Wolfgang E. Nagel · Willi Jäger · Michael Resch
Editors

High Performance Computing in Science and Engineering '06

Transactions of the High Performance Computing Center Stuttgart (HLRS) 2006

With 312 Figures, 101 in Colour, and 46 Tables

Springer

Editors

Wolfgang E. Nagel
Zentrum für Informationsdienste
und Hochleistungsrechnen (ZIH)
Technische Universität Dresden
Willers-Bau, A-Flügel
Zellescher Weg 12
01069 Dresden, Germany
wolfgang.nagel@tu-dresden.de

Willi Jäger
Interdisziplinäres Zentrum
für Wissenschaftliches Rechnen (IWR)
Universität Heidelberg
Im Neuenheimer Feld 368
69120 Heidelberg, Germany
jaeger@iwr.uni-heidelberg.de

Michael Resch
Höchstleistungsrechenzentrum
Stuttgart (HLRS)
Universität Stuttgart
Nobelstraße 19
70569 Stuttgart, Germany
resch@hlrs.de

Front cover figure: Convective flows in the interior of a proto neutron star which were obtained during a core collapse supernova simulation, Max-Planck-Institut für Astrophysik, Garching

ISBN 978-3-642-07160-7 e-ISBN 978-3-540-36183-1

Mathematics Subject Classification (2000): 65Cxx, 65C99, 68U20

Springer is a part of Springer Science+Business Media

springer.com

Cover design: WMXDesign, Heidelberg

Printed on acid-free paper 46/3100/YL - 5 4 3 2 1 0

Nagel · Jäger · Resch (Eds.)
High Performance Computing in Science and Engineering ’06

Preface

The last two years have been great for high performance computing in Baden-Württemberg and beyond. In July 2005, the new building for HLRS as well as Stuttgart's new NEC supercomputer – which is still leading edge in Germany – have been inaugurated. In these days, the SSC Karlsruhe is finalizing the installation of a very large high performance system complex from HP, built from hundreds of Intel Itanium processors and more than three thousand AMD Opteron cores. Additionally, the fast network connection – with a bandwidth of 40 Gbit/s and thus one of the first installations of this kind in Germany – brings the machine rooms of HLRS and SSC Karlsruhe very close together. With the investment of more than 60 Million Euro, we – as the users of such a valuable infrastructure – are not only thankful to science managers and politicians, but also to the people running these components as part of their daily business, on a 24-7 level.

Since about 18 months, there are lots of activities on all scientific, advisory, and political levels to decide if Germany will install an even larger European supercomputer, where the hardware costs alone will be around 200 Million Euro for a five year period. There are many good reasons to invest in such a program because – beyond the infrastructure – such a scientific research tool will attract the best brains to tackle the problems related to the software and methodology challenges. Within the last six months, under the guidance of Professor Andreas Reuter (EML) the German HPC community has made a proposal to reshape High Performance Computing in Germany and to form a German HPC Alliance, with the goal to improve and guarantee competitiveness for the coming years. There is a very good chance that our Federal Ministry for Education and Research (BMBF) – together with the colleagues from the local countries – will support the proposed actions. Beyond the stabilization and strengthening of the existing German infrastructures – including the necessary hardware at a worldwide competitive level – a major request is a related software research and support program to enable Computational Science and Engineering on the required level of expertise and performance which means: running Petascale applications on more than 100,000 processors. It is recommended that for the next years 20 Million Euro are spend – on a yearly basis – for projects to develop algorithms, methods and tools in many areas. As we all know, we do not only need competitive hardware but also excellent software and methods to approach – and solve – the most demanding problems in science and engineering. To achieve this challenging goal every three-year project supported by that program will need to integrate excellent research groups at the universities with colleagues from the competence network of HPC centers in Germany. The success of this approach is of utmost importance for our community and also will strongly influence the development of new technologies and industrial products; beyond that, this will finally determine if Germany will be an accepted partner among the leading technology and research nations.

Since 1996, HLRS is supporting the scientific community as part of its official mission. Like in the years before, the major results of the last 12 months were reported at the Tenth Results and Review Workshop on High Performance Computing in Science and Engineering, which was held October 19–20, 2006 at Stuttgart University. This volume contains the written versions of the research work presented. The papers have been selected from all projects running at HLRS and at SSC Karlsruhe during the one year period beginning October 2005. Overall, 39 papers have been chosen from Physics, Solid State Physics, Computational Fluid Dynamics, Chemistry, and other topics. The largest number of contributions, as in many other years, came from CFD with 17 papers. To a certain extend, the selected papers demonstrate the state of the art in high performance computing in Germany. The authors were encouraged to emphasize computational techniques used in solving the problems examined. The importance of the newly computed results for the specific disciplines, as interesting as they may be from the scientific point of view, were not the major focus of this volume.

We gratefully acknowledge the continued support of the Land Baden-Württemberg in promoting and supporting high performance computing. Grateful acknowledgement is also due to the Deutsche Forschungsgemeinschaft (DFG): many projects processed on the machines of HLRS and SSC could not have been carried out without the support of the DFG. Also, we thank the Springer Verlag for publishing this volume and thus helping to position the local activities into an international frame. We hope that this series of publications is contributing to the global promotion of high performance scientific computing.

Stuttgart, September 2006

Wolfgang E. Nagel
Willi Jäger
Michael Resch

Contents

Physics
H. Ruder and R. Speith 1

Gravitational Wave Signals from Simulations of Black Hole Dynamics
B. Brügmann, J. Gonzalez, M. Hannam, S. Husa, P. Marronetti, U. Sperhake, and W. Tichy 3

The SuperN-Project: Understanding Core Collapse Supernovae
A. Marek, K. Kifonidis, H.-Th. Janka, and B. Müller 19

MHD Code Optimizations and Jets in Dense Gaseous Halos
V. Gaibler, M. Vigelius, M. Krause, and M. Camenzind 35

Anomalous Water Optical Absorption:
Large-Scale *First-Principles* Simulations
W.G. Schmidt, S. Blankenburg, S. Wippermann, A. Hermann, P.H. Hahn, M. Preuss, K. Seino, and F. Bechstedt 49

The Electronic Structures of Nanosystems:
Calculating the Ground States of Sodium Nanoclusters and the Actuation of Carbon Nanotubes
B. Huber, L. Pastewka, P. Koskinen, M. Moseler 59

Object-Oriented SPH-Simulations with Surface Tension
S. Ganzenmüller, A. Nagel, S. Holtwick, W. Rosenstiel, and H. Ruder . . 69

Simulations of Particle Suspensions at the Institute for Computational Physics
J. Harting, M. Hecht, and H. Herrmann 83

Solid State Physics
W. Hanke ... 93

Nano-Systems in External Fields and Reduced Geometry: Numerical Investigations
P. Henseler, C. Schieback, K. Franzrahe, F. Bürzle, M. Dreher, J. Neder, W. Quester, M. Kläui, U. Rüdiger, and P. Nielaba ... 97

Signal Transport and Finite Bias Conductance in and Through Correlated Nanostructures
P. Schmitteckert and G. Schneider ... 113

Atomistic Simulations of Dislocation – Crack Interaction
E. Bitzek and P. Gumbsch ... 127

Monte Carlo Simulations of Strongly Correlated and Frustrated Quantum Systems
C. Lavalle, S.R. Manmana, S. Wessel, and A. Muramatsu ... 137

Chemistry
C. van Wüllen ... 153

Characterization of Catalyst Surfaces by STM Image Calculations
R. Kovacik, B. Meyer, and D. Marx ... 155

Theoretical Investigation of the Self-Diffusion on Au(100)
K. Pötting, T. Jacob, and W. Schmickler ... 171

TrpAQP: Computer Simulations to Determine the Selectivity of Aquaporins
M. Dynowski and U. Ludewig ... 187

Computational Fluid Dynamics
S. Wagner ... 199

Direct Numerical Simulation and Analysis of the Flow Field Around a Swept Laminar Separation Bubble
T. Hetsch and U. Rist ... 205

Direct Numerical Simulation of Primary Breakup Phenomena in Liquid Sheets
W. Sander and B. Weigand ... 223

Direct Numerical Simulation of Mixing and Chemical Reactions in a Round Jet into a Crossflow – a Benchmark
J.A. Denev, J. Fröhlich,and H. Bockhorn ... 237

Numerical Simulation of the Bursting
O. Marxen and D. Henningson ... 253

Parallel Large Eddy Simulation with UG
A. Hauser and G. Wittum 269

LES and DNS of Melt Flow and Heat Transfer in Czochralski Crystal Growth
A. Raufeisen, M. Breuer, V. Kumar, T. Botsch, and F. Durst 279

Efficient Implementation of Nonlinear Deconvolution Methods for Implicit Large-Eddy Simulation
S. Hickel and N.A. Adams 293

Large-Eddy Simulation of Tundish Flow
N. Alkishriwi, M. Meinke, and W. Schröder 307

Large Eddy Simulation of Open-Channel Flow Over Spheres
T. Stoesser and W. Rodi 321

Prediction of the Resonance Characteristics of Combustion Chambers on the Basis of Large-Eddy Simulation
F. Magagnato, B. Pritz, H. Büchner, and M. Gabi 331

Investigations of Flow and Species Transport in Packed Beds by Lattice Boltzmann Simulations
T. Zeiser 343

Rheological Properties of Binary and Ternary Amphiphilic Fluid Mixtures
J. Harting and G. Giupponi 355

The Effects of Vortex Generator Arrays on Heat Transfer and Flow Field
C.F. Dietz, M. Henze, S.O. Neumann, J. von Wolfersdorf, and B. Weigand 365

Investigation of the Influence of the Inlet Geometry on the Flow in a Swirl Burner
M. García-Villalba, J. Fröhlich, and W. Rodi 381

Numerical Investigation and Simulation of Transition Effects in Hypersonic Intake Flows
M. Krause, B. Reinartz, and J. Ballmann 391

Aeroelastic Simulations of Isolated Rotors Using Weak Fluid-Structure Coupling
M. Dietz, M. Kessler, and E. Krämer 407

Computational Study of the Aeroelastic Equilibrium Configuration of a Swept Wind Tunnel Wing Model in Subsonic Flow
L. Reimer, C. Braun, and J. Ballmann 421

Structural Mechanics
P. Wriggers 435

Numerical Prediction of the Residual Stress State after Shot Peening
M. Klemenz, M. Zimmermann, V. Schulze, and D. Löhe 437

Computer-Aided Destruction of Complex Structures by Blasting
S. Mattern, G. Blankenhorn, and K. Schweizerhof 449

Wave Propagation in Automotive Structures Induced by Impact Events
S. Mattern and K. Schweizerhof 459

Miscellaneous Topics
W. Schröder 471

Continental Growth and Thermal Convection in the Earth's Mantle
U. Walzer, R. Hendel, and J. Baumgardner 473

Efficient Satellite Based Geopotential Recovery
O. Baur, G. Austen, and W. Keller 499

Molecular Modeling of Hydrogen Bonding Fluids: Monomethylamine, Dimethylamine, and Water Revised
T. Schnabel, J. Vrabec, and H. Hasse 515

The Application of a Black-Box Solver with Error Estimate to Different Systems of PDEs
T. Adolph and W. Schönauer 527

Scalable Parallel Suffix Array Construction
F. Kulla and P. Sanders 543

Physics

Prof. Dr. Hanns Ruder and Dr. Roland Speith

Institut für Astronomie und Astrophysik, Abteilung Theoretische Astrophysik,
Universität Tübingen, Auf der Morgenstelle 10, D-72076 Tübingen

The articles in this section represent a selection of those projects currently running at the HLRS which are related to physical research. Although the presented work covers a wide range of Physics, all the different projects have in common that they require state-of-the-art, if not even future, super-computers to achieve their research objectives. In particular this seems to hold for research in Astrophysics, because as in the previous years, nearly half of all Physics projects contribute to that field of research.

A particular highlight in this field is the project by Brügmann et al., which deals with the simulation of gravitational wave signals generated by two orbiting black holes. To solve this problem in full General Relativity is a long-standing goal that has turned out to be extremely difficult and that is, in spite of large international efforts, far from being achieved. It is therefore very remarkable that Brügmann et al. managed to extend the period that currently can be simulated from a fraction of the binary orbit to the length of several orbits.

A similarly complex problem is addressed by Marek et al., who attempt to model core collapse supernovae. Several details of the mechanism of a supernova are also in spite of decades of research still not sufficiently understood, in particular the initiation of the explosion. In their work, Marek et al. extend their model to two dimensions including detailed neutrino transport and other relevant physical effects.

Gaibler et al. report on the progress of modelling the propagation of magnetised astrophysical jets into dense and clumpy halos and the interaction with those environments, a project that is already ongoing for several years. In particular they further optimised and adapted their parallel codes to the super-computers available at the HLRS to be able to simulate very light jets which require a particularly high number of time-steps.

In the non-astrophysical fields, the work by Schmidt et al. has to be mentioned, who perform large-scale simulations to model the optical spectrum of water from first principles. Thus they were able to reproduce experimental results, develop a better understanding of the involved processes, and clarify the molecular interactions. Huber et al. have calculated the electronic structure of two large-scale nanosystems, namely sodium nanoclusters and carbon nanotubes. Hipp et al. contribute with a computer science related work investigating the implementation of surface tension into the Smoothed Particle Hydrodynamics scheme. And finally Harting, Hecht & Herrman report on two types of simulations of particle suspensions, in particular clay like colloids described by molecular dynamics and creeping shear flow modelled with a lattice-Boltzmann algorithm.

All the above mentioned projects again demonstrate that high performance computing has in general become a matter of routine. However, several projects also had to report difficulties of technical as well as administrative nature. The porting of existing codes was complicated by compiler bugs or library conflicts, which in a few cases even rendered the whole project unfeasible. Occasionally it was wished for a faster response to enquiries by e-mail. And in one case very long delays in the refereeing process occurred, so that a final decision for computing time had not been made even when the first status report was requested.

Gravitational Wave Signals from Simulations of Black Hole Dynamics

Bernd Brügmann[1], Jose Gonzalez[1], Mark Hannam[1], Sascha Husa[1], Pedro Marronetti[2], Ulrich Sperhake[1], and Wolfgang Tichy[2]

[1] Theoretisch Physikalisches Institut, Universität Jena, Max-Wien-Platz 1, 07743 Jena, Germany, sascha.husa@uni-jena.de

[2] Department of Physics, Florida Atlantic University, Boca Raton FL 33431, USA

1 Introduction

In this paper we present a status report on our work on the numerical simulation of sources of gravitational waves, which is being carried out within project BBH at HLRS Stuttgart. The wider context of this research is provided by a large international effort to detect gravitational waves and start a new field of astrophysical research: "Gravitational Wave Astronomy".

The theory of general relativity predicts the emission of gravitational waves in non-spherical dynamical interactions of masses. This is analogous to the production of electromagnetic waves by accelerated charges in electromagnetism. So far all our knowledge about astrophysics and cosmology is based on electromagnetic observations, and as such the observation of gravitational waves will open a new window into the universe and provide information about phenomena hitherto not accessible to direct observation, such as black holes, dark matter, or the very early universe. Gravitational wave signals are different in several ways from electromagnetic signals: they are not shielded by interstellar masses, they represent the bulk motion of fast compact objects instead of the incoherent superposition of signals from large numbers of individual particles, and current gravitational wave detectors are sensitive to the amplitude rather than the intensity of the waves. The signal strength thus scales inversely proportional with the distance, instead of the distance squared, and the scaling of the observed volume of the universe is thus much better than with electromagnetic observations.

The detection of gravitational waves has not yet been accomplished, but a growing network of gravitational wave detectors is pushing the sensitivity envelope. In the last years a new generation of interferometer-based gravitational wave detectors has come on-line, such as LIGO [1, 25], GEO [16, 17] and VIRGO [29]. Some of these detectors have recently reached their design sensitivities, and LIGO currently performs its main science run. In order to actually extract information about the sources from observations, and to enhance the detection probability, accurate signal templates are needed for various types of sources as input for data analysis efforts underway.

In order to produce detectable signals, astrophysical systems require dynamics involving large masses at relativistic speeds within small regions of spacetime. An ideal source is the final phase of a binary system of two black holes. Systems involving black holes of a few dozen solar masses constitute one of the most likely sources for the current generation of gravitational-wave detectors on earth, based on the frequency dependent sensitivity of these interferometric detectors. The planned space-based detector LISA will detect gravitational waves from the mergers of supermassive black holes at the center of galaxies, to which our simulations apply equally well.

The time evolution of two orbiting black holes can be divided into three phases. At sufficiently large separation the black holes' motion approximately follows Kepler's laws and can be described by post-Newtonian expansions. Over the course of millions of years the orbits tighten due to the emission of energy and angular momentum in the form of gravitational waves, leading finally to the collision and merger of the two black holes. The final orbit is reached when the radial infall motion overtakes the orbital motion. Once the black holes have merged, the resulting single distorted black hole enters the ring-down phase during which the black hole settles down to a stationary, axisymmetric black hole. The late phase of the inspiral requires the numerical solution of the Einstein equations of general relativity in the regime of highly dynamic and highly non-linear gravitational fields. The main focus of this project is to study the last few orbits of a binary black hole coalescence during which the inspiral deviates substantially from quasi-adiabatic motion.

The numerical solution of the full Einstein equations represents a very complex problem, and for two black holes the spacetime singularities that are encountered in the interior of black holes pose an additional challenge. For a long time, typical runs had been severely limited by the achievable evolution time before the simulations become too inaccurate or before the computer code became unstable and produced numerical infinities, and there was serious concern, whether numerical relativity techniques can produce gravitational wave templates, at least in the near future. This picture has drastically changed ever since the first simulations of a complete black hole orbit were obtained in early 2004 [9]. In the course of the last 12 months, several groups have managed to extract gravitational waves from merger simulations that extended over up to several orbits.

The research group in Jena was established just over one year ago, and is currently building up a long-term effort on binary black-hole evolutions. Starting from the code that had been used for the first orbit simulation [9], over recent months several additions and technical improvements have enabled us to produce state of the art simulations lasting for several orbits.

The rest of the paper is organized as follows: In Sect. 2 we describe in more detail the problem setting and our solution strategies, in Sect.3 we describe computational aspects of the work, including our setup for runs at the Cray Opteron cluster at HLRS Stuttgart. First results from our work are presented in Sect. 4, and we conclude with a discussion in Sect. 5.

2 The Binary Black Hole Problem in Numerical Relativity and Solution Strategies

2.1 Numerical Relativity

A central problem of computational physics is to manifestly capture physical features in a discrete system, e.g. to preserve structure during a time integration. This is particularly hard in numerical relativity (NR) – the numerical solution of the Einstein equations, which form a coupled set of nonlinear partial differential equations. Writing the Einstein equations in the form of an initial value problem, the number of computational degrees of freedom (d.o.f.) is larger than the physical d.o.f., because it is not known in general how to separate physical from gauge and constraint violating d.o.f. Due to the complicated nonlinear structure of the Einstein equations it has not yet been possible to directly carry over techniques that solve conceptually related problems in other gauge theories such as the Maxwell equations. Consequently, numerical relativity simulations have typically been plagued by instabilities, which are often rooted in the continuum formulation of the problem.

Furthermore, the study of wave emission from compact objects requires resolution on different scales: a code needs to resolve the compact objects, their orbits, emitted waves and a slowly varying background. In order to obtain accurate results both the use of mesh refinement techniques *and* a good choice of coordinate gauge is essential. As is common in our field, energy, time and length are measured in units of a mass parameter, denoted M, which is set to approximately the total mass of the system. Together with the complicated structure of the equations – a typical code has between ten and several dozen evolution variables, and, when expanded, the right hand sides of the evolution equations have thousands of terms – this yields a computationally very complex and mathematically very subtle problem.

Extracting gravitational wave signals from simulations is difficult for various reasons. First, even though some gravitational wave sources will be among the brightest objects in the universe, gravitational waves are typically only a small effect in the energy balance of such systems. Second, in GR such fundamental quantities as energy, momentum, or emitted gravitational radiation can only be defined unambiguously in terms of asymptotic limits. Consequently, it also becomes particularly difficult to formulate “outgoing radiation boundary conditions” at a finite distance from the sources.

2.2 Problems Specific to Black Hole Binary Evolutions

While stable and accurate numerical evolutions of single black-hole systems have essentially become routine problems in the field, until recently accurate binary evolutions have remained elusive. One of the reasons is that it is much easier to find coordinate gauge conditions that do not trigger instabilities for a single black hole. In a binary evolution, there are two principal choices: a co-rotating coordinate system, or to move the black holes across the grid. The latter choice requires much more sophisticated computational infrastructure, i.e. mesh refinement regions that move with respect to coarser grids, and thus co-rotation approaches have dominated the field until recently. However, this technique typically leads to super-luminal motion of grid points far away from the sources, which makes it hard to implement boundary conditions and extract the wave content. Also, experience has shown that this approach tends to trigger instabilities, and it took until the beginning of 2004 to present the first numerical simulations of binary black-hole systems that last for about one orbital period for close but still separate black holes. The numerical accuracy was, however, insufficient for the determination of gravitational waves. Since then several groups have implemented computational infrastructure for moving black holes through the grid, following quite different approaches [12, 11, 7, 6, 22], with simulations lasting several orbits.

Due to the presence of constraints in the theory, there is no simple relation between the freely specifiable data and their astrophysical interpretation. This makes it a very difficult problem to set up astrophysically relevant initial data, i.e., data that correspond to the actual inspiral of two black holes. Much work has been done in this direction, e.g. work has been done on energy minimization procedures and the use of Post-Newtonian methods. For brevity we only refer to the literature [15], except for mentioning that a significant part of our computational effort is spent on comparing results from different initial data sets, and showing that indeed they produce consistent physics.

A particular problem is posed by the fact that numerical codes cannot handle the appearance of singularities that form inside black holes. One solution is to identify a pure outflow boundary, where the inside region can be excised from the grid ("black hole excision") since no boundary conditions are to be specified on such a surface. An alternative is to "fill" the black hole with a topological defect in the form of an interior space-like asymptotic end, and to freeze the evolution of the asymptotic region through a judicious choice of coordinate gauge. The latter approach, combined with a setup where the topological defect is allowed to move across the grid ("moving puncture" approach) has lead to a giant leap in the field [12, 11, 7, 6].

It is this approach we are following, based on our previous work. In a recent paper [21] some of us have contributed to a clarification of some open theoretical issues with the "moving punctures" approach, in particular regarding the nature of the coordinate singularity associated with the topological defect, through a combination of numerical and analytical methods The implementation of a fourth-order accurate finite differencing scheme together with a mesh refinement scheme based on moving boxes centered at the topological defects through a hierarchy of mesh refinement levels yields a code that is more efficient than traditional adaptive mesh refinement approaches, and still very accurate. With our approach we can currently run production simulations on as few as 8–32 processors, which require up to several hundred processors for alternative approaches [13].

2.3 Equations and Numerical Algorithms in the BAM Code

The most straightforward way to get PDEs (partial differential equations) out of the Einstein equations is to introduce a coordinate system of four functions $\{x^a\}$ that label events in spacetime, and write out Einstein's equations explicitly in this coordinate system. The character of the resulting PDEs depends on the geometry of the coordinate system. The most usual choice here is to choose coordinates $\{x^i, t\}$ $(i = 1, 2, 3)$ such that the metric can be written in the form

$$ds^2 = -(\alpha^2 - \gamma_{ij}\beta^i\beta^j)dt^2 + 2h_{ij}\beta^j dt\, dx^i + \gamma_{ij} dx^i\, dx^j, \tag{1}$$

where γ_{ij} is a positive definite metric on the 'slices' $t = const.$. This corresponds to a 3+1 splitting of spacetime, where by singling out a privileged time direction, the spacetime gets foliated by 3-dimensional space-like hypersurfaces corresponding to slices of constant time t. The field $h_{ij}(x^k, t)$ is the Riemannian metric on the 3-spaces of constant time. In order to obtain a Lorentzian metric, one demands that $\alpha \neq 0$. The diffeomorphism invariance of general relativity is then encoded in the fact that the four functions $\alpha(x^i, t)$ and $\beta^i(x^j, t)$ turn out to be freely specifiable, and merely steer the coordinate system through spacetime as the time evolution proceeds – and that the physical result is independent of this choice.

Writing the Einstein equations as an initial value problem in this way yields a coupled system of second differential order elliptic constraints and hyperbolic evolution equations that preserve the constraints. In the free evolution approach, which is most common in the field, the constraints are only solved initially, and later only the hyperbolic equations are used to construct the solution. For a long time, the standard choice of variables for writing the Einstein equations was based on the so-called "ADM equations" [5] (essentially a first order in time reduction resulting from the procedure sketched above). It is known now, however, that the hyperbolic subsystem thus obtained is not well posed; specifically, it leads to a weakly hyperbolic set of equations. Improved evolution systems are provided by the so called BSSN family [8, 27]. This formulation is characterized by introducing a contracted connection term as a new variable, a conformal decomposition of the metric and extrinsic curvature variables, and adding constraints to the evolution equations. Detailed discussions of well-posedness for the BSSN family have been given by Gundlach and Martin-Garcia [18, 19, 20]. The set of evolved variables are the logarithm of the conformal factor φ, the conformally rescaled three-metric $\tilde{\gamma}_{ij}$, the trace of the extrinsic curvature K, the conformally rescaled traceless extrinsic curvature $\tilde{A}_{ij}$, and the contracted Christoffel symbols $\tilde{\Gamma}^i$:

$$
\begin{aligned}
\varphi &= \log(\det\gamma_{ij})/12, && (2)\\
\tilde{\gamma}_{ij} &= e^{-4\varphi}\gamma_{ij}, && (3)\\
K &= \gamma^{ij}K_{ij}, && (4)\\
\tilde{A}_{ij} &= e^{-4\varphi}(K_{ij} - (1/3)\gamma_{ij}K), && (5)\\
\tilde{\Gamma}^i &= \tilde{\Gamma}^i_{jk}\tilde{\gamma}^{jk}. && (6)
\end{aligned}
$$

This immediately leads to one differential and two algebraic constraints:

$$
\mathcal{G} = \tilde{\Gamma}^i - \tilde{\gamma}^{jk}\tilde{\Gamma}^i_{jk} = 0, \qquad \mathcal{S} = \det\gamma_{ij} - 1 = 0, \qquad \mathcal{A} = \tilde{A}^i_i = 0, \tag{7}
$$

which are again propagated by the evolution equations. Note that densitized quantities (those with a tilde) have their indices raised and lowered with the conformally rescaled three-metric $\tilde{\gamma}_{ij}$. The standard Hamiltonian and momentum constraints of general relativity take the form

$$
\begin{aligned}
\mathcal{H} &= e^{-4\varphi}\tilde{R} - 8e^{-4\varphi}\tilde{D}^j\tilde{D}_j\varphi - 8e^{-4\varphi}(\tilde{D}^j\varphi)(\tilde{D}_j\varphi)\\
&\quad + (2/3)K^2 - \tilde{A}_{ij}\tilde{A}^{ij} - (2/3)\mathcal{A}K,\\
\mathcal{M}_i &= 6\tilde{A}^j{}_i(\tilde{D}_j\varphi) - 2\mathcal{A}(\tilde{D}_i\varphi) - (2/3)(\tilde{D}_iK) + \tilde{\gamma}^{kj}(\tilde{D}_j\tilde{A}_{ki}).
\end{aligned}
$$

The BSSN evolution equations, which are obtained from the Einstein equations by using the definitions (2) and making a standard choice for adding constraints, become

$$
\begin{aligned}
\mathcal{L}_n\varphi &= -(1/6)\alpha K,\\
\mathcal{L}_n\tilde{\gamma}_{ij} &= -2\alpha\tilde{A}_{ij},\\
\mathcal{L}_nK &= -D^iD_i\alpha + \alpha\tilde{A}_{ij}\tilde{A}^{ij} + (1/3)\alpha K^2,\\
\mathcal{L}_n\tilde{A}_{ij} &= -e^{-4\varphi}(D_iD_j\alpha)^{TF} + e^{-4\varphi}\alpha(R_{ij})^{TF} + \alpha K\tilde{A}_{ij} - 2\alpha\tilde{A}_{ik}\tilde{A}^k{}_j,\\
\mathcal{L}_n\tilde{\Gamma}^i &= -2(\partial_j\alpha)\tilde{A}^{ij} + 2\alpha\big(\tilde{\Gamma}^i_{jk}\tilde{A}^{kj} - (2/3)\tilde{\gamma}^{ij}(\partial_jK) + 6\tilde{A}^{ij}(\partial_j\varphi)\big).,
\end{aligned}
$$

where $\tilde{D}_i$ is covariant derivative associated with $\tilde{\gamma}_{ij}$, and $\mathcal{L}_n = \partial_t - \mathcal{L}_\beta$ is the Lie derivative along the unit normal. The Ricci curvature in terms of the BSSN variables takes the form

$$
\begin{aligned}
R_{ij} &= \tilde{R}_{ij} + R^\varphi_{ij},\\
R^\varphi_{ij} &= -2\tilde{D}_i\tilde{D}_j\varphi - 2\tilde{\gamma}_{ij}\tilde{D}^k\tilde{D}_k\varphi + 4(\tilde{D}_i\varphi)(\tilde{D}_j\varphi) - 4\tilde{\gamma}_{ij}(\tilde{D}^k\varphi)(\tilde{D}_k\varphi),\\
\tilde{R}_{ij} &= -(1/2)\tilde{\gamma}^{lk}\partial_l\partial_k\tilde{\gamma}_{ij} + \tilde{\gamma}_{k(i}\partial_{j)}\tilde{\Gamma}^k + \tilde{\Gamma}^k\tilde{\Gamma}_{(ij)k} + 2\tilde{\gamma}^{lm}\tilde{\Gamma}^k_{l(i}\tilde{\Gamma}_{j)km} + \tilde{\gamma}^{lm}\tilde{\Gamma}^k_{im}\tilde{\Gamma}_{klj}.
\end{aligned}
$$

The trace-free part of the Ricci tensor is computed by the projection operation $R_{ij}^{TF} = R_{ij} - R\gamma_{ij}i/3$. The algebraic constraints are solved at every time step by setting $\tilde{\gamma}_{ij} \to \tilde{\gamma}_{ij}\det\tilde{\gamma}^{-1/3}$ and $\tilde{A}_{ij} \to \tilde{A}_{ij} - \frac{1}{3}\tilde{A}_{lm}\tilde{\gamma}^{il}\tilde{\gamma}^{jm}$.

In spite of the fact that much experience has been gained in recent times in constructing improved gauge conditions, this issue still represents an active area of research. Several of our numerical experiments are devoted to testing the effects of different gauge choices.

Currently, our standard choice for evolving the lapse function is given by the so-called Bona-Masso family of slicing conditions:

$$\partial_t \alpha = -\alpha K f(\alpha), \tag{8}$$

in particular the choices $f = 1$, corresponding to harmonic time slicing, and $f = 2/\alpha$, which is usually termed "$1 + \log$" slicing, see [3].

In order to obtain long-term stable numerical simulations, it is equally important to construct a suitable shift vector field β^i. Here we report on evolutions where we evolve the shift vector according to a Gamma-freezing prescription [3]. A key feature of this particular choice is to drive the dynamics of the variable $\tilde{\Gamma}^i$ towards a stationary state. As a "side effect" this choice creates a coordinate motion that drags the black holes along an inspiral orbit. A crucial effect of this method is that the resulting coordinate motion which corresponds to the naive physical intuition reduces artificial distortions in the geometry, which otherwise could easily trigger instabilities

In summary, the solution procedure for the equations is as follows. First, we specify free data motivated by quasi-equilibrium arguments, then solve the 9 components of the constraint equations $(\mathcal{H}, \mathcal{M}_i, \mathcal{G}^i, \mathcal{A}, \mathcal{S})$ to obtain initial data for the 17 evolution variables $(\varphi, \tilde{\gamma}_{ij}, K, \tilde{A}_{ij}, \tilde{\Gamma}^i)$. The evolution system is completed by specifying evolution equations for the four gauge quantities (α, β^i), which yields a hyperbolic system that is second order in space and first order in time, and which determines the evolution of all 21 components of the "state vector" describing the geometry of spacetime.

In finite difference codes for the solution of nonlinear hyperbolic equations, it is common practice to either use high-resolution finite volume shock capturing schemes or standard finite difference methods with artificial dissipation terms added to all right-hand-sides of the time evolution equations, schematically written as

$$\partial_t \mathbf{u} \to \partial_t \mathbf{u} + Q\mathbf{u}. \tag{9}$$

Since shocks in the standard sense do not appear in general relativity due to the linear degeneracy of the evolution equations [2], we use a standard finite difference approach combined with a Kreiss–Oliger dissipation [24] operator (Q) of order $2r$

$$Q = \sigma(-h)^{2r-1}(D_+)^r \rho (D_-)^r / 2^{2r}, \tag{10}$$

for a $2r - 2$ accurate scheme, with σ a parameter regulating the strength of the dissipation, and ρ a weight function that we typically set to unity.

Discretization in space is performed with standard second- or fourth-order accurate stencils. In particular, symmetric stencils are used, with the exception of the advection terms associated with the shift vector, where we use lop-sided upwind stencils, see e.g. [31]. Time integration is performed by standard Runge-Kutta type methods, in particular 3rd and 4th order Runge-Kutta and second order accurate three-step iterative Crank-Nicholson integrators as described in [10], where Courant limits and stability properties are discussed for the types of equations used here.

The new technical contribution in our work is a fully fourth-order accurate box-based mesh refinement algorithm, which is tuned to the problems we consider and indeed very efficient for our work. Details and results are described in the next sections.

2.4 BAM Code Structure

The BAM code that we are developing is designed to solve partial differential equations on structured meshes, in particular a coupled system of (typically hyperbolic) evolution equations and elliptic equations. The peculiarities of numerical relativity are that the equations are rather complex and that the Einstein equations do not correspond to a unique set of partial differential equations – an optimal formulation of the continuum equations for the purpose of numerical simulations is thus an active field of research. This leads to the requirement of a short development cycle when modifying the equations, which we address by using a Mathematica package integrated into the code, which produces C-code from high level problem descriptions in Mathematica notebooks.

Furthermore, the code is organized as a "framework", similar in spirit to the Cactus code [4, 28], which is frequently used in our community, but dropping much of its complexity. The core idea is that code is organized into a central core and code modules, called "projects", which interact with the core through a well-defined interface of registering parameters, variables and functions. User-defined functions are called by registering them into "time-bins", e.g. for initial data, the evolution step, or analysis of a time step.

A typical production run for a black hole binary requires at least roughly 100^3 grid points on 10 refinement levels, with 100–160 grid functions defined, which amounts to roughly 10 GByte of storage. The computational domain is decomposed into cubes, following standard domain-decomposition algorithms, and is parallelized with MPI.

Mesh refinement techniques are handled to resolve the different scales of the problem, and to follow the motion of the black holes. The relevant spatial scales of the binary black-hole problem are the scales of the holes, their orbital motion, the typical wave lengths of the ring-down of the individual and merged black holes, the typical wavelength of the merger waveform and the asymptotic falloff of the fields. All of these scales can be estimated from the initial data and vary relatively slowly with time. It is thus very efficient to essentially use a fixed mesh refinement strategy, with inner level refinement boxes following the motion of the black holes.

The theoretical gain in efficiency is partially offset by a significant increase in code complexity and by significant overhead for parallelized computation. The initial goal of the simulations at the HLRS is therefore to implement and optimize parallelization strategies, before moving on to physics production runs.

3 Simulations Performed at HLRS Stuttgart

3.1 Infrastructure

We use C code compiled with the Portland Group compiler and MPICH for interprocessor communication. Quick visualizations during runs and a first analysis are typically performed with gnuplot or ygraph [30], and for detailed analysis data are downloaded to local workstations where 2D and 3D visualizations are performed with VTK-based software, in particular a VTK-based C++ code our group is developing.

3.2 Performance and Scaling

We have only very recently improved our numerical evolution scheme from second to fourth order accuracy and have also enhanced the capabilities of our mesh-refinement engine to allow for multiple boxes at each refinement level. While these developments are still ongoing, we have already achieved vastly improved accuracy in our simulations while increasing the computational efficiency. In the following we give the preliminary performance results.

The only code that we can compare efficiency results with, is a code developed at NASA-Goddard [23, 14, 7, 6], which uses adaptive mesh refinement based on the PARAMESH package [26]. The main differences to our approach, apart from the adaptivity in the Goddard code, is that in their approach the boundaries of mesh refinement levels are only treated with second order accuracy, and that in the Goddard approach all levels use the same time step. The Goddard group reports that a typical production run requires 256–512 processors [13]. Our main focus in the last weeks has been to gain comparable accuracy with runs that require only 32 processors and fit into a 48-hour queue limit. We have now achieved this goal.

Our current typical choice of initial data (equal-mass black holes and suitably aligned spins) possess reflection symmetries, which allow us to evolve on one quadrant of the full 3-space by applying appropriate symmetry-boundary conditions as necessary. In the context of moving black holes, this leads to two configurations of refinement levels that we work with: in BOX mode, we only evolve one box at each refinement level. At times when both holes are close to the symmetry plane, this leads to bar-like grid-configurations as shown in

Fig. 1, and to corresponding peaks in memory requirements. A more efficient alternative approach that we have recently implemented evolves several boxes at the fine levels (typically one for each moving hole). This approach leads to drastically reduced but still significant fluctuations in memory requirements as the holes come very close and merge, but has become our standard mode of evolution during the last weeks.

Scaling with larger numbers of processors has been best tested with the second order accurate version of our code, and non-moving boxes of refinement. In such cases we typically reach 80–90 % of scaling on up to 128 processors, which has been tested on different machines. With the new fourth-order code we have so far only tested scaling on up to 32 processors. We get 77 % scaling on the Cray Opteron cluster strider going from 4 to 32 processors, and increasing the problem size by a factor close to 8. The 32 processor run corresponds to the initial $10M$ of runtime of a typical production run with 14.4 GByte memory usage (i.e. approximately 450 MByte per processor).

3.3 Numerical Experiments

At HLRS we essentially perform 3 types of runs:

- Small debugging runs, typically 2–4 processors, with run times below 8 hours.
- Single black hole numerical experiments, typically using 8–16 processors for 10–20 hours.
- Binary black hole numerical experiments, typically using 16–32 processors for 40–48 hours.

Small debugging runs are mostly used to test our parallel infrastructure and our newly developed fourth-order accurate mesh refinement technique, e.g. regarding the numerical stability of certain options for the restriction and prolongation operations. Since binary black hole evolutions are very resource intensive, a significant part of the testing is therefore done in single black hole situations. In these runs we either consider black holes that are not moving, or which only have linear momentum and move across straight lines.

Memory requirements vary significantly, depending on whether we use BOX or BOXES configurations for moving black holes, or fixed mesh refinement for non-moving black holes, and there can be significant variation during runs. Typical runs require several hundred MByte of memory per processor, with peaks close to 2 GByte.

Most of the computer time on strider has been devoted to testing and optimizing our choices of time integrator, resolution, gauge conditions and mesh refinement strategy, and to get first physics results which are sufficiently accurate to develop and test analysis tools such as algorithms to extract the gravitational wave content and derived quantities such as the radiated energy, and linear and angular momentum.

4 Results

In this section we present preliminary results the dynamics and gravitational wave signals from inspiralling black holes, which give a first glimpse into the type of results we plan to produce at HLRS in the near future.

Typical grid structures appearing in the BOX and BOXES modes are shown in Fig. 1. The BOX mode leads to the formation of computationally expensive bar configurations when the holes are close to the symmetry plane. The grid function displayed is the conformal factor ϕ, which is sharply peaked at the position of the topological defect which marks the position of the black hole. The BOXES mode, where we treat the two black holes with individual refinement boxes around each object is much more efficient and flexible, and will allow us to also treat the very interesting configurations with unequal masses or non-aligned spins, which would not be consistent with a quadrant symmetry.

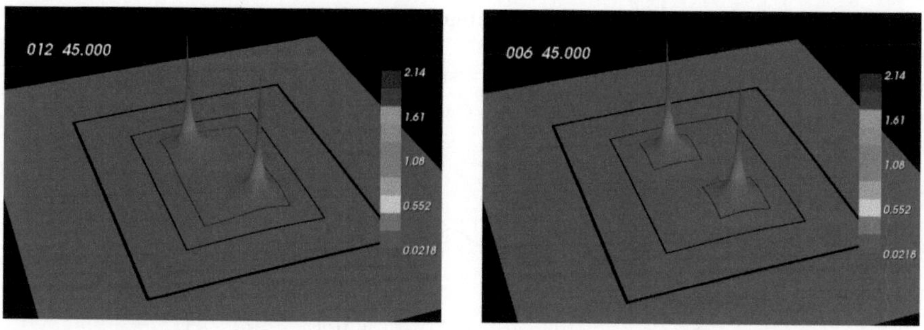

Fig. 1. Comparison of grid structures in BOX (*left*) and BOXES (*right*) modes. BOX mode leads to the formation of computationally expensive bar configurations when the holes are close to the symmetry plane. The grid function displayed here is the conformal factor ϕ. *Dark lines* have been introduced to mark the boundary of mesh refinement regions

The orbital motion of the black holes is shown in Fig. 2. Note that the motion does depend on the choice of coordinates and, thus, is gauge dependent, and is essentially determined by our choice of shift vector field β^a. However, as it turns out, the judicious choice made for the shift does in fact correspond quite accurately with the physical motion expected from the waveform shown in Fig. 3. Figure 3 displays typical results for the gravitational wave signal from an equal mass binary from an initial separation of $\approx 3.2M$. We plot the real part of the ψ_4 component of the Weyl tensor, which in vacuum equals the Riemann curvature tensor, rescaled by the radius of the extraction sphere. This rescaled ψ_4 component of the Weyl tensor is a complex quantity that asymptotically characterizes the 2 polarizations of the gravitational wave signal, and can be converted to a prediction of a detector signal, and thus to

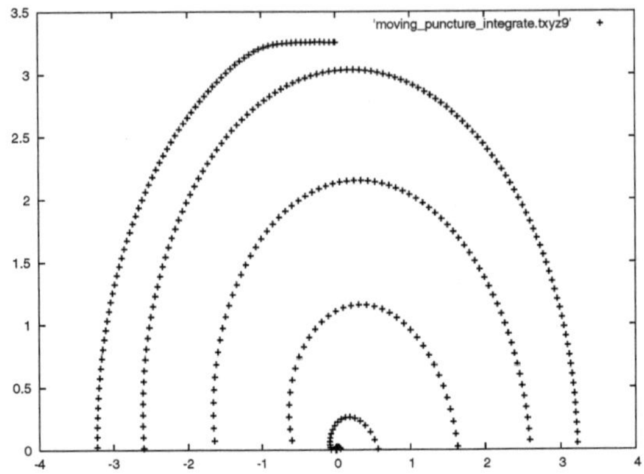

Fig. 2. Orbital motion of one of the black holes in a binary system. As a result of our implementation of quadrant symmetry, we swap the hole displayed every time the holes cross the symmetry plane. The initial flat stretch is an artifact of suboptimal initial data for the gauge quantities α and β^i (lapse and shift), which quickly relax to appropriate values by virtue of the gauge evolution equations we have chosen

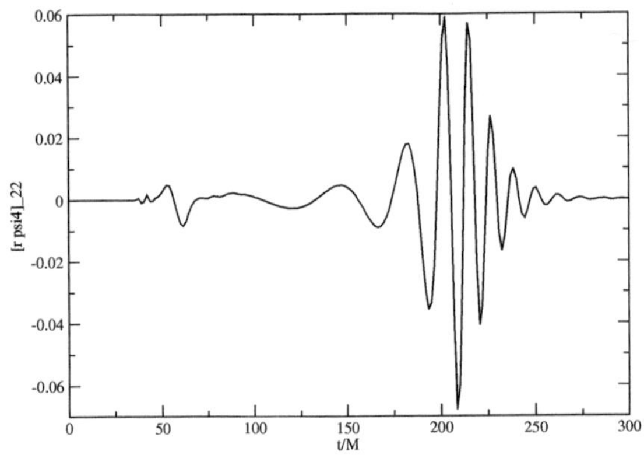

Fig. 3. Waveform from the inspiral of an equal mass binary from an initial separation of $\approx 3.2M$. The initial pulse up to approximately $100M$ is an artifact of the choice of initial data. We plot the real part of the ψ_4 component of the Weyl tensor, rescaled by the radius of the extraction sphere, which asymptotically corresponds to the value of the area coordinate or luminosity distance

a search template for gravitational wave data analysis. At late times the solution approaches a stationary state which corresponds to a Kerr black hole. The fact that our computational quantities also reach a stationary state relies again on a beneficial choice of coordinate gauge, see [21].

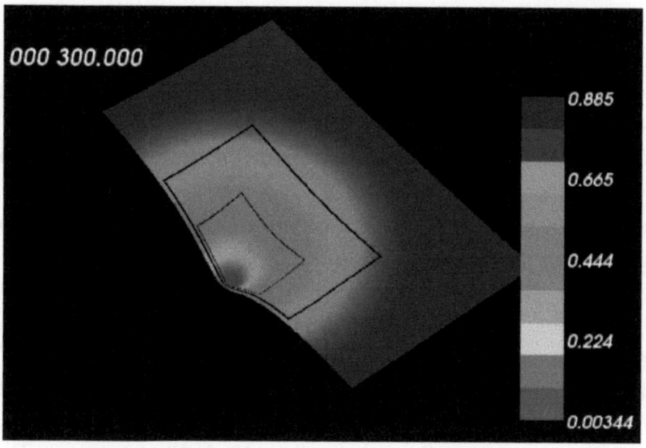

Fig. 4. Late time stationary profile of the lapse function α

5 Discussion

Since we have started our project at HLRS, there have been drastic changes in our code and in the status of the field. We have made a transition from an essentially single-processor second order accurate fixed mesh refinement code that was able to track roughly one orbit of a black-hole binary, to a fourth-order accurate code, where fine grid levels track the motion of the black holes, giving us robust waveform calculations from evolutions lasting several orbits, and which scales well at least up to 32 processors.

These developments provide us with a tool that we believe will outperform the best methods for binary black hole simulations currently described in the literature, and will allow us to start extensive parameter studies of astrophysically interesting initial data sets. A "downside" of our progress is that since current evolutions are significantly more stable and can and should be run 2–3 times longer than what we had hoped to be possible, this leads to a corresponding increase in our needs for computer time.

For the future we plan to document our current work in several publications, and to start parameter studies of astrophysically interesting binary black hole systems, e.g. with large mass ratios and large spins, but also to progress with the theoretical understanding of our methods and with interfacing our results to the gravitational wave data analysis community.

Acknowledgements

This work was supported by the SFB/Transregio 7 on *"Gravitational Wave Astronomy"* of the DFG.

References

1. A. A. Abramovici, W. Althouse, R. P. Drever, Y. Gursel, S. Kawamura, F. Raab, D. Shoemaker, L. Sievers, R. Spero, K. S. Thorne, R. Vogt, R. Weiss, S. Whitcomb, and M. Zuker. Ligo: The laser interferometer gravitational-wave observatory. *Science*, 256:325–333, 1992.
2. Miguel Alcubierre. The appearance of coordinate shocks in hyperbolic formulations of general relativity. *Phys. Rev. D*, 55:5981–5991, 1997.
3. Miguel Alcubierre, Bernd Brügmann, Peter Diener, Michael Koppitz, Denis Pollney, Edward Seidel, and Ryoji Takahashi. Gauge conditions for long-term numerical black hole evolutions without excision. *Phys. Rev. D*, 67:084023, 2003.
4. G. Allen, T. Goodale, J. Massó, and E. Seidel. The cactus computational toolkit and using distributed computing to collide neutron stars. In *Proceedings of Eighth IEEE International Symposium on High Performance Distributed Computing, HPDC-8, Redondo Beach, 1999*. IEEE Press, 1999.
5. Richard Arnowitt, Stanley Deser, and Charles W. Misner. The dynamics of general relativity. In L. Witten, editor, *Gravitation: An introduction to current research*, pages 227–265. John Wiley, New York, 1962.
6. John G. Baker, Joan Centrella, Dae-Il Choi, Michael Koppitz, and James van Meter. Binary black hole merger dynamics and waveforms. *Phys. Rev. D*, 73:104002, 2006.
7. John G. Baker, Joan Centrella, Dae-Il Choi, Michael Koppitz, and James van Meter. Gravitational wave extraction from an inspiraling configuration of merging black holes. *Phys. Rev. Lett.*, 96:111102, 2006.
8. Thomas W. Baumgarte and Stuart L. Shapiro. On the numerical integration of Einstein's field equations. *Phys. Rev. D*, 59:024007, 1999.
9. Bernd Brügmann, Wolfgang Tichy, and Nina Jansen. Numerical simulation of orbiting black holes. *Phys. Rev. Lett.*, 92:211101, 2004.
10. Gioel Calabrese, Ian Hinder, and Sascha Husa. Numerical stability for finite difference approximations of Einstein's equations. *J. Comp. Phys*, 2005. in press.
11. Manuela Campanelli, C. O. Lousto, and Y. Zlochower. The last orbit of binary black holes. *Phys. Rev. D*, 73:061501(R), 2006.
12. Manuela Campanelli, Carlos O. Lousto, Pedro Marronetti, and Yosef Zlochower. Accurate evolutions of orbiting black-hole binaries without excision. *Phys. Rev. Letter*, 96:111101, 2006.
13. Dae-Il Choi. Recent results on binary black hole simulations. Talk given at Penn State Sources and Simulations Seminar Seroes, April 11 2006, 2006.
14. Dae-Il Choi, J. David Brown, Breno Imbiriba, Joan Centrella, and Peter MacNeice. Interface conditions for wave propagation through mesh refinement boundaries. *J. Comput. Phys.*, 193:398–425, 2004.
15. Gregory B. Cook. Initial data for numerical relativity. *Living Rev. Rel.*, 3:5, 2000.
16. K. Danzmann. The geo project: a long baseline laser interferometer for the detection of gravitational waves. *Lecture Notes in Physics*, 410:184–209, 1992.
17. GEO600 – http://www.geo600.uni-hannover.de/.
18. Carsten Gundlach and Jose M. Martin-Garcia. Symmetric hyperbolicity and consistent boundary conditions for second-order Einstein equations. *Phys. Rev. D*, 70:044032, 2004.

19. Carsten Gundlach and Jose M. Martin-Garcia. Hyperbolicity of second-order in space systems of evolution equations. 2005.
20. Carsten Gundlach and Jose M. Martin-Garcia. Well-posedness of formulations of the einstein equations with dynamical lapse and shift conditions. 2006.
21. Mark Hannam, Sascha Husa, Denis Pollney, Bernd Bruegmann, and Niall O'Murchadha. Geometry and regularity of moving punctures. 2006.
22. Frank Herrmann, Deirdre Shoemaker, and Pablo Laguna. Unequal-mass binary black hole inspirals. 2006.
23. Breno Imbiriba, John Baker, Dae-Il Choi, Joan Centrella, David R. Fiske, J. David Brown, James R. van Meter, and Kevin Olson. Evolving a puncture black hole with fixed mesh refinement. *Phys. Rev. D*, 70:124025, 2004.
24. Heinz-Otto Kreiss and Joseph Oliger. Methods for the approximate solution of time dependent problems. *Global atmospheric research programme publications series*, 10, 1973.
25. LIGO – http://www.ligo.caltech.edu/.
26. Peter MacNeice, Kevin M. Olson, Clark Mobarry, Rosalinda de Fainchtein, and Charles Packer. Paramesh: A parallel adaptive mesh refinement community toolkit. *Computer Physics Communications*, 126(3):330–354, 11 April 2000.
27. Masaru Shibata and Takashi Nakamura. Evolution of three-dimensional gravitational waves: Harmonic slicing case. *Phys. Rev. D*, 52:5428, 1995.
28. Cactus Computational Toolkit. `http://www.cactuscode.org`.
29. VIRGO – http://www.virgo.infn.it/.
30. The xgraph and ygraph Home Pages
`http://jean-luc.aei-potsdam.mpg.de/Codes/xgraph`,
`http://www.aei.mpg.de/~pollney/ygraph`.
31. Y. Zlochower, J. G. Baker, M. Campanelli, and C. O. Lousto. Accurate black hole evolutions by fourth-order numerical relativity. *Phys. Rev. D*, 72:024021, 2005.

The SuperN-Project: Understanding Core Collapse Supernovae

A. Marek, K. Kifonidis, H.-Th. Janka, and B. Müller

Max-Planck-Institut für Astrophysik, Karl-Schwarzschild-Strasse 1, Postfach 1317, D-85741 Garching bei München, Germany
amarek@mpa-garching.mpg.de

Summary. We give an overview of the problems and the current status of (core collapse) supernova modeling, and discuss the system of equations and the algorithm for its solution that are employed in our code. We also report on our recent progress, and focus on the ongoing calculations that are performed on the SX-8 at the HLRS Stuttgart.

1 Introduction

A star more massive than about 8 solar masses ends its live in a cataclysmic explosion, a supernova. Its quiescent evolution comes to an end, when the pressure in its inner layers is no longer able to balance the inward pull of gravity. Throughout its life, the star sustained this balance by generating energy through a sequence of nuclear fusion reactions, forming increasingly heavier elements in its core. However, when the core consists mainly of iron-group nuclei, central energy generation ceases. The fusion reactions producing iron-group nuclei relocate to the core's surface, and their "ashes" continuously increase the core's mass. Similar to a white dwarf, such a core is stabilized against gravity by the pressure of its degenerate gas of electrons. However, to remain stable, its mass must stay smaller than the Chandrasekhar limit. When the core grows larger than this limit, it collapses to a neutron star, and a huge amount ($\sim 10^{53}$ erg) of gravitational binding energy is set free. Most ($\sim 99\%$) of this energy is radiated away in neutrinos, but a small fraction is transferred to the outer stellar layers and drives the violent mass ejection which disrupts the star in a supernova.

Despite 40 years of research, the details of how this energy transfer happens and how the explosion is initiated are still not well understood. Observational evidence about the physical processes deep inside the collapsing star is sparse and almost exclusively indirect. The only direct observational access is via measurements of neutrinos or gravitational waves. To obtain insight into the events in the core, one must therefore heavily rely on sophisticated numerical simulations. The enormous amount of computer power required for this purpose has led to the use of several, often questionable, approximations and numerous ambiguous results in the past. Fortunately, however, the development of numerical tools and computational resources has meanwhile advanced to a point, where it is becoming possible to perform multi-dimensional simulations with unprecedented accuracy. Therefore there is hope that the physical processes which are essential for the explosion can finally be unraveled.

An understanding of the explosion mechanism is required to answer many important questions of nuclear, gravitational, and astro-physics like the following:

- How do the explosion energy, the explosion timescale, and the mass of the compact remnant depend on the progenitor's mass? Is the explosion mechanism the same for all progenitors? For which stars are black holes left behind as compact remnants instead of neutron stars?
- What is the role of the – poorly known – equation of state (EoS) for the proto neutron star? Do softer or stiffer EoSs favor the explosion of a core collapse supernova?
- What is the role of rotation during the explosion? How rapidly do newly formed neutron stars rotate?
- How do neutron stars receive their natal kicks? Are they accelerated by asymmetric mass ejection and/or anisotropic neutrino emission?
- What are the generic properties of the neutrino emission and of the gravitational wave signal that are produced during stellar core collapse and explosion? Up to which distances could these signals be measured with operating or planned detectors on earth and in space? And what can one learn about supernova dynamics from a future measurement of such signals in case of a Galactic supernova?

2 Numerical Models

2.1 History and Constraints

According to theory, a shock wave is launched at the moment of "core bounce" when the neutron star begins to emerge from the collapsing stellar iron core. There is general agreement, supported by all "modern" numerical simulations, that this shock is unable to propagate directly into the stellar mantle and envelope, because it looses too much energy in dissociating iron into free nucleons while it moves through the outer core. The "prompt" shock ultimately stalls. Thus the currently favored theoretical paradigm needs to exploit the fact that a huge energy reservoir is present in the form of neutrinos, which are abundantly emitted from the hot, nascent neutron star. The absorption of electron neutrinos and antineutrinos by free nucleons in the post shock layer is thought to reenergize the shock, and lead to the supernova explosion.

Detailed *spherically symmetric* hydrodynamic models, which recently include a very accurate treatment of the time-dependent, multi-flavor, multi-frequency neutrino transport based on a numerical solution of the Boltzmann transport equation [1, 2, 3, 4], reveal that this "delayed, neutrino-driven mechanism" does not work as simply as originally envisioned. Although in principle able to trigger the explosion (e.g., [5], [6], [7]), neutrino energy transfer to the postshock matter turned out to be too weak. For inverting the infall of the stellar core and initiating powerful mass ejection, an increase of the efficiency of neutrino energy deposition is needed.

A number of physical phenomena have been pointed out that can enhance neutrino energy deposition behind the stalled supernova shock. They are all linked to the fact that the real world is multi-dimensional instead of spherically symmetric (or one-dimensional; 1D) as assumed in the work cited above:

(1) Convective instabilities in the neutrino-heated layer between the neutron star and the supernova shock develop to violent convective overturn [8]. This convective overturn is helpful for the explosion, mainly because (a) neutrino-heated matter rises and increases the pressure behind the shock, thus pushing the shock further out, and (b) cool matter is able to penetrate closer to the neutron star where it can absorb neutrino energy more efficiently. Both effects allow multi-dimensional models to explode easier than spherically symmetric ones [9, 10, 11].
(2) Recent work [12, 13, 14, 15] has demonstrated that the stalled supernova shock is also subject to a second non-radial instability which can grow to a dipolar, global deformation of the shock [15].
(3) Convective energy transport inside the nascent neutron star [16, 17, 18] might enhance the energy transport to the neutrinosphere and could thus boost the neutrino luminosities. This would in turn increase the neutrino-heating behind the shock.

This list of multi-dimensional phenomena awaits more detailed exploration in multi-dimensional simulations. Until recently, such simulations have been performed with only a grossly simplified treatment of the involved microphysics, in particular of the neutrino transport and neutrino-matter interactions. At best, grey (i.e., single energy) flux-limited diffusion schemes were employed. All published successful simulations of supernova explosions by the convectively aided neutrino-heating mechanism in two [9, 10, 19, 20] and three dimensions [21, 22] used such a radical approximation of the neutrino transport.

Since, however, the role of the neutrinos is crucial for the problem, and because previous experience shows that the outcome of simulations is indeed very sensitive to the employed transport approximations, studies of the explosion mechanism require the best available description of the neutrino physics. This implies that one has to solve the Boltzmann transport equation for neutrinos.

2.2 Recent Calculations and the Need for TFlop Simulations

We have recently advanced to a new level of accuracy for supernova simulations by generalizing the VERTEX code, a Boltzmann solver for neutrino transport, from spherical symmetry [23] to multi-dimensional applications [24, 25]. The corresponding mathematical model, and in particular our method for tackling the integro-differential transport problem in multi-dimensions, will be summarized in Sect. 3.

Results of a set of simulations with our code in 1D and 2D for progenitor stars with different masses have recently been published by [25], and with respect to the expected gravitational-wave signals from rotating and convective supernova cores by [26]. The recent progress in supernova modeling was summarized and set in perspective in a conference article by [24].

Our collection of simulations has helped us to identify a number of effects which have brought our two-dimensional models close to the threshold of explosion. This makes us optimistic that the solution of the long-standing problem of how massive stars explode may be in reach. In particular, we have recognized the following aspects as advantageous:

- Stellar rotation, even at a moderate level, supports the expansion of the stalled shock by centrifugal forces and instigates overturn motion in the neutrino-heated postshock matter by meridional circulation flows in addition to convective instabilities.
- Changing from the current "standard" and most widely used EoS for stellar core-collapse simulations [27] to alternative descriptions [28, 29], we found in 1D calculations that a higher incompressibility of the supranuclear phase yields a less dramatic and less rapid recession of the stalled shock after it has reached its maximum expansion [24]. This finding suggests that the EoS of [29] might lead to more favorable conditions for strong postshock convection, and thus more efficient neutrino heating, than current 2D simulations with the EoS of [27].

All these effects are potentially important, and some (or even all of them) may represent crucial ingredients for a successful supernova simulation. So far no multi-dimensional calculations have been performed, in which two or more of these items have been taken into account simultaneously, and thus their mutual interaction awaits to be investigated. It should also be kept in mind that our knowledge of supernova microphysics, and especially the EoS of neutron star matter, is still incomplete, which implies major uncertainties for supernova modeling. Unfortunately, the impact of different descriptions for this input physics has so far not been satisfactorily explored with respect to the neutrino-heating mechanism and the long-time behavior of the supernova shock, in particular in multi-dimensional models.

From this it is clear that rather extensive parameter studies using multi-dimensional simulations are required to identify the physical processes which are essential for the explosion. Since on a dedicated machine performing at a sustained speed of about 30 GFlops already a single 2D simulation has a turn-around time of more than half a year, these parameter studies are not possible without TFlop simulations.

3 The Mathematical Model

The non-linear system of partial differential equations which is solved in our code consists of the following components:

- The Euler equations of hydrodynamics, supplemented by advection equations for the electron fraction and the chemical composition of the fluid, and formulated in spherical coordinates;
- the Poisson equation for calculating the gravitational source terms which enter the Euler equations, including corrections for general relativistic effects;
- the Boltzmann transport equation which determines the (non-equilibrium) distribution function of the neutrinos;
- the emission, absorption, and scattering rates of neutrinos, which are required for the solution of the Boltzmann equation;
- the equation of state of the stellar fluid, which provides the closure relation between the variables entering the Euler equations, i.e. density, momentum, energy, electron fraction, composition, and pressure.

In what follows we will briefly summarize the neutrino transport algorithms. For a more complete description of the entire code we refer the reader to [25], and the references therein.

3.1 "Ray-by-Ray Plus" Variable Eddington Factor Solution of the Neutrino Transport Problem

The crucial quantity required to determine the source terms for the energy, momentum, and electron fraction of the fluid owing to its interaction with the neutrinos is the neutrino distribution function in phase space, $f(r, \vartheta, \phi, \epsilon, \Theta, \Phi, t)$. Equivalently, the neutrino intensity $I = c/(2\pi\hbar c)^3 \cdot \epsilon^3 f$ may be used. Both are seven-dimensional functions, as they describe, at every point in space (r, ϑ, ϕ), the distribution of neutrinos propagating with energy ϵ into the direction (Θ, Φ) at time t (Fig. 1).

The evolution of I (or f) in time is governed by the Boltzmann equation, and solving this equation is, in general, a six-dimensional problem (as time is usually not counted as a separate dimension). A solution of this equation by direct discretization (using an S_N scheme) would require computational resources in the PetaFlop range. Although there are attempts by at least one group in the United States to follow such an approach, we feel that, with the currently available computational resources, it is mandatory to reduce the dimensionality of the problem.

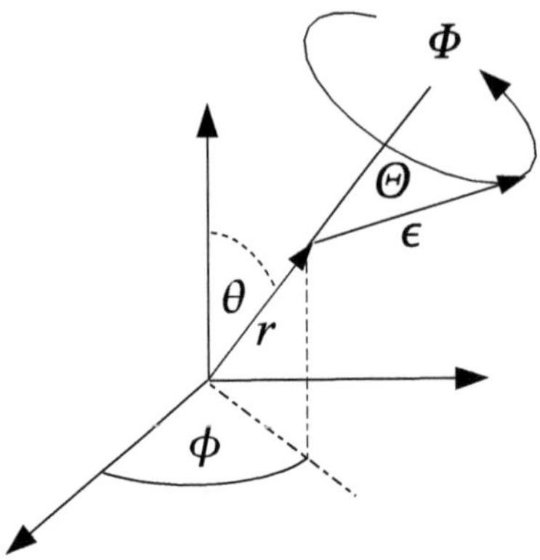

Fig. 1. Illustration of the phase space coordinates (see the main text)

Actually this should be possible, since the source terms entering the hydrodynamic equations are *integrals* of I over momentum space (i.e. over ϵ, Θ, and Φ), and thus only a fraction of the information contained in I is truly required to compute the dynamics of the flow. It makes therefore sense to consider angular moments of I, and to solve evolution equations for these moments, instead of dealing with the Boltzmann equation directly. The 0th to 3rd order moments are defined as

$$J, \boldsymbol{H}, \mathbf{K}, \mathbf{L}, \ldots (r, \vartheta, \phi, \epsilon, t) = \frac{1}{4\pi} \int I(r, \vartheta, \phi, \epsilon, \Theta, \Phi, t)\, \boldsymbol{n}^{0,1,2,3,\ldots}\, \mathrm{d}\Omega \qquad (1)$$

where $\mathrm{d}\Omega = \sin\Theta\, \mathrm{d}\Theta\, \mathrm{d}\Phi$, $\boldsymbol{n} = (\cos\Theta, \sin\Theta\cos\Phi, \sin\Theta\sin\Phi)$, and exponentiation represents repeated application of the dyadic product. Note that the moments are *tensors* of the required rank.

This leaves us with a four-dimensional problem. So far no approximations have been made. In order to reduce the size of the problem even further, one needs to resort to assumptions on its symmetry. At this point, one usually employs azimuthal symmetry for the stellar matter distribution, i.e. any dependence on the azimuth angle ϕ is ignored, which implies that the hydrodynamics of the problem can be treated in two dimensions. It also implies $I(r, \vartheta, \epsilon, \Theta, \Phi) = I(r, \vartheta, \epsilon, \Theta, -\Phi)$. If, in addition, it is assumed that I is even independent of Φ, then each of the angular moments of I becomes a *scalar*, which depends on two spatial dimensions, and one dimension in momentum space: $J, H, K, L = J, H, K, L(r, \vartheta, \epsilon, t)$. Thus we have reduced the problem to three dimensions in total.

The System of Equations

With the aforementioned assumptions it can be shown [25], that in order to compute the source terms for the energy and electron fraction of the fluid, the following two transport equations need to be solved:

$$\left(\frac{1}{c}\frac{\partial}{\partial t} + \beta_r \frac{\partial}{\partial r} + \frac{\beta_\vartheta}{r}\frac{\partial}{\partial \vartheta}\right) J + J\left(\frac{1}{r^2}\frac{\partial(r^2\beta_r)}{\partial r} + \frac{1}{r\sin\vartheta}\frac{\partial(\sin\vartheta\beta_\vartheta)}{\partial\vartheta}\right)$$
$$+\frac{1}{r^2}\frac{\partial(r^2 H)}{\partial r} + \frac{\beta_r}{c}\frac{\partial H}{\partial t} - \frac{\partial}{\partial\epsilon}\left\{\frac{\epsilon}{c}\frac{\partial\beta_r}{\partial t}H\right\} - \frac{\partial}{\partial\epsilon}\left\{\epsilon J\left(\frac{\beta_r}{r} + \frac{1}{2r\sin\vartheta}\frac{\partial(\sin\vartheta\beta_\vartheta)}{\partial\vartheta}\right)\right\}$$
$$-\frac{\partial}{\partial\epsilon}\left\{\epsilon K\left(\frac{\partial\beta_r}{\partial r} - \frac{\beta_r}{r} - \frac{1}{2r\sin\vartheta}\frac{\partial(\sin\vartheta\beta_\vartheta)}{\partial\vartheta}\right)\right\} + J\left(\frac{\beta_r}{r} + \frac{1}{2r\sin\vartheta}\frac{\partial(\sin\vartheta\beta_\vartheta)}{\partial\vartheta}\right)$$
$$+K\left(\frac{\partial\beta_r}{\partial r} - \frac{\beta_r}{r} - \frac{1}{2r\sin\vartheta}\frac{\partial(\sin\vartheta\beta_\vartheta)}{\partial\vartheta}\right) + \frac{2}{c}\frac{\partial\beta_r}{\partial t}H = C^{(0)}, \quad (2)$$

$$\left(\frac{1}{c}\frac{\partial}{\partial t} + \beta_r \frac{\partial}{\partial r} + \frac{\beta_\vartheta}{r}\frac{\partial}{\partial \vartheta}\right) H + H\left(\frac{1}{r^2}\frac{\partial(r^2\beta_r)}{\partial r} + \frac{1}{r\sin\vartheta}\frac{\partial(\sin\vartheta\beta_\vartheta)}{\partial\vartheta}\right)$$
$$+\frac{\partial K}{\partial r} + \frac{3K-J}{r} + H\left(\frac{\partial\beta_r}{\partial r}\right) + \frac{\beta_r}{c}\frac{\partial K}{\partial t} - \frac{\partial}{\partial\epsilon}\left\{\frac{\epsilon}{c}\frac{\partial\beta_r}{\partial t}K\right\}$$
$$-\frac{\partial}{\partial\epsilon}\left\{\epsilon L\left(\frac{\partial\beta_r}{\partial r} - \frac{\beta_r}{r} - \frac{1}{2r\sin\vartheta}\frac{\partial(\sin\vartheta\beta_\vartheta)}{\partial\vartheta}\right)\right\}$$
$$-\frac{\partial}{\partial\epsilon}\left\{\epsilon H\left(\frac{\beta_r}{r} + \frac{1}{2r\sin\vartheta}\frac{\partial(\sin\vartheta\beta_\vartheta)}{\partial\vartheta}\right)\right\} + \frac{1}{c}\frac{\partial\beta_r}{\partial t}(J+K) = C^{(1)}. \quad (3)$$

These are evolution equations for the neutrino energy density, J, and the neutrino flux, H, and follow from the zeroth and first moment equations of the comoving frame (Boltzmann) transport equation in the Newtonian, $\mathcal{O}(v/c)$ approximation. The quantities $C^{(0)}$ and $C^{(1)}$ are source terms that result from the collision term of the Boltzmann equation, while $\beta_r = v_r/c$ and $\beta_\vartheta = v_\vartheta/c$, where v_r and v_ϑ are the components of the hydrodynamic velocity, and c is the speed of light. The functional dependences $\beta_r = \beta_r(r,\vartheta,t)$, $J = J(r,\vartheta,\epsilon,t)$, etc. are suppressed in the notation. This system includes four unknown moments (J,H,K,L) but only two equations, and thus needs to be supplemented by two more relations. This is done by substituting $K = f_K \cdot J$ and $L = f_L \cdot J$, where f_K and f_L are the variable Eddington factors, which for the moment may be regarded as being known, but in our case is indeed determined from a separate simplified ("model") Boltzmann equation.

A finite volume discretization of Eqs. (2–3) is sufficient to guarantee exact conservation of the total neutrino energy. However, and as described in detail in [23], it is not sufficient to guarantee also exact conservation of the neutrino number. To achieve this, we discretize and solve a set of two additional equations. With $\mathcal{J} = J/\epsilon$, $\mathcal{H} = H/\epsilon$, $\mathcal{K} = K/\epsilon$, and $\mathcal{L} = L/\epsilon$, this set of equations reads

$$\left(\frac{1}{c}\frac{\partial}{\partial t}+\beta_r\frac{\partial}{\partial r}+\boldsymbol{\frac{\beta_\vartheta}{r}\frac{\partial}{\partial\vartheta}}\right)\mathcal{J}+\mathcal{J}\left(\frac{1}{r^2}\frac{\partial(r^2\beta_r)}{\partial r}+\boldsymbol{\frac{1}{r\sin\vartheta}\frac{\partial(\sin\vartheta\beta_\vartheta)}{\partial\vartheta}}\right)$$
$$+\frac{1}{r^2}\frac{\partial(r^2\mathcal{H})}{\partial r}+\frac{\beta_r}{c}\frac{\partial\mathcal{H}}{\partial t}-\frac{\partial}{\partial\epsilon}\left\{\frac{\epsilon}{c}\frac{\partial\beta_r}{\partial t}\mathcal{H}\right\}-\frac{\partial}{\partial\epsilon}\left\{\epsilon\mathcal{J}\left(\frac{\beta_r}{r}+\boldsymbol{\frac{1}{2r\sin\vartheta}\frac{\partial(\sin\vartheta\beta_\vartheta)}{\partial\vartheta}}\right)\right\}$$
$$-\frac{\partial}{\partial\epsilon}\left\{\epsilon\mathcal{K}\left(\frac{\partial\beta_r}{\partial r}-\frac{\beta_r}{r}-\boldsymbol{\frac{1}{2r\sin\vartheta}\frac{\partial(\sin\vartheta\beta_\vartheta)}{\partial\vartheta}}\right)\right\}+\frac{1}{c}\frac{\partial\beta_r}{\partial t}\mathcal{H}=\mathcal{C}^{(0)}, \quad (4)$$

$$\left(\frac{1}{c}\frac{\partial}{\partial t}+\beta_r\frac{\partial}{\partial r}+\boldsymbol{\frac{\beta_\vartheta}{r}\frac{\partial}{\partial\vartheta}}\right)\mathcal{H}+\mathcal{H}\left(\frac{1}{r^2}\frac{\partial(r^2\beta_r)}{\partial r}+\boldsymbol{\frac{1}{r\sin\vartheta}\frac{\partial(\sin\vartheta\beta_\vartheta)}{\partial\vartheta}}\right)$$
$$+\frac{\partial\mathcal{K}}{\partial r}+\frac{3\mathcal{K}-\mathcal{J}}{r}+\mathcal{H}\left(\frac{\partial\beta_r}{\partial r}\right)+\frac{\beta_r}{c}\frac{\partial\mathcal{K}}{\partial t}-\frac{\partial}{\partial\epsilon}\left\{\frac{\epsilon}{c}\frac{\partial\beta_r}{\partial t}\mathcal{K}\right\}$$
$$-\frac{\partial}{\partial\epsilon}\left\{\epsilon\mathcal{L}\left(\frac{\partial\beta_r}{\partial r}-\frac{\beta_r}{r}-\boldsymbol{\frac{1}{2r\sin\vartheta}\frac{\partial(\sin\vartheta\beta_\vartheta)}{\partial\vartheta}}\right)\right\}$$
$$-\frac{\partial}{\partial\epsilon}\left\{\epsilon\mathcal{H}\left(\frac{\beta_r}{r}+\boldsymbol{\frac{1}{2r\sin\vartheta}\frac{\partial(\sin\vartheta\beta_\vartheta)}{\partial\vartheta}}\right)\right\}-\mathcal{L}\left(\frac{\partial\beta_r}{\partial r}-\frac{\beta_r}{r}-\boldsymbol{\frac{1}{2r\sin\vartheta}\frac{\partial(\sin\vartheta\beta_\vartheta)}{\partial\vartheta}}\right)$$
$$-\mathcal{H}\left(\frac{\beta_r}{r}+\boldsymbol{\frac{1}{2r\sin\vartheta}\frac{\partial(\sin\vartheta\beta_\vartheta)}{\partial\vartheta}}\right)+\frac{1}{c}\frac{\partial\beta_r}{\partial t}\mathcal{J}=\mathcal{C}^{(1)}. \quad (5)$$

The moment equations (2–5) are very similar to the $\mathcal{O}(v/c)$ equations in spherical symmetry which were solved in the 1D simulations of [23] (see Eqs. 7,8,30, and 31 of the latter work). This similarity has allowed us to reuse a good fraction of the one-dimensional version of VERTEX, for coding the multi-dimensional algorithm. The additional terms necessary for this purpose have been set in boldface above.

Finally, the changes of the energy, e, and electron fraction, Y_e, required for the hydrodynamics are given by the following two equations

$$\frac{\mathrm{d}e}{\mathrm{d}t}=-\frac{4\pi}{\rho}\int_0^\infty\mathrm{d}\epsilon\sum_{\nu\in(\nu_\mathrm{e},\bar\nu_\mathrm{e},\dots)}C_\nu^{(0)}(\epsilon), \quad (6)$$

$$\frac{\mathrm{d}Y_\mathrm{e}}{\mathrm{d}t}=-\frac{4\pi\, m_\mathrm{B}}{\rho}\int_0^\infty\mathrm{d}\epsilon\left(C_{\nu_\mathrm{e}}^{(0)}(\epsilon)-C_{\bar\nu_\mathrm{e}}^{(0)}(\epsilon)\right) \quad (7)$$

(for the momentum source terms due to neutrinos see [25]). Here m_B is the baryon mass, and the sum in Eq. (6) runs over all neutrino types. The full system consisting of Eqs. (2–7) is stiff, and thus requires an appropriate discretization scheme for its stable solution.

Method of Solution

In order to discretize Eqs. (2–7), the spatial domain $[0, r_\mathrm{max}]\times[\vartheta_\mathrm{min},\vartheta_\mathrm{max}]$ is covered by N_r radial, and N_ϑ angular zones, where $\vartheta_\mathrm{min}=0$ and $\vartheta_\mathrm{max}=\pi$ correspond to the north and south poles, respectively, of the spherical grid.

(In general, we allow for grids with different radial resolutions in the neutrino transport and hydrodynamic parts of the code. The number of radial zones for the hydrodynamics will be denoted by N_r^{hyd}.) The number of bins used in energy space is N_ϵ and the number of neutrino types taken into account is N_ν.

The equations are solved in two operator-split steps corresponding to a lateral and a radial sweep.

In the first step, we treat the boldface terms in the respectively first lines of Eqs. (2–5), which describe the lateral advection of the neutrinos with the stellar fluid, and thus couple the angular moments of the neutrino distribution of neighbouring angular zones. For this purpose we consider the equation

$$\frac{1}{c}\frac{\partial \Xi}{\partial t} + \frac{1}{r \sin\vartheta}\frac{\partial(\sin\vartheta\, \beta_\vartheta\, \Xi)}{\partial \vartheta} = 0\,, \tag{8}$$

where Ξ represents one of the moments J, H, $\mathcal{J}$, or $\mathcal{H}$. Although it has been suppressed in the above notation, an equation of this form has to be solved for each radius, for each energy bin, and for each type of neutrino. An explicit upwind scheme is used for this purpose.

In the second step, the radial sweep is performed. Several points need to be noted here:

- terms in boldface not yet taken into account in the lateral sweep, need to be included into the discretization scheme of the radial sweep. This can be done in a straightforward way since these remaining terms do not include derivatives of the transport variables (J, H) or $(\mathcal{J}, \mathcal{H})$. They only depend on the hydrodynamic velocity v_ϑ, which is a *constant* scalar field for the transport problem.
- the right hand sides (source terms) of the equations and the coupling in energy space have to be accounted for. The coupling in energy is non-local, since the source terms of Eqs. (2–5) stem from the Boltzmann equation, which is an integro-differential equation and couples all the energy bins
- the discretization scheme for the radial sweep is *implicit* in time. Explicit schemes would require very small time steps to cope with the stiffness of the source terms in the optically thick regime, and the small CFL time step dictated by neutrino propagation with the speed of light in the optically thin regime. Still, even with an implicit scheme $\gtrsim 10^5$ time steps are required per simulation. This makes the calculations expensive.

Once the equations for the radial sweep have been discretized in radius and energy, the resulting solver is applied ray-by-ray for each angle ϑ and for each type of neutrino, i.e. for constant ϑ, N_ν two-dimensional problems need to be solved.

The discretization itself is done using a second order accurate scheme with backward differencing in time according to [23]. This leads to a non-linear system of algebraic equations, which is solved by Newton-Raphson iteration with explicit construction and inversion of the corresponding Jacobian matrix.

4 Results and Ongoing Work

With the computer power available at the HLRS we try to answer some of the important questions in SN theory (see Sect. 1) with 2D-simulations. At the HLRS, we typically run our code on one node (8 processors) of the SX-8 with 98.3% of vector operations and 22000 MFLOPS per second and one simulation roughly needs 75000 CPU-hours.

In the following we present some of our (preliminary) results from these simulations that are currently performed at the HLRS. All of these simulations are calculated on a 180 degree (north pole–south pole) plane and have an angular resolution of 0.94 degree for the non-rotating models, and 1.41 degree for the rotating model, respectively. As neutrino interaction rates we use the full set as described in [30] and general relativistic effects are taken into account according to [31]. These simulations are the best resolved and computationally most expensive calculations ever performed by the Garching supernova group.

4.1 Supernova Explosions for Low Mass Progenitor Stars

Successful supernova explosions with Boltzmann neutrino transport are – even in multidimensional simulations – not routinely obtained. However, [32, 25] recently reported such successful explosions with low-mass progenitor stars, i.e. 8.8 $M_\odot$ and 11.2 $M_\odot$, whereas more massive stars fail to explode in simulations with detailed neutrino transport. This suggests that these explosions were either obtained by "chance" (i.e. triggered by very special combinations of progenitor properties), or – more likely – it suggests that the neutrino heating mechanism reliably works for progenitors in a certain low mass range. This

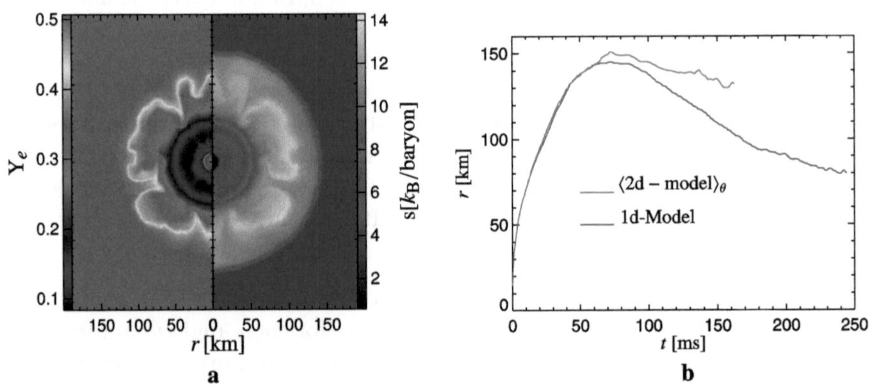

Fig. 2. **a**: A snapshot of entropy (*left*) and the electron fraction (*right*) at a time of 130 ms after the shock formation for a n10.2 $M_\odot$ progenitor (see text). **b**: The shock positions as functions of time for the corresponding 1D-model and the laterally averaged 2D-model. Time is normalized to the moment of core bounce

is an important question since it is nowadays speculated by a few experts that some other physical mechanisms (like energy transport by sound waves [33] or the presence of magnetic fields e.g. [34]) are necessary for the explosion of a core collapse supernova. By taking a progenitor model of $10.2\,M_\odot$, we have started to simulate the supernova evolution for another progenitor between $8\,M_\odot$ and $11.2\,M_\odot$ in order to further investigate the supernova problem in the low mass progenitor range. This simulation is not finished yet, and we cannot confirm or rule out another successful explosion for a low mass progenitor star. In Fig. 2 a we show a snapshot of the entropy and proton-to-baryon-ratio ("electron fraction") at a time of 130 ms after the shock formation. In Fig. 2b we compare the (lateraly averaged) shock positions of our 2D and 1D model. As one can see, convective flows are clearly present at this time, but have only moderate influence on the shock position. However, the subsequent evolution of these convective instabilities will be crucial for a success of the explosion.

4.2 The Nuclear EoS for Proto-Neutron Stars

As we have have already pointed out, the knowledge of the EoS for neutron stars is still incomplete. Spherically symmetric simulations indicate that the EoS determines the expansion of the shock, and affects the luminosity of emitted neutrinos, and may have important influence on the growth of convective instabilities. In a set of two simulations we have started to explore the effects of the EoS on the evolution of a $15\,M_\odot$ star. For these simulations we make use of the Wolff-EoS [29] and the standard EoS for supernova simulations according to [27]. Though these simulations are not yet finished, we can report different dynamical behavior in these simulations: we find indeed that the EoS strongly determines the timescales and the strength of the development of hydrodynamic instabilities. In Figs. 3a,b,c,d we depict snapshots of both simulations with different EoSs at a time of 17 ms after the shock formation (upper panels) and at a time of 170 ms after the shock formation (middle panels). At early times, one can clearly see convective patterns in the simulations with the Wolff-EoS, whereas the simulation with the L&S-EoS does not show any convective motions at this time. As the snapshot at later times reveals, we also find different convective activity in the neutrino heating region below the shock front. This activity clearly influences the position of the shock front at this time. Furthermore the entropies in the model calculated with the Wolff-EoS are higher, which indicates a larger energy content in the heating region in this model. It is important that the EoS influences the development of convective motion, since this may also influence the development of low mode ($l = 1, 2, 3\ldots$) non-radial instabilities of the region below the shock front, see [15]. One can also see in Figs. 3c,d that the EoS determines – by its stiffness – the size of the compact inner core (roughly the green circular region in the inner cores of Figs. 3c,d). This is an important fact, because the size of the shrinking inner core influences the shock position, see e.g. [24], and may be important for an successful supernova explosion.

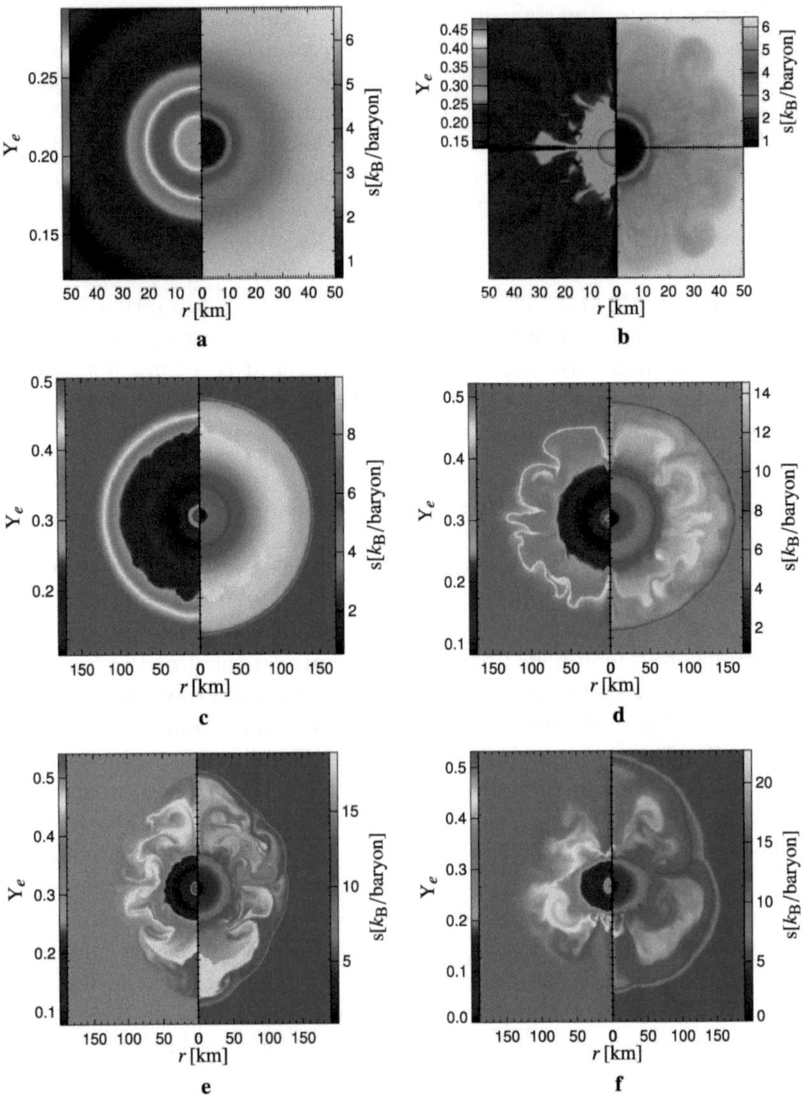

Fig. 3. Snapshots of entropy and electron fraction for different simulations of a 15 $M_\odot$ progenitor. Note the different axis scales on all plots. **a**: A snapshot for a calculation with the L&S-EoS at a time of 17 ms after the shock formation. **b**: The same snapshot as in panel a for a calculation with the Wolff-EoS. Note that at this time the simulation was performed from pole to equator and later mirrored at the equator plane. **c**: A snapshot for the same model as in panel a (L&S-EoS), but at a time of 170 ms after the shock formation. **d**: A snapshot for the same model as in panel b (Wolff-EoS), but at a time of 170 ms after the shock formation. **e**: A snapshot of the calculation with the L&S-EoS at a time of 220 ms after shock formation. **f**: The same snapshot as in panel e, however in this simulation the star is rotating (see text)

4.3 Effects of Rotation

It is a well known fact from observations that stars do rotate. However, rotation is not routinely included in calculations with Boltzmann neutrino transport. The reason for this is that the exact rotation frequencies and rotation profiles for the *inner* core of a star are poorly known from theory as well as from observations. With a first set of simulations we have started to look into the effects of rotation on the evolution of a $15\,M_\odot$ progenitor star. In this model we have chosen an initial rotation frequency of 0.5 rad/s for the inner core, which extremizes the effects of rotation since it is higher than the value predicted by state-of-the-art stellar evolution calculations shortly before the onset of gravitational instability. This still ongoing simulation – it was partially performed at the SX-8 at the HLRS – is now the longest multi-dimensional Boltzmann neutrino transport simulation worldwide. The reason for pushing this simulation to such long times is that rotation and angular momentum will become more and more important at later times as matter will have fallen from larger radii to the shock position. It is interesting to note that we clearly find the development of low-mode ($l = 1, 2, 3$) non-radial instabilities, as was already seen in models with approximative neutrino transport (see Sect. 2.1). In Figs. 3e,f we show snapshots of this rotating and a non rotating model. Clearly, rotation produces oblate-shaped proto-neutron stars (which is roughly the green region in the inner cores of Figs. 3e,f), but also the convective pattern of the postshock region is clearly different. We currently try to push the rotating and the non-rotating models to as large times as possibly in order to be able to study in detail the effects of rotation on the supernova evolution at late post-bounce stages.

5 Conclusions and Outlook

We have started to simulate well resolved 2D models of core collapse supernovae with detailed neutrino transport and varied input physics (e.g. progenitor models, EoS, and rotation). Preliminary results reveal interesting differences that depend on these variations and may be bearing on the supernova exlposion mechanism. This calls for systematic parameter studies which continue to much later post-bounce times, which in turn requires the use of a code with Teraflop capability. Such a code is developed right now by the Garching supernova group for the SX-8 of the HLRS and promises interesting results in the future.

Acknowledgements

Support from the SFB 375 "Astroparticle Physics", SFB/Tr7 "Gravitationswellenastronomie" of the Deutsche Forschungsgemeinschaft, and computer time at the HLRS and the Rechenzentrum Garching are acknowledged. We also thank M. Galle and R. Fischer for performing the benchmarks on the NEC machines.

References

1. Rampp, M., Janka, H.T.: Spherically Symmetric Simulation with Boltzmann Neutrino Transport of Core Collapse and Postbounce Evolution of a 15 $M_\odot$ Star. Astrophys. J. **539** (2000) L33–L36
2. Mezzacappa, A., Liebendörfer, M., Messer, O.E., Hix, W.R., Thielemann, F., Bruenn, S.W.: Simulation of the Spherically Symmetric Stellar Core Collapse, Bounce, and Postbounce Evolution of a Star of 13 Solar Masses with Boltzmann Neutrino Transport, and Its Implications for the Supernova Mechanism. Phys. Rev. Letters **86** (2001) 1935–1938
3. Liebendörfer, M., Mezzacappa, A., Thielemann, F., Messer, O.E., Hix, W.R., Bruenn, S.W.: Probing the gravitational well: No supernova explosion in spherical symmetry with general relativistic Boltzmann neutrino transport. Phys. Rev. D **63** (2001) 103004–+
4. Thompson, T.A., Burrows, A., Pinto, P.A.: Shock Breakout in Core-Collapse Supernovae and Its Neutrino Signature. Astrophys. J. **592** (2003) 434–456
5. Bethe, H.A.: Supernova mechanisms. Reviews of Modern Physics **62** (1990) 801–866
6. Burrows, A., Goshy, J.: A Theory of Supernova Explosions. Astrophys. J. **416** (1993) L75
7. Janka, H.T.: Conditions for shock revival by neutrino heating in core-collapse supernovae. Astron. Astrophys. **368** (2001) 527–560
8. Herant, M., Benz, W., Colgate, S.: Postcollapse hydrodynamics of SN 1987A – Two-dimensional simulations of the early evolution. Astrophys. J. **395** (1992) 642–653
9. Herant, M., Benz, W., Hix, W.R., Fryer, C.L., Colgate, S.A.: Inside the supernova: A powerful convective engine. Astrophys. J. **435** (1994) 339
10. Burrows, A., Hayes, J., Fryxell, B.A.: On the nature of core-collapse supernova explosions. Astrophys. J. **450** (1995) 830
11. Janka, H.T., Müller, E.: Neutrino heating, convection, and the mechanism of Type-II supernova explosions. Astron. Astrophys. **306** (1996) 167–+
12. Thompson, C.: Accretional Heating of Asymmetric Supernova Cores. Astrophys. J. **534** (2000) 915–933
13. Foglizzo, T.: Non-radial instabilities of isothermal Bondi accretion with a shock: Vortical-acoustic cycle vs. post-shock acceleration. Astron. Astrophys. **392** (2002) 353–368
14. Blondin, J.M., Mezzacappa, A., DeMarino, C.: Stability of Standing Accretion Shocks, with an Eye toward Core-Collapse Supernovae. Astrophys. J. **584** (2003) 971–980
15. Scheck, L., Plewa, T., Janka, H.T., Kifonidis, K., Müller, E.: Pulsar Recoil by Large-Scale Anisotropies in Supernova Explosions. Phys. Rev. Letters **92** (2004) 011103–+
16. Keil, W., Janka, H.T., Mueller, E.: Ledoux Convection in Protoneutron Stars – A Clue to Supernova Nucleosynthesis? Astrophys. J. **473** (1996) L111

17. Burrows, A., Lattimer, J.M.: The birth of neutron stars. Astrophys. J. **307** (1986) 178–196
18. Pons, J.A., Reddy, S., Prakash, M., Lattimer, J.M., Miralles, J.A.: Evolution of Proto-Neutron Stars. Astrophys. J. **513** (1999) 780–804
19. Fryer, C.L.: Mass Limits For Black Hole Formation. Astrophys. J. **522** (1999) 413–418
20. Fryer, C.L., Heger, A.: Core-Collapse Simulations of Rotating Stars. Astrophys. J. **541** (2000) 1033–1050
21. Fryer, C.L., Warren, M.S.: Modeling Core-Collapse Supernovae in Three Dimensions. Astrophys. J. **574** (2002) L65–L68
22. Fryer, C.L., Warren, M.S.: The Collapse of Rotating Massive Stars in Three Dimensions. Astrophys. J. **601** (2004) 391–404
23. Rampp, M., Janka, H.T.: Radiation hydrodynamics with neutrinos. Variable Eddington factor method for core-collapse supernova simulations. Astron. Astrophys. **396** (2002) 361–392
24. Janka, H.T., Buras, R., Kifonidis, K., Marek, A., Rampp, M.: Core-Collapse Supernovae at the Threshold. In Marcaide, J.M., Weiler, K.W., eds.: Supernovae, Procs. of the IAU Coll. 192, Berlin, Springer (2004)
25. Buras, R., Rampp, M., Janka, H.T., Kifonidis, K.: Two-dimensional hydrodynamic core-collapse supernova simulations with spectral neutrino transport. I. Numerical method and results for a 15 Mȯ star. Astron. Astrophys. **447** (2006) 1049–1092
26. Müller, E., Rampp, M., Buras, R., Janka, H.T., Shoemaker, D.H.: Toward Gravitational Wave Signals from Realistic Core-Collapse Supernova Models. Astrophys. J. **603** (2004) 221–230
27. Lattimer, J.M., Swesty, F.D.: A generalized equation of state for hot, dense manner. Nuclear Physics A **535** (1991) 331–+
28. Shen, H., Toki, H., Oyamatsu, K., Sumiyoshi, K.: Relativistic Equation of State of Nuclear Matter for Supernova Explosion. Progress of Theoretical Physics **100** (1998) 1013–1031
29. Hillebrandt, W., Wolff, R.G.: Models of Type II Supernova Explosions. In Arnett, W.D., Truran, J.W., eds.: Nucleosynthesis: Challenges and New Developments, Chicago, University of Chicago Press (1985) 131
30. Marek, A., Janka, H.T., Buras, R., Liebendörfer, M., Rampp, M.: On ion-ion correlation effects during stellar core collapse. Astron. Astrophys. **443** (2005) 201–210
31. Marek, A., Dimmelmeier, H., Janka, H.T., Müller, E., Buras, R.: Exploring the relativistic regime with Newtonian hydrodynamics: an improved effective gravitational potential for supernova simulations. Astron. Astrophys. **445** (2006) 273–289
32. Kitaura, F.S., Janka, H.T., Hillebrandt, W.: Explosions of O-Ne-Mg Cores, the Crab Supernova, and Subluminous Type II-P Supernovae. astro-ph/0512065, A&A in press (2005)
33. Burrows, A., Livne, E., Dessart, L., Ott, C., Murphy, J.: A New Mechanism for Core-Collapse Supernova Explosions. ArXiv Astrophysics e-prints (2005)
34. Akiyama, S., Wheeler, J.C., Meier, D.L., Lichtenstadt, I.: Feedback Effects of the Magnetorotational Instability on Core Collapse Supernovae. Bulletin of the American Astronomical Society **34** (2002) 664–+

MHD Code Optimizations and Jets in Dense Gaseous Halos

Volker Gaibler[1], Matthias Vigelius[2], Martin Krause[3], and Max Camenzind[4]

[1] Landessternwarte Königstuhl, 69117 Heidelberg, Germany
V.Gaibler@lsw.uni-heidelberg.de
[2] School of Physics, University of Melbourne, Victoria 3010, Australia
mvigeliu@physics.unimelb.edu.au
[3] Astrophysics Group, Cavendish Laboratory, Madingley Road, Cambridge CB3 0HE, United Kingdom, M.Krause@mrao.cam.ac.uk
[4] Landessternwarte Königstuhl, 69117 Heidelberg, Germany
M.Camenzind@lsw.uni-heidelberg.de

Summary. We have further optimized and extended the 3D-MHD-code *NIRVANA*. The magnetized part runs in parallel, reaching 19 Gflops per SX-6 node, and has a passively advected particle population. In addition, the code is MPI-parallel now – on top of the shared memory parallelization. On a 512^3 grid, we reach 561 Gflops with 32 nodes on the SX-8. Also, we have successfully used *FLASH* on the Opteron cluster. Scientific results are preliminary so far. We report one computation of highly resolved cocoon turbulence. While we find some similarities to earlier 2D work by us and others, we note a strange reluctancy of cold material to enter the low density cocoon, which has to be investigated further.

1 Introduction

Radio galaxies host powerful jets, which are the source of the intense radio emission. These jets, highly collimated, bipolar plasma streams, are believed to be launched by supermassive black holes by accretion processes with speeds very close to the speed of light and can extend far into the surrounding medium of the galaxy cluster.

Supermassive black holes with enormous masses of up to 10^{10} solar masses are generally present in the bright central galaxies of galaxy clusters. According to our present understanding, these galaxies, mainly ellipticals, are not formed by late mergers of smaller galaxies. Radio galaxies at high redshifts might be the progenitors of these objects, as suggested by HST images showing individual gas clumps and a space density of local galaxy clusters similar to high redshift radio galaxies. Furthermore, the space density of bright ellipticals seems to be constant over a wide range in redshift [1]. While powerful radio sources were quite common at high redshift, local galaxies with huge supermassive black holes generally only show weak radio emission, Cygnus A being an exception to this.

But not only jet activity was a powerful source of energy in the early universe. At that time, formation of galaxies was still under way and the birth and death of lots of massive stars probably lead to the formation of galactic winds. High supernova rates powered the formation of global outflows, creating shocks and sweeping up matter into a dense shell [2].

Optical emission line regions and continuum associated with the radio galaxies changes drastically with cosmic epoch from small narrow line regions on the 100 parsec scale nearby to about 1 000 times larger at a redshift of about one [3]. Giant gas haloes are associated with the highest redshift radio galaxies [4]. Similar haloes have been detected without a contained radio source [5; 6]. These haloes, now frequently called Lyman α blobs, are somehow related to the formation of galaxies in the young universe. An understanding of the differences of these similar objects is surely linked to the interaction of the jets with such environments and promises new insights into the conditions out of which galaxies form. Based on computations with the NEC SX-6, we have already proposed to explain the frequently observed Lyman α absorbers around high redshift radio galaxies by galactic wind shells [7]. Similar absorbers have now been found against Lyman α blobs [8], with the same interpretation. This strengthens our model even more.

We concentrate on two scenarios: the propagation of jets into the dense and clumpy medium of the galaxy clusters (very light jets) and the interaction of such a jet with the dense shell created by a galactic wind. The propagation of jets at these early times is examined with hydrodynamical and magnetohydrodynamical (MHD) simulations on the SX-6/8 and the CRAY Opteron cluster at the HLRS.

Very light jets are especially different from heavier jets in that they have a much smaller propagation speed. While the velocity of the jet plasma is not different, the jet head, seen on radio images as "hot spots", moves very slowly, because of the little momentum the underdense jet carries. As the CFL timestep is limited by the high speed of sound and high Mach numbers, the number of timesteps to be carried out is very large and such computations are only feasible on supercomputers. Especially for jet-wind-interaction and cocoon turbulence simulations, sufficient resolution is necessary.

2 Simulations of Very Light Jets

2.1 Computational Technique

These simulations of very light jets were carried out using the *NIRVANA* code [11]. It is a non-relativistic finite-differences code to solve the magnetohydrodynamic equations in two or three dimensions. It was vectorized and parallelized by Martin Krause. For a more detailed description, please see [9; 12].

Code Optimization Results

As reported in [9], the magnetic parts of *NIRVANA* were additionally optimized by Volker Gaibler. During the last year, this was extended to reach optimum performance on multiple CPUs and we now report on the performance results. Simulations with the optimized magnetic routines were still done on the SX-6 because of the more relaxed limits on the simulation wall-clock time.

A speedup of 6.9 was reached for runs with 8 CPUs (1 node). When smart domain handling is switched on, the average vector length (usually above 254) is worse because of shorter loops, but the overall CPU time decreased up to a factor of 2. Typical performance values with smart domain handling are shown in Table 1.

Table 1. A typical 2D-MHD run with the optimized code on 8 CPUs

Real Time (sec)	69380.275892	Max Concurrent Proc.	8
User Time (sec)	477814.162355	Conc. Time(>= 1)(sec)	63817.206581
Sys Time (sec)	4046.758538	Conc. Time(>= 2)(sec)	59590.022448
Vector Time (sec)	403174.461913	Conc. Time(>= 7)(sec)	59213.994233
MOPS	6658.524164	Conc. Time(>= 8)(sec)	57416.318422
MFLOPS	2570.727031	Lock Busy Count	98910156
MOPS (concurrent)	49853.908001	Lock Wait (sec)	9161.936500
MFLOPS (concurrent)	19247.626916	I-Cache (sec)	500.862733
A.V. Length	248.060178	O-Cache (sec)	6117.597996
V. Op. Ratio (%)	99.487482	Bank (sec)	3353.742165
Memory Size (MB)	752.000000		

We noticed, that the turbulence in the jet backflow is quite sensitive to compiler optimization. The results of three simulations with identical initial conditions, two on a Intel-CPU based workstation with gcc 3.2 and gcc 3.3.5, and one on the SX-6 (sxcc rev. 063), are shown in Fig. 1. While a certain sensitivity of turbulence on floating-point arithmetics would be expected, it is interesting to see that after 210 000 timesteps the differences already are quite obvious. Differences in the length of the jet are mainly caused by "pumping" of the jet, which occurs at slightly different times for the simulations. The differences in the extent of the bow shock are 1.5% in axial and 0.7% in radial direction on average. Figure 2 shows the time evolution – there doesn't seem to be a systematical trend or increase with time.

Implementation of a Particle Population

To be able to improve our emission models, we decided to implement a particle population for the code and follow the evolution of the physical variables along the trajectories of passively advected particles, through regions of different density and magnetic fields. For these particle locations, it will be possible

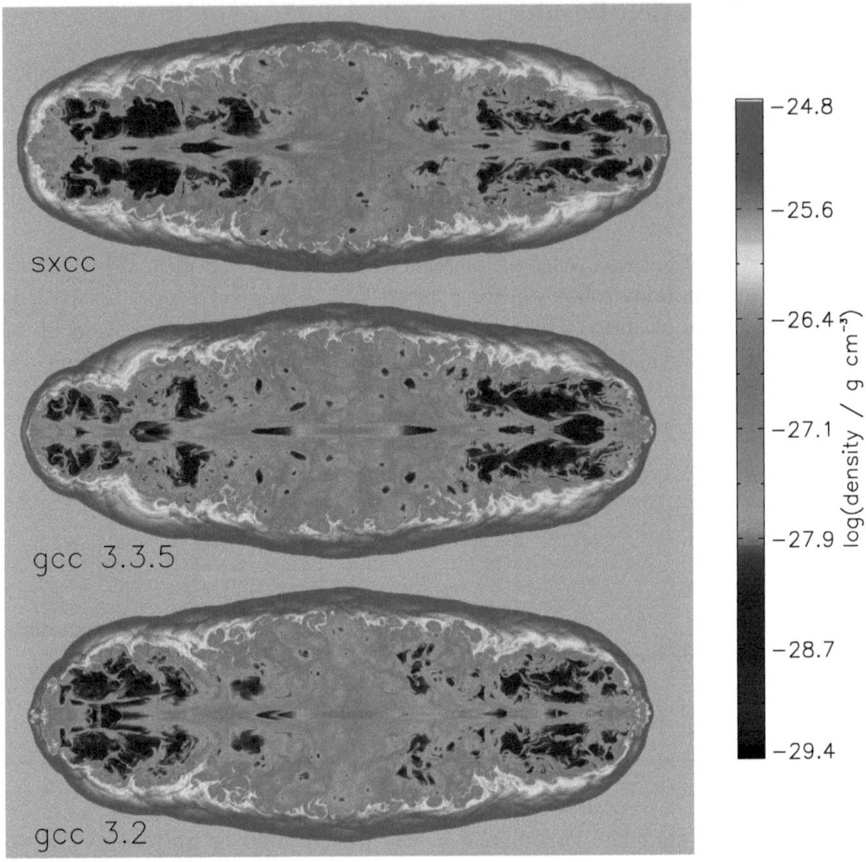

Fig. 1. Comparison between three 2D-MHD runs with identical initial conditions, compiled on different systems. This snapshot shows the logarithmically scaled density after 6.0 Myr with a resolution of $4\,000 \times 800$ cells (200×40 kpc^2) and a density contrast of $\eta = 10^{-1}$ to save computation time. After some Myrs, the differences become visible, as the turbulence is very sensitive to numerical inaccuracies, but the global behaviour does not seem to be affected

to determine the energy distributions of the electrons, given some initial energy distributions and the known history of these particles. As this can be done after the long simulation run, various initial energy distributions can be probed and resulting emission of the electrons can be computed for the particle locations. Simulations including a particle population have already been run, but evaluation of the particle data is not yet finished. Some preliminary results are presented in 2.2 for further information.

In this implementation, a fixed number of particles is placed into the jet nozzle at regular intervals over the whole simulation time. These particles carry the information of their position and motion as well as the fluid variables

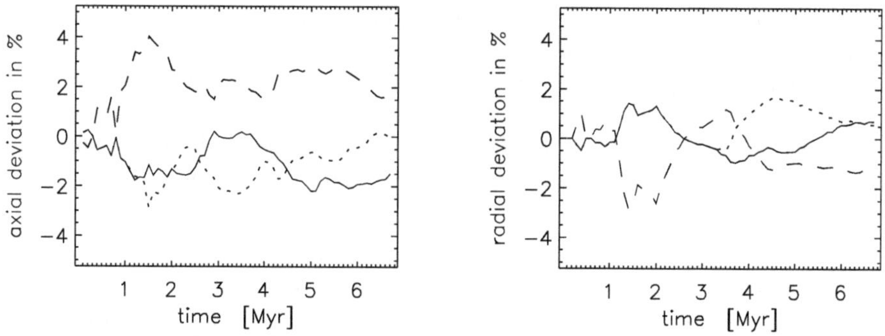

Fig. 2. Time evolution of the differences in the axial and radial extent of the bow shock for runs with different compilers (*solid*: gcc 3.2, *dotted*: gcc 3.3.5, *dashed*: sxcc). The deviation in percents is expressed relative to the common average of the three runs. Aside from sxcc producing slightly longer bow shocks, there doesn't seem to be a systematical increase of the deviations with time

at their position. In addition, div(**v**) and a shock indicator variable (as in [13]) are saved to find shock encounters for the particle. The particles are advected with the velocity of the grid fluid. This leads to the problem that the particle velocity has discontinuities exactly when the particles moves into a neighbouring grid cell. To avoid this, we use bilinear interpolation of the grid variables for particle variables.

As only a very limited number of grid data files can be written during the simulation, the need to save the trajectories of the particles including all physical variables faced us with the serious problem, how this data can be saved in a memory-efficient way. We came up with the solution that each particle is only saved when at least one of its variables changed by a specified percentage compared to its last-saved state. While 5% gives a really good precision, even 50% gives fairly good results and "compression" because usually at least one variable is very sensitive. For our already finished simulations with 5% sensitivity, the disk space requirements for the particle population was still clearly below the disk space for the grid variables.

MPI Version

We have developed an MPI version of *NIRVANA*. The code is intended for 3D simulations. So far, the first dimension (X) was vectorized, and the second one (Y) parallelized by automatic shared memory parallelization via compiler directives. In the new MPI-version, the computational domain is sliced in the Y- and Z-dimension. MPI-parallelization in the Y-dimension on top of the shared memory parallelization becomes advantageous for problems with more than $n_Y = 256$ cells in the Y-dimension.

The adopted MPI parallelization strategy is simple: in the part of the code that makes the grid, each processor computes the limits of its particular share of the computational domain. In the following, this processor solves the problem in his domain. Communication is necessary when the boundary conditions are called – here the boundary cells are exchanged with the corresponding processors. We also use one global timestep everywhere. No communication is needed for data-dumping. Each MPI-process dumps its data to a separate file, which are joined together later by the visualization software.

Table 2. Typical MPIPROGINF output (512^3 cells)

Global Data of 32 processes	Min	[U,R]	Max	[U,R]	Average
Real Time (sec)	66.499	[0,15]	66.516	[0,12]	66.512
User Time (sec)	309.403	[0,28]	332.470	[0,5]	325.880
System Time (sec)	2.122	[0,2]	2.982	[0,4]	2.602
Vector Time (sec)	147.328	[0,3]	154.457	[0,26]	151.650
Instruction Count	27183311328	[0,28]	28901999224	[0,5]	28411664374
Vector Instruction Count	7942472144	[0,3]	8311479771	[0,26]	8181309056
Vector Element Count	1974781136436	[0,3]	2068061463258	[0,25]	2036197214990
FLOP Count	809168128634	[0,3]	846711904021	[0,18]	834065906220
MOPS	6084.053	[0,3]	6476.024	[0,30]	6311.554
MFLOPS	2467.181	[0,3]	2625.667	[0,30]	2559.886
Average Vector Length	248.243	[0,31]	249.323	[0,6]	248.881
Vector Operation Ratio (%)	98.967	[0,3]	99.040	[0,29]	99.016
Memory size used (MB)	1760.000	[0,0]	1776.000	[0,1]	1774.000
Global Memory size used (MB)	64.000	[0,0]	64.000	[0,0]	64.000
MIPS	86.589	[0,20]	88.326	[0,31]	87.190
Instruction Cache miss (sec)	1.491	[0,2]	1.834	[0,27]	1.739
Operand Cache miss (sec)	24.799	[0,28]	26.531	[0,2]	25.824
Bank Conflict Time (sec)	14.864	[0,0]	16.330	[0,10]	15.757
Max. Concurrent Processes	8	[0,0]	8	[0,0]	8
MOPS (concurrent)	40350.753	[0,31]	43079.353	[0,22]	41957.596
MFLOPS (concurrent)	16348.186	[0,31]	17472.222	[0,22]	17017.573
MIPS (concurrent)	554.814	[0,31]	593.148	[0,6]	579.654
Event Busy Count	0	[0,0]	0	[0,0]	0
Event Wait (sec)	0.000	[0,0]	0.000	[0,0]	0.000
Lock Busy Count	53391	[0,2]	72759	[0,11]	69061
Lock Wait (sec)	4.813	[0,0]	6.229	[0,11]	5.970
Barrier Busy Count	0	[0,0]	0	[0,0]	0
Barrier Wait (sec)	0.000	[0,0]	0.000	[0,0]	0.000

Overall Data	
Real Time (sec)	66.516
User Time (sec)	10428.173
System Time (sec)	83.258
Vector Time (sec)	4852.808
GOPS (rel. to User Time)	201.932
GFLOPS (rel. to User Time)	81.902
GOPS (concurrent)	1342.465
GFLOPS (concurrent)	544.490
Memory size used (GB)	55.438
Global Memory size used (GB)	2.000

There are several complications when combining MPI with shared memory parallelization. First, the MPI routines are *not thread safe*. Consequently, the application crashes as soon as any two nodes do the same MPI-call simultaneously. Therefore, we embedded these calls typically in a serial section. Then, only the master node does the MPI-call. This is also important for communication calls, when the MPI-rank is addressed directly. We note in particular

Table 3. Typical MPICOMMINF output (512^3 cells, nonzero entries only)

Real MPI Idle Time (sec)	7.167 [0,6]	12.737 [0,31]	9.042
User MPI Idle Time (sec)	7.134 [0,6]	12.707 [0,31]	9.018
Total real MPI Time (sec)	19.048 [0,5]	22.557 [0,0]	20.129
Send count	19911 [0,1]	26511 [0,4]	24966
Recv count	19912 [0,1]	26512 [0,4]	24966
Barrier count	218 [0,0]	218 [0,0]	218
Number of bytes sent	7273306488 [0,3]	11072794488 [0,5]	10089021421
Number of bytes recv	7273329301 [0,3]	11072817301 [0,5]	10089021421

that an MPI_Barrier can apparently be released when a certain number of microtasks of the same node reach it, i.e. there is no runtime error at this point. It took some time to find out these details, and the documentation was not always helpful. The boundary cells are exchanged via the buffered mode (MPI_Bsend). The documentation recommends global memory allocation for the MPI-buffer via MPI_Alloc_mem. We allocate this memory once, at the beginning of the simulation and release at the end. The order of the commands seems to be critical – the following works:

```
bsize=2.*sizeof(double)*18*((g->ny+1)*(g->nx+1)-1);
mpi_buffer=dvector(0,bsize);
if (i=MPI_Alloc_mem(bsize, MPI_INFO_NULL, mpi_buffer))
printf("MKGRD :: ERROR IN GLOBAL ALLOCATION\n");
if (i=MPI_Buffer_attach(mpi_buffer,bsize))
printf("MKGRD :: ERROR IN BUFFER ATTACH\n");
```

The buffer is freed at the end of the program by the corresponding commands in reverse order.

The code was then thoroughly tested and improved for MPI-parallelization in the Z-direction until agreement with the non-parallel mode was down to machine accuracy. We are confident that the combined Y-Z-MPI-parallelization is fine, too (compare below).

As recommended in the documentation, we measured the performance of the MPI-parallelized code by setting the environment variables MPIPROGINF=DETAIL (Table 2), and MPICOMMINF=YES (Table 3), and linking the profiling libraries for the test cases. MPICOMMINF indicates a data transfer rate of about 0.5 GB/s (real time). Subtracting idle time, we still end up below 1 GB/s. That is not very much, given the nominal data transfer rate of 16 GB/s. We must therefore conclude that the data transfer rate is dominated by waiting and latency. The timings indicate that $\approx 50\%$ performance may still be gained by improving the data transfer. This is consistent with the direct FLOP measurements: while the MPI run gives 2.6 Gflops per CPU on average, the single processor performance for that loop length is 4 Gflops. However, this will involve a considerable effort in coding and testing which we did not choose to undertake so far.

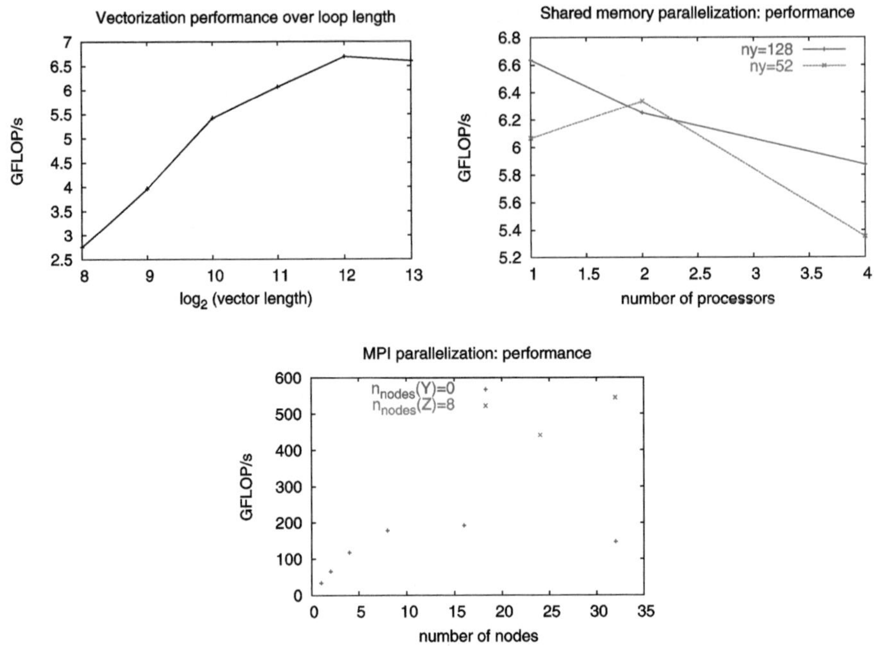

Fig. 3. The vectorization has been tested on a $nx \times 52 \times 12$ grid. For the shared memory parallelization, a $2048 \times ny \times 12$ grid has been used. For the MPI-parallelization, the 512^3 grid of the real simulations has been used. Parallelization in Z-direction only is limited by a bout 200 Gflops for this problem size (*red pluses*). We could increase the performance further by also parallelizing the already shared memory parallelized Y-dimension (*green crosses*)

Since we did not report on performance details for the SX-8 yet, we do it here in full (Fig. 3). Vectorization reaches its maximum performance at a loop length of 4096 with 6.7 Gflops. For the shared memory parallelization, the performance on eight processors is still acceptable for Y-loop lengths of about 128. The MPI-part yields a good speedup for slices of more than 64 cells. This puts no additional constraint to the Y-slicing. For the 512^3 periodic box hydrodynamic turbulence simulation described below, we reached a performance of 561 Gflops, using 32 nodes, where the Z-direction was split into 8 parts.

2.2 Scientific Results

Very Light Magnetic Jets

The optimized code enables us to run simulations of very light magnetic jets with density contrasts of up to $\eta = 10^{-4}$. The resolution was set to $(n_Z \times n_R) = (4\,000 \times 1\,600)$ cells (physical scale: 200×80 kpc^2) and a particle number of

1 000 was chosen. Analysis of the data is still under way, so only preliminary results can be presented here.

Figure 4 shows the density evolution for a member of the particle population in the $\eta = 10^{-1}$ simulation over a time span of 6 Myrs. It starts inside the jet nozzle, but is soon pushed out of the jet beam and then moves away from it towards the midplane at $Z = 0$. Changes in density (and the other variables) are more violent at early times, near the jet beam, and decrease later on.

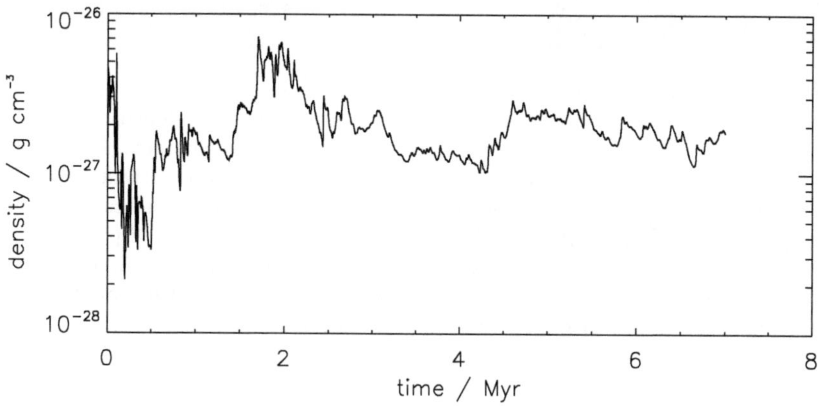

Fig. 4. Evolution of a sample particle in a MHD simulation with density contrast $\eta = 10^{-1}$ and a sensitivity for particle data saving of 5% for all variables. About 53 000 points out of 210 000 timesteps were saved for this particle. Even with a sensitiviy of only 50%, this evolution can be reproduced

High-Resolution Cocoon Turbulence

In order to study the interaction of jets with their environment in greater detail, we have performed simulations of the turbulent mixing at the contact surface between jet cocoon and shocked ambient gas in the presence of denser clouds. This is essentially a Kelvin-Helmholtz instability in a 3D periodic box, with the addition of a cooling cloud. The initial density ratio is 10 000, appropriate for such contact surfaces. We add an elliptical cloud to the denser phase, overdense by another factor of 10 (100).

The first runs have just completed and so we can report on the very first impressions from the simulations only. We performed hydrodynamic simulations with cooling for a 512^3 data cube. The density distribution in two slices ($X = 50$ pc, $Y = 50$ pc) is shown in Fig. 5. The simulation run in MPI-parallel mode on $(Y \times Z) = (4 \times 8)$ nodes of the NEC SX-8 for 8.5 realtime hours, using 1 431 CPU hours at 561 Gflops. This is 14% of the nominal peak performance

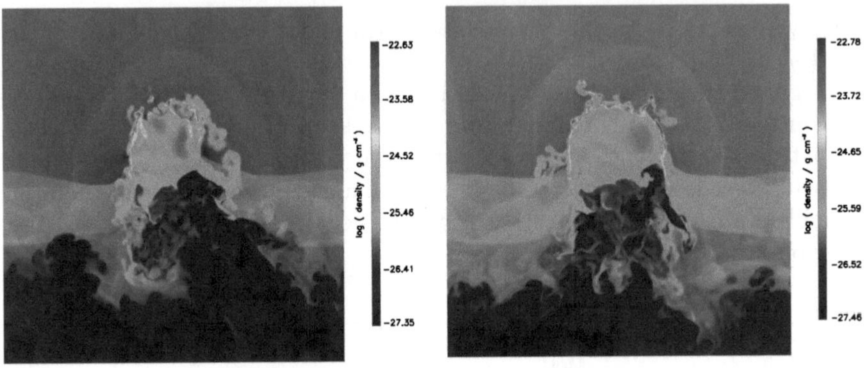

Fig. 5. Slices through the midplane (*left*: $Y = 50$ pc, *right*: $X = 50$ pc) of the (100 pc)^3 simulation box. We simulate a shear instability with density ratio 10 000 and a dense (factor ten) cloud in the denser (*upper*) part of the box. The elliptical cloud collapses due to thermal instability. The shear instability drives shocks into the upper medium, which are reflected at boundaries and the cloud surface, where they produce small filaments. The vertical dimension was cut into 8 parts and distributed among multiple SX-8 nodes for evolution. The horizontal direction in the *right plot* was separated into 4 parts. The absence of suspect features at the boundaries demonstrates the successful MPI-parallelization of our magnetohydrodynamics code *NIRVANA*

(16 Gflops per CPU). The simulation needed 51 000 timesteps in total. Therefore, we need 4 nanoseconds per cell and timestep, using 753 nanoseconds of CPU time.

Our earlier 2D results suggest that due to the rapid cooling, the cold and dense phase survives in the turbulent multiphase jet cocoon, gets considerably stirred up and interacts strongly with the other components, which causes particular (power law) density and temperature distribution functions and enhanced cooling and cold cloud dropout for the warmer phase. We would like to test these findings also with the 3D simulations. Yet, these runs are literally just completed at the time of writing, wherefore we content ourselves by making four points: First, we observe the turbulent mixing at the shear instability, as expected. Second, the shocks expected due to the shear instability in the upper (denser) part, shred the cloud into smaller but denser fragments, in agreement with other work in the literature [10]. Third, the dense fragments avoid the mixing zone of the other two phases, unlike to what had happened in the similar 2D simulations. As has been mentioned before, the process of data reduction and interpretation is not yet finished. Finally, we note that no artifacts from the MPI parallelization can be spotted at the boundaries, demonstrating the success of the code adaptation.

3 Simulations of Jet – Galactic Wind – Interaction

3.1 Computational Technique

The simulations were performed using *FLASH* [14], a parallelized hydrodynamic solver based on the second-order piecewise parabolic method (PPM). *FLASH* solves the inviscid hydrodynamic equations in conservative form. It manages the adaptive mesh with the *Paramesh* library [15]. The mesh is refined and coarsened in response to the second-order error in the dynamical variables (for details, see [16]).

The scaling properties of *FLASH* are extensively discussed in [14]. Generally, the scaling is nearly ideal for a uniform mesh up to $\sim 1\,000$ processors. For an adaptive mesh, the scaling is not so good if more than ~ 100 processors are involved, due to the increased communication costs.

We converted the existing *NIRVANA* setup [17] to *FLASH.* Up to now, we have performed only 2.5D runs in cylindrical coordinates with a maximum refinement level of 7, resulting in a maximum of $(R \times Z) = (512 \times 1024)$ cells.

The simulations start with an isothermal King atmosphere with $\rho(r) = 0.3\, m_p\, \mathrm{cm}^{-3} \,/\, (1 + (r \,/\, 10\,\mathrm{kpc}))$ at $T = 10^6$ K. The atmosphere is stabilized by an associated dark matter halo gravitational field which is implemented in *FLASH* by a source term in the momentum equation. The galactic wind is set up by a mass and energy injection of 10 $\mathrm{M}_\odot$/yr and 10^{51} erg/yr, respectively, exponentially distributed over a region of 3 kpc around the origin. We use a temperature-dependent cooling function (implemented as optically thin cooling with an energy sink in the energy equation) according to [18]. Both of these are added to the cooling function of *FLASH.* After 80 Myrs, the jet is started. It is implemented by updating a region of 1 kpc around the origin with a z velocity of $2c/3$ and a density of $10^{-5}\, m_p\, cm^{-3}$.

We plan to drastically increase the resolution of the 2.5-dimensional simulations. This will allow us to study the clumping and the time evolution of the grade of covering of the neutral hydrogen clouds, examine, how effective a light jet can destroy a wind shell and how this evolution affects the optical spectrum. Furthermore, we plan to extend our code to 3 dimensions, since 2.5D simulations artificially stabilize the jet–shell configuration. It is obvious, that these aims require parallel computations as well as the use of the adaptive mesh capability of *FLASH.*

3.2 Scientific Goals

In addition to the large energy output in the radio band, high-z (~ 2) radio galaxies are usually associated with an extended Ly α halo, often exceeding the size of the radio emission region. However, for radio galaxies smaller than $\sim$ 50 kpc, Lyα radiation is effectively absorbed, mostly in the blue wing. These absorbers can be associated with the radio emitting galaxy [19].

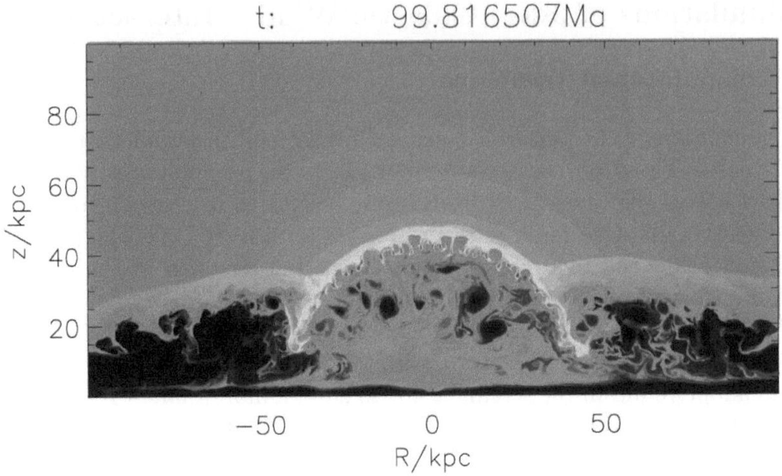

Fig. 6. Interaction between jet and galactic wind with *FLASH*: Density (logarithmic scale) after 100 Myr. The jet has destroyed the wind shell

[17] proposed a model to explain these associated absorbers. In the early stage of galactic evolution, stellar winds and supernova explosions inject gas and energy into the interstellar medium. This galactic wind results in a spherical expanding bow shock. The Lyman α emission is due to gas ionized either by stellar radiation or collisional excitation. If the cooling time scale is comparable with the propagation time scale, the material behind the bow shock will cool and form a dense shell, effectively absorbing Ly α emission at a narrow velocity range. Later ($\sim$ 80 Myr), accretion onto the supermassive black hole in the center of the galaxy will feed a jet. This jet will finally destroy the absorbing shell.

The result of a 2.5-dimensional simulation with *FLASH* is shown in Fig. 6. The jet has destroyed the wind shell almost completely. This run took 3.65 hrs on 20 processors. The number of evolved zones per second was 673 600 or 558 450 cells/sec per processor and timestep.

4 Summary

This year, we have much concentrated on code development. This includes both performance optimization and new physics. The magnetic parts of our traditional code *NIRVANA* have now been fully parallelized – we reach a performance of 2.6 Gflops per CPU on the SX-6. We report completion of the MPI-parallelization, too. This version enables us to evolve the hydrodynamic equations at 560 Gflops, using 32 nodes of the SX-8. We report the development of a new tracer particle module for *NIRVANA* that will much improve our synchrotron emission diagnostics for jet simulations. The particles are passively advected with the grid plasma, and the trajectories and evolution of physical variables is efficiently saved for postprocessing. We also started computations on the CRAY Opteron cluster, employing another hydrodynamics code – *FLASH*. It is a well-maintained parallel code, and we reached the expected performance.

In addition, we report some preliminary results. A magnetized jet was run on the SX-6, including the new passive tracers. The turbulence seems to be very sensitive to compiler versions and optimisations, but global properties such as the bow shock size differ only on the percent level. Some high resolution runs of jet cocoon turbulence are reported. These show some behaviour familiar from our earlier 2D work and the literature, but a surprising reluctancy for the cold matter to mix into the cocoon, which should be investigated further. Using *FLASH*, we could reproduce earlier 2D work with *NIRVANA*. This is soon to be extended to 3D simulations.

Acknowledgements

This work was also supported by the Deutsche Forschungsgemeinschaft (Sonderforschungsbereich 439, MK's fellowship: KR 2857/1-1).

References

[1] P. Saracco, et al. The density of very massive evolved galaxies to $z \simeq 1.7$. *MNRAS*, 357:L40–L44, 2005.
[2] V. Gaibler, M. Camenzind, and M. Krause. Evolution of the ISM and Galactic Activity. astro-ph/0502403
[3] P.J. McCarthy. High redshift radio galaxies. *A&AReview*, 31:639–688, 1993.
[4] M. Reuland, et al. Giant Ly α Nebulae Associated with High-Redshift Radio Galaxies. *ApJ*, 592:755–766, 2003.
[5] C. Steidel et al. Ly α Imaging of a Proto-Cluster Region at z=3.09. *ApJ*, 532:170–182, 2000.
[6] A. Dey, et al. Discovery of a Large ~ 200 kpc Gaseous Nebula at $z \sim 2.7$ with the Spitzer Space Telescope. *ApJ*, 629:654–666, 2005.
[7] M. Krause. Very light jets II: Bipolar large scale simulations in King atmospheres. *A&A*, 431:45–64, 2005.
[8] R.J. Wilman. The discovery of a galaxy-wide superwind from a young massive galaxy at redshift $z \sim 3$. *Nature*, 436:227–229, 2005.
[9] M. Krause, V. Gaibler, and M. Camenzind. Simulations of Astrophysical Jets in Dense Environments. In *High Performance Computing in Science and Engeneering '05, eds.: W.E. Nagel, W. Jäger, and M. Resch, Springer*, 2005.
[10] G. Mellema, J. Kurk, H. Röttgering. Evolution of clouds in radio galaxy cocoons. *A&A*, 395:L13–L16, 2003.

[11] U. Ziegler and H. W. Yorke. A nested grid refinement technique for magnetohydrodynamical flows. *Computer Physics Communications*, 101:54, 1997.

[12] M. Krause and M. Camenzind. Interaction of Jets with Galactic Winds. In *High Performance Computing in Science and Engeneering '04, eds.: E. Krause, W. Jäger, and M. Resch, Springer*, 2004.

[13] A. Mignone, G. Bodo. An HLLC Solver for Relativistic Flows – II. Magnetohydrodynamics. astro-ph/0601640

[14] B. Fryxell, et al. 2000, ApJS , 131, 273

[15] K.M. Olson, et al. 1999, Bulletin of the American Astronomical Society, 31, 1430

[16] R. Loehner. 1987, Comp. Meth. App. Mech. Eng., 61, 323

[17] M. Krause. Galactic Wind Shells and High Redshift Radio Galaxies On the Nature of Associated Absorbers. *A&A*, 436:845–851, 2005.

[18] R.S. Sutherland and M.A. Dopita. *ApJ Supplement*, 88, 253, 1993.

[19] R. van Ojik, H. J. A. Röttgering, G. K. Miley, and R. W. Hunstead. The gaseous environments of radio galaxies in the early Universe: kinematics of the Lyman α emission and spatially resolved H I absorption. *A&A*, 317:358–384, 1997.

Anomalous Water Optical Absorption: Large-Scale *First-Principles* Simulations

W.G. Schmidt[1,3], S. Blankenburg[1], S. Wippermann[1], A. Hermann[2], P.H. Hahn[3], M. Preuss[3], K. Seino[3], and F. Bechstedt[3]

[1] Theoretische Physik, Universität Paderborn, 33095 Paderborn, Germany
[2] Institute of Fundamental Sciences, Massey University, Auckland, New Zealand
[3] Institut für Festköpertheorie und -optik, Friedrich-Schiller-Universität Jena, Germany

Summary. The optical spectrum of water is not well-understood. For example, the main absorption peak shifts *upwards* by 1.3 eV upon condensation of gas-phase water monomers, which is contrary to the behaviour expected from aggregation-induced broadening of molecular levels. We investigate theoretically the effects of electron-electron and electron-hole correlation, finding that condensation leads to delocalisation of the exciton onto nearby hydrogen-bonded molecules. This reduces its binding energy and has a dramatic impact on the line shape. The calculated spectrum is in excellent agreement with experiment.

1 Introduction

Despite the apparent simplicity of the water molecule, the hydrogen bonds between aggregated water molecules cause many peculiarities that still elude understanding. The thorough, microscopic understanding of water in its various phases is not only of fundamental interest, but forms the prerequisite for any serious attempt to address the many chemical, biological and technological processes related to it. Here we address the water optical absorption, which is not well-understood, despite decades of effort. The spectra of amorphous, hexagonal as well as cubic ice are dominated by a very pronounced first absorption peak at about 8.7 eV [1, 2]. The absorption spectrum of the liquid shows a similar structure [3, 4]. In the case of cubic ice, the first absorption peak was attributed to the calculated density of states [5]. However, the universal occurrence of the 8.7 eV peak in a variety of water phases suggests its molecular origin, as opposed to being due to specific transitions between electronic states arising from a crystalline structure with a particular symmetry. However, the lowest-lying excitation of the H_2O molecule occurs at 7.4 eV [6]. Intermolecular interactions are expected to induce a significant broadening of the energy levels relative to the isolated molecule. This should lead to a decrease of the transition energies relative to the molecular values [7, 8, 9]. Numerical studies of the optical properties of ice [10] based on the local density approximation, *i.e.*, neglecting many-body effects beyond a mean-field description, indeed find the absorption onset at just below 6 eV.

The blueshift observed experimentally upon condensation of water molecules has been alternatively attributed to excitons of *molecular* origin [8, 9] or to solvation and Rydbergisation effects (destabilisation of highly excited states in condensed phases by interactions with nearby molecules) [11]. However, accurate description of optical spectra requires inclusion of many-body correlation effects from *first principles*, which is a difficult and computationally expensive task for complex disordered systems.

We exploit the recent progress in accurate modelling of optical properties using the many-body Green's function approach [12] to explain and quantitatively account for the peculiar absorption properties of solid water [13] and single water monomers [14]. The results highlight the major role played by nearby hydrogen-bonded water molecules in affecting the spacial extent and the energetics of the exciton dominating the main absorption peak. It is clear that the exciton state "sees" nearby molecules due to its spacial extent and will be affected by changes in the first and even the second coordination shell.

Naturally occurring hexagonal ice (ice-Ih) is chosen as a model system. The spectra of cubic and amorphous ice, however, are very similar. We proceed in three steps: (i) We use density-functional theory in generalized gradient approximation (DFT-GGA) to obtain the structurally relaxed ground state configuration of ice-Ih and the Kohn-Sham eigenvalues and eigenfunctions that enter the single- and two-particle Green's functions, (ii) the electronic quasiparticle spectrum is obtained within the GW approximation (GWA) [15] to the exchange-correlation self-energy, and (iii) the Bethe-Salpeter equation (BSE) is solved for coupled electron-hole excitations [16, 17, 18], thereby accounting for the screened electron-hole attraction and the unscreened electron-hole exchange [19, 20, 21].

The oxygen atoms in the ice-Ih structure lie on a hexagonal wurtzite lattice. Hydrogen atoms occupy the sites between neighbouring oxygens in a disordered fashion but subject to the ice rules, *i.e.*, each oxygen is covalently bonded to two hydrogen atoms with the constraint that only one H atom can lie between two neighboring O atoms. We model the proton disorder within a periodically repeated supercell consisting of 16 molecules [22].

2 Computational Method

We start from *first-principles* pseudopotential calculations within the DFT-GGA [23]. The mean-field effects of exchange and correlation in GGA are modelled using the PBE functional [24]. This functional is known to accurately reproduce the ice-Ih ground-state properties, including hydrogen bonding [25]. The Kohn-Sham energies calculated for the 16-molecule supercell are shown

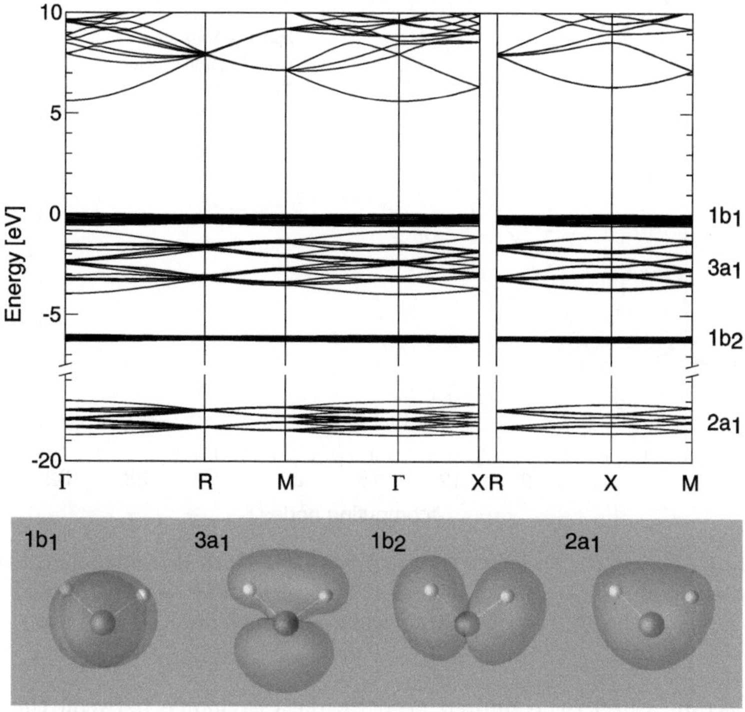

Fig. 1. Valence and conduction energy bands calculated in DFT-GGA for the 16 molecule supercell. The occupied molecular states, from which the valence bands are derived, are shown below

in Fig. 1. Due to intermolecular interactions, the occupied molecular states $2a_1$, $1b_2$, $3a_1$, and $1b_1$ broaden into energy bands. We calculate a band gap of 5.6 eV, 0.6 eV smaller than the lowest molecular transition energy. Earlier DFT studies found an energy gap of about 6 eV for cubic ice [10] and 4.6 eV for the liquid [7].

The ground-state DFT calculations were parallelised for different bands and sampling points in the Brillouin zone using the message passing interface (MPI). Fig. 2 shows benchmark calculations to determine the electronic ground state of a water dimer starting from randomised wave functions. The calculations were performed on the Cray Opteron cluster of the Höchstleistungs-Rechenzentrum Stuttgart (2GHz AMD Opteron, Myrinet 2000 node-node interconnect), the Arminius cluster of the Paderborn Center for Parallel Computing (hpcline, Xeon EM64T 3.2 GHz, Infiniband) and a local Cray XD1 machine (2Ghz Dual Core AMD Opteron, RapidArray Interconnect). As can be seen, for systems of the size studied here, an efficient parallelisation is possible using up to 16 nodes. It is interesting to note that the details of the node-node interconnect seem to be of minor importance for

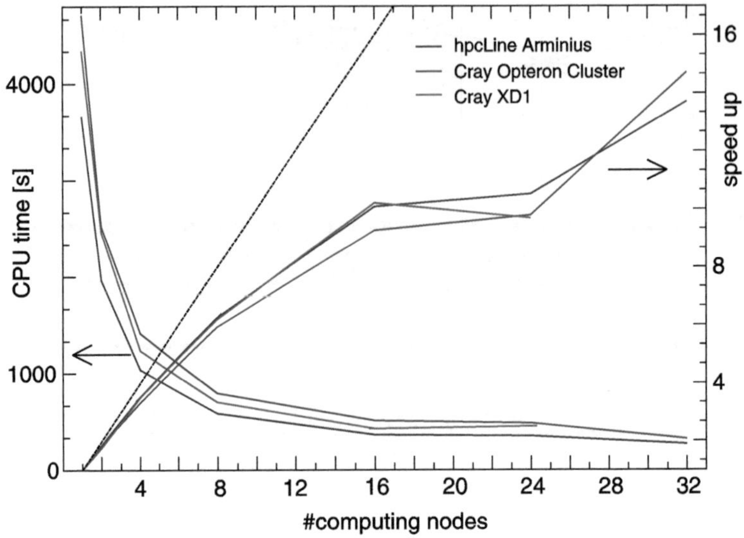

Fig. 2. CPU time per node and speed-up of a DFT-GGA ground-state calculation for the water dimer vs. number of nodes. The calculations were performed on the HLRS Opteron cluster, the PC^2 Xeon cluster and the Cray XD1 (see text)

the speedup, due to the limited need for data exchange between the nodes during the calculations.

In the second step we include electronic self-energy effects. This requires the replacement of the GGA exchange and correlation potential by the non-local and energy-dependent self-energy operator $\Sigma(\mathbf{r}, \mathbf{r}'; E)$. We calculate Σ in the GW approximation [15], where it is expressed as a convolution of the single-particle propagator G and the dynamically screened Coulomb interaction W. For single water monomers, the screening was obtained from *first principles*, using the random-phase approximation. Numerical details are given in [14]. For bulk water, a further approximation had to be introduced to cope with the numerical load: We use a model dielectric function [26] to calculate W. This speeds up the calculations substantially and results in quasiparticle energies that are within about 0.1–0.2 eV of the complete calculations [27, 28].

The electron-hole interaction is taken into account in the third step. The two-particle Hamiltonian

$$H_{vc\mathbf{k},v'c'\mathbf{k}'} = (\epsilon_{c\mathbf{k}} - \epsilon_{v\mathbf{k}})\delta_{vv'}\delta_{cc'}\delta_{\mathbf{k},\mathbf{k}'} + \tag{1}$$
$$2\iint d\mathbf{r}d\mathbf{r}'\psi^*_{c\mathbf{k}}(\mathbf{r})\psi_{v\mathbf{k}}(\mathbf{r})\bar{v}(\mathbf{r}-\mathbf{r}')\psi_{c'\mathbf{k}'}(\mathbf{r}')\psi^*_{v'\mathbf{k}'}(\mathbf{r}') -$$
$$\iint d\mathbf{x}d\mathbf{r}'\psi^*_{c\mathbf{k}}(\mathbf{r})\psi_{c'\mathbf{k}'}(\mathbf{r})W(\mathbf{r},\mathbf{r}')\psi_{v\mathbf{k}}(\mathbf{r}')\psi^*_{v'\mathbf{k}'}(\mathbf{r}') ,$$

describes the interaction of pairs of electrons in conduction states $|c\mathbf{k}\rangle$ and holes in valence states $|v\mathbf{k}\rangle$ [19, 20, 21]. The diagonal first part is given by the quasiparticle energies obtained in GW approximation. The second, the electron-hole exchange term, where the short-range part of the bare Coulomb potential $\bar{v}$ enters, reflects the influence of local fields. Finally, the third part, which describes the screened electron-hole attraction, is calculated using the same approximations for W as in the self-energy.

The eigenvalues and eigenvectors of the two-particle Hamiltonian (1) can be used to calculate the macroscopic dielectric function and thus the absorption spectrum. In the case of ice Ih, the Hamiltonian has been set up from 64 valence and 64 conduction bands using 40 **k** points. In order to bypass the diagonalisation of the Hamiltonian, we follow Glutsch *et al.* [29, 30]: If the energy dependence of the macroscopic polarisability on the eigenvalues of the exciton Hamiltonian is Fourier transformed, the polarisability can be obtained from the solution of an initial-value problem for the vector $|\mu(t)\rangle$. Its time evolution is driven by the pair Hamiltonian (1)

$$i\hbar|\dot{\mu}(t)\rangle = H|\mu(t)\rangle \ . \tag{2}$$

The initial values of the vector elements are given by

$$\mu^i_{cv\mathbf{k}}(0) = \frac{\langle c\mathbf{k}|v_i|v\mathbf{k}\rangle}{\epsilon^{LDA}_{c\mathbf{k}} - \epsilon^{LDA}_{v\mathbf{k}}} \ , \tag{3}$$

where v_i is the $i(= x, y, z)$ component of the velocity operator. The macroscopic dielectric function with the broadening parameter γ is then obtained by the Fourier transform of $e^{-\gamma t} \cdot \langle\mu(0)|\mu(t)\rangle$.

3 Results

The ice optical spectra calculated according to the three levels of theory described above are compared with experiment [2] in Fig. 3. The spectrum obtained within DFT-GGA agrees with earlier independent-particle results [10]. The calculated onset of absorption occurs at too low energies (below 6 eV), and the first absorption maximum at about 8 eV is far less pronounced than the corresponding feature in experiment. The inclusion of the many-body electron-electron interaction in GWA, *i.e.*, the electronic self-energy, leads to a nearly rigid blueshift of the spectrum, severely overestimating the energy positions of the measured peaks. The Coulomb correlation of electrons and holes, accounted for by solving the BSE, leads to the appearance of a sharp excitonic peak below the onset of the vertical quasiparticle transition energies. The peak positions and the line shape obtained from the BSE agree well with experiment. Since no input parameters to the calculation have been taken from experiment or fitted, the agreement between theory and measurement is truly impressive.

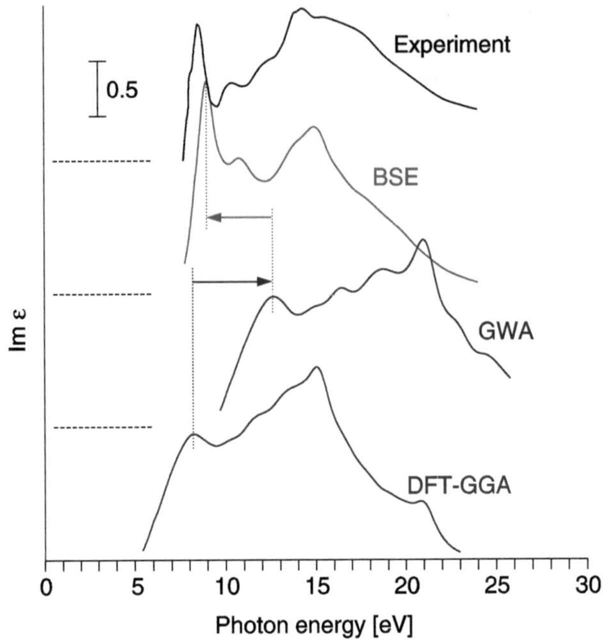

Fig. 3. Imaginary part of the dielectric function of hexagonal ice measured at 80 K (from Ref. [2]), and calculated within the DFT-GGA, the GWA, and from the BSE (see text)

After having reproduced the main features of the optical absorption of ice, we explore the origin of the prominent first peak. To this end we calculate the optical transitions of gas-phase H_2O molecules. The water monomer was modelled in a periodic (simple cubic) supercell with with a lattice constant of 10 Å. The DFT-GGA ground-state calculation yield a oxygen-hydrogen bond length of 0.966 Å and an angle between the two bonds of 104.49°. The corresponding experimental values for gas-phase molecules are 0.957 Å and 104.47°. According to the molecule symmetry, the atomic orbitals are derived from six linear combinations of a_1, a_2, b_1, and b_2 symmetry with non-degenerate energy levels. The eight valence electrons occupy the four lowest states. The dependence of the self-energy on the singe-particle eigenvalues is shown in Fig. 4. It is most important for the occupied molecular states, but nearly vanishes for the unoccupied states. The quasiparticle shifts vary between –7.7 and –4.7 eV (0.4 and 0.8 eV) for the occupied (empty) states. At the single-particle level of theory, *i.e.*, in DFT-GGA, an energy of 6.2 eV is obtained for the lowest singlet pair excitation. The electronic self-energy blueshifts this value by 6.3 eV to 12.5 eV. This is partially compensated by an exciton binding energy of 5.3 eV, leading finally to an optical absorption at 7.2 eV, in excellent agreement with the experimental value of 7.4 eV [6].

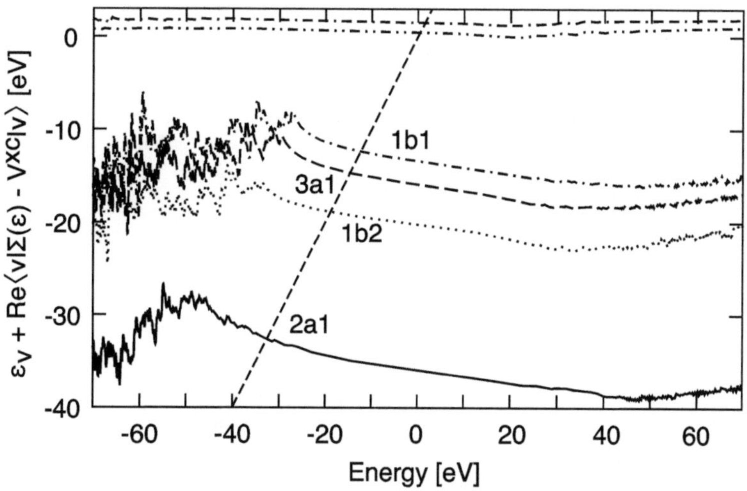

Fig. 4. Real part of the H_2O self-energy differences added to the corresponding KS eigenvalues versus single-particle excitation energy. The frequency-dependent dielectric matrix has been computed including 260 states. The *straight short-dashed line* represents the linear function ε. The crossings with the other lines define the quasiparticle energies $\varepsilon_\nu^{\mathrm{QP}}$

These results can be compared with those for ice, shown in Fig. 3, where we find for the first absorption peak a self-energy shift of about 4.5 eV (blue arrow) and an exciton binding energy (related to the position of the first major peak of the vertical quasiparticle transitions, red arrow) of 3.2 eV. Self-energy and excitonic effects in ice are thus reduced by 29 and 40%, respectively, compared to gas-phase molecules. The reduction of the self-energy and the exciton binding energy upon condensation of the molecules is expected, due to the larger screening in the solid. The fact, however, that the exciton binding energy is affected more than the self-energy points to an additional effect, namely a change in the localisation of the exciton. The attractive interaction between an electron and a hole is inversely proportional to their average distance, the so-called exciton radius

$$R = \int d\mathbf{r}_e \int d\mathbf{r}_h |\mathbf{r}_e - \mathbf{r}_h| |\Psi(\mathbf{r}_e, \mathbf{r}_h)|^2, \tag{4}$$

where $\Psi(\mathbf{r}_e, \mathbf{r}_h)$ is the electron-hole pair wave function depending on the positions $\mathbf{r}_e$ and $\mathbf{r}_h$ of electrons and holes, respectively. The spatial distribution of electron-hole pairs therefore influences their energy.

In the case of an isolated molecule, the localisation of the electron-hole pair is immediately obtained from the spatial extension of the respective molecular orbitals, which relax upon optical excitation. The highest occupied molecular orbital (HOMO) of water is a nonbonding oxygen-localized $1b_1$ orbital. It barely changes upon excitation of one electron into the lowest unoccupied

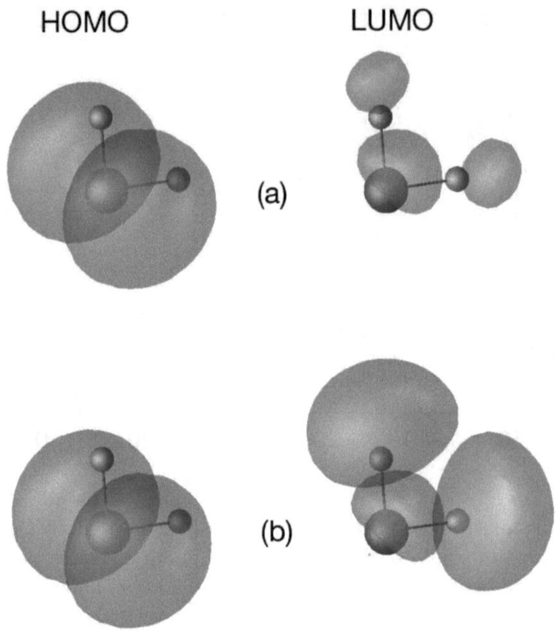

Fig. 5. HOMO and LUMO of the gas-phase water molecule in the ground state (**a**) and upon optical excitation (**b**)

molecular orbital (LUMO). The latter, however, is strongly affected by partial occupation. The LUMO shows a significant probability density in the proximity of the O atom, as well as two lobes that protrude from the molecule in the proton directions, see Fig. 5(a). Optical excitation of one electron from the HOMO into the LUMO leads to a considerable expansion of these lobes, as shown in Fig. 5(b). Nevertheless, due to the attractive interaction with the hole at the oxygen atom, the electron remains close to the molecule. We calculate an average electron-hole distance R of 2.27 Å.

How does the spatial distribution of the excited electrons change upon condensation of the H_2O molecules? The solution of the BSE yields correlated electron-hole pair states. The pair wave functions $\Psi(\mathbf{r}_e, \mathbf{r}_h)$ are scalar functions in the two-particle space. The center of gravity of the hole belonging to the exciton wave function responsible for the first absorption peak is close to an oxygen position τ_{oxygen}. In Fig. 6 we show the probability density $|\Psi(\mathbf{r}_e, \mathbf{r}_h = \tau_{oxygen})|^2$ corresponding to the electron distribution of that particular exciton. It is still reminiscent of the molecular LUMO shown in Fig. 5, but it expanded considerably farther towards the nearest-neighbour molecules. The calculated mean distance R between the electron and the hole is 4.02 Å, much larger than in the case of a gas-phase water molecule.

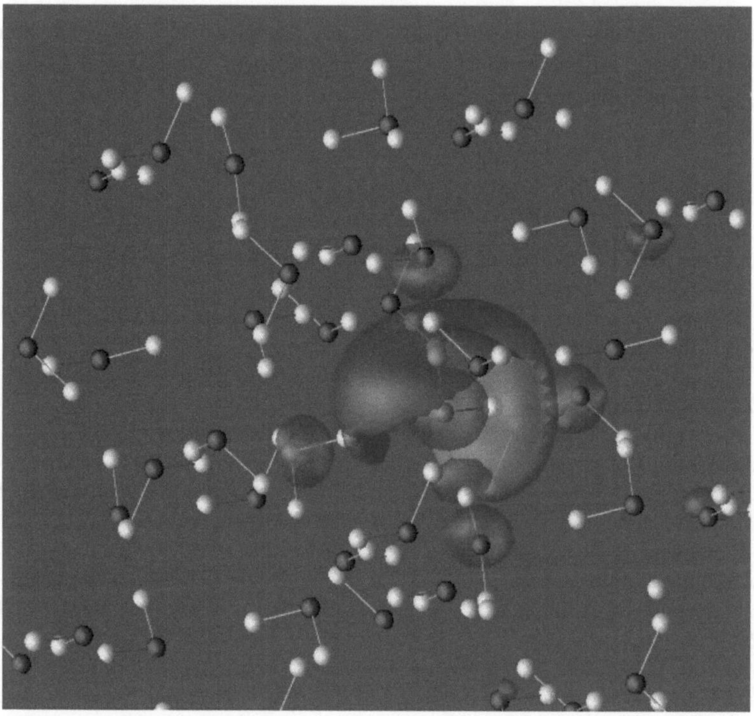

Fig. 6. Spatial distribution of the electron in the exciton associated with the lowest optical absorption peak of ice-Ih

4 Conclusions

Our work shows that (i) the three-step approach consisting of DFT-GGA ground-state calculations, quasiparticle calculations within the GWA and the solution of the Bethe-Salpeter equation for the polarizability is appropriate to investigate excited-state properties of water structures. (ii) The water monomer exciton is extremely sensitive to hydrogen-bonded neighbour molecules. This suggests the possibility of detecting changes in the solvation shell due to missing molecules or distorted hydrogen bonds by means of optical spectroscopy. The proposed calculations should thus be very helpful to clarify the interaction of water molecules with each other and also with foreign molecules. (iii) The simulations provide an intuitive explanation for the surprising energy shift discovered in experiments: The matrix of surrounding H_2O molecules enables the optically excited electron to move farther away from the hole left behind at the oxygen atom, thus reducing the binding energy of the electron-hole pair from 5.3 to 3.2 eV. This effect overcompensates the narrowing of the HOMO-LUMO energy gap resulting from the condensation of the H_2O molecules and leads to the observed blueshift of the optical absorption.

Generous grants of computer time from the Höchstleistungs-Rechenzentrum Stuttgart (HLRS) and the Paderborn Center for Parallel Computing (PC^2) are gratefully acknowledged.

References

1. R Onaka and T. Takahashi, J. Phys. Soc. Jpn. **24**, 548 (1968).
2. K Kobayashi, J. Phys. Chem. **87**, 4317 (1983).
3. G D Kerr, R N Hamm, M W Williams, R D Birkhoff, and L R Painter, Phys. Rev. A **5**, 2523 (1972).
4. H Hayashi, N Watanabe, Y Udagawa, and C-C Kao, Proc. Natl. Acad. Sci. USA **97**, 6264 (2000).
5. V F Petrenko and I A Ryzhkin, Phys. Rev. Lett. **71**, 2626 (1993).
6. W F Chan, G Cooper, and C E Brion, Chem. Phys. **178**, 387 (1993).
7. K Laasonen, M Sprik, M Parrinello, and R Car, J. Chem. Phys. **99**, 9080 (1993).
8. G P Parravicini and L Resca, Phys. Rev. B **8**, 3009 (1973).
9. L Resca and R Resta, phys. stat. sol. (b) **81**, 129 (1977).
10. W Y Ching, M-Z Huang, and Y-N Xu, Phys. Rev. Lett. **71**, 2840 (1993).
11. B D Bursulaya, J Jeon, C-N Yang, and H J Kim, J. Phys. Chem. A **104**, 45 (2000).
12. G Onida, L Reining, and A Rubio, Rev. Mod. Phys. **74**, 601 (2002).
13. P H Hahn, W G Schmidt, K Seino, M Preuss, F Bechstedt, and J Bernholc, Phys. Rev. Lett. **94**, 037404 (2005).
14. P H Hahn, W G Schmidt, and F Bechstedt, Phys. Rev. B **72**, 245425 (2005).
15. M S Hybertsen and S G Louie, Phys. Rev. B **34**, 5390 (1986).
16. S Albrecht, L Reining, R Del Sole, and G Onida, Phys. Rev. Lett. **80**, 4510 (1998).
17. L X Benedict, E L Shirley, and R B Bohn, Phys. Rev. Lett. **80**, 4514 (1998).
18. M Rohlfing and S G Louie, Phys. Rev. Lett. **83**, 856 (1999).
19. L J Sham and T M Rice, Phys. Rev. **144**, 708 (1966).
20. W Hanke and L J Sham, Phys. Rev. B **12**, 4501 (1975).
21. W Hanke and L J Sham, Phys. Rev. B **21**, 4656 (1980).
22. I Morrison, J-C Li, S Jenkins, S S Xantheas, and M C Payne, J. Phys. Chem. B **101**, 6146 (1997).
23. E L Briggs, D J Sullivan, and J Bernholc, Phys. Rev. B **54**, 14362 (1996).
24. J P Perdew, K Burke, and M Ernzerhof, Phys. Rev. Lett. **77**, 3865 (1996).
25. D R Hamann, Phys. Rev. B **55**, R10157 (1997).
26. F Bechstedt, R Del Sole, G Cappellini, and L Reining, Solid State Commun. **84**, 765 (1992).
27. J E Northrup, Phys. Rev. B **47**, R10032 (1993).
28. W G Schmidt, S Glutsch, P H Hahn, and F Bechstedt, Phys. Rev. B **67**, 085307 (2003).
29. S Glutsch, D S Chemla, and F Bechstedt, Phys. Rev. B **54**, 11592 (1996).
30. P H Hahn, W G Schmidt, and F Bechstedt, Phys. Rev. Lett. **88**, 016402 (2002).

The Electronic Structures of Nanosystems: Calculating the Ground States of Sodium Nanoclusters and the Actuation of Carbon Nanotubes

B. Huber[1,2], L. Pastewka[2], P. Koskinen[1,2], and M. Moseler[1,2]

[1] Freiburg Materials Research Center, Stefan-Meier-Str. 21, 79104 Freiburg
[2] Fraunhofer Institut für Werkstoffmechanik, Wöhlerstr. 11, 79108 Freiburg
mos@iwm.fhg.de

Summary. This contribution reports large scale electronic structure calculations for two important classes of nanosystems, namely metal nanoclusters and carbon nanotubes. In the former case we focus on the ground-state structure of sodium cluster anions which is obtained by solving the Kohn-Sham equations of density functional theory (DFT) in combination with an efficient genetic optimisation algorithm. The accuracy and efficiency of this approach is reflected in the electronic density of states of the lowest lying isomer showing excellent agreement with experimental photoelectron spectra. In the case of the carbon nanotubes (CNTs), the swelling and shrinking of the CNT paper due to electron or hole injection (bond strength induced actuation) has been studied. Experimentally, the charging of the tubes is caused by a small applied voltage and it is compensated by a electrochemical double layer formed in a surounding liquid electrolyte. Here we focus on models describing a singel nanotubes and the double layer within density functional and density functional based tight-binding band structure calculations. We show that the frequently used representation of the double layer by a jellium background reveals an unphysical dependence of the CNT actuation on the size of the computational box. As an alternative a cylindrical charge compensation, not suffering of this shortcoming, is suggested.

1 Theoretical Method and Computational Details

Accurate quantum chemistry calculations of many electron systems requires heavy computer power. In this context Kohn's density functional theory [1] plays a dominant role, since it reduces the complicated many electron system to a more tractable picture of a single electron in the mean field of the other electrons resulting in a three-dimensional eigenvalue equation, the so-called Kohn-Sham equation. For large systems, this equation can be solved with great accuracy and efficiency using a plane wave basis set for the single electron wave functions [2]. However, the memory and CPU requirement still exceeds modern serial hardware and thus massive parallel computing is the only way to solve the Kohn-Sham equations for a large number of atoms.

In DFT we search for the solution of the Kohn-Sham equation for the electrons

$$\left(-\frac{1}{2}\nabla^2 + v_{eff,\sigma}(\mathbf{r}) \right) \phi_{i,\sigma}(\mathbf{r}) = \epsilon_{i,\sigma}\phi_{i,\sigma}(\mathbf{r}). \tag{1}$$

Here the $\phi_{i,\sigma}$ are a set of a single particle electronic wave function, $\epsilon_{i,\sigma}$ their energies and the effective potential is given by

$$v_{eff,\sigma}(\mathbf{r}) = v(\mathbf{r}) + \int d^3r' \frac{n(\mathbf{r}')}{|\mathbf{r} - \mathbf{r}'|} + v_{xc,\sigma}(\mathbf{r}). \tag{2}$$

The electron density n of the system, as the central quantity of density functional theory, derives from the occupied Kohn-Sham orbitals

$$n(\mathbf{r}) = \sum_{i,\sigma}^{occ} |\phi_{i,\sigma}(\mathbf{r})|^2. \tag{3}$$

In order to make the computations less expensive, only chemically active electrons are considered and therefore a pseudo potential v is used for the confinement of the valence electrons representing the influence of the naked ions and the core electrons [3]. The exchange-correlation potential v_{xc} takes into account many-body effects that are not included in the classical Coulomb field $\int d^3r' \frac{n(\mathbf{r}')}{|\mathbf{r}-\mathbf{r}'|}$ in the above equation. It is treated in the framework of the generalized gradient approximation [4]. The spin of the system is explicitely taken into account by calculating the wave functions of both spin manifolds $\sigma = \uparrow, \downarrow$, thus making the description of magnetism possible within this formalism. For more details on spin density functional theory, the reader is refered to standard textbooks [5].

The method for the numerical solution of eq. (1) utilizes the Born-Oppenheimer-local-spin-density-molecular-dynamics (BO-LSD-MD) approach of Barnett and Landman [2] and benefits from the fact that the differential operator $-\frac{1}{2}\nabla^2$ is diagonal and given by by $-\frac{1}{2}k^2$ for the Fourier transform $\phi_{\mathbf{k}}$ of the wave function. An iterative Block-Davidson eigenvalue solver needs the action of the Hamiltonian $-\frac{1}{2}\nabla^2 + v_{eff}$ onto the wave function and therefore a dual space technique treating the kinetic energy in Fourier and the potential energy part in real space provides a very efficient scheme to solve eq. (1). A domain decomposition of both spaces and an efficient parallelisation of the fast fourier transform (FFT) connecting k- and real space results in a very good parallel efficiency on massively parallel machines like the HP XC6000. The FFT is also used to calculate the Coulomb field $\int d^3r' \frac{n(\mathbf{r}')}{|\mathbf{r}-\mathbf{r}'|}$ since it satisfies Poissons equation which is algebraic and thus easily solvable in k-space. For more details on the numerical aspects of the method see [2].

After the solution of the Kohn-Sham equations, the forces on the ions are calculated employing the Hellmann-Feynman-Theorem [5], and the molecular dynamics of the respective system can be studied with very high accuracy. In certain cases where finite temperature infomation are needed a Langevin thermostat is used in order to simulate a canonical ensemble.

For our DFT calculations on the HP XC6000 Cluster we used up to 8 CPUs for the global optimisation algorithm and up to 32 CPUs for local optimisations of large sodium cluster with more then 300 electrons. Bechmark runs showed an almost linear scaling for the runtime. We are particularly delighted that the HP XC6000 gives us the opportunity for long jobs (3 days) which is very important for the efficiency of our global optimisation algorithm.

2 Electronic and Geometric Structure of Sodium Nanoclusters

Since the experiments of Knight *et al.* [6] and their interpretation in terms of the jellium model [7, 8], sodium clusters have attracted great attention of scientists, both experimentalists and theoreticians [9, 10, 11]. One reason for that is the electronic structure of sodium. The valence electrons in sodium are almost completely free and therefore bulk sodium is the model case of an ideal free electron metal. This results in a very simple density of states for the electrons, the electronic shell structure.

The experimental determination of the geometric structure of sodium clusters is a big challenge. Often the only way to find the ground-state structure is by calculating the total energy of a large set of possible configurations with an acurate theoretical method. Since the number of stable isomers increases exponentially with the number of atoms this is a huge numerical task and limited to small clusters. The standard method for the determination of the ground-state structures in the last decades was the simulated annealing technique where the Born-Oppenheimer energy surface is scanned by thermal molecular dynamics and some configurations of the trajectory gets optimised to the next local energy minimum. This method often fails in systems with high energy-barriers which trap the simulation in one of the numerous metastable configurations. Thus an algorithm is needed which can hop from one minimum to another. Recently some new promising methods were introduced using genetic algorithms (GA). The basic concept of GA is derived from the biological evolution. Parent structures are combined to form offspring, which then compete with other structures. The property of the "fitness" is here the total energy of the system. The method for the search for the global minimum we use is essentially based on the single-parent genetic algorithm [12]. Two genetic-like mating operations are applied to the parent structure, "piece reflection" and "piece rotation', which divides the cluster in two equal size parts and then creates a new child out of one part and its reflection or rotation. This new structure gets optimised and, if lower in energy, is established as

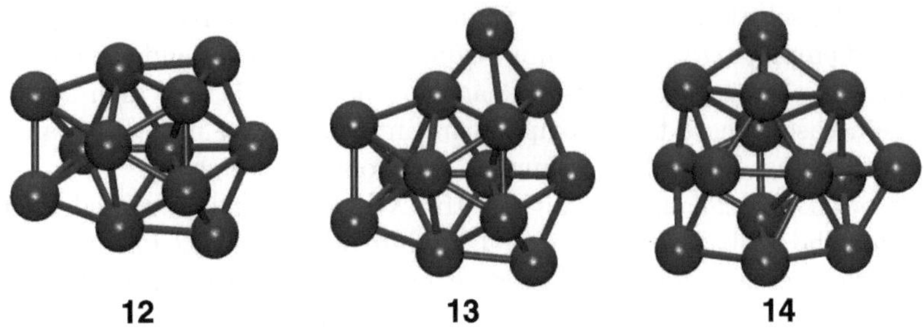

Fig. 1. The new groudstate structures found by the single-parent evolution algorithm for Na^-_{12-14}

a new parent. When the single-parent method is used in combination with a slightly modified version of another method called big bang [13], one gets a rather effective algorithm to search for the ground state and explore many of the low potential energy isomers.

First calculations on the sizes $N < 20$ confirmed the structures found in an earlier study using simulated annealing [9], but we also discovered for Na^-_{12}, Na^-_{13}, Na^-_{14} and Na^-_{16} lower energy structures (see Fig. 1). The efficiency of this algorithm is quite astonishing. For small clusters it often takes less than fifty local optimisation to reach the ground-state structure.

Regarding only the energies of the lowest isomers one can never be sure that the single-parent evolution algorithm has already converged to the "real" ground-state structure. Therefore it would be helpful to find a property which allows to compare experiment and theoretical calculations. A reasonable way to compare theoretical predictions on sodium cluster ground-state structures to experimental results is a comparision of the calculated DOS to the measured photoelectron spectra (PES) [9]. Since the electronic structures of clusters are very sensitive to the arrangement of the atoms in the cluster it is possible to assign for each size the ground-state configuration.

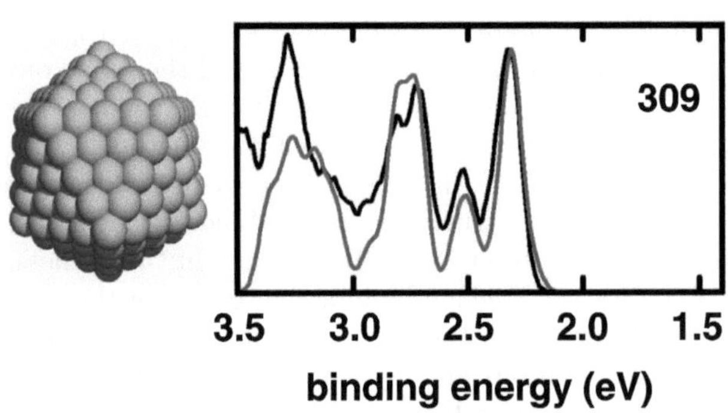

Fig. 2. Electron density of states (*red curve*) and photoelectron spectra (*black curve*) of the Na^-_{309} Mackay icosahehron

The success of this method on much larger systems than in Ref. [9] is described in Ref. [14] including the results of our calculations on the HP XC6000 Cluster. We will here just summarise the main points of this paper. First of all we showed that sodium clusters prefer Mackay icosahedral structures for the complete shells Na_{55}^-, Na_{147}^- and Na_{309}^-. In Fig. 2 the DOS and PES of Na_{309}^- are compared and they show a very good agreement. Even at this size the clusters show a clear electronic shell structure and the DOS is still far away from bulk behaviour. For the sizes between closed Mackay icosahedron shells we saw a good agreement between experiment and theory for clusters with complete twisted Mackay-caps on the top of the icosahedrons. This indicates a Mackay-like growth motif for sodium clusters. For further informations we want to refer again to Ref. [14].

3 Actuation of Charged Carbon Nanotubes

Since their discovery in 1992 carbon nanotubes have been proposed for a wealth of applications. Certainly one of the most exciting ones is the use of carbon nanotubes as actuators as proposed by Baughman *et al.* in 1999 [15]. Baughman immersed nanotube paper in a liquid electrolyte and found the paper to expand or contract depending on the applied voltage. This application shows the possibility to have huge technological impact because of the high work densities realized. Despite some research in this field the actuation mechanism is basically ununderstood: According to Baughman, repulsion of the double layer of mutual nanotubes or quantum mechanical expansion of the C-C bond length might be responsible for the observed actuation.

Here, we investigate the latter effect. Previous studies [16, 17, 18] based on density functional theory (DFT) have proven unsatisfying because the charged tubes were modelled in a uniform Jellium background which turns out to render the results for effective one-dimensional systems such as nanotubes unusable. We present a similar study using a cylindrical compensating background. Our study is based on density functional tight-binding (DFTB) [19] within the self-consistent charge (SCC) [20] level of theory. Within this level, the electronic state of the system is described by the total energy

$$E = \sum_{I\mu J\nu} \rho^{I\mu J\nu} H_{J\nu I\mu} + \sum_{IJ} (E_{IJ}^{\mathrm{rep}} + \gamma_{IJ} \Delta q_I \Delta q_J) \tag{4}$$

where ρ denotes the density matrix, H the tight-binding Hamiltonian, E^{rep} the repulsive energy and Δq the excess charge. In this notation, the indices I and J denote atoms, whereas the indices μ and ν denote orbitals. Excess charge is considered up to first order perturbation theory, where in our case we assume the radial decay of the excess charge to be of Gaussian nature [21]. Minimisation of the total energy E leads to the equivalent generalised eigenvalue problem

$$\sum_{J\nu} C^i_{J\nu} \left\{ H_{I\mu J\nu} + \frac{1}{2} S_{I\mu J\nu} \sum_{\xi} \left(\gamma_{I\xi} + \gamma_{J\xi} \right) \Delta q_\xi - \epsilon_i S_{I\mu J\nu} \right\} = 0 \qquad (5)$$

which is solved directly in our code. The symbol S denotes the overlap matrix elements and the excess charges Δq_I are determined by a Mulliken charge analysis [22]. In addition, we consider Brillouin zone sampling [22] to be able to treat very large tubes. The extension of Eq. (5) to include Brillouin zone sampling is straighforward. Note that this theory can be regarded as a first order perturbation expansion in the charge density within density functional theory using a minimal localised basis set. The use of the SCC-DFTB approach now allows for the treatment of systems much larger than previously possible.

3.1 Simulation Procedure and Results

For each tube the charge per atom is varied between $-0.03e$ and $0.03e$. This corresponds approximately to the charge per atom extracted from experimentally determined capacities and applied voltages [15]. For each tube chirality and each charge the simulation box size is varied and the energy vs. box size curve is extracted. From this it is possible to determine the minimum energy configuration and the elastic modulus of the tube at different charging levels.

First, the influence of the jellium on the obtained actuation is investigated. It is found that the actuation of the tubes does crucially depend on the box size perpendicular to the tubes. A larger box size leads to a diluted compensating jellium which in return leads to more actuation due to simple Coulomb repulsion. Therefore, we employ the more natural model of a compensating charged cylinder located at a distance of 8 a.u. from the nanotubes. This cyclinder mimicks the electrostatic double layer in a natural way and does lead to results that are independent of box size. A comparison of the jellium and the double layer model is depicted in Fig. 3. These results render previous studies [16, 17, 18] useless because of the difficulty to relate the jellium approximation to physically realisable situations.

The actuation and elastic modulus of chosen zig-zag tubes is shown in Fig. 4 and Fig. 5, respectively. It can be seen that each tube has its very individual actuation signature, the total actuation does however stay well below 1%. Furthermore, the change of the elastic modulus does show a similar individual signature. The strengthening and weakening of the tubes lies within 5%.

3.2 Outlook

Our SCC-DFTB simulation code has been parallelised using OpenMP. However, for the solution of the eigenvalue problem (5) it does rely on the routines

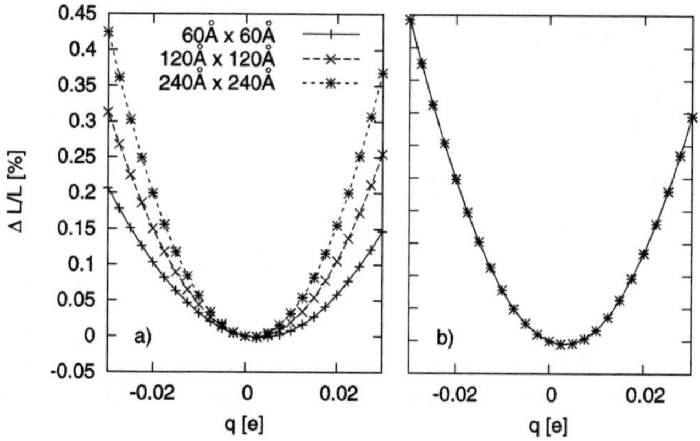

Fig. 3. Comparison of the jellium and the double layer model for charge compensation. Within the jellium approach a), a variation of the box size perpendicular to the tube leads to a different actuation behavior due to different jellium concentration. Within the double layer model b), the curves for the three box sizes do perfectly coincide. The only free parameter within the double layer approach is thus the distance of the layer from the tube which is chosen to be 4 a.u. in all simulations. The chiralty of the tube used in these simulation is 5×5 using a Brillouin zone sampling after the method of Monkhorst and Pack [23] with a k-space mesh of $1 \times 1 \times 20$. The lines connect calculated points and are intended to guide the eye

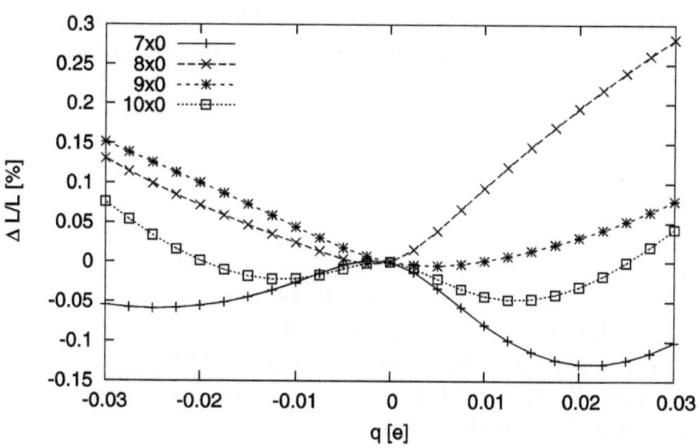

Fig. 4. Actuation of 7×0, 8×0, 9×0 and 10×0 tubes. Each tube has an individual actuation signature. The total actuation does always stay well below 1% for all the charging levels investigated. An all simulations the Brillouin zone was sampled after the method of Monkhorst and Pack [23] with a k-space mesh of $1 \times 1 \times 20$. The lines connect calculated points and are intended to guide the eye

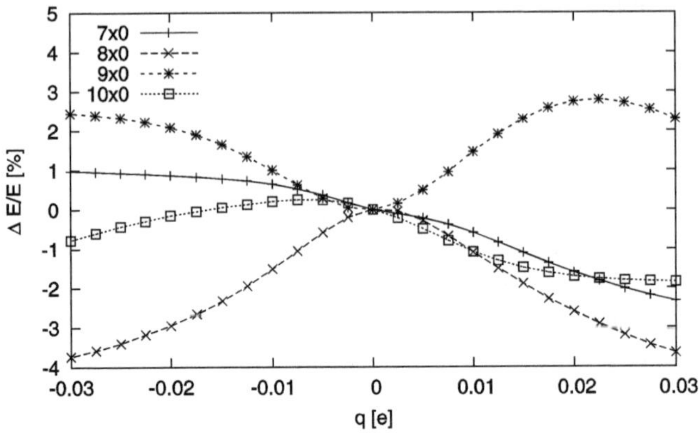

Fig. 5. Change of the elastic modulus of charged 7×0, 8×0, 9×0 and 10×0 tubes. In all simulations the Brillouin zone was sampled after the method of Monkhorst and Pack [23] with a k-space mesh of $1 \times 1 \times 20$. The lines connect calculated points and are intended to guide the eye

provided by LAPACK which do only scale up to 2 processors for systems of the size investigated here. Due to the fact that for each charging level the total energy for about 10 box sizes needs to be determined, the whole computational procedure does still scale very well. However, our research is directed into the investigation of much larger systems. Thus in the future, $O(N)$ methods and parallelised diagonalization techniques will be explored to be able to treat even larger tubes.

References

1. Kohn, W., Sham, L.J., Phys. Rev. A **140**, 1133, (1965)
2. Barnett, R., Landman, U., Phys. Rev. B **48**, 2081 (1993)
3. Troullier, N., Martins, J.L., Phys. Rev. B **43**, 1993 (1991)
4. Perdew, J.P., *et al.*, Phys. Rev. Lett. 77, 3865 (1996)
5. Parr, R., Yang, W.: Density functional theory of atoms and molecules. Oxford University Press, New York (1989)
6. Knight, W.D., *et al.*, Phys. Rev. Lett. **52**, 2141 (1984)
7. Ekardt, W., Phys. Rev. Lett. **52**, 1925 (1984)
8. Beck, D.E., Phys. Rev. B **30**, 6935 (1984)
9. Moseler, M., *et al.*, Phys. Rev. B **68**, 165413 (2003)
10. Kostko, O., *et al.*, Eur. Phys. J. D **34**, 133 (2005)
11. Haberland, H., *et al.*, Phys. Rev. Lett. **94**, 035701 (2005)
12. Rata, I., *et al.*, Phys. Rev. Lett. **85**, 546 (2000)
13. Jackson, K.A., *et al.*, Phys. Rev. Lett. **93**, 013401 (2004)
14. Kostko, O., von Issendorff, B., Huber, B., Moseler, M., Phys. Rev. Lett. submitted
15. Baughman, R.H. et al., Science, **284**, 1340 (1999)

16. Sun, G., Kürti, J., Kertesz, M., Baughman, R.H., J. Am. Chem. Soc., **124**, 15076 (2002)
17. Sun, G., Kertesz, M., Kürti, J., Baughman, R.H., Phys. Rev. B, **68**, 125411 (2003)
18. Verissimo-Alves, M., Koiller, B., Chacham, H., Capaz, R.B., Phys. Rev. B, **67**, R161401 (2003)
19. Porezag, D., Frauenheim, T., Köhler, T., Seifert, G., Kaschner, R., Phys. Rev. B, **51**, 12947 (1995)
20. Elstner, M., *et al.*, Phys. Rev. B, **58**, 7260 (1998)
21. Bernstein, N., Mehl, J., Papaconstantopoulos, D.A., Phys. Rev. B, **66**, 075212 (2002)
22. Finnis, M.: Interatomic forces in condensed matter. Oxford University Press, New York (2003)
23. Monkhorst, H.J., Pack, J.D., Phys. Rev. B, **13**, 5188 (1976)

Object-Oriented SPH-Simulations with Surface Tension*

S. Ganzenmüller[1], A. Nagel[1], S. Holtwick[2], W. Rosenstiel[1], and H. Ruder[2]

[1] Wilhelm-Schickard-Institut für Informatik, Universität Tübingen
[2] Institut für Astronomie und Astrophysik, Universität Tübingen

Summary. Today object-oriented software development is well established in industry and research and has replaced procedural techniques. Nevertheless there is a lack of integration of object-oriented concepts in parallel scientific applications and the underlying parallelization libraries. To benefit from advantages of object-oriented programming like easy configurability and good extensibility, we implemented a framework for SPH simulations to support the ongoing development of the SPH method. We used the high performance systems at HLRS during the last year to test our object-oriented approach and the influences of several configurations on the runtimes and speedup on different machines. We integrated parallel I/O and surface tension in our framework and show first results. The implementation of object-oriented parallel I/O improved the performance significantly. The next step is to optimize and advance the surface tension model.

1 Introduction

Smoothed Particle Hydrodynamics (SPH) is a widespread particle simulation method in parallel scientific computing. It is a grid-free Lagrangian method for solving the system of hydrodynamic equations for compressible and viscous fluids. SPH was introduced in 1977 [5], [8] and has become a widely used numerical method for astrophysical problems. Although its main application is still located in astrophysics, the SPH approach is nowadays also used in fields of material sciences, for modeling multiphase flows such as diesel injection [10], and the simulation of brittle solids [1].

In the Collaborative Research Center (CRC) 382 physicists, mathematicians and computer scientists work together to research new aspects of astrophysics and the motion of multiphase flows, evolve them to models and run parameter studies to verify these models. Several particle codes are used to simulate these physical problems.

* This project is funded by the DFG within CRC 382: *Verfahren und Algorithmen zur Simulation physikalischer Prozesse auf Höchstleistungsrechnern* (Methods and algorithms to simulate physical processes on supercomputers).

Resolution and accuracy of a simulation depend on the number of used particles and interaction partners. Actual physical problems like the simulation of surface tension, turbulence or cavitation need several millions of particles to achieve reasonable results. Thus, high-performance computers are indispensable.

Our group has a strong effort to develop fast particle libraries which are portable to all important parallel platforms. Therefore, we developed a parallel object-oriented SPH framework which is clearly structured, easy to configure, maintain and extendable.

In the second chapter we introduce our object-oriented framework and present a comparison of several configurations on the Kepler Cluster in Tübingen, the Cray Opteron Cluster and the HP zx6000 at HLRS. The third chapter explains parallel I/O in sph2000 together with performance results. The fourth chapter presents the model and implementation of surface tension in SPH together with physical results.

2 Object-oriented SPH

2.1 Design Goals

The benefit of an object-oriented view is in the contiguousness of the model and the modeled system. The object-oriented approach unifies analysis, design and implementation of applications. Our main goal was to develop a parallel object-oriented SPH framework with extensibility, maintainability and reusability of the code. A main concern in the design was the strict decoupling of parallelization and physics. Another goal was to prove the feasibility of the object-oriented approach in the performance critical domain of particle simulations without loosing efficiency. The result is a particle simulation framework written in C++, called sph2000. Classes modeling the elements of the problem domain generate a well structured design. The use of several design patterns [4] helped to organize the classes clearly and efficient. They are introduced to structure the class library, to separate and group the application elements, as well as to define uniform interfaces. Additionally, they allow to insert extensions, like the newly implemented surface tension model, more simply because of decoupled elements. The independent elements can be extended, causing no changes in other classes.

2.2 Configuration and Initialization of Simulation Runs

Another goal was to simplify the configuration of simulations, which mainly means to configure a simulation run after the compilation at runtime by reading a parameter file. The key design pattern of the framework is the strategy pattern, which is based on the object-oriented concept of polymorphism. Every configurable element of the framework, like the hydrodynamic equations, the timing and integration or the parallelization method, defines an abstract base class with a uniform interface. The classes within an element can easily be exchanged because of these interfaces.

The complete configuration with all parameters of a simulation is read from a text file and stored in an object of the class ParameterMap (configuration table pattern) and spread to all nodes. Every object which needs some configuration parameters owns a reference to the ParameterMap object, thus all objects can access the configuration uncomplicatedly.

Mainly the initialization objects (builder pattern) access the ParameterMap to realize the exchangeability of the components. Every strategy offers an accordant parameter to determine which concrete implementation is used in the simulation. In every simulation run, the builders create and initialize only the needed objects.

For example two different communication classes are implemented, SingleCommunicator for stand-alone workstations to test the application and several configurations easily and TpoCommunicator to run simulations in parallel. Due to the configuration file, the appropriate communicator object will be created.

2.3 Performance Results

The SPH framework sph2000 was used so far on workstations with Solaris and different Linux systems and parallel in a network of workstations and on the Kepler cluster. To analyze the runtime behaviour of the framework, it was ported on other machines, mainly the Cray Opteron Cluster and the HP zx6000 in Stuttgart [7]. On all machines the GNU C++ compiler is used, so the results could improve with the Portland Group compiler on Cray or Intel compiler on HP.

An experimental porting to the Hitachi SR8000 in Munich failed due to compilers with a lacking support for the STL. The porting to the NEC SX-8 in Stuttgart is not yet completed. The SX-8 compiler has problems with inline and template methods of the frameworks classes.

Table 1 compares the application on Kepler Cluster (Intel Pentium III, 650 MHz), Cray Opteron Cluster (AMD Opteron, 2 GHz) and HP zx6000 (Intel Itanium 2, 900 MHz). All measurements are executed with 2 processors per node. The speedup of 2.2 on 3 nodes is a consequence of the few particles used in the comparing measurements which results in a relative high administration effort. The initialization and administration of the objects and the structures to organize the communication are designed for a high throughput. But in order to make all results comparable, a small common ground must be found. To obtain an optimal speedup, the entire memory of the machines should be utilized. The fact that the speedup of 1.3 from 2 to 3 nodes is lower than the speedup 1.6 from 1 to 2 nodes is caused in the smaller communication overhead of 2 nodes, which has quadratic subdomains in the present configuration.

Table 1. Results of simulations on 1, 2 and 3 nodes. The relative speedup is based on the Kepler Pentium results

Nodes	1	2	3	Speedup
Execution time (in s)				
– Kepler Pentium	78.1 s	47.9 s	36.4 s	2.1
– Cray Opteron	13.9 s	8.9 s	6.5 s	2.1
– HP zx6000	47.2 s	28.3 s	21.4 s	2.2
Relative Speedup				
– Cray Opteron	5.6	5.4	5.6	
– HP zx6000	1.7	1.7	1.7	

Number of Interactions

The interaction search takes a considerable part of the computing time of an SPH simulation. Table 2 compares simulation sets with a different amount of interactions per particle. Whereas Kepler and HP zx6000 need 1.7 times longer for a double number of interactions, Cray shows a better factor of 1.5. With a higher number Cray grows above average, so the dimension of the problem is to small to take advantage of the machine.

Table 2. Results of simulations with an average of 50, 75 and 100 interaction partners per particle

Interactions	50	75	100	Runtime extension
Execution time (in s)				
– Kepler Pentium	25.0 s	35.0 s	43.6 s	1.7
– Cray Opteron	6.6 s	8.6 s	10.2 s	1.5
– HP zx6000	14.7 s	20.5 s	25.7 s	1.7
Relative Speedup				
– Cray Opteron	3.8	4.0	4.3	
– HP zx6000	1.7	1.7	1.7	

Time Integration Method

The choice of the integration method leads to a convenient comparison of runtimes. The communication increases proportional with the order of the integration method, since the interaction search and the right hand side calculation must be done for every gradient. While the Euler method calculates only one gradient, the standard Runge-Kutta method is a four-step integrator. Table 3 shows, that the runtime extension is lower than expected. Cray needs 3.3 times longer, Kepler and HP zx6000 perform better with only 2.5 times. All machines compensate the enhanced communication and calculation overhead.

Table 3. Results of simulations with Euler and Runge-Kutta as time integration method

Integrator	Euler	Runge-Kutta	Runtime extension
Execution time (in s)			
– Kepler Pentium	17.4 s	44.3 s	2.5
– Cray Opteron	3.2 s	10.6 s	3.3
– HP zx6000	10.7 s	26.9 s	2.5
Relative Speedup			
– Cray Opteron	5.4	4.2	
– HP zx6000	1.6	1.6	

Number of Particles

With the number of particles, involving a higher number of interactions, the calculation cost increases disproportionate. There are more possible interactions to prove, more right hand sides must be calculated and the amount of proxy particles to communicate between the nodes rises. Table 4 shows, that the runtime on HP zx6000 increases approximately linear with the particles. Kepler with an extension factor of 6.4 and Cray with only 5.7 work noticeably better.

Table 4. Results of simulations with an increasing particle count

Particles	5000	20000	50000	Runtime extension
Execution time (in s)				
– Kepler Pentium	19.7 s	56.4 s	125.6 s	6.4
– Cray Opteron	4.0 s	10.5 s	22.7 s	5.7
– HP zx6000	7.1 s	32.9 s	73.5 s	10.4
Relative Speedup				
– Cray Opteron	4.9	5.4	5.5	
– HP zx6000	2.8	1.7	1.7	

Conclusion

The results demonstrate, that the object-oriented application works well on different machines. With different configurations and thus increasing communication and computing power all three systems keep a constant balance to each other. Based on the Kepler Pentium nodes, HP zx6000 shows an average speedup of 1.8, Cray runs 4.9 times faster. The framework compensates the increasing demands very good, the runtime extension is notedly less then expected. The slow speedup with identical configurations on more nodes is due to a small particle density and computing complexity per processor. The results indicate, that the speedup gets better with a better utilization of the memory of the machines.

3 sph2000 with Parallel I/O

3.1 Parallelization and Domain Decomposition

For an efficient parallelization we implemented a domain decomposition, dividing up the simulation area into several rectangular subdomains. These dynamically change their size during runtime to keep the load balanced between all processors.

In every time step each subdomain has to process the same tasks, like preparing the calculation, communicating particles to neighbouring subdomains, computing lists which contain the interaction partners, calculating the right hand side, or integrating the equations of motion. Each subdomain therefore contains classes and objects respectively for these tasks. From an object-oriented view each subdomain is a group of objects, which represent this area geometrically.

All subdomains are independent from each other and communicate through a special communicator which is needed to pass objects from one subdomain to another. The implementation of this Communicator class (inter node mediator) follows the strategy and mediator design patterns. It encapsulates the information about the whole communication structure and defines the interface between intra node and inter node communication. Thus all scientific objects are decoupled form the parallelization method. The major advantage of this concept is, that it enables the user to easily divide up the simulation domain into as many subdomains as processors are available. For communication between the processors the message passing object TpoCommunicator is generated which utilizes TPO++ as object-oriented message passing library.

To communicate objects, the corresponding class must implement member functions `serialize()` and `deserialize()` to determine which data of a dynamic structure should be transmitted and which can be reconstructed from the receiver. The `TPO_MARSHALL_DYNAMIC` macro allows TPO++ to generate an efficent communication at compile time:

```
#include<tpo++.H>

class Particle
{
  public:
    Particle();

    // make a particle object sendable ... and receivable
    void serialize(TPO::Message_data& msg) const;
    void deserialize(TPO::Message_data& msg);
    ...
};
TPO_MARSHALL_DYNAMIC(Particle);
```

To specify the data to transmit inside the `serialize()` method, the member attributes must be inserted in the message stream, complete STL containers are inserted via their iterators:

```
void
Particle::serialize(TPO::Message_data& msg) const
{
    msg.insert(scalarQuantities.begin(), scalarQuantities.end());
    msg.insert(vectorQuantities.begin(), vectorQuantities.end());

    msg.insert(id);
    msg.insert(fluid);
}
```

To deserialize an object `msg.insert()` must be replaced by `msg.extract()`.

3.2 Parallel I/O

The first results of sph2000 showed a significant lack in performance due to sequential I/O. The constantly growing gap between processor and hard disk performance during the last decades makes this problem even worse and requires the use of parallel I/O systems. The most often used interface for parallel I/O is the standard MPI-IO [2].

However, MPI 2 only supports procedural interfaces for parallel I/O, making it impossible to be used in our object-oriented application. Therefore, we extended TPO++ by an object-oriented interface for parallel I/O [11]. The major goals were to provide an efficient object-oriented interface, which supports standard datatypes like STL containers and provides the same functionality and be conform to MPI-IO.

So far, in sph2000 I/O was treated sequentially. The master process gathers the part results from the other processes and saves the whole file in ASCII format to disk. Figure 1 shows the sequential implementation. To save the data, all subdomains get the `collectParticles()` message. Node 0 writes its particles to disk and waits to receive and write the particles of the other nodes:

```
void
TpoCommunicator::collectParticles(const string& savingNumber) const
{
    const ParticleContainer& myparticles = mediator->collectParticles();
    if (group == 0) // receiver
        particleIO.saveDataFile(savingNumber, myparticles);
        for (int i = 1; i < groupCount; ++i)
        {
            ParticleContainer otherparticles;
            TPO::net_back_insert_iterator<ParticleContainer> pit(otherparticles);
            TPO::CommWorld.recv(pit);
            particleIO.saveDataFile(savingNumber, otherparticles);
        }
    else // senders
        TPO::CommWorld.send(myparticles.begin(), myparticles.end(), 0);
}
```

Because the amount of particles of the other domains is not known, the `TPO::net_back_insert_iterator` is used to fill the receiving STL container. The particleIO object uses `std::ofstream` to save the relevant part of the particles. A particle object stores all quantities which are needed for the simulation, which includes helper variables for the time integrator. The ParticleIO member `writeFile()` excludes these values from saving. The format of the data files is predetermined by the configuration parameters, so there is no meta-data needed:

```
void
ParticleIO::saveDataFile(const ParticleContainer& particles, const string& filename)
{
    std::ofstream file(filename.c_str());
    writeFile(file, particles);
    file.close();
}
```

To provide sph2000 with parallel I/O, the ParticleIO class had to be adapted. With the implementation of collective I/O, all processes can access the same file in parallel, see Fig. 2. This improves the performance significantly, since the communication to the master process is needless and the whole parallel I/O bandwidth can be used for transferring the data to and from disk. In addition, the library internally calculates the correct offsets within the file where each process has to place its part, avoiding any extra implementation by the user.

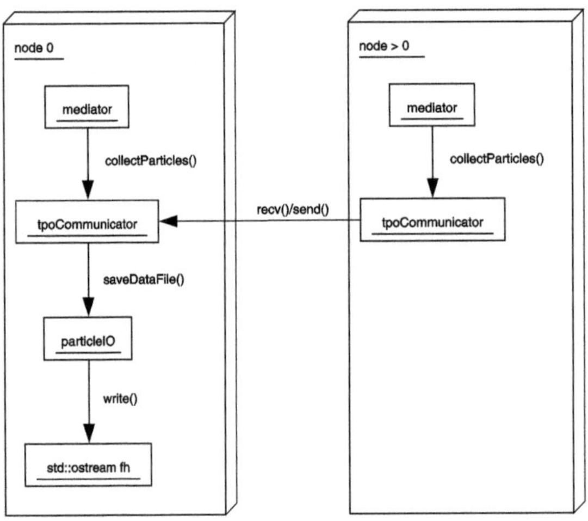

Fig. 1. Object diagram for sequential file I/O with one master node collecting all particles

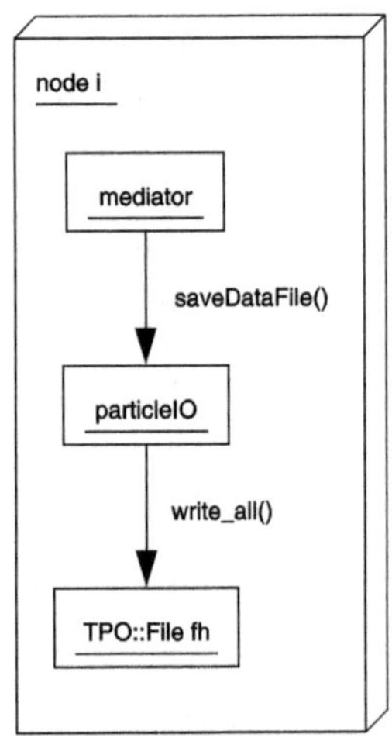

Fig. 2. Object diagram with parallel file I/O

The following listing shows the adapted method `saveDataFile()` and represents the simple usage of the interface. This implementation of using a single collective call with `fh.write_all()` reduces the size of the original code by about 100 lines of code and saves the containers of particles of each processor to disk simultaneously:

```
#include<tpo++.H>

void ParticleIO::saveDataFile(const ParticleContainer& particles, string name)
{
    TPO::File fh;
```

```
    int code = fh.open(TPO::CommWorld, name, TPO_MODE_CREATE);
    fh.write_all(particles.begin(), particles.end());
    fh.close();
}
```

3.3 Performance Results

Performance meassurements with a synthetic benchmark on the Cray Opteron Cluster [6] show, that TPO++ can handle data quantities from 64 k up without overhead. As a consequence the usage of object-oriented communication has no disadvantages in production runs. While these measurements took place on Cray, we switched to Kepler to run long simulation with many particles to test parallel I/O in our application. With 1 Mio. particles, a Runge-Kutta integrator with 5th order and a writing to disk in every second time step, the configuration corresponds to production runs.

We measured two different simulation setups: The first running on the Pentium nodes with disabled I/O and the second on the Athlons with I/O enabled in every second time step, to compare the performance of sequential and parallel I/O. We always used only one processor per node.

Figure 3 shows the performance results of the setups. Due to memory shortage of the Pentium nodes, the measurements of the second setup were made starting with 2 processors. The parallelization scales very good up to 64 processors leading to a remarkable speedup of 44.10.

The results of the second simulation setup show the significant effect of parallel I/O on the performance results. The I/O part could be improved by a factor of 20 when using 32 processors working on a parallel file system with 32 distributed disks. Since I/O is only a small proportion of the whole simulation, the overall gain using parallel I/O reduces to a – still remarkable – factor of 3.5 (64 processors). Note that even the sequential simulation with parallel I/O is faster than without parallel I/O, due to changing from ASCII file format to binary format and less code overhead for saving the particles.

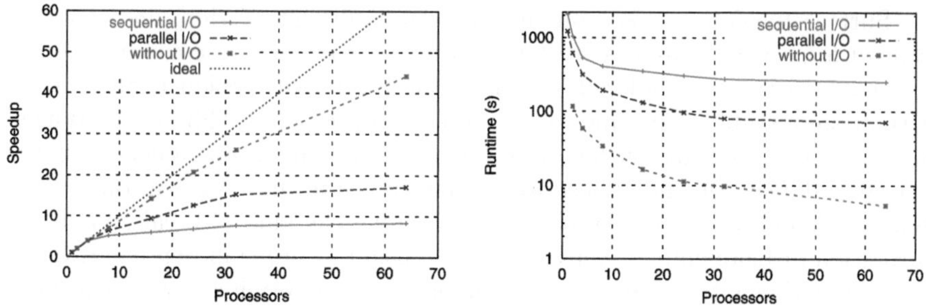

Fig. 3. Results of both simulation setups with 1 million particles (*left*) and corresponding runtimes of the simulation (*right*)

Table 5. Results of the second simulation setup with 1 million particles on Athlon showing the relative speedup when switching from sequential to parallel I/O

Processors	1	2	4	8	16	24	32	64
Execution time (in s)								
– sequential I/O	2090.2	1050.9	530.4	407.0	349.4	305.1	272.4	251.6
– parallel I/O	1223.5	617.2	312.5	191.4	130.9	96.5	79.8	71.3
Relative Speedup	1.7	1.7	1.7	2.1	2.7	3.2	3.4	3.5

4 Surface Tension

4.1 Model with Exterior Forces

The objectives of the current workings are to expand the SPH method by the possibility to simulate surface tension and cavitation. We examine the primary breakup of a compact liquid jet as well as collisions of droplets within dense sprays.

We apply the smoothed particle hydrodynamics method. This method provides the numerical algorithm to solve a system of coupled partial differential equations. SPH is an entirely grid free method. The real fluid is represented by an ensemble of SPH particles which are moving corresponding to the flow of the fluid, according to the lagrangian perception. The formalism essentially consists of three parts: the kernel smoothing, the Monte Carlo integration and the integration in time. The method is parallelized by domain decomposition.

The principle applicability of the SPH method for the simulation of compact fluid jets was revealed in former workings. To be able to simulate the unsteady atomization at very high pressures several enhancements of the method are necessary. The current simulations do not consider the divergence of the fluid jet, the breakup into ligaments and droplets due to surface tension as well as cavitation as significant source of turbulence within the fluid jet and the resulting structures.

Surface tension is based on the Van der Walls interaction, an attractive central force with comparative large coverage, particularly compared to the other interactions considered by the method. Surface tension results in the formation of stable drops and to the increase of pressure within drops. Within the simulated systems surface tension has a fundamental influence on the energy balance. During procedures of intermixture and segregation surface tension also plays a decisive role by determining the kind of breakup and the formation of ligaments as well as at the collision of drops in dense sprays.

Of practical interest are atomization processes at very high pressures above 1000 bar. The entire spray consists of some 10 billion droplets. The distribution of the sizes of the droplets corresponds to a standardized normal distribution around 5 μm. The relevant domain for the simulation contains about some 10.000 droplets. The surface energy within these systems is about two orders of magnitude greater than the kinetic energy and gives a lower limit for the distribution of drop sizes.

Because inner forces of the fluids have to be expressed as exterior forces acting upon the SPH particles, the physical equations can not simply be assumed. First an elementary derivation for a stress tensor was found and surface tension was formulated as singular force density on the surface, which implements the Continuum Surface Force method. The introduction of an artificial separation force proved to be partially necessary to prevent intermixture of the fluids. A variety of SPH equations is now available to compute the surface forces.

With the introduction of the surface tension into the SPH method resulted a couple of problems which partly still have to be solved. The SPH method shows in principle an instability to tractive forces. The sufficiently accurate computation of the local curvature from a distribution of smoothed particles is challenging. The resulting surface forces are being smoothed themselves and act on different kinds of particles with highly different masses. This results in huge acceleration of particles of the lighter kind and therefore leads to immense fluctuation of the step sizes of the integration.

The surface force does not act symmetrically, therefore the conservation of momentum is not warranted. Several particles can easily pass through a faintly curved surface. The intermixture of phases is energetically favored and made the introduction of an artificial separation force necessary.

As problems for testing the approach and implementation a freely oscillating boundary layer, a freely oscillating round drop, a freely oscillating extremely deformed drop, the disintegration of a thin bar by constriction and the collision of drops have been chosen. The increase of pressure, frequencies of oscillation, devolution of momentum, particles crossing the boundary layer, conservation of momentum and energy as well as the sizes of timesteps of the integrator are evaluated.

Figures 4 and 5 show a two dimensional simulation with surface tension of a deformed drop. The first simulation turned out to be instable due to the selected kernel function. The second simulation displays a stable drop even after a long time of simulation. The oscillation is dying out and the drop evolves to a round shape.

The hitherto results indicate, that not all kernel functions are applicable to simulate surface tension. Partly but not at all times an artificial separation force is required. The increase of pressure is not throughout correct and the conservation of energy and momentum are not always simultaneously good.

Simulations in three dimensions with much larger numbers of particles and higher accuracy are required to determine the advantages and disadvantages of the different equations and the effects of other numerical and physical parameters. It is also necessary to investigate the boundary layer of fluids with very different sound velocities and to stabilize the SPH method to huge density varieties.

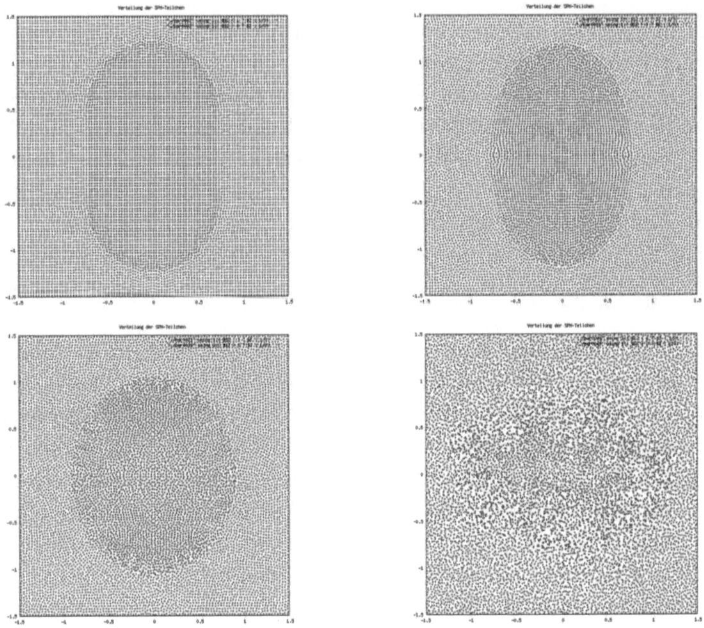

Fig. 4. Instable simulation of a droplet with an unsuitable kernel

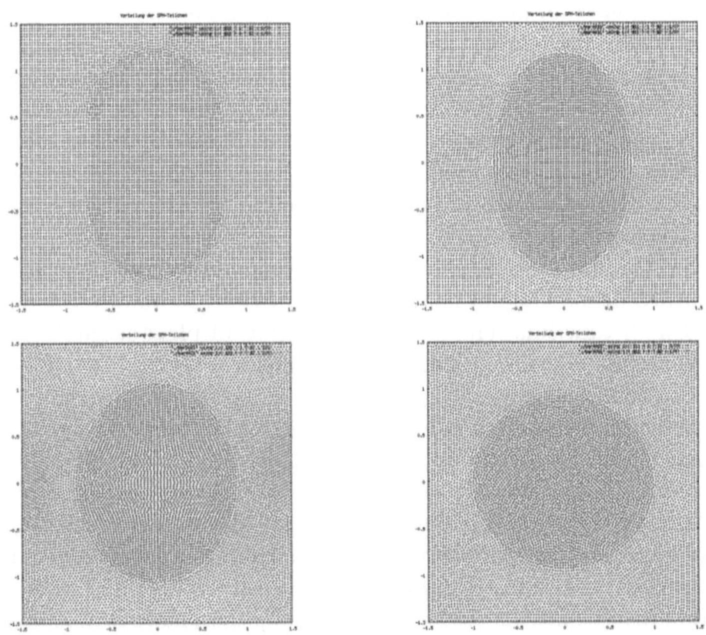

Fig. 5. Stable simulation of a droplet with a suitable kernel

4.2 A new Approach with Intrinsic Forces

The models with exterior forces has the disadvantage, that it works only well, if the normal vector and the curvature can be calculated consistent. This is not allways possible because of the mixture of the fluids, rough surface, etc. A new approach is to implement the effect of surface tension directly, see [3] for a related model. Figure 6 compares the simulations of diesel injection. The left side shows a former simulation without any surface tension, the right side is the result of a current simulation with intrinsic surface tension, which is more realistic.

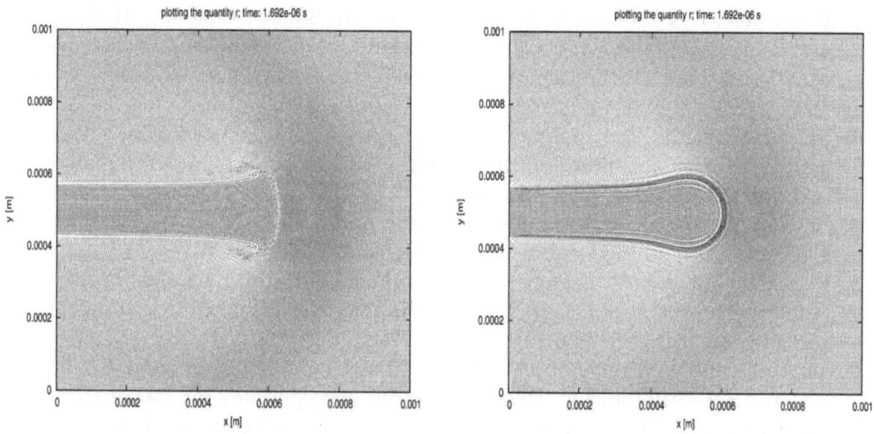

Fig. 6. Injection of diesel in an air filled chamber, simulation with 100.000 particles. The left picture shows the injected diesel without surface forces. On the right the intrinsic surface tension is enabled

5 Summary

The usage of object-oriented development methods has improved the quality of our simulation codes. The implementation is very easy to maintain and extend, e.g. to add the physics of surface tension or turbulence. The result of object-oriented techniques with design patterns is a framework, in which classes have clear and strictly separated responsibilities. Different methods can be interchanged without influencing other code. The use of our parallel I/O library and optimizations for communication reduced the sequential parts of the framework to a minimum which leads to a well scaling parallel performance.

In the future, we focus on the improvement of the surface tension model and the development of models for simulating turbulences. Due to an increased number of particles which is needed to simulate these effects meaningful, we are investigating solutions to decrease the amount of calculated interactions without increasing the runtime of the simulations.

At the moment we port our application to the NEC SX-8, the Portland Group compiler on Cray and the Intel compiler at HP zx6000 and try to utilize OpenMP for our object-oriented SPH framework.

References

1. W. Benz, E. Asphaug. Catastrophic Disruptions Revisited. In *Icarus*, 142: 5–20, 1999
2. P. Corbett, D. Feitelson, Y. Hsu, J.-P. Prost, M. Snir, S. Fineberg, B. Nitzberg, and B. Traversatand Parkson Wong. MPI-IO: A Parallel File I/O Interface for MPI Version 0.3. Technical Report NAS-95-002, NASA Ames Research Center, 1995.
3. M. Desbrun, M. Cani. Smoothed Particles: A new paradigm for animating highly deformable bodies. Eurographics Workshop on Computer Animation and Simulation (EGCAS), page 61–76 - Aug 1996.
4. E. Gamma and R. Helm and R. Johnson and J. Vlissides. Design Patterns: Elements of Reusable Object-Oriented Software. Addison-Wesley, 1995
5. R.A. Gingold, J.J. Monaghan. Smoothed Particle Hydrodynamics: Theory and Application to Non-Spherical Stars. In *Monthly Notices of the Royal Astronomical Society*, 181: 375–389, 1977
6. M. Hipp and S. Pinkenburg and S. Holtwick and S. Kunze and C. Schäfer and W. Rosenstiel and H. Ruder. Libraries and Methods for Parallel Particle Simulations. In *Proceedings of High Performance Computing in Science and Engineering '05*, W.E. Nagel, W. Jäger, M. Resch (ed), Springer, 2006.
7. A. Jahnke. Portierung und Laufzeitmessungen einer physikalischen Simulationbibliothek auf Kepler Cray und HP. Studienarbeit, University of Tübingen, 2006.
8. L.B. Lucy. A Numerical Approach to Testing the Fission Hypothesis. In *The Astronomical Journal*, 82(12): 1013–1024, 1977
9. Message Passing Interface Forum. MPI-2: Extensions to the Message-Passing Interface. *Online. URL: http://www.mpi-forum.org/docs/mpi-20-html/mpi2-report.html*, July 1997.
10. F. Ott, E. Schnetter. A modified SPH approach for fluids with large density differences. In *ArXiv Physics e-prints*, 3112-+, 2003
11. S. Pinkenburg and W. Rosenstiel. Parallel I/O in an Object-Oriented Message-Passing Library. In *Proceedings of the 11th European PVM/MPI Users' Group Meeting*, 2004.

Simulations of Particle Suspensions at the Institute for Computational Physics

Jens Harting[1], Martin Hecht[1], and Hans Herrmann[2]

[1] Institut für Computerphysik, Pfaffenwaldring 27, 70569 Stuttgart, Germany
[2] Institute for Building Materials, ETH Hönggerberg, HIF E 12, 8093 Zürich, Switzerland

Summary. In this report we describe some of our projects related to the simulation of particle-laden flows. We give a short introduction to the topic and the methods used, namely the Stochastic Rotation Dynamics and the lattice Boltzmann method. Then, we show results from our work related to the behaviour of claylike colloids in shear flow as well as structuring effects of particles in the vicinity of rigid walls.

1 Introduction

Simulating the flow of suspensions is an extremely difficult and demanding problem. Suspensions are mixtures of fluid and granular materials, and each component alone is a challenge to numerical modelers. When they are combined in a suspension, neither component can be neglected, so all the difficulties of both fluids and grains must be solved, in addition to the new problem of describing the interaction between the two. These suspensions are ubiquitous in our daily life, but are not well understood due to their complexity. During the last twenty years, various simulation methods have been developed in order to model these systems. Due to varying properties of the solved particles and the solvents, one has to choose the simulation method properly in order to use the available compute resources most effectively with resolving the system as well as needed. Various techniques for the simulation of particle suspensions have been implemented at the Institute for Computational Physics allowing us to study the properties of clay-like systems, where Brownian motion is important, more macroscopic particles like glass spheres or fibers solved in liquids, or even the pneumatic transport of powders in pipes.

Computer simulation methods are indispensable for such many-particle systems, for the inclusion of inertia effects (Reynolds numbers > 1) and Brownian motion (Peclet number of order 1). These systems often contain a large number of important time scales which differ by many orders of magnitude, but nevertheless have to be resolved by the simulation, leading to a large numerical effort. However, simulations have the potential to increase our knowledge of elementary processes and to enable us to find the aforementioned relations from simulations instead of experiments.

In this paper we will present two approaches which have been applied during the last year: A combined Molecular Dynamics and Stochastic Rotation Dynamics method has been applied to the simulation of claylike colloids and the lattice-Boltzmann method was used as a fluid solver to study the behaviour of glass spheres in creeping shear flows.

2 Simulation of Claylike Colloids: Stochastic Rotation Dynamics

We investigate properties of dense suspensions and sediments of small spherical Al_2O_3 particles by means of a combined Molecular Dynamics (MD) and Stochastic Rotation Dynamics (SRD) simulation. Stochastic Rotation Dynamics is a simulation method developed by Malevanets and Kapral [17, 18] for a fluctuating fluid. The work this chapter is dealing with is presented in more detail in references [7, 6].

We simulate claylike colloids, for which in many cases the attractive Van-der-Waals forces are relevant. They are often called "peloids" (Greek: claylike). The colloidal particles have diameters in the range of some nm up to some μm. The term "*peloid*" originally comes from soil mechanics, but particles of this size are also important in many engineering processes. Our model systems of Al_2O_3-particles of about half a μm in diameter suspended in water are often used ceramics and play an important role in technical processes. In soil mechanics [22] and ceramics science [20], questions on the shear viscosity and compressibility as well as on porosity of the microscopic structure which is formed by the particles, arise [26, 16]. In both areas, usually high volume fractions ($\Phi > 20\%$) are of interest. The mechanical properties of these suspensions are difficult to understand. Apart from the attractive forces, electrostatic repulsion strongly determines the properties of the suspension. Depending on the surface potential, one can either observe formation of clusters or the particles are stabilized in suspension and do sediment only very slowly. The surface potential can be adjusted by the pH-value of the solvent. Within Debye-Hückel theory one can derive a so-called $2pK$ charge regulation model which relates the simulation parameters with the pH-value and ionic strength I adjusted in the experiment. In addition to the static interactions hydrodynamic effects are also important for a complete description of the suspension. Since typical Peclet numbers are of order one in our system, Brownian motion cannot be neglected.

The colloidal particles are simulated with molecular dynamics (MD), whereas the solvent is modeled with stochastic rotation dynamics (SRD).

In the MD part of our simulation we include effective electrostatic interactions and van der Waals attraction, a lubrication force and Hertzian contact forces. In order to correctly model the statics and dynamics when approaching stationary states, realistic potentials are needed. The interaction between the particles is described by DLVO theory [9, 24, 16]. If the colloidal particles are suspended in a solvent, typically water, ions move into solution, whereas their counter ions remain in the particle due to a different resolvability. Thus, the colloidal particle carries a charge. The ions in solution are attracted by the charge on the particles and form the electric double layer. It has been shown (see [24]), that the resulting electrostatic interaction between two of these particles can be described by an exponentially screened Coulomb potential

$$V_{\rm Coul} = \pi \varepsilon_r \varepsilon_0 \left[\frac{2+\kappa d}{1+\kappa d} \cdot \frac{4k_{\rm B}T}{ze} \tanh\left(\frac{ze\zeta}{4k_{\rm B}T}\right)\right]^2 \times \frac{d^2}{r} \exp(-\kappa[r-d]) \,, \quad (1)$$

where d denotes the particle diameter and r is the distance between the particle centers. e is the elementary charge, T the temperature, $k_{\rm B}$ the Boltzmann constant, and z is the valency of the ions of added salt. Within DLVO theory one assumes linear screening, mainly by one species of ions with valency z (e.g. $z = +1$ for NH_4^+). The first fraction in equation 1 is a correction to the original DLVO potential, which takes the surface curvature into account and is valid for spherical particles [2]. The effective surface potential ζ is the electrostatic potential at the border between the diffuse layer and the compact layer, it may therefore be identified with *the* ζ-potential. It includes the effect of the bare charge of the colloidal particle itself, as well as the charge of the ions in the Stern layer, where the ions are bound permanently to the colloidal particle. In other words, DLVO theory uses a renormalized surface charge. This charge can be related to the pH value of the solvent within Debye-Hückel theory [6]. ε_0 is the permittivity of the vacuum, ε_r the relative dielectric constant of the solvent. κ is the inverse Debye length defined by $\kappa^2 = 8\pi\ell_B I$, with the ionic strength I and the Bjerrum length ℓ_B. We use ε_r=81 for water, which implies $\ell_B = 7\,$Å.

The Coulomb term of the DLVO potential competes with the attractive van der Waals term

$$V_{\rm VdW} = -\frac{A_{\rm H}}{12}\left[\frac{d^2}{r^2-d^2} + \frac{d^2}{r^2} + 2\ln\left(\frac{r^2-d^2}{r^2}\right)\right] . \quad (2)$$

$A_{\rm H} = 4.76 \cdot 10^{-20}\,$J is the Hamaker constant [8] which involves the polarizability of the particles.

For the integration of the translational motion we utilize a velocity Verlet algorithm [1] to update the velocity and position of particle i according to the equations

$$\mathbf{x}_i(t+\delta t) = \mathbf{x}_i(t) + \delta t\, \mathbf{v}_i(t) + \delta t^2 \frac{F_i(t)}{m} \,, \quad (3)$$

$$\mathbf{v}_i(t+\delta t) = \mathbf{v}_i(t) + \delta t\, \frac{F_i(t) + F_i(t+\delta t)}{2m} \,. \quad (4)$$

For the rotation, a simple Euler algorithm is applied:

$$\omega_i(t+\delta t) = \omega_i(t) + \delta t\, \mathbf{T}_i \; , \tag{5}$$

$$\vartheta_i(t+\delta t) = \vartheta_i(t) + F(\vartheta_i, \omega_i, \delta t) \; , \tag{6}$$

where $\omega_i(t)$ is the angular velocity of particle i at time t, $\mathbf{T}_i$ is the torque exerted by non central forces on the particle i, $\vartheta_i(t)$ is the orientation of particle i at time t, expressed by a quaternion, and $F(\vartheta_i, \omega_i, \delta t)$ gives the evolution of ϑ_i of particle i rotating with the angular velocity $\omega_i(t)$ at time t.

The concept of quaternions [1] is often used to calculate rotational motions in simulations, because the Euler angles and rotation matrices can easily be derived from quaternions. Using Euler angles to describe the orientation would give rise to singularities for the two orientations with $\vartheta = \pm 90°$. The numerical problems related to this fact and the relatively high computational effort of a matrix inversion can be avoided using quaternions.

The Stochastic Rotation Dynamics method (SRD) introduced by Malevanets and Kapral [17, 18] is a promising tool for a coarse-grained description of a fluctuating solvent, in particular for colloidal and polymer suspensions. The method is also known as "Real-coded Lattice Gas" [10] or as "multi-particle-collision dynamics" (MPCD) [23]. It can be seen as a "hydrodynamic heat bath", whose details are not fully resolved but which provides the correct hydrodynamic interaction among embedded particles [15]. SRD is especially well suited for flow problems with Peclet numbers of order one and Reynolds numbers on the particle scale between 0.05 and 20 for ensembles of many particles. The method is based on so-called fluid particles with continuous positions and velocities. Each time step is composed of two simple steps: One streaming step and one interaction step. In the streaming step the positions of the fluid particles are updated as in the Euler integration scheme known from Molecular Dynamics simulations:

$$\mathbf{r}_i(t+\tau) = \mathbf{r}_i(t) + \tau\, \mathbf{v}_i(t) \; , \tag{7}$$

where $\mathbf{r}_i(t)$ denotes the position of the particle i at time t, $\mathbf{v}_i(t)$ its velocity at time t and τ is the time step used for the SRD simulation. After updating the positions of all fluid particles they interact collectively in an interaction step which is constructed to preserve momentum, energy and particle number. The fluid particles are sorted into cubic cells of a regular lattice and only the particles within the same cell are involved in the interaction step. First, their mean velocity $\mathbf{u}_j(t') = \frac{1}{N_j(t')} \sum_{i=1}^{N_j(t')} \mathbf{v}_i(t)$ is calculated, where $\mathbf{u}_j(t')$ denotes the mean velocity of cell j containing $N_j(t')$ fluid particles at time $t' = t + \tau$. Then, the velocities of each fluid particle in cell j are updated as:

$$\mathbf{v}_i(t+\tau) = \mathbf{u}_j(t') + \boldsymbol{\Omega}_j(t') \cdot [\mathbf{v}_i(t) - \mathbf{u}_j(t')] \; . \tag{8}$$

$\boldsymbol{\Omega}_j(t')$ is a rotation matrix, which is independently chosen randomly for each time step and each cell. The mean velocity $\mathbf{u}_j(t)$ in the cell j can be seen as streaming velocity of the fluid at the position of the cell j at the time t, whereas the difference $[\mathbf{v}_i(t) - \mathbf{u}_j(t')]$ entering the interaction step can be interpreted as a contribution to the thermal fluctuations.

To couple the two parts of the simulation, MD on the one hand and SRD on the other one, the colloidal particles are sorted into the SRD boxes and their velocities are included in the rotation step. This technique has been used to model protein chains suspended in a liquid [4, 27]. Since the mass of the fluid particles is much smaller than the mass of the colloidal particles, one has to use the mass of each particle – colloidal or fluid particle – as a weight factor when calculating the mean velocity.

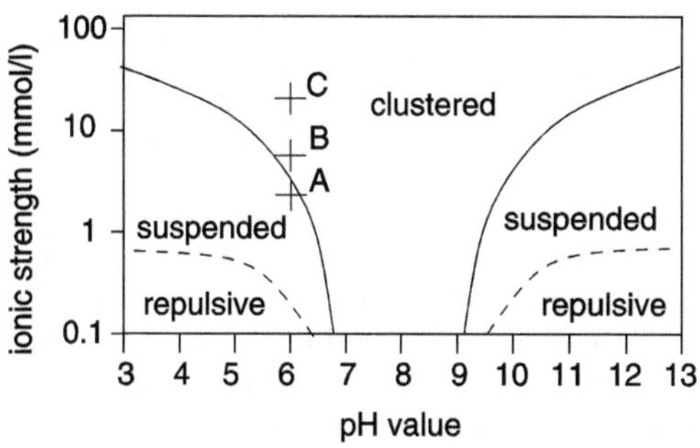

Fig. 1. Schematic phase diagram for volume fraction $\Phi = 35\%$ in terms of pH-value and ionic strength involving three different phases: a clustering regime due to van der Waals attraction, stable suspensions where the charge of the colloidal particles prevents clustering, and a repulsive structure for further increased electrostatic repulsion. This work concentrates on state A (pH $= 6$, $I = 3\,\mathrm{mmol/l}$) in the suspended phase, state B (pH $= 6$, $I = 7\,\mathrm{mmol/l}$) close to the phase border but already in the clustered phase, and state C (pH $= 6$, $I = 25\,\mathrm{mmol/l}$) in the clustered phase [6]

Depending on the experimental conditions, one can obtain three different phases in a Al_2O_3 suspension: A clustered region, a suspended phase, and a repulsive structure. These phases can be reproduced in the simulations and we can quantitatively relate interaction potentials to certain experimental conditions. A schematic picture of the phase diagram is shown in Fig. 1. Close to the isoelectric point (pH $= 8.7$), the particles form clusters for all ionic strengths since they are not charged. At lower or higher pH values one can prepare a stable suspension for low ionic strengths because of the charge, which is carried by the colloidal particles. At even more extreme pH values, one can obtain a repulsive structure due to very strong electrostatic potentials (up to $\zeta = 170\,\mathrm{mV}$ for pH $= 4$ and $I = 1\,\mathrm{mmol/l}$, according to our model). The repulsive structure is characterized by an increased shear viscosity. In the following we focus on three states: State A (pH $= 6$, $I = 3\,\mathrm{mmol/l}$) is in the suspended phase, state B (pH $= 6$, $I = 7\,\mathrm{mmol/l}$) is a point already in

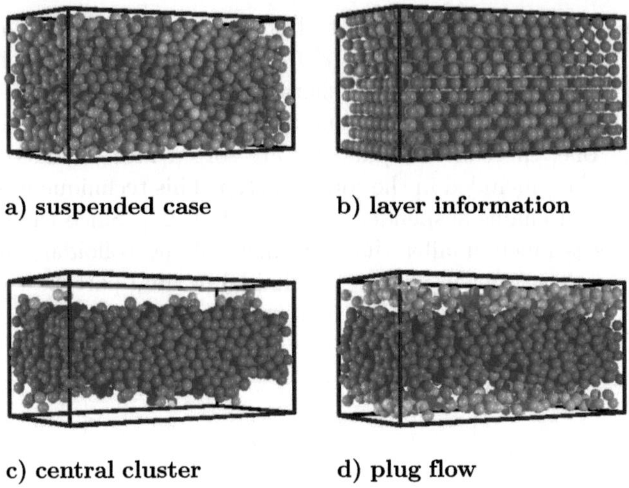

Fig. 2. Images of four different cases. For better visibility we have chosen smaller systems than we usually use for the calculation of the viscosity. The *colors* denote velocities: *Dark particles* are slow, *bright ones* move fast. The potentials do not correspond exactly to the cases A–C in Fig. 1, but they show qualitatively the differences between the different states: **a**) Suspension like in state A, at low shear rates. **b**) Layer formation, which occurs in the repulsive regime, but also in the suspension (state A) at high shear rates. **c**) Strong clustering, like in state C, so that the single cluster in the simulation is deformed. **d**) Weak clustering close to the phase border like in state B, where the cluster can be broken into pieces, which follow the flow of the fluid (plug flow)

the clustered phase but still close to the phase border, and state C ($p\mathrm{H} = 6$, $I = 25\,\mathrm{mmol/l}$) is located well in the clustered phase.

Some typical examples for the different phases are shown in Fig. 2a)–d). These examples are meant to be only illustrative and do not correspond exactly to the cases A–C in Fig. 1 denoted by uppercase letters. In the suspended case (a), the particles are mainly coupled by hydrodynamic interactions. One can find a linear velocity profile and a slight shear thinning. If one increases the shear rate $\dot{\gamma} > 500/\mathrm{s}$, the particles arrange in layers. The same can be observed if the Debye-screening length of the electrostatic potential is increased (b), which means that the solvent contains less ions ($I < 0.3\,\mathrm{mmol/l}$) to screen the particle charges. On the other hand, if one increases the salt concentration, electrostatic repulsion is screened even more and attractive van der Waals interaction becomes dominant ($I > 4\,\mathrm{mmol/l}$). Then the particles form clusters (c), and viscosity rises. A special case, called "plug flow", can be observed for high shear rates, where it is possible to tear the clusters apart and smaller parts of them follow with the flow of the solvent (d). This happens in our simulations for $I = 25\,\mathrm{mmol/l}$ (state C) at a shear rate of $\dot{\gamma} > 500/\mathrm{s}$. However, as long as there are only one or two big clusters in the system, it is too small to expect quantitative agreement with experiments. In these cases we have to focus on state B ($I = 7\,\mathrm{mmol/l}$) close to the phase border.

Within this project we have investigated many details of the phase diagram and also compared our simulations to experimental data of a group in Karlsruhe. Furthermore, a large number of simulations has been carried out to investigate the properties of the simulation method and to improve it where needed. We are currently studying the dependence of autocorrelation functions, structure factors, and the fractality of particle clusters. Until now, two papers have been published describing the results of this projects in much greater detail than possible within this report [7, 6].

3 Transport Phenomena and Structuring in Suspensions: Lattice-Boltzmann Simulations

The object of this project is a theory describing the influence of solid boundary walls in the creeping shear flow of suspensions. To analyze the various effects, the results of theories, simulations, and experiments are compared.

We are interested in structuring effects which might occur in the solid fraction of the suspension. Such effects are known from dry granular media resting on a plane surface or gliding down an inclined chute [21, 25]. In addition, the wall causes a demixing of the solid and fluid components which might have an unwanted influence on the properties of the suspension. Near the wall one finds a thin lubrication layer which contains almost no particles and causes a so-called "pseudo wall slip". We expect structuring close to a rigid wall at much smaller concentrations than in granular media because of long-range hydrodynamic interactions. In [11], we study these effects by the means of particle volume concentrations versus distance to the wall.

We model the fluid in by means of a lattice-Boltzmann algorithm. The lattice-Boltzmann method is a simple scheme for simulating the dynamics of fluids. By incorporating solid particles into the model fluid and imposing the correct boundary condition at the solid/fluid interface, colloidal suspensions can be studied. Pioneering work on the development of this method has been done by Ladd et al. [12, 13, 14] and we use their approach to model sheared suspensions near solid walls.

The lattice-Boltzmann (hereafter LB) simulation technique which is based on the well-established connection between the dynamics of a dilute gas and the Navier-Stokes equations [3]. We consider the time evolution of the one-particle velocity distribution function $n(\mathbf{r}, \mathbf{v}, t)$, which defines the density of particles with velocity $\mathbf{v}$ around the space-time point $(\mathbf{r}, t)$. By introducing the assumption of molecular chaos, i.e. that successive binary collisions in a dilute gas are uncorrelated, Boltzmann was able to derive the integro-differential equation for $n(\mathbf{r}, \mathbf{v}, t)$ named after him [3]. The LB technique arose from the realization that only a small set of discrete velocities is necessary to simulate the Navier-Stokes equations [5]. Much of the kinetic theory of dilute gases can be rewritten in a discretized version. The time evolution of the distribution functions n is described by a discrete analogue of the Boltzmann equation [14]:

$$n_i(\mathbf{r} + \mathbf{c}_i \Delta t, t + \Delta t) = n_i(\mathbf{r}, t) + \Delta_i(\mathbf{r}, t) \, , \tag{9}$$

where Δ_i is a multi-particle collision term. Here, $n_i(\mathbf{r}, t)$ gives the density of particles with velocity $\mathbf{c}_i$ at $(\mathbf{r}, t)$. In our simulations, we use 19 different discrete velocities $\mathbf{c}_i$.

To simulate the hydrodynamic interactions between solid particles in suspensions, the lattice-Boltzmann model has to be modified to incorporate the boundary conditions imposed on the fluid by the solid particles. Stationary solid objects are introduced into the model by replacing the usual collision rules at a specified set of boundary nodes by the "link-bounce-back" collision rule [19].When placed on the lattice, the boundary surface cuts some of the links between lattice nodes. The fluid particles moving along these links interact with the solid surface at boundary nodes placed halfway along the links. Thus, a discrete representation of the surface is obtained, which becomes more and more precise as the surface curvature gets smaller and which is exact for surfaces parallel to lattice planes.

The particle position and velocity are calculated using Newton's equations in a similar manner as in the section on SRD simulations. However, particles do not feel electrostatic interactions, but behave like hard spheres in the case presented in this section.

The purpose of our simulations is the reproduction of rheological experiments on computers. We simulate a representative volume element of the experimental setup and compare our calculations with experimentally accessible data, i.e. density profiles, time dependence of shear stress and shear rate. We also get experimentally inaccessible data from our simulations like translational and rotational velocity distributions, particle-particle and particle-wall interaction frequencies. The experimental setup consists of a rheoscope with two spherical plates. The upper plate can be rotated either by exertion of a constant force or with a constant velocity, while the complementary value is measured simultaneously. The material between the rheoscope plates consist of glass spheres suspended in a sugar-water solution. The radius of the spheres varies between 75 and 150 μm.

We are currently investigating the occurrence of non-Gaussian velocity distributions of particles for higher particle densities and higher shear rates. For this, improvements of the method are mandatory in order to prevent instabilities of the simulation. By utilizing an implicit scheme for the update of the particle velocities [14, 19] we are able to overcome artefacts caused by numerical inaccuracies at high volume fractions or shear rates. Figure 3 shows a snapshot of a system containing 1536 particles after 28000 timesteps.

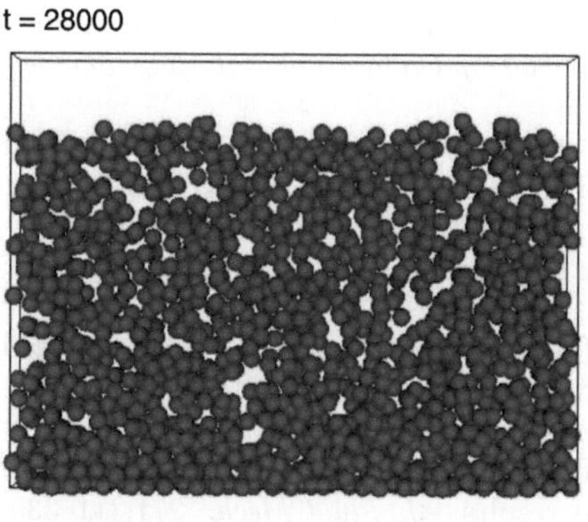

Fig. 3. A snapshot of a suspension with 1536 spheres after 28000 timesteps used to gain statistics of particle velocity distributions

4 Conclusions

We have reviewed two of the projects at the Institute for Computational Physics concerning particle suspensions. Both projects have developed simulations that can be directly compared to experiments. In this way, our numerical work contributes to answering important questions in both physics and engineering. Until now, a number of papers have been published describing the applied models and the results in much more detail [7, 6, 11]. The reader is referred to those in order to gain a better understanding of the implementation details.

References

1. M. P. Allen and D. J. Tildesley. *Computer simulation of liquids.* Oxford Science Publications. Clarendon Press, 1987.
2. L. Bocquet, E. Trizac, and M. Aubouy. Effective charge saturation in colloidal suspensions. *J. Chem. Phys.*, 117:8138, 2002.
3. S. Chapman and T. G. Cowling. *The Mathematical Theory of Non-uniform Gases.* Cambridge University Press, second edition, 1952.
4. E. Falck, J. M. Lahtinen, I. Vattulainen, and T. Ala-Nissila. Influence of hydrodynamics on many-particle diffusion in 2d colloidal suspensions. *Eur. Phys. J. E*, 13:267–275, 2004.
5. U. Frisch, B. Hasslacher, and Y. Pomeau. Lattice-gas automata for the Navier-Stokes equation. *Phys. Rev. Lett.*, 56(14):1505–1508, 1986.
6. M. Hecht, J. Harting, M. Bier, J. Reinshagen, and H. J. Herrmann. Shear viscosity of clay-like colloids: Computer simulations and experimental verification. *submitted to Phys. Rev. E*, 2006. cond-mat/0601413.
7. M. Hecht, J. Harting, T. Ihle, and H. J. Herrmann. Simulation of claylike colloids. *Physical Review E*, 72:011408, 2005.
8. M. Hütter. *Brownian Dynamics Simulation of Stable and of Coagulating Colloids in Aqueous Suspension.* PhD thesis, Swiss Federal Institute of Technology Zurich, 1999.

9. M. Hütter. Local structure evolution in particle network formation studied by brownian dynamics simulation. *Journal of Colloid and Interface Science*, 231:337–350, 2000.
10. Y. Inoue, Y. Chen, and H. Ohashi. Development of a simulation model for solid objects suspended in a fluctuating fluid. *J. Stat. Phys.*, 107(1):85–100, 2002.
11. A. Komnik, J. Harting, and H. J. Herrmann. Transport phenomena and structuring in shear flow of suspensions near solid walls. *J. Stat. Mech: Theor. Exp.*, P12003, 2004.
12. A. J. C. Ladd. Numerical simulations of particulate suspensions via a discretized boltzmann equation. part 1. theoretical foundation. *J. Fluid Mech.*, 271:285–309, 1994.
13. A. J. C. Ladd. Numerical simulations of particulate suspensions via a discretized boltzmann equation. part 2. numerical results. *J. Fluid Mech.*, 271:311–339, 1994.
14. A. J. C. Ladd and R. Verberg. Lattice-boltzmann simulations of particle-fluid suspensions. *J. Stat. Phys.*, 104(5):1191, 2001.
15. A. Lamura, G. Gompper, T. Ihle, and D. M. Kroll. Multi-particle-collision dynamics: Flow around a circular and a square cylinder. *Eur. Phys. Lett*, 56:319, 2001.
16. J. A. Lewis. Colloidal processing of ceramics. *J. Am. Ceram. Soc.*, 83:2341–59, 2000.
17. A. Malevanets and R. Kapral. Mesoscopic model for solvent dynamics. *J. Chem. Phys.*, 110:8605, 1999.
18. A. Malevanets and R. Kapral. Solute dynamics in mesoscale solvent. *J. Chem. Phys.*, 112:7260, 2000.
19. N. Q. Nguyen and A. J. C. Ladd. Lubrication corrections for lattice-boltzmann simulations of particle suspensions. *Phys. Rev. E*, 66(4):046708, 2002.
20. R. Oberacker, J. Reinshagen, H. von Both, and M. J. Hoffmann. Ceramic slurries with bimodal particle size distributions: Rheology, suspension structure and behaviour during pressure filtration. *Ceramic Transactions*, 112:179–184, 2001.
21. P. Mijatović. Bewegung asymmetrischer Teilchen unter stochastischen Kräften. Master-thesis, Universität Stuttgart, 2002.
22. S. Richter and G. Huber. Resonant column experiments with fine-grained model material – evidence of particle surface forces. *Granular Matter*, 5:121–128, 2003.
23. M. Ripoll, K. Mussawisade, R. G. Winkler, and G. Gompper. Low-reynolds-number hydrodynamics of complex fluids by multi-particle-collision dynamics. *Europhys. Lett.*, 68:106–12, 2004.
24. W. B. Russel, D. A. Saville, and W. Schowalter. *Colloidal Dispersions.* Cambridge Univ. Press., Cambridge, 1995.
25. T. Pöschel. Granular material flowing down an inclined chute: a molecular dynamics simulation. *J. Phys. II*, 3(1):27–40, 1993.
26. G. Wang, P. Sarkar, and P. S. Nicholson. Surface chemistry and rheology of electrostatically (ionically) stabillized allumina suspensions in polar media. *J. Am. Ceram. Soc.*, 82(4):849–56, 1999.
27. R. G. Winkler, K. Mussawisade, M. Ripoll, and G. Gompper. Rod-like colloids and polymers in shear flow: a multi-particle-collision dynamics study. *J. of Physics-Condensed Matter*, 16(38):S3941–54, 2004.

Solid State Physics

Prof. Dr. Werner Hanke

Institut für Theoretische Physik und Astrophysik, Universität Würzburg,
Am Hubland, D-97074 Würzburg

The contributions stemming from solid-state physics concerning high-performance computing in Stuttgart are devided in several topical areas:

The first one is a research project by Prof. P. Nielaba from the Physics Department of the University of Konstanz. It deals with nanostructures in reduced geometry, which have become an interesting research topic in the last years. In this demanding field, computer simulations have become more and more important. This is due to the fact that nanosystems in reduced geometry contain about tenthousand particles, a number which is nearly ideal for the application of computer simulation techniques. Many important results have been obtained by the Nielaba group in the last years based on the strong support of high-performance computing centers, in particular the HLRS. In the last year, new insights have in particular been obtained into the elastic properties and phase transitions of model colloids in external periodic fields, the dynamics in micro-channels, spin-structures in nanosystems with latteral constrictions, quantum effects in nanowires and structures in electronic properties of clusters. Simulations are based on the path integral formulation of the partition function, where a quantum particle is represented by a classical chain of so-called "Totter-particles" in the limit of infinite numbers of these particles. This algorithm, which has been optimized for parallel computing, is described very nicely in the longer paper by P. Nielaba in this volume.

Another example of solid-state physics, where the use of supercomputing is nicely demonstrated to the advantage of the research project, is the report by P. Schmitteckert and G. Schneider from the Institute for Theoretical Physics in Karlsruhe. This topic concerns the signal transport and finite bias conductance in and through correlated nanostructures. It is devoted to the description of non-equilibrium transport properties like the signal transport or the finite bias conductance of an interacting nanostructure attached to leads. The increasing theoretical interest in these sytems is based on the fact that during the past decade, improved experimental techniques have made the production and the corresponding experimental investigation on one-dimensional systems possible. However, the theoretical description is a very challenging task. For non-interacting particles, the conductance can be extracted from the transition of the single-particle levels, which are easily obtained. But in principle, the actual physical process is a complicated many-body problem: The screening of electrons is reduced by reducing the size of structures under investigation and, as a consequence, electron correlations, which also build up the screening, can no longer be neglected. Therefore, an adequate description had to be worked out by P. Schmitteckert and G. Schneider to be able to treat the strong correlations and, in particular, also the non-equilibrium behavior in these systems in a rigorous manner. The authors used the so-called density-matrix renormalization group method, a powerful method for low-dimensional systems, which was here extended to the treatment of non-equilibrium transport properties. The authors (in their corresponding reports) describe nicely that they are now heading towards actual applications for which they need higher energy resolution and corresponding longer simulation times. Basically, the aim is to enable a guided search for novel nanostructure devices based on correlation effects, which would replace the so-far mostly used empirical search for these devices.

Another project on the boarderline between solid-state physics and more mechanically oriented material science is the project by E. Blitzek and P. Gumbsch from the Institute of Applied Science, University of Karlsruhe. This topic deals with the question of how dislocations or pre-existing dislocations in a material influence the low-temperature fracture toughness and the brittle-to-ductile-transition. Silicon has been widely used as a model material to study this transition because it can be grown as a nearly defect-free single crystal. Experiments then on single crystal show that the dislocation nucleation can be highly inhomogeneous, as single dislocation intersecting the crack front can then stimulate the emission of other dislocations in an avalanche-type multiplication process. Thus, the dislocation sources created at the crack front can play an important role in the fracture behavior of all brittle material. The subject of this high-performance computer study of the Karlsruhe group is, therefore, the investigation of the detailed mechanism of the stimulated emission and multiplication of disclocations at the crack tip. The evolution of the system is then calculated using standard molecular dynamics, also making use of atomic interaction potentials, which are modelled in accordance to experimental informations. The basic point of these simulations is that they are, for the first time, three-dimensional large scale atomistic simulations of dislocation crack interactions. This requires high-performance algorithmic developments, which have been undertaken by the Karlsruhe group.

The last project in the solid-state physics series is by the Stuttgart group C. Lavalle, S.R. Manmana, W. Wessel and A. Muramatsu, dealing with Quantum-Monte-Carlo (QMC) simulations of strongly correlated and frustrated quantum systems. The importance of this study is that many topical material properties are nowadays known to be determined, or at least essentially influenced, by strong electronic correlations. Topical examples for this are high-temperature superconductors, the manganites, which are important for magnetic data recording, or heavy fermion compounds and quantum magnetic systems. Also ultra-cold atomic gases in optical lattices have recently become very fashionable. They provide a new exciting bridge between the physics of a quantum condensed-matter system and the field of quantum optics. The basic route to an understanding of these many-body systems is to choose a model Hamiltonian, which govers the kinetic energy and Coloumb correlation processes of the underlying particles. A typical model is the so-called t-J model which is considered to be the effective Hamiltonian for the low energy physics of the high-T_c materials. In spite of its formal simplicity the model has turned out to be extremely difficult to approach with both analytical and numerical techniques. The Stuttgart group has used a newly developed QMC algorithm, the so-called hybrid-loop QMC, to study, in particular, different dynamical observables, in particular the spectral functions, which can be directly compared to experiment. Again using large-scale QMC simulations, also the properties of ultra-cold atomic gases in optical lattices have been studied. This is a very exciting topic since recently novel quantum phenomena have been realized in such systems. The Stuttgart group has found, for example, that (on not a rectangular but) on a triangular lattice, the presence of the resulting frustration in the underlying lattice leads to the emergence of the so-called super-solid state of matter. Furthermore, the possibility of randomness in the interaction strength has let to the formation of a Bose-glass phase of the atoms, another exciting phase to be compared with experimental observation. This group has also started to investigate non-equilibrium dynamics of strongly correlated systems using a new variant of the already above-mentioned density matrix renormalization group (DMRG).

Nano-Systems in External Fields and Reduced Geometry: Numerical Investigations

P. Henseler, C. Schieback, K. Franzrahe, F. Bürzle, M. Dreher, J. Neder, W. Quester, M. Kläui, U. Rüdiger, and P. Nielaba

Physics Department, University of Konstanz, 78457 Konstanz, Germany
peter.nielaba@uni-konstanz.de

Summary. Properties of magnetic domain walls have been studied as well as flow properties and phase transitions of model colloids in external potentials and structural and electronic properties of nano-wires and Si clusters. In the following sections an overview is given on the results of our recent computations on quantum effects, structures and phase transitions in such systems.

1 Introduction and General Remarks

Nanostructures in reduced geometry and external fields have become interesting research fields in the last years. Despite the fact that by experimental techniques many structural-, elastic-, electronic-, and phase-properties of systems in the size of a few nanometers have been obtained, the theoretical investigations and analyses are still in an initial stage. This is partly due to the fact that systems which are far away from the thermodynamic limit (with infinitely many particles) due to their finite size are difficult to handle by analytical methods which are suitable for systems with either few particles (2–5) or in the limit of infinitely many particles. In this field computer simulations have become more and more important since nano-systems in reduced geometry contain about 10–10.000 particles, which is nearly ideal for the application of computer simulation methods. Many important results have been obtained by the support of HPC centres (HLRS, SSC, NIC) [1–4].

In this paper we report on new insights into properties of magnetic domain walls, flow properties and phase transitions of model colloids in external potentials, structural and electronic properties of nano-wires and Si clusters.

2 Simulations of Spin Structures in Nano-Structures with Lateral Constrictions

Nano-structured magnetic materials form the basic building blocks for several devices of the next generation, biosensors [5] and magneto- and spin-electronics [6, 7]. Due to their small size, the reduced dimensionality and the mutual interaction, new properties result, which are not known from the corresponding bulk materials [8, 9, 10, 11]. Magnetic nano-structures reveal new properties and phenomena, when the system sizes are comparable to characteristic lengths as the spin diffusion length, the mean free path length or the domain wall (DW) size. The spin structure and the magnetisation reversal process of magnetic nano-structures, for example, differs drastically, if the lateral extent is below the domain wall size. Not only the lateral structural size, but also the shape of constrictions in magnetic nano-structures, for example, has a lasting effect on the magnetisation reversion process, such that by selection of structural size and shape the static and dynamic properties can be designed [8–14].

We investigate configurations of magnetisation and spin structures in constrictions. The local geometry has a crucial influence on the spin structure of domain walls. Therefore, we systematically examine the influence of the size and the shape of the constriction and the influence of thermal fluctuations on the DW by means of computer simulations.

So far numerical investigations were carried out [15, 13] using the simulation package OOMMF [16]. These simulations assume the magnetisation to be a continuous function in space. Therefore the program only yields good results, if the angle between neighbouring spins is small. In this research project, however, atomically sharp DWs in constrictions will be investigated where large spin angles occur. In addition experimental measurements are all performed at room temperature. In order to consider all thermal excitations (spin waves) in the simulation, the cell size has to match the interatomic distance (Ångstroms). To circumvent the large angle problem we have chosen an atomistic Heisenberg model which also permits the correct description of thermal excitations [17].

Model, Simulation

A classical Heisenberg model has been considered. The magnetic moments are located on a cubic lattice with ferromagnetic exchange coupling between nearest neighbours.

Hamiltonian of the Heisenberg Model

$$\mathcal{H} = -J \sum_{\langle ij \rangle} \mathbf{S}_i \cdot \mathbf{S}_j - \mu_s \mathbf{B} \cdot \sum_i \mathbf{S}_i - d_x \sum_i S_{x,i}^2 - \frac{w}{2} \sum_i \sum_{i \neq j} \frac{3(\mathbf{S}_i \cdot \mathbf{e}_{i,j})(\mathbf{e}_{i,j} \cdot \mathbf{S}_j) - \mathbf{S}_i \cdot \mathbf{S}_j}{r_{i,j}^3}$$

$\mathbf{S}_i = \underline{\mu}_i / \mu_s$ three-dimensional magnetic moment of unit length with $\mu_s = |\underline{\boldsymbol{\mu}}_i|$, J with $J > 0$ ferromagnetic exchange coupling constant, w strength of the dipole-dipole interaction, $r_{i,j}$ distance between the magnetic moments i and

j, $\mathbf{e}_{i,j}$ unit vector in the direction of $\mathbf{r}_{i,j}$, d_x anisotropy constant, for $d_x > 0$ the x-axis is a so called magnetically "easy" axis.

The first term of the Hamiltonian describes the isotropic exchange interaction between neighbouring magnetic moments. With a ferromagnetic exchange interaction the magnetic moments align parallel at zero temperature. The second term of the Hamiltonian describes the interaction of magnetic moments with an external magnetic field **B**. The third term describes an uniaxial anisotropy. The magnetic moments preferentially align themselves along a magnetic "easy" axis. The last term of the Hamiltonian describes the long ranged magnetic dipole-dipole interaction, also known as shape-anisotropy. These interactions compete and determine the order of the magnetic moments in a system.

Landau-Lifshitz-Gilbert Equation

The dynamics of the system respectively the equation of motion of the magnetic moments $\mathbf{S}_i$ is given by the Landau-Lifshitz-Gilbert (LLG) equation [18]:

$$\frac{\partial \mathbf{S}_i}{\partial t} = \underbrace{-\frac{\gamma}{(1+\alpha^2)\cdot\mu_s}\mathbf{S}_i \times \mathbf{H}_i}_{\text{precession}} \underbrace{-\frac{\gamma\cdot\alpha}{(1+\alpha^2)\cdot\mu_s}\mathbf{S}_i \times (\mathbf{S}_i \times \mathbf{H}_i)}_{\text{damping}} \tag{1}$$

$\mathbf{H}_i$ are the effective fields, α is the dimensionless damping constant, $\gamma = g\mu_B/\hbar$ is the gyromagnetic ratio, and μ_s is the saturation magnetization.

The first term of the LLG equation describes the precession of the magnetic moment $\mathbf{S}_i$ within the effective field $\mathbf{H}_i$. The second term describes the relaxation of the magnetic moment. The numerical integration of the LLG equation has been carried out using a Heun-method [19, 20].

Effective Fields and Thermal Excitations

The effective field $\mathbf{H}_i(t)$ results as the derivative of Heisenberg model Hamiltonian $\mathcal{H}$ with respect to the magnetic moment. To take thermal fluctuations into account a time dependent thermal noise term $\boldsymbol{\zeta}_i(t)$ can be added into the effective field $\mathbf{H}_i$ [21].

$$\mathbf{H}_i(t) = -\frac{\partial \mathcal{H}}{\partial \mathbf{S}_i} + \underline{\zeta}_i(t) = J\sum_j \mathbf{S}_j + w\cdot\mathbf{H}_{i,dip} + 2\cdot\underline{\underline{\mathbf{d}}}\cdot\mathbf{S}_i + \mu_s\mathbf{B} + \underline{\zeta}_i(t) \tag{2}$$

$\underline{\zeta}_i(t)$ possesses the properties of white noise: $\langle\underline{\zeta}_i(t)\rangle = 0$, $\langle\zeta_i^\nu(t)\zeta_j^\Theta(t')\rangle = 2(\alpha\mu_s/\gamma)k_BT\delta_{i,j}\delta_{\nu,\Theta}\delta(t-t')$ where Θ, ν are the Cartesian coordinates, k_B is the Boltzmann constant, and T is the temperature. In the simulation thermal fluctuations are represented by Gaussian distributed random numbers.

In order to test the proper implementation of thermal excitations, the paramagnetic-ferromagnetic phase transition of a Heisenberg model in a simple cubic system has been determined and compared to results known from the literature [22]. The phase transition was computed here in very good agreement with results from the literature.

In the computer simulations [23] as well as in the experimental results, both a transverse (Fig. 1 upper part) and a vortex (Fig. 1 lower part) DW can be observed. The transverse DW was found inside the constriction and the vortex DW was found adjacent to it [24]. In the case of the transverse DW, an abrupt head-to-head DW in the middle of the constriction was used as initial configuration and in the case of the vortex wall a random distribution was used. The magnetic moments at the right and left border were fixed. In the experiment a saturation field parallel to the constriction was applied to position the DWs.

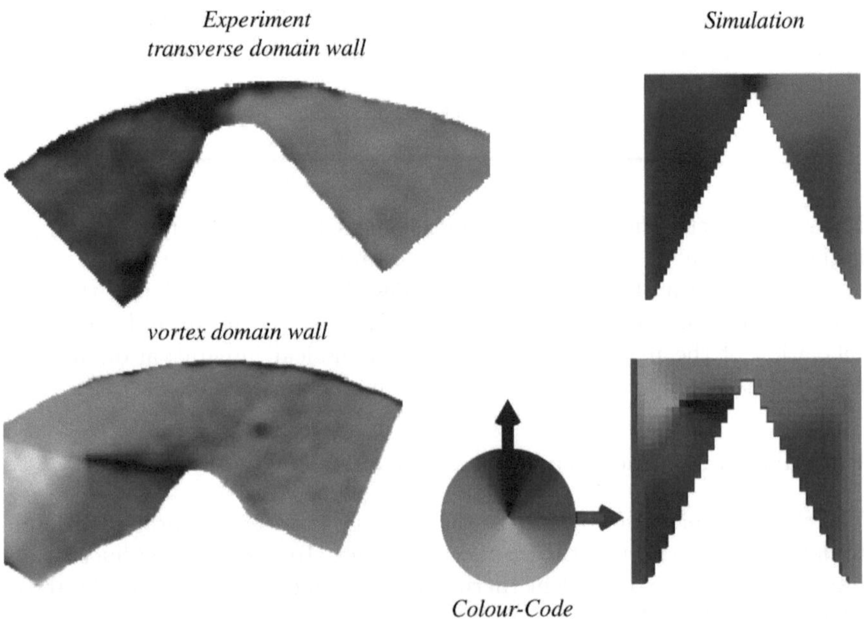

Fig. 1. Comparison between experimental results (*left side*) [13] and computer simulations (without thermal fluctuations) (*right side*). The line width of the shown experimental structure is 200 nm. In case of the transverse domain wall the constriction has a width of 35 nm, in case of the vortex domain wall 140 nm. In the simulation a system of size 256 nm × 256 nm × 32 nm has been studied with the parameters of permalloy. The constriction has a width of 16 nm. In case of the transverse domain wall the cell size is 4 nm, and the system was started from a head-to-head domain wall. In case of the vortex domain wall the cell size is 8 nm, and the system has been started from a random initial configuration, the system parameters for permalloy are standard literature values

Comparing the configurations [23] for constrictions with different widths (16 nm and 56 nm), it can be seen that the shape and size of the transverse DW depend on the width of the constriction. The smaller the constriction width the smaller the DW width.

Simulations [23] have been done for permalloy rings of different thicknesses. In good qualitative agreement with experiments it turns out, that in case of thick rings vortex DW occur, in case of thin rings transverse domain walls [25, 24]. A publication with our results is in preparation [26].

Spin-Torque-Effects

Interesting effects of electric current on the dynamics of domain walls have been studied theoretically recently [27, 28, 29]. In these studies the interactions between the electron spins and the magnetic moments have been treated by additional terms in the LLG equation:

$$\begin{aligned}\frac{\partial \mathbf{S_i}}{\partial t} = &- \frac{\gamma}{(1+\alpha^2)\mu_s} \cdot \mathbf{S_i} \times \mathbf{H_i} - \frac{\gamma \cdot \alpha}{(1+\alpha^2)\mu_s} \cdot \mathbf{S_i} \times (\mathbf{S_i} \times \mathbf{H_i}) \\ &- \frac{1+\beta\alpha}{(1+\alpha^2)} \cdot \left[(\mathbf{u} \cdot \mathbf{\nabla})\mathbf{S_i}\right] - \frac{\alpha - \beta}{(1+\alpha^2)} \cdot \mathbf{S_i} \times \left[(\mathbf{u} \cdot \mathbf{\nabla})\mathbf{S_i}\right] \end{aligned} \quad (3)$$

Here $\mathbf{u}$ is the effective velocity ($|\mathbf{u}| = JPg\mu_B/(2eM_s)$, J the current density, P the polarisation, and β a nonadiabaticity parameter.

In Ref. [27] a current along the x-direction is applied to a wire with a standard in-plane Neel wall separating two head to head domains along the wire direction (x-direction). Analytical and numerical results for the velocity, the reversible displacement x_{dw} and the deformation of the domain-wall have been obtained for the case of a low current u_x below the Walker breakdown [27] and $\beta = 0$. In particular, $x_{dw} \propto u_x$. In Fig. 2 we show results of our simulations for the position of the domain wall as a function of current. Good

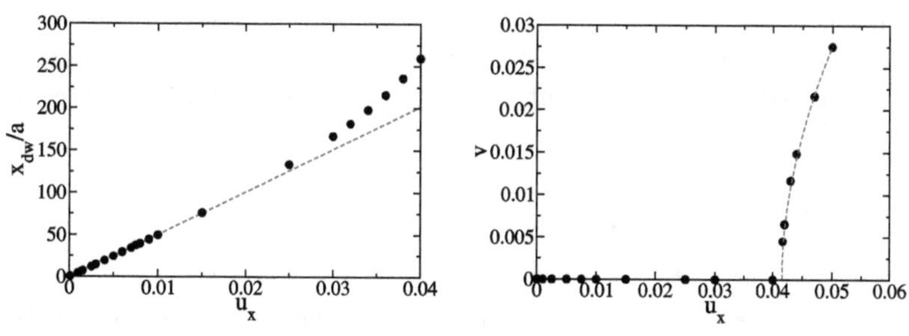

Fig. 2. *Left*: Simulation results [23] (*symbols*) for the final x-displacement x_{dw} of the domain wall as a function of effective velocity. The *line* shows a linear fit for small u_x. *Right*: Simulation results [23] (*symbols*) for the long term domain-wall velocity as function of effective velocity. The *line* shows a fit to the analytical prediction [28]. Parameters: $\alpha = 0.02, \beta = 0$

agreement with the low current predictions is found, for larger currents we find deviations. In [28] an approximate analytical prediction for the long term domain-wall velocity v as a function of effective velocity u_x has been derived: $v = \sqrt{u_x^2 - u_c^2}/(1+\alpha^2)$ for effective velocities exceeding a critical effective velocity u_c. These predictions have been reproduced by our simulations, see Fig. 2.

The effects of non-adiabatic spin-torque contributions ($\beta \neq 0$) have been investigated as well. In Fig. 3 we present results for the long term domain-wall velocity as a function of effective velocity u_x for different values of β. Good agreement with the results of [29] is obtained.

In future studies we plan the investigation of the influence of the constriction geometry and constriction size on the DW structure and further simulations with thermal fluctuations. Here the lattice constant has to be chosen in the range of Ångstroms to take thermal excitations into account. This reduction of the lattice constant leads to an increase of the number of magnetic moments and an increase in system size, respectively. In addition we plan to extend our investigations of the influence of the spin torque on DWs.

In 2004 Manfred Albrecht an coworkers have shown that evaporating strongly anisotropic Co/Pd-layers on a structured substrates of spherical polystyrene particles results in magnetically isolated nanocaps [30]. These caps are single-domain and their switching behaviour depends on the angle of the applied external magnetic field. We investigated [31] especially the latter aspect by Monte Carlo simulations for caps ranging from 20 nm to 100nm.

The model used consists of classical magnetic moments located on a simple cubic lattice with lattice constant a [17]. The Hamiltonian contains contributions from exchange, dipole-dipole interaction and an external magnetic field, as given in Eq. (1), the anisotropy H_a, however, is given by $H_a = -d\sum_i(\mathbf{S}_i \cdot \mathbf{A}_i)^2$. It is constructed in such a way to match the spherical anisotropy of the film deposited on the cap. $\mathbf{A}_i$ is a unit vector at lattice site i pointing radially away from the center of the magnetic cap and $\mathbf{S}_i$ is the normalized magnetic moment. The material parameters are taken from the literature [32].

The hysteresis loops are recorded after the system has been carefully thermalized. The systems analyzed at, e.g. 300K, are cooled down in steps of 10K

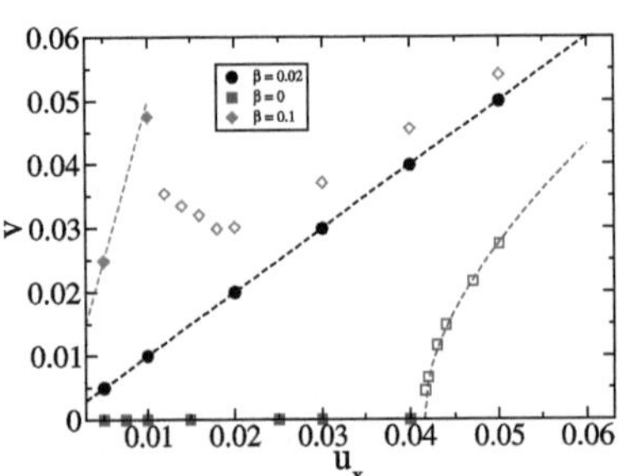

Fig. 3. Simulation results [23] (*symbols*) for the long term domain-wall velocity as function of effective velocity for different values of β. The *line* shows a fit to the analytical prediction [29]. Parameters: $\alpha = 0.02$

beginning at 500K. At each step 10^5MCS are performed, the external magnetic field is at its maximum value of $b_{max} = \mu_s B_{max}/J$ and its tilt against the normal of the substrate is α. At the final temperature the magnetic field is reduced by Δb every 10^5MCS until the $-b_{max}$ is reached and then increased again. The magnetization M is defined as the projection of the magnetic moments $\mathbf{S}_i$ onto the direction of the maximum magnetic field $\mathbf{E}_\alpha = \mathbf{b}_{max}/b_{max}$ field: $M = (1/N)\sum_{i=1}^{N} \mathbf{S}_i \cdot \mathbf{E}_\alpha$, averages are taken over 10^5MCS. Since the numerical effort of calculating the dipole-dipole interaction in Monte Carlo simulation exactly is immense, an approximative procedure for the sum has been used and tested against results obtained with the complete sum: The dipolar field is recalculated after every 10 MCS (one MC step consists of one sweep through the whole lattice). In particular for the cap of diameter 50 nm it was found, that the hysteresis loops differ only within the error bars, if the dipolar field is newly calculated every 1MCS, 5MCS and 10MCS, respectively.

In Fig. 4 we show simulation results [31] for a cap of diameter 50 nm. Hysteresis loops are found, and for increasing tilt angles the hysteresis regions decrease. Future studies shall involve the analysis of the dynamics of the systems by the methods described above (LLG).

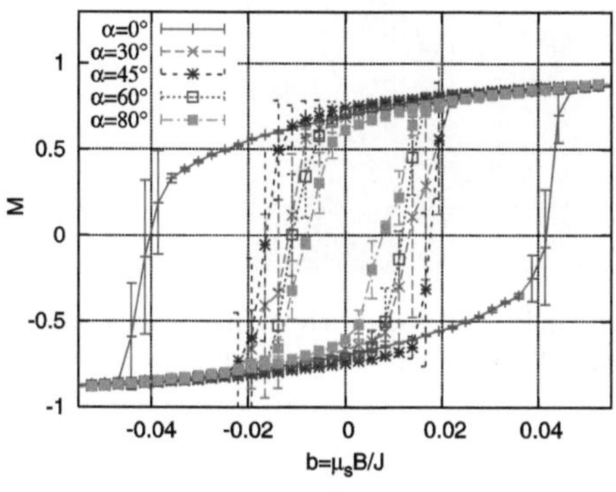

Fig. 4. Hysteresis loops for different tilt angles α. (Parameters: diameter = 50 nm d_{max} = 10 nm, temperature = 300 K, cell size = 25 Å, number of cells = 846)

3 Simulations of Soft Matter Systems

Bi-disperse colloidal crystals can be characterised by their structural and elastic properties. Monte-Carlo Simulations are an effective means for a systematic analysis of these properties. We are interested in the dependence of these properties on the mixing ratio and the size ratio of the components [33, 34] as well as on the phase diagram in external periodic light fields [33].

MC simulations for hard disk mixtures with different diameter ratios d_B/d_A have been performed in the NPT- and the NVT-ensemble [33] in order to analyse the structural properties and phase transition parameters in such systems. Interesting high pressure phases have been found (for nomenclature see ref. [35]). An example for such an high pressure phase is shown in Fig. 5 for a diameter ratio of $d_B/d_A = 0.2$. In order to compute a phase diagram [33], the transition densities have to be computed by cumulant intersection methods for various diameter ratios.

The lack or long range order in confined two-dimensional colloidal crystals has been investigated recently [36]. The interesting effect of "laser induced freezing" and "laser induced melting" has been studied in the HPC projects by Monte Carlo simulations in two dimensions using commensurate [37–42] and incommensurate [43] potentials. For the case of a lattice constant twice the size of the (sin-) potential wave length, a transition has been investigated [44] from a phase, in which every lane with a potential minimum is occupied on the average, to a phase, in which only every second lane is occupied, see Fig. 6.

The effect of finite mass on the phase diagram has been quantified by path integral Monte Carlo simulations [45, 46], and qualitative differences were found. We plan to explore this interesting topic for systems with different particle masses by PIMC studies and finite-size-scaling methods. Besides

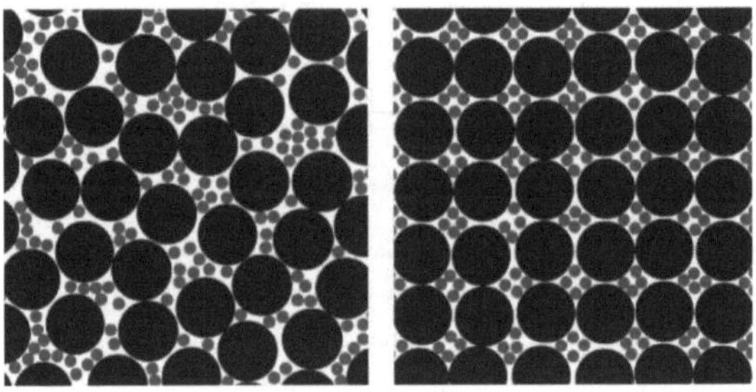

Fig. 5. Details of configurations of hard disk mixtures with a diameter ratio of $d_B/d_A = 0.2$ for low ($P^* = 100$) and high ($P^* = 102$) pressures. At high pressures the $S_2(AB_4)$ phase is stable

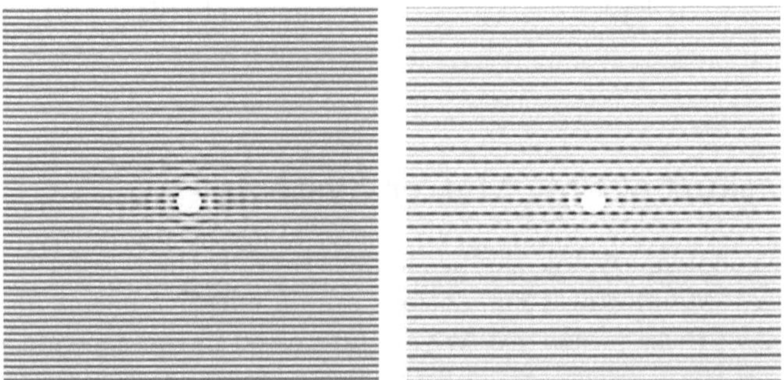

Fig. 6. Two dimensional pair correlation function of a system of hard disks in an external periodic (sin-) potential with amplitude $V_0 = 5$ and a wave length of half the lattice constant of the triangular lattice at density $\rho = 0.83$ (*left*) and $\rho = 0.86$ (*right*)

this we plan to analyse the order of the phase transition at high potential amplitudes by an application of our new method for the computation of elastic constants. The influence of a potential with higher symmetry shall be explored as well as the effect of the potential amplitude on the Bose condensation in case of systems with Bose statistics.

The effect of external periodic fields on the phase transition scenario in bi-disperse hard sphere mixtures was analysed [33]. In case of a 50% mixture for a diameter ratio of $d_b/d_a = 0.414$ a freezing transition was detected, see Fig. 7.

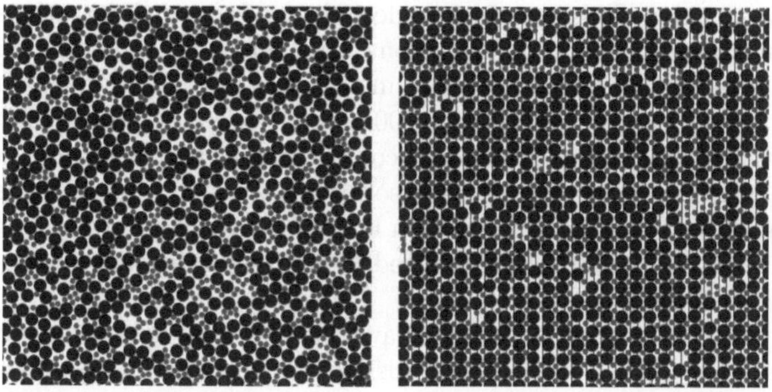

Fig. 7. Details of configurations of hard disk mixtures with a diameter ratio of $d_B/d_A = 0.414$ at density $\varrho^* = 1.5$ in an external periodic laser field with amplitude $V_0 = 0$ (*left side*) and $V_0 = 15$ (*right side*)

The computation of a full phase diagram requires the computation of the transition density for various field amplitudes. A typical computation (with $N = 882$ particles) of a point in the phase diagram requires 200 CPU hours on a standard processor.

We have developed a new configurations-based method for the computation of elastic constants. This method has been applied to models of colloidal systems containing quenched point impurities (and to colloidal mixtures) [4, 37, 33, 47, 43, 48, 38, 34]. A substantial hardening of the material with impurities has been detected [33]. A typical run for a fixed system size and a quenched configuration of impurities requires about 100 CPU h, an average over at least 30 configurations is required to perform the quenched average.

4 Dynamics in Micro Channels

In the context of micro-fluidics and "lab-on-chip" devices one is interested in non-equilibrium transport and mixing phenomena on the microscopic scale [49, 50]. Computer simulations [47, 51] and experiments [51] show lane formation and lane reduction in a system of gravitationally driven colloidal particles. The following experimental observations [51] are made: a) occurrence of ordered structures, b) density gradient along the channel, c) "lane transitions", i.e. changes between triangular lattice structures with different lattice parameters.

We carried out non-equilibrium computer simulations [47, 51] in order to investigate the transport of classical particles through channels of various configurations. The particles are driven by externally applied potential gradients. In the corresponding experiment [51] on super-paramagnetic colloidal particles the micro-channel is tilted, so that the particles are driven by the gravitational field. The computer simulations are based on an overdamped Langevin equation. This approach neglects hydrodynamic interactions as well as the short-time momentum relaxation of the particles. Both approximations are fully justified in the current experimental context. Typical momentum relaxation times are on the order of 100 µs and therefore much shorter than the repetition rate of the video microscopy setup (10s). The channel walls act like ideal hard walls: At $x = 0$ we also have a hard wall, and at the end of the channel ($x = L_x$) we have an open boundary. To keep the overall particle number fixed a new particle is inserted in the reservoir each time a particle drops out of the channel.

The particles are driven by the gravitational force due to the inclination of the micro channel. The channel walls are modelled as ideal hard walls, and open boundary conditions are applied in the flow direction, i.e. particles which drop out of the channel at the lower end will be inserted at a random position at the beginning of the channel.

The colloidal trajectories $\mathbf{r}_i(t) = (x_i(t), y_i(t))$ $(i = 1, ..., N)$ are given by the stochastic position Langevin equations [52] with the friction constant ξ,

$$\xi \frac{d\mathbf{r}_i}{dt} = -\nabla_{\mathbf{r}_i} \sum_{i \neq j} V_{ij}(r_{ij}) + \mathbf{F}_i^{\text{ext}}(t) + \tilde{\mathbf{F}}_i(t) \tag{4}$$

with the external force $\mathbf{F}_i^{ext}(t) = mg\cos(\alpha)\hat{\mathbf{x}}$ and the random forces $\tilde{\mathbf{F}}_i(t)$, given by random numbers with variance $\langle \tilde{\mathbf{F}}_{i\alpha}(t)\tilde{\mathbf{F}}_{i\beta}(0)\rangle = 2k_BT\xi\delta(t)\delta_{ij}\delta_{\alpha\beta}$ and zero mean $\langle \tilde{\mathbf{F}}_i(t)\rangle = 0$. The point particles interact via the dipolar potential $V_{ij}(r_{ij}) = \mu_0 M^2/4\pi r_{ij}^3$, with the magnetization M. A typical simulation run for a fixed inclination angle requires about 1000 CPU h, an average over at least 10 runs is required for statistical reasons.

A formation of lanes in the motion of the particles is detected along the channel. A similar layering phenomenon has been observed in channels with non-moving particles. The number of lanes decreases along the channel. In between areas of well defined number of lanes a point exists in which the particles are not well-ordered and is called the point of lane-reduction. The reduction of the number of lanes originates from a density gradient along the channel. The density decreases linearly along the direction of the motion of the particles. The density profile does not change even for the longest times. The first 10% of the channel act as a reservoir and the channel end is coupled to an empty reservoir. Hence the density gradient is solely driven by the dynamics due to the particle motion along the channel.

The lane-reduction transition can be described by an appropriate order parameter. The conventional coordination order parameter (ψ_6) is not suitable for this system, as it is sensitive to any perturbation of the hexagonal order. We therefore define the lane order parameter in the following way: $\psi_{lane} = |\frac{1}{n_{\rm bin}} \sum_{j=1}^{n_{\rm bin}} e^{i\mathbf{k}\cdot\mathbf{r}}| = |\frac{1}{n_{\rm bin}} \sum_{j=1}^{n_{\rm bin}} e^{i\frac{2\pi n_l y}{L_y}}|$, with $\mathbf{k} = \frac{2\pi n_l}{L_y}\hat{\mathbf{y}}$, ($n_l \in \mathbf{N}$), L_y is the channel width, $n_{\rm bin}$ the number of particles in bin, the number of lanes is $n_l + 1$. The lane order parameter along the channel in x-direction is shown in Fig. 8. Clearly a discontinuous behaviour across the lane-reduction is found. For this system of channels with ideal hard walls we find:

a) the particles form ordered structures and one observes transitions between different triangular lattice structures,
b) The lane reduction always coincides with a defect,
c) the drift velocity increases monotonically,
d) the particle flow j_x increases linearly with the packing fraction η.

We compared the experimental results gained from video microscopy with simulations of particle flow through constrictions under very similar conditions. The results of both cases show qualitative and quantitative agreement [51]. In future studies we plan to extend our studies to explore the flow behaviour in dependence of the particle interaction range, the characteristics of the channel walls, and the channel geometry (bottlenecks, barriers).

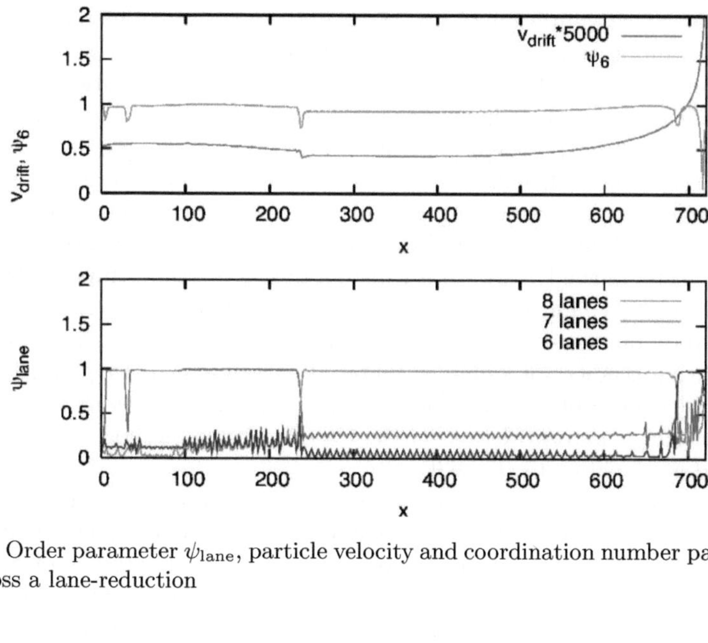

Fig. 8. Order parameter $\psi_{\rm lane}$, particle velocity and coordination number parameter ψ_6 across a lane-reduction

5 Electronic and Structural Properties of Nano Wires and Clusters

In this section we report on our computations of structural and electronic properties of atomic wires. Such systems were studied by experimental methods [53, 54, 55], where wires have been stretched down to single atom contacts.

Many experiments have shown that the conductance histograms of metallic atomic-sized contacts exhibit a peak structure, which is characteristic of thc corresponding material. The origin of these peaks still remains as an open problem. In order to shed some light on this issue, we have computed conductance histograms of atomic contacts of a variety of systems (Au, Fe, Pt, Co, Ni,...). In our HPC project we have combined classical molecular dynamics simulations of the breaking of nanocontacts with conductance calculations based on a tight-binding model. This combination gives us access to crucial information such as contact geometries, forces, minimum cross-section, total conductance and transmission coefficients of the individual conduction channels. The conductance as function of stretching distance of the electrode has been computed [56, 58] for Pt-wires by our method described in Ref. [57] with model parameters for Pt taken from the literature [59, 60]. We find the formation of single atom wires (see Fig. 9), containing about 5 atoms for large stretching distances. In contrast to Au-nano contacts the conductance has values larger than $1G_0$. From Fig. 9 the stretching force behaviour can be analysed. When the stretching force increases we are in the elastic region, in inelastic regions with constant force or decreasing force we find atomic rearrangements with resulting conductance changes.

A typical computation of the structural and conductance evolution requires a computation with $1.1 \cdot 10^6$ MD steps where after the equilibration of the system every $4 \cdot 10^3$ MD steps a full conductance compilation is required. The resulting numerical effort is about 400 CPU-hours (single processor) for a single stretching process. The total CPU-time of our studies was about $20 \cdot 10^3$ CPU-hours. Conduction histograms can be obtained as the result of averaging over many stretching processes at different temperatures. In order to compute such histograms and to be able to analyse the experimentally observed effects, in the HLRS project histograms have been computed by about 100 molecular dynamics simulations of single stretching processes.

Based on our results [57, 58] further detailed numerical investigations and comparative studies for different materials are planned. In parallel an improved treatment of the electronic components of the system at the single atom contact is planned by use of the Car-Parrinello-method and the results obtained at the SSC [61, 62, 63].

In addition Car-Parrinello-studies are performed for clusters at surfaces [64, 63], where in particular the cluster stability and their usefulness as building blocks as cluster material are analyzed. Following our studies of Si_4 clusters [61, 63], we focussed on the structural and electronic properties of Si_7 clusters. For the simulations the implementation of DFT available at [65] was

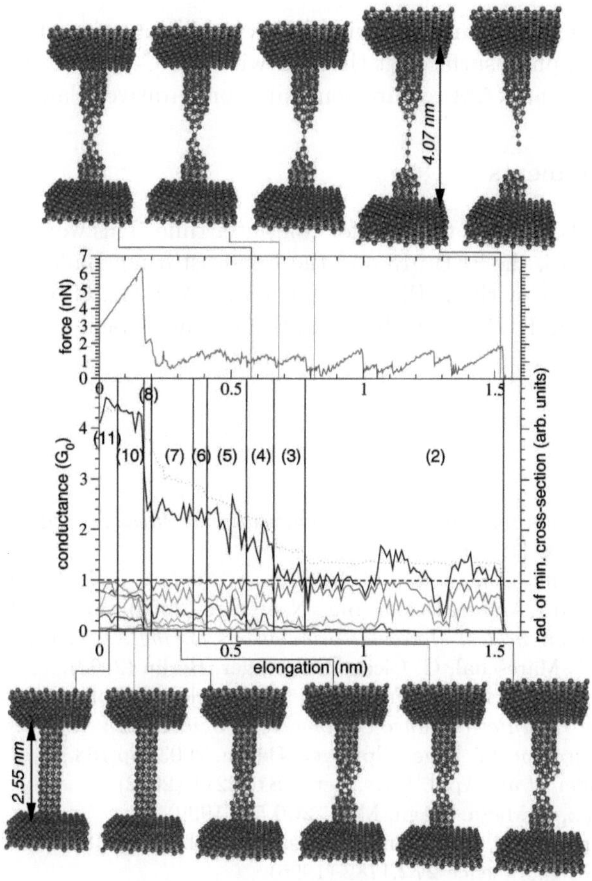

Fig. 9. *Upper panel* (*red line*): strain force on nano contact (in nN) as function of stretching distance. *Middle panel*: conductance in units of $G_0 = 2e^2/h$ (*black line*); radius of minimum cross section in arbitrary units (*dashed orange line*). *Lower panel*: single channel contributions to the conductance (in units of G_0). *Vertical lines*: separate regions with different channel numbers (number of channels is given in the *brackets in the upper part*). Below and above the graph: configuration pictures of Platin-wires

used. Calculations were performed using norm conserving pseudo potentials of the Trouiller-Martins type and PBE [66] Exchange Correlation functionals. As the interaction of the clusters with the (HOPG) surface is very weak, the surface was initiated by not allowing certain atoms to move out of the xy-plane. Additional studies involved the microscopic treatment of the surface [64]. In summary, we have all the knowledge to perform simulations of the adsorption of atoms and clusters on graphite surfaces. Further studies will consider additional reaction channels and the deposition process, investigate the role of oxygen in the experiments and other possible cluster material building blocks, for example C_{10} rings. A typical simulation of a single configuration of a graphite surfaces at the SSC with 108 C-particles required about 32 CPU hours, the RAM requirements are comparatively large.

Acknowledgements

We thank the HLRS and the SSC for computer time. This work was supported by the SFB TR6, the SFB 513 and the Landesstiftung Baden-Württemberg. Useful discussions with K. Binder, C. Cuevas, A. Erbe, G. Ganteför, P. Leiderer, U. Nowak, F. Pauly, A. Ricci, E. Scheer and S. Sengupta are gratefully acknowledged.

References

1. P. Nielaba, in *Annual Reviews of Computational Physics V*, edited by D. Stauffer, p. 137–199 (1997).
2. P. Nielaba, in: *Computational Methods in Surface and Colloid Science*, M. Borowko (Ed.), Marcel Dekker Inc., New York (2000), pp.77–134.
3. *Bridging Time Scales: Molecular Simulations for the Next Decade*, edited by P. Nielaba, M. Mareschal, G. Ciccotti, Springer, Berlin (2002).
4. M. Dreher, D. Fischer, K. Franzrahe, P. Henseler, J. Hoffmann, W. Strepp, P. Nielaba, in *High Performance Computing in Science and Engineering 02*, edited by E. Krause and W. Jäger, Springer, Berlin, 2003, pp.168.
5. M. M. Miller *et al.*, Appl. Phys. Lett., **81**, 2211 (2002).
6. G. A. Prinz, J. Magn. Magn. Mat., **200**,57 (1999).
7. B. D. Terris and T. Thomson, J. Phys. D: Appl. Phys. **38**,R199 (2005).
8. M. Hehn *et al.*, Science **272**,1782 (1996).
9. J. Jorzick *et al.*, Phys. Rev. Lett. **88**, 047204 (2002).
10. T. Shinjo, T. Okuno, R. Hassdorf, K. Shigeto, T. Ono, Science **289**, 930 (2000).
11. A. Wachowiak *et al.*, Science **298**,577 (2002).
12. Y. G. Yoo, M. Kläui *et al.*, Appl. Phys. Lett. **82**, 2470 (2003).
13. M. Kläui, H. Ehrke, U. Rüdiger, T. Kasama, R. E. Dunin-Borowski, D. Backes, L. J. Heyderman, C. A. F. Vaz, J. A. C. Bland, G. Faini, E. Cambril, and W. Wernsdorfer, Appl. Phys. Lett. **87**, 102509 (2005).
14. M. Kläui, C. A. F. Vaz, J. A. C. Bland, W. Wernsdorfer, G. Faini, E. Cambril, L. J. Heyerman, F. Nolting, and U. Rüdiger, Phys. Rev. Lett. **94**, 106601 (2005).
15. M. Kläui *et al.*, Phys. Rev. Lett. **90**, 97202 (2003).
16. The free OOMMF package is available at http://math.nist.gov/oommf.
17. U. Nowak, Ann. Rev. of Comp. Phys. **9**, 105 (2001).
18. L. D. Landau und E. M. Lifshitz, On the Theory of the Dispersion of Magnetic Permeabiltity in Ferromagnetic Bodies, Phys. Z. Sowjetunion **8**, 153 (1935).
19. J. L. Garcia-Palacios and F. J. Lázaro, Phys. Rev. **B 58**, 14937 (1998).
20. C. W. Gardiner, *Handbook of Stochastic Methods*, Springer-Verlag, Berlin, 1990.
21. W. F. Brown, Phys. Rev., **130**, 1677 (1963).
22. D.P. Landau, K. Binder; *A Guide to Monte Carlo Simulations in Statistical Physics*, Cambridge University Press, Cambridge (2000).

23. Chr. Schieback, Doktorarbeit, U. Konstanz (in preparation).
24. M. Laufenberg, D. Backes, W. Bührer, D. Bedau, M. Kläui, U. Rüdiger, C. A. F. Vaz, J. A. C. Bland, L. J. Heyderman, F. Nolting, S. Cherifi, A. Locatelli, R. Belkhou, S. Heun, and E. Bauer, Appl. Phys. Lett. **88**, 052507 (2006).
25. R. D. McMichael and M. J. Donahue, IEEE Trans. Mag. **33**, 4167 (1997).
26. C. Schieback, M. Kläui, D. Backes, R. Dunin-Borkowski, U. Rüdiger, P. Nielaba, *Domain wall widths in nanoscale constrictions* (in preparation).
27. Z. Li *et al.*, Phys. Rev. **B70**, 024417 (2004).
28. A. Thiaville, Y. Nakatani, J. Miltat, N. Vernier, J. Appl. Phys. **95**, 7049 (2004).
29. A. Thiaville, Y. Nakatani, J. Miltat, Y. Suzuki, Europhys. Lett. **69**, 990 (2005).
30. M. Albrecht *et al.*, Nature Materials 4, 203–206 (2005).
31. J. Neder, Diplomarbeit, U. Konstanz (2005).
32. T. C. Ulbrich *et al.*, Phys. Rev Lett. **96**, 77202 (2006).
33. K. Franzrahe, Doktorarbeit, U. Konstanz (in preparation).
34. K. Franzrahe, P. Henseler, A. Ricci, W. Strepp, S. Sengupta, M. Dreher, Chr. Kircher, M. Lohrer, W. Quester, K. Binder, P. Nielaba, Comp. Phys. Commun. **169**, 197 (2005).
35. C.N. Likos, C.L. Henley, Phil. Mag. **B68**, 85 (1993).
36. A. Ricci, P. Nielaba, S. Sengupta, K. Binder, Phys. Rev. **E**, in press.
37. P. Nielaba, K. Binder, D. Chaudhuri, K. Franzrahe, P. Henseler, M. Lohrer, A. Ricci, S. Sengupta, W. Strepp, J. Phys. Cond. Mat.: **16**, S4115 (2004).
38. M. Dreher, D. Fischer, K. Franzrahe, G. Günther, P. Henseler, J. Hoffmann, W. Strepp, P. Nielaba, in *NIC Symposium 2004*, edited by D. Wolf, G. Münster, M. Kremer, pp. 291 (2004).
39. W. Strepp, S. Sengupta, P. Nielaba, Phys. Rev. **E63**, 046106 (2001).
40. W. Strepp, S. Sengupta, P. Nielaba, Phys. Rev. **E66**, 056109 (2002).
41. W.Strepp,S.Sengupta,M.Lohrer,P.Nielaba,Comp.Phys.Commun.**147**,370(2002).
42. W.Strepp,S.Sengupta,M.Lohrer,P.Nielaba,Math.Comp. in Simul.**62**,519(2003).
43. Chr. Kircher, Diplomarbeit, U. Konstanz (2004).
44. F. Bürzle, Diplomarbeit, U. Konstanz (2006).
45. M. Dreher, D. Fischer, K. Franzrahe, P. Henseler, Chr. Kircher, M. Lohrer, W. Quester, A. Ricci, S. Sengupta, W. Strepp, K. Binder, P. Nielaba, Phase Transitions **78**, 751–772 (2005).
46. W.Strepp, P. Nielaba, draft-preprint.
47. P. Henseler, Doktorarbeit, U. Konstanz (in preparation).
48. W. Quester, Diplomarbeit, U. Konstanz (2004).
49. H.A.Stone, A.D.Strook, A.Ajdari, Ann.Rev.Fluid.Mech.**36**,381(2004).
50. T.M. Squires and S.R. Quake, Rev. Mod. Phys. **77**, 977 (2005).
51. M. Köppl, P. Henseler, A. Erbe, P. Nielaba, P.Leiderer, preprint.
52. M.P. Allen, D.J. Tildesley, *Computer Simulations of Liquids*, Oxford (1987).
53. E. Scheer *et al.*, Phys. Rev. Lett.**78**, 3535 (1997).
54. E. Scheer *et al.*, Nature **394**, 154 (1998).
55. E. Scheer *et al.* Phys. Rev. Lett. **86**, 284 (2000).
56. M. Dreher, Doktorarbeit, U. Konstanz (in preparation).
57. M. Dreher, F. Pauly, J. Heurich, J.C. Cuevas, E.Scheer, P. Nielaba, Phys. Rev **B 72**, 075435 (2005).
58. F. Pauly, M. Dreher, J.K. Viljas, M. Häfner, J.C. Cuevas, P. Nielaba, preprint.
59. K.W. Jacobsen, P. Stoltze, J.K. Norskov; Surf. Sci. **366**, 394 (1996).
60. M.J. Mehl, D.A. Papaconstantopoulos; Phys. Rev. **B54**, 4519 (1996).

61. D. Fischer, Dissertation, Univ. Konstanz (2002).
62. D. Fischer *et al.*, Chem. Phys. Lett. **361**, 389 (2002).
63. M. Grass, D. Fischer, M. Mathes, G. Ganteför, P. Nielaba; Appl. Phys. Lett. **81**, 3810 (2002).
64. W. Quester, Dissertation, Univ. Konstanz (in preparation).
65. The homepage of the CPMD consortium, http://www.cpmd.org.
66. J.P. Perdew et al., Phys. Rev. Lett. **77**, 3865 (1996).

Signal Transport and Finite Bias Conductance in and Through Correlated Nanostructures

Peter Schmitteckert[1,2] and Günter Schneider[1,3]

[1] Institut für Theorie der Kondensierten Materie, Wolfgang Gaede Straße 1, D-76128 Karlsruhe
[2] peter@tkm.physik.uni-karlsruhe.de
[3] gs@tkm.physik.uni-karlsruhe.de

During the past decade improved experimental techniques have made production of and measurements on one-dimensional systems possible, and hence led to an increasing theoretical interest in these systems. However, the description of non-equilibrium transport properties, like the signal transport or the finite bias conductance of an interacting nanostructure attached to leads, is a challenging task. For non-interacting particles, the conductance can be extracted from the transmission of the single particle levels. However, the screening of electrons is reduced by reducing the size of structures under investigation and electron-electron correlations can no longer be neglected. Therefore, an adequate description has to be able to treat strong correlations and non-equilibrium in a rigorous manner.

While several methods have been developed to calculate the zero bias conductance of strongly interacting nanostructures, there are no general methods available to obtain rigorous results for the finite bias conductance. While the problem has been formally solved by Meir and Wingreen using Keldysh Greens functions [1], the evaluation of these formulas for interacting systems is generally based on approximate schemes. In this project we apply the real time density matrix renormalization group method (RT-DMRG) to simulate the signal transport in one-dimensional, interacting quantum systems, and the conductance of interacting nanostructures attached to one-dimensional, non-interacting leads, where the nanostructure and the leads are described with a time dependent many particle wavefunction.

1 The Density Matrix Renormalization Group Method

The density matrix renormalization group method is now a well established method to treat strongly correlated quantum systems in low dimensions. Originally developed for static properties, it has recently been extended to simulate the time evolution of quantum systems. The method is an iterative diagonalization scheme, in which one tries to optimize a subspace of the complete Hilbert space of the problem. This optimization is performed by a sweeping technique, where one sweeps through a real space partitioning of the system [2, 3]. The optimized subspace represents the desired states, which one calls target states. These states typically consist of the low lying states of the system of interest, and states needed to solve resolvent equations. In the case of the time evolution schemes in DMRG one has to include the time dependent wavefunctions. For an introduction to the method we refer to reviews in the literature [4].

1.1 Sparse Matrix Representation

In order to describe our parallelization scheme it is necessary to introduce the data structure of our DMRG implementation. In the DMRG procedure one divides the system into a left (A) and a right (B) block, where the vector space of the complete system is given by the tensor product space of block A and B. In our implementation we explicitly block for the inserted sites (σ, τ) leading to the $A \cdot \cdot B$ blocking displayed in Fig. 1.

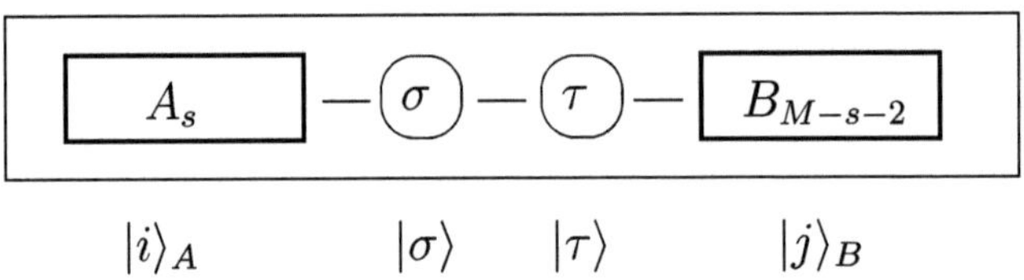

Fig. 1. The $A \cdot \cdot B$ blocking scheme used in our DMRG implementation

A state $|\Psi\rangle_C$ of the super block $C = A \cdot \cdot B$ is given by a tensor product state of the the four individual blocks A, σ, τ and B

$$|\Psi\rangle_C = \sum_{\substack{q_A, q_B \\ i,j,\sigma,\tau}} \Psi_{i,j}^{q_A,q_B,\sigma,\tau} |q_A, i, \sigma; q_B, j, \tau\rangle_C \tag{1}$$

where

$$|q_A, i, \sigma; q_B, j, \tau\rangle_C = |q_A, i\rangle_A \otimes |\sigma\rangle \otimes |\tau\rangle \otimes |q_B, j\rangle_B \tag{2}$$

are the basis states of the product space $\mathcal{V}$ of $C = A \cdot \cdot B$.

Since we can utilize symmetries, like particle or spin conservation, the Hilbert space $\mathcal{V}$ is not just a simple tensor product, but is given by a direct sum of tensor product spaces, where each subspace $\mathcal{V}_{q_A,q_B,\sigma,\tau}$ has to obey the global symmetries $\mathcal{Q}$

$$\mathcal{V} = \oplus \, \mathcal{V}_{q_A,q_B,\sigma,\tau}\Big|_{(q_A,q_B,\sigma,\tau)\in\mathcal{Q}} . \tag{3}$$

For example, if we want to look at a system with only one electron, we have four subspaces, where the electron can be in one of the four blocks, while the other three blocks correspond to a system without an electron, and the quantum number q_ι is given by the electron number.

A state $|\gamma\rangle = |\alpha\rangle_{q_A} \otimes |\beta\rangle_{q_B} \otimes |\sigma,\tau\rangle \in \mathcal{V}_{q_A,q_B,\sigma,\tau}$ of the super block C is now represented by a tuple of dyadic vectors $\gamma^{q_A,q_B,\sigma,\tau}$

$$\gamma^{q_A,q_B,\sigma,\tau} = \gamma = \boldsymbol{\alpha}_{\mathbf{q_A},\boldsymbol{\sigma}} \cdot \boldsymbol{\beta}^{\top}_{q_B,\tau} \cdot . \tag{4}$$

The action of an operator $\hat{C} = \hat{A} \otimes \hat{B}$ is given by

$$\hat{C}_{ik,j\ell}\,\gamma_{j\ell} = \hat{A}_{i,j}\,\boldsymbol{\alpha}_j\,\boldsymbol{\beta}_\ell^{\top}\,\hat{B}^{\top}_{\ell,k} = \left(\hat{A}\cdot\gamma\cdot\hat{B}^{\top}\right)_{i,k} . \tag{5}$$

For instance the Hamiltonian $\mathcal{H}$ is given by a set of operators $\hat{A}^\ell$ and $\hat{B}^\ell$

$$\mathcal{H} = \sum_\ell \hat{A}^\ell \otimes \hat{B}^\ell \qquad \hat{A}^\ell \otimes \hat{B}^\ell : \mathcal{V}_{q_A,q_B,\sigma,\tau} \rightarrow \mathcal{V}_{q'_A,q'_B,\sigma',\tau'} \tag{6}$$

where $\hat{A}^\ell$ ($\hat{B}^\ell$) acts only on the left (right) of a sub-block $\gamma_{j\ell}$ of a vector γ. This structure uses the fact, that the representation of the inserted sites consists of the direct sum of one-dimensional subspaces, which is the reason for employing the $A \cdot \cdot B$ blocking scheme.

The big advantage of this representation is that the sparse matrix-vector product $\mathcal{V} \rightarrow \mathcal{V}$ is now implemented by a set of BLAS-3 matrix-matrix products $\mathcal{V}_{q_A,q_B,\sigma,\tau} \rightarrow \mathcal{V}_{q'_A,q'_B,\sigma',\tau'}$. As it turns out, the operators $\hat{A}^\ell$ ($\hat{B}^\ell$) are represented by dense matrices. Therefore the matrix-matrix multiplications in equation 6 consist of a set of BLAS-3 xgemm calls. Using the hardware counters of the Itanium2 processor we measured 2.7 FLOP/cycle over a complete medium sized DMRG run. Since MKL dgemm does not deliver more than 3.6 FLOP/cycle, this is already 75% of the maximally achievable performance, and 68% of the theoretically possible performance of 4FLOP/cycle. We expect that the large scale calculations presented below had an even better performance.

1.2 Parallelization Using Posix Threads

While the blocked representation of the matrices described in the previous section leads to a highly efficient representation, an efficient implementation for distributed systems is a challenging task. However, for the conductance calculations (see Sect. 3), a single run results in a single point of a current voltage ($I(V)$) characteristics, and many runs are needed to get a complete $I(V)$ curve, which will then allow the calculation of the differential conductance. We therefore decided to implement a SMP parallelization to use multiple cores of a compute node, while distributing the calculation of different data points in an embarrassingly parallel manner.

To this end we designed a master-worker parallelization using Posix threads, where the distribution of the work load is encapsulated in a master class. The advantage of Posix threads in comparison to an OpenMP implementation lies in the higher flexibility of Posix threads, which allow for a more efficient concurrent programming model. In our implementation we took care that the master class can be extended by an MPI interface in the future.

Implementation

We have implemented a master worker class which serves as an interfacc to the linear algebra operations, e.g. it encapsulates the calls to BLAS and LAPACK and provides some extended routines, which are needed to guarantee atomic calculation of combined operations, like the matrix-matrix product $\hat{C} = \hat{A} \cdot \hat{X} \cdot \hat{B}$, which also facilitates thread local memory for the intermediate result of $\hat{X} \cdot \hat{B}$, if supported by the OS kernel.

The master class encapsulates all operations needed for the parallelization and would enable alternative implementations, e.g. one that utilizes MPI parallelization. It provides

- changing the number of worker threads,
- allocation of task indices,
- global and task index specific waits,
- BLAS, LAPACK and combined operations,
- scheduling of arbitrary functions,
- accepts single work entries and complete work queues for scheduling.

Threading Performance

We investigated the threading performance by a test program, which schedules matrix-matrix multiplications and compares the wall time for multithreaded execution against a serial version. In Fig. 2 we compare the threading performance for scheduling the matrix-matrix multiplication for matrices of dimension 100x100. It shows perfect scaling for the Linux 2.6 kernel, (the Power5 system utilizes simultaneous multithreading), while the Linux 2.4 kernel cannot achieve linear scaling on the XC.

In Fig. 3 we show the results for the scheduling of small (20x20) matrices. The results show that the Linux 2.4 kernel has a performance problem with the threading overhead, while the 2.6 kernel manages to provide a reasonable speedup even for these small work units. While the typical workload of our DMRG calculation corresponds to at least the 100x100 example, there are always some small work units involved, which can spoil the scalability. From these timings we expect a big improvement in scalability of our code, once the XC has moved to the 2.6 kernel[1].

[1] After the transition to kernel 2.6.9 we found an performance boost up to 30% on large jobs on the fat nodes. However, the kernel is still not NUMA aware. So the performance gain comes from improved threading and IO.

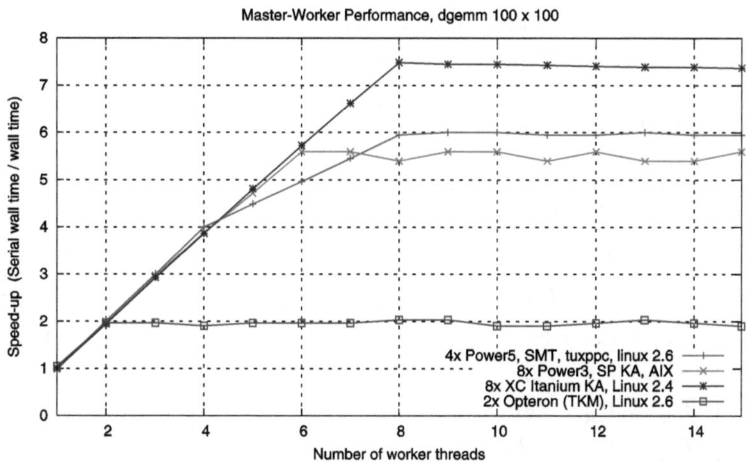

Fig. 2. Threading performance for 100 × 100 dgemm for different architectures

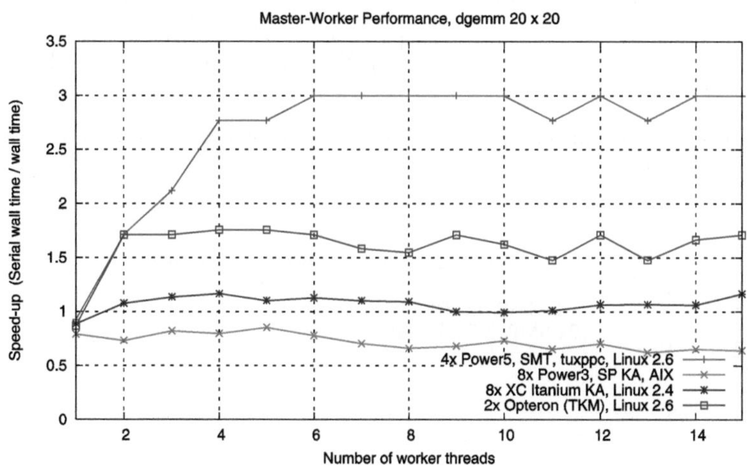

Fig. 3. Threading performance for 20 × 20 dgemm for different architectures

Concurrency

The flexibility of our parallelization is exemplified in Fig. 4 where we describe the scheduling of the modified Gram-Schmidt orthogonalization. It shows that the main thread can always schedule scalar products while the worker queue is still occupied with the previous step. This concurrency is achieved by explicitly splitting the complex operations into real and imaginary parts, which can be scheduled independently. The scalability of the complete sparse matrix exponential is shown in Fig. 5. On a thin node we find an efficiency of our parallelization of 84% (72%) compared to a run using a single worker thread (complete serial run). On a the thick node the efficiency drops from 86% (81%)

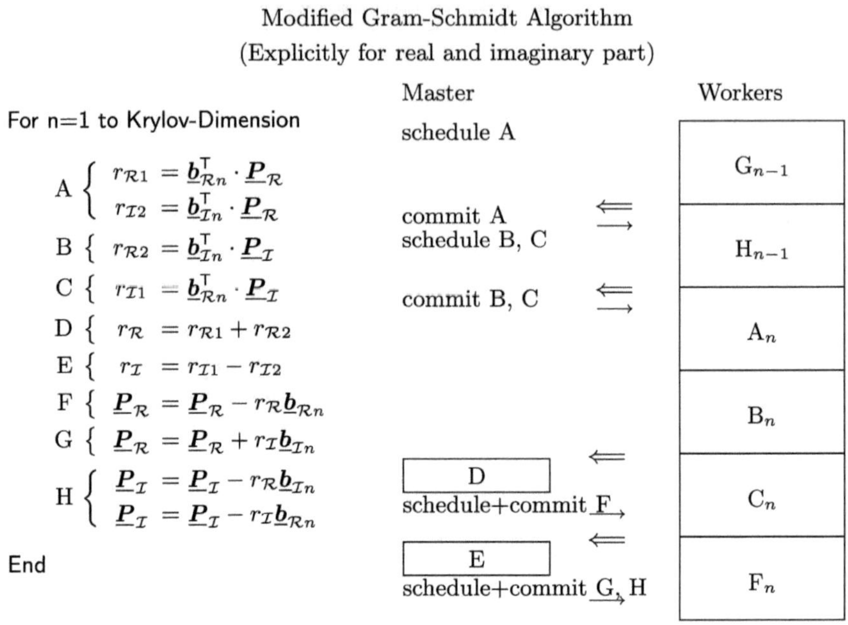

Fig. 4. Concurrency in the modified Gram-Schmidt orthogonalization of the sparse matrix exponential

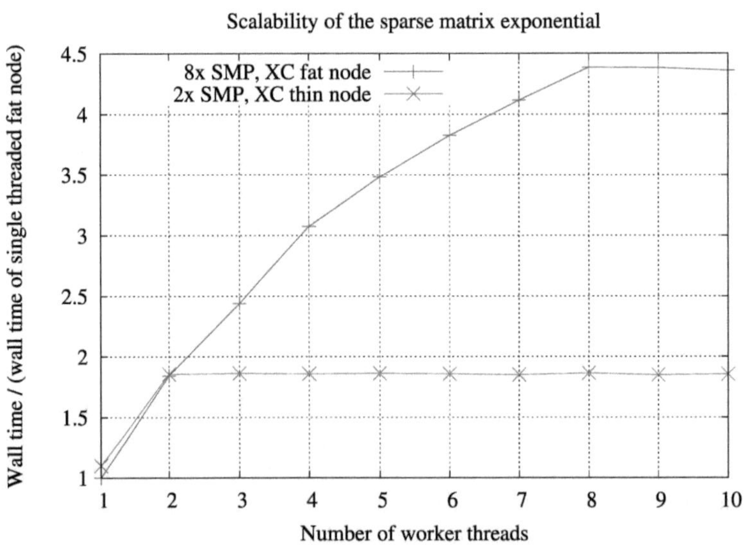

Fig. 5. Scalability of the sparse matrix exponential for a spin 1/2 electron system with $n_{\mathrm{Cut}} = 5000$

using two threads, 77% (71%) using four threads, down to 59% (51%) using eight threads compared to a run using a single worker thread (complete serial run). From comparison with dual opteron systems at our institute running a Linux 2.6 kernel we conclude that most of the missing efficiency can be attributed to the threading deficiencies of the Linux 2.4 kernel. Therefore we shifted a detailed threading analysis until after the 2.6 kernel is available on the XC. In addition, the current kernel is not NUMA aware and always interleaves memory over all CPUs, which interferes with multithreading and does not exploit the thread local memory.

The Senior Server Concept

The advantage of the Posix thread approach is that we can schedule arbitrary subroutines. We exploit this feature by reusing the master class for scheduling larger work units. For this purpose we create a second instance, which we call senior server, which takes extended subroutines, like a sparse matrix exponential, a sparse matrix diagonalization, or the evaluation of observables as argument. We can therefore not only exploit the parallelism of the algorithms, but include the concurrent evaluation of large work units to reduce idling due to inherently serial parts of a subroutine. In addition we are in the process of moving the IO load into a third instance of the server, in order to perform our IO operations asynchronously.

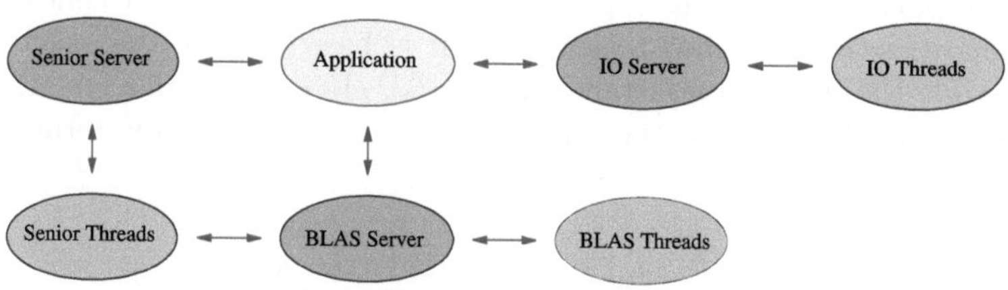

Fig. 6. The three master approach to DMRG. All three servers are instances of the same master class. However, they are used for different tasks

Performance Issues

Memory Latency

In our calculation we see a typical performance loss of 20% to 30% when comparing runs on the eight way fat nodes relative to the performance of the same runs on the two way thin nodes. This corresponds to the increased memory latencies due to the interleaved memory. We expect that a switch to a NUMA aware kernel will remove this problem. In addition to the general improvement of the threading behavior, it would also enable us to use thread local matrix temporaries, leading to an additional increase of scalability.

IO

During this project it turned out that the Lustre file system has performance problems with our IO load. In order to address this problem, we implemented an IO buffer, which buffers the book keeping data and small matrices into a single IO block. In addition we utilize scattered IO for the remaining memory blocks to maximize IO performance and to decrease interference of read/write operations from different nodes.

Restart

Since our calculations need longer execution times than provided by the available job limit of three days we implemented a restart feature, which allows us to split a single job into several jobs. In addition, this allows us to perform an initial run on the dual nodes, and restarting it on fat nodes, once the problem size is sufficiently large. This approach utilizes the fact that in the first few DMRG sweeps one does not need to use the full number of desired states, as the DMRG needs a few sweeps to adapt to the problem. Therefore, we increase the number of states kept during the finite lattice sweeps.

2 Linear Response

Linear response calculations [6] provide a method to calculate the conductance of a nanostructure attached to leads. As it is based on the the exact Kubo formula for the linear conductance $g \equiv \frac{e^2}{h}\langle\tilde{J}\rangle/V_{SD}$ it is valid for arbitrary interaction. In the DC limit the conductance can be expressed in terms of two different correlators,

$$g_{J_jN} = -\frac{e^2}{h}\langle\psi_0|\hat{J}_{n_j}\frac{4\pi i\eta}{(\hat{H}_0 - E_0)^2 + \eta^2}\hat{N}|\psi_0\rangle, \tag{7}$$

$$g_{JJ} = \frac{e^2}{h}\langle\psi_0|\hat{J}_{n_1}\frac{8\pi\eta(\hat{H}_0 - E_0)}{\left[(\hat{H}_0 - E_0)^2 + \eta^2\right]^2}\hat{J}_{n_2}|\psi_0\rangle, \tag{8}$$

where the positions n_j are in principle arbitrary. However, the positions n_1 and n_2 should be placed close to the nanostructure to minimize finite size effects. As described in [6] one has to introduce exponentially reduced hopping terms close to the boundary of the leads to minimize finite size effects, which in return leads to ill-conditioned linear systems. In order to solve these equations, we start with a system without damping at the boundaries, and then turn on the damping in additional sweeps (iterations). In [6] we typically used four steps each consisting of four finite lattice sweeps for each reduction of the hopping, e.g. we decreased the damping factor to 0.95, 0.9, 0.85 and finally 0.8. In each DMRG step of these 16 additional sweeps the full DMRG calculation, including exact diagonalization and solving the resolvent equation have to be performed. Due to the damping, the level splitting of the low lying states is typically of the order of 10^{-6}, while the kinetic energy is of the order of 1 and the interaction is up to 30, which leads to a high condition number [6]. Even worse, for the current-current correlation one has to solve for $(\mathcal{H} - E_0 - i\eta)^2$. This resolvent equation is hard to solve, e.g. conjugate gradient does not converge, and even preconditioned minimal residual methods tend to shoot of. Even if these methods converge, iteration counts larger than 1000 are needed, which render these solvers unpractical. We solved this problem by implementing a diagonal block preconditioned full orthogonalization scheme, which converges typically within 20 to 50 iterations. Finally, the typical run time for the systems in [6] is of the order of a few days up to a week on a dual thin node for systems on or near a conductance resonance, while systems off resonance converge much faster.

Due to the sharp resonances found in strongly interacting systems, many data points are needed to resolve all features of the conductance curves. As each conductance point requires a full run, we have used a distributed approach, where we typically utilized 10 nodes simultaneously, with each node performing an independent conductance calculation. With these calculations we proofed that for strongly interacting systems, the resonance width is not just given by the contact parameter, but is strongly influenced by the interaction strength. Therefore commonly used methods based on the Landauer-Büttiker formula are not appropriate for these systems. For the detailed results we refer to [6].

In addition to the interesting results of the linear response calculations, they serve as a benchmark for our finite bias calculations in the regime of small voltages.

3 Time Evolution

The time evolution of a wavefunction corresponding to a time independent Hamiltonian is given by

$$|\Psi(t')\rangle = \mathrm{e}^{i\mathcal{H}(t'-t)}|\Psi(t)\rangle . \tag{9}$$

Although the calculation of the time evolution operator is infeasible as it is given by a dense matrix, the action of the time evolution operator on a wavevector can be efficiently calculated by a sparse matrix exponential using a Krylov space approximation [7].

We have implemented an adaptive DMRG scheme based on the sparse matrix exponential [7, 5]. In our implementation of the sparse matrix exponential we employ a full orthogonalization Arnoldi procedure as we found that the simplified Lanczos scheme is vulnerable to the loss of orthogonality, especially in the linear regime of conductance calculations.

In our implementation of the DMRG scheme we do not fix the number of states n_{Cut} kept per block. Instead we take n_{Cut} as a lower limit and increase the number of states kept per block in order to keep the Hilbert space dimension constant in the asymmetric block configuration, which would otherwise drop by a factor of three to five. This can lead to dramatically increased block sizes. e.g. $\sim$ 26000 states in an otherwise 5000 state calculation at the turn of the sweep direction.

In Fig. 7 we show the time evolution of a Gaussian density excitation in a 2/3 filled Hubbard model consisting of 33 sites and periodic boundary condition. In this calculation we used at least 10000 states per DMRG block, with slightly increased block sizes for the asymmetric configurations (which we limited to max. 12500 states due to memory restrictions). The size of the target space in this calculation was of the order of 70 million states and we performed two finite lattice sweeps for each time step of $\Delta t = 0.5$. The run time of the initial time step was $\sim$ 24h and increased up to $\sim$ 45h for the seventh time steps. After the upgrade to kernel 2.6.9, the run time dropped to 30h per time step, compare Fig. 8. Memory consumption was around 45 GB and the scratch files occupied 50 GB. The initial static DMRG was first performed on a thin node up to $n_{\text{Cut}} = 2500$ and then restarted on a fat node to perform the large block size calculation.

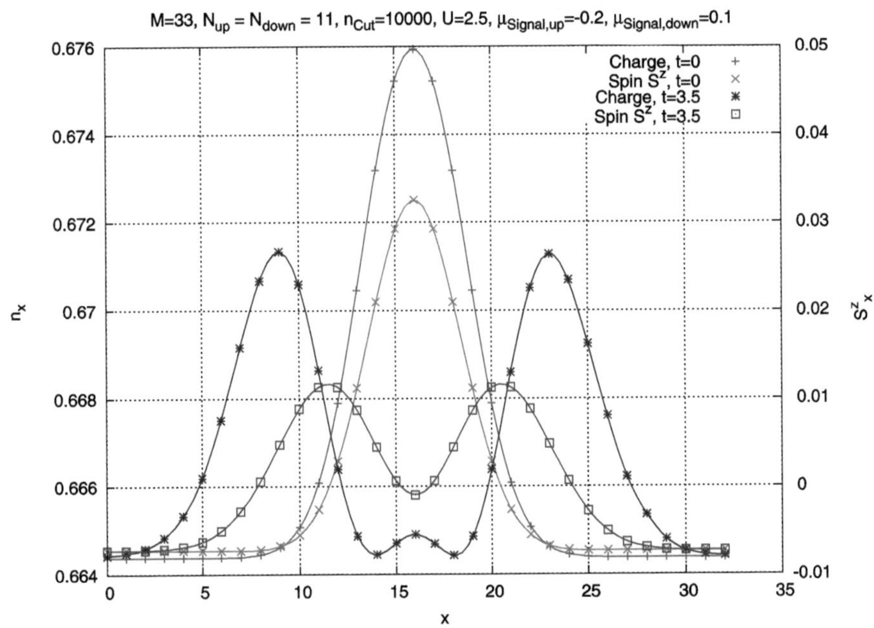

Fig. 7. Time evolution of a density excitation created with a gaussian perturbation of width $\sigma = 2.5$ and signal strength $\mu_{\uparrow} = -0.2$ and $\mu_{\downarrow} = 0.1$. The system is a 2/3 filled Hubbard model at $S^z = 0$ and $M = 33$ sites and periodic boundary conditions. At least $n_{\text{Cut}} = 10000$ states per block have been used

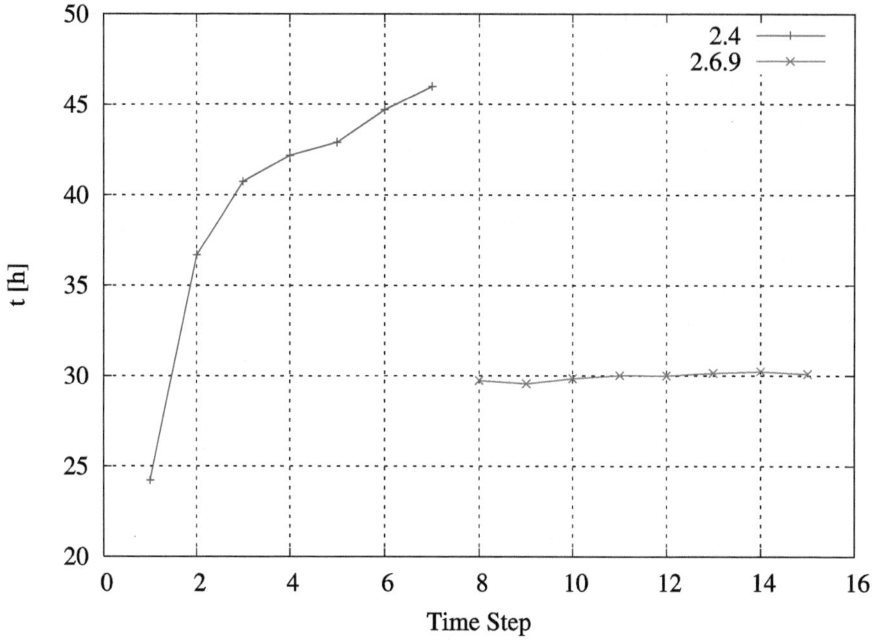

Fig. 8. Timing of a spin charge separation run of a 33 site Hubbard model, $N_{\rm up} = N_{\rm down} = 11$ using at least $n_{\rm Cut} = 10000$ states per block. A single time consisted of two sweeps. Compare Fig. 7

While the results could have been obtained with a 3000 state calculation, we only know that by checking our result with such a large calculation. Interestingly this calculation shows, that even 1000 states per block are not sufficient to get an accurate description of the dynamics. Even for the 10000 state calculation the discarded entropy is only slightly below 10^{-6} in the last sweep of the initial static DMRG run, increasing up to $1.1 \cdot 10^{-5}$ in the seventh time step, which shows that this calculation is still limited by the DMRG truncation scheme and that such large dimensions are indeed needed to make definite statements. On the XC fat nodes we have been able to increase the block size to 12000 states, which still leads to a discarded entropy of the order of $4 \cdot 10^{-7}$.

We would like to point out that we are not aware of any other research group worldwide being able to perform real-time dynamics within DMRG which such large vector spaces.

Differential Conductance

In the main part of our project we are addressing the problem of calculating the finite bias differential conductance of a nanostructure attached to leads in the presence of strong correlations. Due to the quantum mechanical nature of the problem it is important that the leads are treated on equal footing with the nanostructure, which in our approach is achieved by simulating the many particle wavefunction of the complete system consisting of the nanostructure and the leads.

We developed the following recipe for calculating the conductance from real time dynamics

- Prepare an initial extended wave packet by an source drain voltage U_{SD}
- Perform time evolution.
- Look at quasi-stationary state.
- Measure current close to nanostructure.
- Linear Response (small bias voltage): $\mathrm{g}(\mu_{Dot}) = \mathrm{I}(\mu_{Dot})/\mathrm{U}_{SD}$
- Nonlinear regime: Differential conductance $g(U_{\mathrm{SD}}) = \partial \mathrm{I}/\partial U_{\mathrm{SD}}$

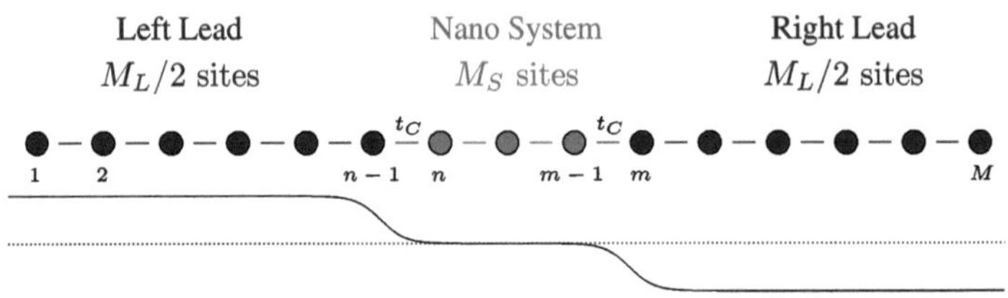

Fig. 9. Nanostructure attached to leads and schematic density profile of the initial wavepacket at $T = 0$

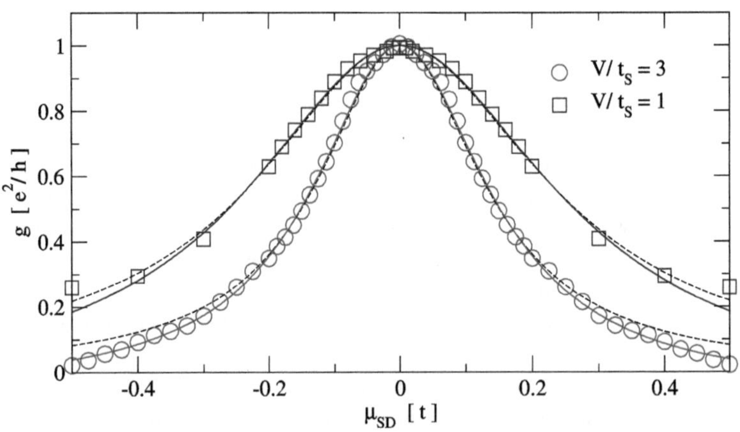

Fig. 10. Differential conductance as a function of bias voltage through a 7 site nanostructure with nearest neighbor interaction. Parameters are $t_{\mathrm{C}} = 0.5t$, $t_{\mathrm{S}} = 0.8t$, and N/M = 0.5. *Squares* (*circles*) denote weak (strong) interaction with $V/t_{\mathrm{S}} = 1$ (3). *Lines* are fits to a Lorentzian with an energy dependent self energy $\Sigma = i\eta_0 + i\eta_1\mu^2$. *Dashed lines*: $\eta_1 = 0$. System size is $M = 144$ ($M = 192$) and 600 (800) states were kept in the DMRG

where the schematics of the system and the initial, extended wave packet are displayed in Fig. 9. It is important to note that this method is currently the only method available to calculate the finite bias conductance for strongly interacting nanostructures attached to leads. Other methods are either based on the free fermion Landauer-Büttiker formula, weak coupling expansions or other approximate schemes. In Fig. 10 we show the differential conductance for a seven site system in the regime of weak $V = 1t_{\text{Dot}}$ and strong $V = 3t_{\text{Dot}}$ nearest neighbor interaction for a spinless fermion system. A careful analysis of the data shows, that our results are accurate enough to demonstrate a small energy dependent line width, which cannot be obtained by a standard Landauer-Büttiker approach.

For a detailed discussion of the results we refer to [5].

4 Outlook

After having established a new method for transport calculations we are now heading toward applications, for which we need higher energy resolution, and correspondingly longer simulation times. In addition we have implemented a new approach to calculate directly the differential conductance avoiding the necessity to perform a numerical derivative of the $I(V)$ curve. To include low temperature effects we implemented the treatment of several low lying states. Both approaches demand for the time evolution of several states at once, which fit perfectly in our parallelization concept using senior and BLAS server threads. Finally we have implemented all ingredients needed to perform quantum chemistry calculations, especially an automatic treatment of the fermion signs which allows us to perform the step from model physics to more realistic simulations, which should enable the search for novel nanostructure devices based on correlation effects.

4.1 Implementation Plans

Since the XC is now upgraded to a 2.6 linux kernel, we will first evaluate, whether our parallelization was limited by the 2.4 threading issues. Should the switch to the 2.6 kernel not significantly increase our scalability, we will test a high/low water queue based on a linear buffer, instead of our current list based approach.

In addition we plan to implement work queue scheduling optimization, e.g. scheduling the large matrices first in order to minimize the time between the end of the first and last thread for a given task. Since the work queues for the time consuming operations, e.g. the Hamiltonian matrix lists, are needed often in the sparse matrix methods, the overhead for presorting them once is expected to be negligible.

References

1. Y. Meir and N. S. Wingreen, Phys. Rev. Lett. **68**, 2512 (1992).
2. S. R. White, Phys. Rev. Lett. **69**, 2863 (1992).
3. S. R. White, Phys. Rev. B **48**, 10345 (1993).
4. Density Matrix Renormalization – A New Numerical Method in Physics, edited by I. Peschel, X. Wang, M.Kaulke, and K. Hallberg (Springer, Berlin, 1999); Reinhard M. Noack and Salvatore R. Manmana, Diagonalization- and Numerical Renormalization-Group-Based Methods for Interacting Quantum Systems, AIP Conf. Proc. 789, 93–163 (2005).
5. Günter Schneider, Peter Schmitteckert: Conductance in strongly correlated 1D systems: Real-Time Dynamics in DMRG, condmat-0601389.
6. Dan Bohr, Peter Schmitteckert, Peter Wölfle: DMRG evaluation of the Kubo formula – Conductance of strongly interacting quantum systems Europhys. Lett., 73 (2), pp. 246–252 (2006).
7. Peter Schmitteckert: Nonequilibrium electron transport using the density matrix renormalization group. Phys. Rev. B **70**, 121302 (2004).

Atomistic Simulations of Dislocation – Crack Interaction

Erik Bitzek[1] and Peter Gumbsch[1,2]

[1] Institut für Zuverlässigkeit von Bauteilen und Systemen (izbs), Universität Karlsruhe (TH), Kaiserstrasse 12, 76131 Karlsruhe, Germany
erik.bitzek@izbs.uni-karlsruhe.de

[2] Fraunhofer-Institut für Werkstoffmechanik IWM, Wöhlerstr. 11, 79108 Freiburg, Germany

Summary. The interaction of dislocations with a static mode I crack is studied by large scale molecular dynamics simulations. The model consists of a blunted [001](110) crack in nickel, to which after relaxation at $K < K_{Ic}$ the displacement field of a dislocation is added. The response of the system is monitored during its evolution in the micro-canonical ensemble. The three dimensional nature of the problem requires the simulation of many millions of atoms. The great demands on the computational resources and data storage can only be met by high performance computing platforms and by the development of appropriate simulation methods. The simulations allowed to identify different characteristic processes during the interaction of the impinging dislocation with the crack. In particular, stimulated dislocation emission and cross slip processes are observed to be important for the development of a plastic zone.

1 Introduction

One of the great unknowns in the description of semi-brittle fracture is the origin of the dislocations near the crack tip. Preexisting dislocations are known to influence the low temperature fracture toughness and the brittle-to-ductile-transition (BDT) dramatically (see [1, 2, 3] and references therein). Silicon has been widely used as model material to study the BDT because it can be grown as a nearly defect free single crystal. Experiments on single crystalline silicon show that dislocation nucleation can be a highly inhomogeneous process if preexisting defects are present at the crack front. A single dislocation intersecting the crack front can stimulate the emission of other dislocations in an avalanche-type multiplication process [1]. Dislocation sources created at the crack front by intersecting dislocations are believed to play an important role in the fracture behaviour of all brittle materials. The detailed mechanism of this stimulated emission and multiplication of dislocations at the crack tip are still largely unexplored and the subject of this study.

Until today, no simple atomic interaction potential can realistically reproduce fracture in silicon [4]. Face-centered-cubic (fcc) metals, for which good potentials are available, have the same slip systems as the the diamond-cubic structure of silicon, but usually show ductile failure. However, cleavage fracture in fcc metals is possible if the crack is forced to propagate along special crystallographic planes [5]. In such a set up fcc metals can be used to study generic features of dislocation–crack interaction. The details of the processes, however, can not be transfered to silicon, as the strong tetrahedral bonding of Si determines the fracture behavior as well as the complex dislocation core structures and the high Peierls stress.

In this article, we report on large scale molecular dynamics (MD) simulations of the interaction of dislocations attracted to the crack tip of a stable, blunted crack in nickel.

2 Simulation Method and Analysis

2.1 Model

Following the crystallographic situation in experiments [6], the simulations are performed on γ-oriented cracks which reside on a (110) plane with crack front along the [001] direction. The orientation of the simulation box containing the crack front and of the available slip planes on the Thompson tetrahedron are shown in Fig. 1.

The samples consists of about 38 million atoms, with dimensions of approximately $75 \times 75 \times 75$ nm. Fixed boundary conditions are used for the atoms in the outermost boundary layers, except along the crack front direction, where the motion of the atoms in the boundary layers is restricted to the xy-plane (2D dynamic boundary conditions). All atoms are displaced according to the

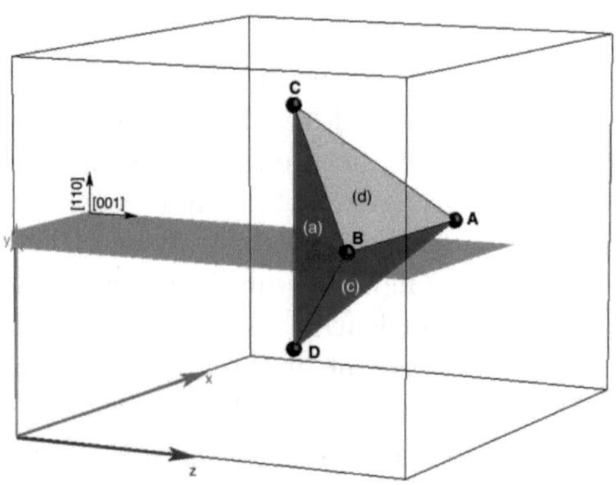

Fig. 1. Orientation of the slip planes with respect to the crack front in the γ orientation. The Thompson tetrahedron notation is applied to identify the Burgers vectors and glide planes

displacement field of a crack in a rigidly clamped thin strip [5]. With these boundary conditions crack propagation in a truly three dimensional set up is possible without enforcing periodicity along the crack front and without the generation of dislocations at the free surfaces.

In this configuration a blunted crack is introduced by removing 1 atomic layer and subsequently relaxing the system at a predetermined strain ϵ_G. The crack can be further loaded or unloaded by appropriately scaling the displacement field obtained from the relaxed crack [5]. The stability regime of the crack can be determined in this way. A Straight dislocation line is then introduced by linear superposition of their predetermined displacement field to that of the crack. A typical starting configuration is shown in Fig. 2.

The critical strain ϵ_G necessary for crack advance is related to the box size (volume $V = x \times y \times z$, area $A = x \times z$) via the Griffith criterion [7] as:

$$\epsilon_G = \sqrt{\frac{4\gamma^{(110)} A}{E^* V}} \sim y^{-\frac{1}{2}} \tag{1}$$

(E^*: Young's modulus in plane stress, $\gamma^{(110)}$: (110) surface energy). The simulation box therefore has to be sufficiently large to reach sensible strain values. The boundary conditions have a significant effect on the values of the surface energy and of the elastic energy. The critical strain can therefore not be directly calculated using eq.(1) and the theoretical values of E^* and $\gamma^{(110)}$, but has to be determined by iterative static calculations on the total simulation box.

Fig. 2. Typical simulation setup, containing a blunted crack in the γ-orientation and a 60° dislocation. Fixed boundary condition are used in all directions except in crack front direction where 2D dynamic boundary conditions are used to allow fracture to take place

The simulations are performed using the well tested embedded atom method (EAM) potential for nickel [8] by Y. Mishin.

2.2 Numerical Methods

The evolution of the system is calculated using standard molecular dynamics with a leap-frog integrator. For details of the implementation and parallelization of the used MD program package IMD see [9, 10, 11, 12, 13]. The implementation of the Embedded Atom Method in IMD is described in [13].

Multiple static calculations have to be performed to determine the stable configurations of the crack and the dislocation, as well as to determine the energy of the system as a function of strain for the calculation of the critical strain. Using standard algorithms for energy minimization like the conjugated gradient method [14] usually require many evaluations of the force and energy calculation routine ('calls to force'). Thus relaxation calculations eat up a large part of the computing time.

The development of a new simple and robust minimization method allowed to significantly reduce the computational cost for structural relaxation of the systems, see Fig. 3. The algorithm FIRE (Fast Inertial Relaxation Engine, patent pending) will be described in detail elsewhere [15].

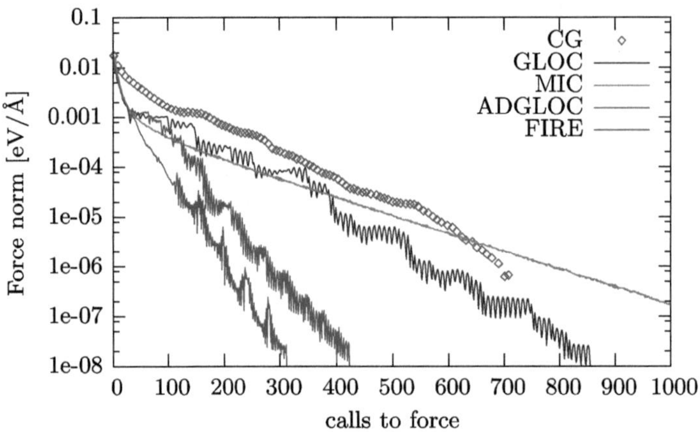

Fig. 3. Relaxation of a crack in a small test system with different algorithms. The newly developed algorithm FIRE reduces the necessary computation time by a factor of about three compared to the formerly used algorithms like conjugated gradient (CG) or GLOC [15]

2.3 Characterization of Crystallographic Defects and Visualization

The analysis and visualization of large scale atomistic simulations containing many millions of atoms is a challenging task because the appropriate tools are often not parallelized but still require the information of the entire sample. The 0K simulations, however, allow to easily identify atoms near defects by their increased potential energy. The thus filtered data can be visualized by the parallel visualization program AtomEye [16]. Crystallographic defects within interesting regions can then be further analyzed using all atoms within these regions.

The determination of the Burgers vector of newly created dislocations is essential to the study of dislocation–crack interactions. For this purpose a tool was developed which allows the extraction of dislocation cores from the large configurations for the calculation of the Nye tensor [17]. With this method it is possible to identify the Burgers vector of dislocations where other methods like the slip vector analysis [18] fail. This kind of topological analysis, however, requires the location of all atoms in the neighborhood of the defect. A significant data reduction of the output generated during the simulation is therefore not feasible.

2.4 Computational Issues

The three dimensional nature of the problem as well as the long range stress field of the crack and the dislocation require the simulation of very large systems. The great demands on the computational resources and data storage can only be met by high performance computing platforms. Most of the calculations were performed on the HP XC6000 cluster in Karlsruhe, smaller test calculations were performed on the 32 CPU cluster of the izbs. Typical jobs on the XC6000 consisted of 24 hour runs on 64 CPUs. With this number of processors a MD step on one processor takes 4.9 μs per atom, 10000 steps (the equivalent of 20 ps simulated time) with the large sample take about 8 h on 64 CPUs. On the izbs cluster the memory requirements of the large samples can only be fulfilled by using all nodes. One MD step per atom takes on a Xeon with 3,06 GHz and Gigabit interconnect 7.9μs, 10 000 MD steps on the izbs cluster require therefore around 26 hours.

The excellent performance of IMD on Itanium processors[1] and its very good scaling behavior up to large CPU numbers has been documented in previous reports [10, 13, 11, 12], and up-to date benchmarks can be found on the homepage of IMD [9].

We therefore want to point the attention to the issue of data storage. The increase in computing power enables the simulation of larger samples which require also more storage space and capacities for analysis. A typical snapshot of our simulation runs requires about 3,5 GB storage space, at typical run thus needs about 175 GB. This data needs to be analyzed (a simple analysis of a run can require up to 10 h), compressed and transfered via Gigabit connection. Together with the generation of a sample (which requires about 1,5 h) the

[1] The performance is, however, very sensitive to the employed compiler version and compiling options, see [9].

time for data processing makes up a significant part of the simulation process. Parallelization of these tasks is not very efficient as the filesystem seems to be the limiting factor and it requires additional time to organize the data for parallel processing. The limit to large simulations on the XC6000 might thus probably not be the available number of processors but the throughput of and the available space on the filesystem.

3 Results

The events during the interaction of an impinging dislocation with the crack of course depend on the character and glide plane of the incoming dislocation and on the applied load. For incoming dislocations on the (a)-plane with Burgers vector DB, three typical processes can be identified, (see also Fig. 4):

1. Change of glide plane of the leading dislocation at the crack front. This mechanism is already active at low loads ($0.85\epsilon_G$).
2. Stimulated nucleation of (partial and full) dislocation loops on the inclined glide planes (c) and/or (d), starting at $0.95\epsilon_G$.
3. Cross slip on (c) of a part of the incoming dislocation which has attained screw character (starting at $0.95\epsilon_G$).

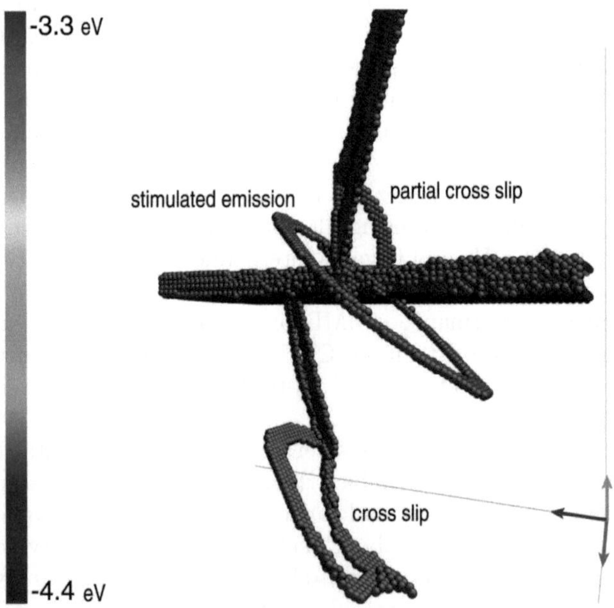

Fig. 4. Snapshot of a dynamic simulation of a 60° dislocation on the (a) plane interacting with a blunted crack at a subcritical load. The typical processes observed in or simulations of dislocation–crack interactions are depicted. Only atoms within the simulation box which have an increased potential energy are shown

The partial cross slip of the leading partial dislocation $D\alpha \rightarrow D\beta + \alpha\beta$ upon contact with the crack front is observed in all studied configurations. The cross slip of the incoming dislocation DB from the (a)-plane to the (c)-plane, however, requires that the dislocation reaches screw orientation within the region of the crack stress field where the Peach-Koehler force on the dislocation is higher on the (c) plane than on the (a)-plane.

It is interesting to note that stimulated emission of dislocations from the crack front does not take place in each case. In addition to the critical loading of the crack, the dislocation nucleation process depends on the velocity of the incoming dislocation. Figure 5 compares two series of snapshots of simulations with the same starting configuration and boundary conditions. In one simulation, however, the velocity of all atoms was set to zero before the dis-

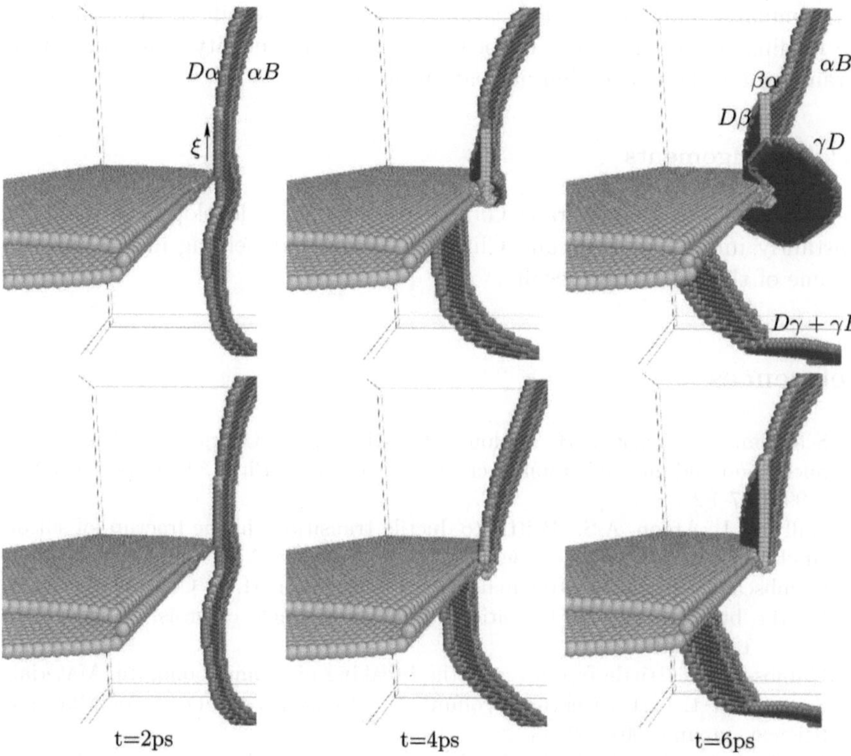

Fig. 5. Time series of the process of stimulated emission. *Upper* and *lower row* show simulations with the same starting configuration and boundary conditions. However, in the *lower row* the velocities of all atoms is set to 0 at $t = 2$ ps. Contrary to the undisturbed simulation (*upper row*), no stimulated dislocation nucleation is observed in this simulation. The defects are identified using the common neighbor analysis [20]. Dislocations are characterized using the Thompson tetrahedron notation (see Fig. 1)

location met the crack front. The impinging velocity v_i of the dislocation is thus reduced ($v_i \approx 8$Å/ps instead of $v_i \approx 22$Å/ps), and no dislocation emission takes place. Similar dynamic effects have also been seen in the interaction of dislocations with localized obstacles [19].

A more detailed description of some of the results can be found in [21], a thorough analysis and discussion of the simulations will be presented elsewhere [22].

4 Conclusions

In this article we have reported on the first three dimensional large scale atomistic simulations of dislocation crack interactions. Typical processes, including the stimulated emission of dislocations, have been identified. In addition to the loading of the crack the dislocation impinging velocity is an important parameter for the dislocation nucleation process.

Acknowledgements

We would like to thank Franz Gähler and the IMD-development team for constantly improving IMD and Christian Brandl for helping in the analysis of some of the simulation results.

References

1. Scandian, C., Azzouzi, H., Maloufi, N., Michot, G., George, A.: Dislocation nucleation and multiplication at crack tips in silicon. Phys. Status Solidi A **171** (1999) 67–82
2. Gally, B.J., Argon, A.S.: Brittle-to-ductile transitions in the fracture of silicon single crystals by dynamic crack arrest. Philos. Mag. A **81** (2001) 699–740
3. Gumbsch, P., Riedle, J., Hartmaier, A., Fischmeister, H.F.: Controlling factors for the brittle-to-ductile transition in tungsten single crystals. Science **282** (1998) 1293–1295
4. Gumbsch, P.: Brittle fracture and the breaking of atomic bond. In: Materials Science for the 21st Century. Volume A. JSMS , The Society of Materials Science, Japan (2001) 50–58
5. Gumbsch, P., Zhou, S., Holian, L.: Molecular dynamics investigation of dynamic crack stability. Phys. Rev. B **55**(6) (1997) 3445–3455
6. Scandian, C.: Conditions d'émission et de multiplication des dislocations à l'extrémité d'une fissure. Application au cas du silicium. PhD thesis, Institut National Polytechnique de Lorraine (2000)
7. Thomson, R.: Physics of fracture. In Ehrenreich, H., Turnbull, D., eds.: Solid State Physics. Volume 39., New York, Academic Press (1986) 1–129
8. Mishin, Y.: Atomistic modeling of the γ and γ'-phases of the Ni-Al system. Acta Metall. **52** (2004) 1451–1467

9. IMD : the ITAP Molecular Dynamics Program. (http://www.itap.physik.uni-stuttgart.de/~imd)
10. Roth, J., et al.: IMD – a massively parallel MD package for classical simulations in condensed matter. In Krause, E., Jäger, W., eds.: High Performance Comput. in Sci. and Eng. '99, Berlin, Springer (2000) 72–81
11. Rudhart, C., Rösch, F., Gähler, F., Roth, J., Trebin, H.R.: Crack propagation in icosahedral model quasicrystals. In Krause, E., Jäger, W., Resch, M., eds.: High Performance Computing in Science and Engineering 2003, Heidelberg, Springer (2004) 107–116
12. Gähler, F., Kohler, C., Roth, J., Trebin, H.R.: Computation of strain distributions in quantum dot nanostructures by means of atomistic simulations. In Krause, E., Jäger, W., eds.: High Performance Computing in Science and Engineering 2002, Heidelberg, Springer (2002) 3–14
13. Bitzek, E., Gähler, F., Hahn, J., Kohler, C., Krdzalic, G., Roth, J., Rudhart, C., Schaaf, G., Stadler, J., Trebin, H.R.: Recent developments in IMD: Interactions for covalent and metallic systems. In Krause, E., Jäger, W., eds.: High Performance Computing in Science and Engineering 2000, Springer , Heidelberg (2001) 37–47
14. Press, W.H., Flannery, B.P., Teukolsky, S.A., Vetterling, W.T.: Numerical Recipes in C. Cambridge University Press, Cambridge (1997)
15. Bitzek, E., Koskinen, P., Gähler, F., Moseler, M., Gumbsch, P.: Fire: Structural relaxation made simple. (to be published)
16. Li, J.: Atomeye: an efficient atomistic configuration viewer. Modelling Simul. Mater. Sci. Eng. **11** (2003) 173–177
17. Hartley, C.S., Mishin, Y.: Characterization and visualization of the lattice misfit associated with dislocation cores. Acta Metall. **53** (2005) 1313–1321
18. Zimmerman, J.A., Kelchner, C.L., Klein, P.A., Hamilton, J.C., Foiles, S.M.: Surface step effects on nanoindentation. Phys. Rev. Lett. **87**(16) (2001) (165507–1)–(165507–4)
19. Bitzek, E., Gumbsch, P.: Dynamics aspects of dislocation motion: atomistic simulations. Mater. Sci. Eng. A **400-401** (2005) 40–44
20. J.D.Honeycutt, H.C.Andersen: Molecular-dynamics study of melting and freezing of small lennard-jones clusters. J. Phys. Chem. **91** (1987) 4950–4963
21. Brandl, C.: Untersuchungen von Versetzungen im Rissspannungsfeld. Diplomarbeit, Universität Karlsruhe (TH) (2006)
22. Bitzek, E., Gumbsch, P.: Atomistic simulation of dislocation–crack interaction. (in preparation)

9. IMD – the ITAP Molecular Dynamics Program, http://www.itap.physik.uni-stuttgart.de/~imd/
10. Roth, J., et al.: IMD – a massively parallel MD package for classical simulations in condensed matter. In: Krause, E., Jäger, W. (eds.) High Performance Computing in Science and Eng. '99. Berlin: Springer (2000) [illegible]

[illegible]

Monte Carlo Simulations of Strongly Correlated and Frustrated Quantum Systems

C. Lavalle[1], S.R. Manmana[1,2], S. Wessel[1], and A. Muramatsu[1]

[1] Institut für Theoretische Physik III, Universität Stuttgart, Pfaffenwaldring 57, D-70550 Stuttgart, Germany
[2] Fachbereich Physik, Philipps Universität Marburg, D-35032 Marburg, Germany

Summary. We study the dynamics of the 1D t-J model with nearest neighbor (n.n) interaction at finite doping using the *hybrid-loop* quantum Monte Carlo. On the basis of the spectral functions of the $1/r^2$ t-J model the excitation content for the one- and two-particle spectral functions for the n.n. t-J model are obtained from the Bethe-Ansatz solution and compared with the Monte Carlo results. We find that the procedure describes with extremely hight accuracy the excitation of the n.n. t-J model. We furthermore use quantum Monte Carlo simulations to analyze the phases of ultra-cold bosonic atom gases in optical lattices, in particular in the presence of frustrated and random interaction strength. We find that such systems display interesting quantum phases, such as supersolid and Bose-glass phases, and discuss possible experimental setups to examine such phenomena.

1 Introduction

Strong electronic correlations have become an active research area of solid state physics in the last decades, due to their relevance for e.g. heavy fermion compounds [1], high-temperature superconductors [2], and quantum magnetism [3]. Furthermore, ultra-cold atomic gases in optical lattices provide a new exciting bridge between the physics of these quantum condensed matter systems and the field of quantum optics [4, 5]. In the project "CorrSys" novel numerical schemes are developed and employed to effectively simulate such systems.

We study low-dimensional highly correlated electron systems on the basis of the n.n. t-J model which is considered to be the effective Hamiltonian for the low energy physics of the copper-oxides materials. In spite of its formal simplicity the model has turned out to be extremely difficult to approach with both analytical and numerical techniques especially concerning its dynamical properties which are directly accessible by spectroscopic experiments. Using a newly developed quantum Monte Carlo (QMC) algorithm, the *hybrid-loop* QMC, we have studied different dynamical observables (spectral function and dynamical spin and charge correlation functions) for large systems at finite doping.

We furthermore employ large-scale quantum Monte Carlo simulations to study the properties of ultra-cold bosonic atom gases in optical lattices. Recent experimental progress made possible the realization of novel quantum phenomena in such systems. In particular, since detailed experimental control of the relevant system parameters can now be archived. Based on the stochastic series expansion technique, we analyze the ground state phase diagram of the effective Bose-Hubbard model describing such a atom gas in the interesting, strongly correlated regime. We in particular examined possible novel phases induced in such systems by a frustrated lattice geometry or randomness in the interatomic interaction strength. We find that on the triangular lattice the presence of frustration in the underlying lattice leads to the emergence of a supersolid state of matter, due to a novel order-by-disorder effect out of a macroscopic degeneracy of the model in the classical limit. Furthermore, we show that the presence of randomness in the interaction strength leads to the formation of a Bose-glass phase of the atoms, and the presence of a tri-critical point in the zero-temperature phase diagram. We also discuss possible experimental realization of these scenarios.

In addition to the work detailed below, our research focuses on the following topics: i) Superconductivity in novel systems like sodium-cobaltates, that can be considered as doped frustrated quantum antiferromagnets on a triangular lattice [6], and ii) nonequilibrium dynamics of strongly correlated systems, where a new variant of density matrix renormalization group (DMRG) is applied to simulate the time evolution of many-body systems far from equilibrium [7].

2 Doped n.n. t-J Model in One-dimension

The n.n. t-J model reads:

$$H_{t-J} = -t \sum_{<i,j>,\sigma} \tilde{c}^{\dagger}_{i,\sigma}\tilde{c}_{j,\sigma} + J \sum_{<i,j>} \left(\mathbf{S}_i \cdot \mathbf{S}_j - \frac{1}{4}\tilde{n}_i\tilde{n}_j \right) . \tag{1}$$

Here, $\tilde{c}^{\dagger}_{i,\sigma}$ are projected fermion operators in the subspace without doubly occupied sites $\tilde{c}^{\dagger}_{i,\sigma} = (1 - c^{\dagger}_{i,-\sigma}c_{i,-\sigma})c^{\dagger}_{i,\sigma}$, $\tilde{n}_i = \sum_{\alpha} \tilde{c}^{\dagger}_{i,\alpha}\tilde{c}_{i,\alpha}$, $\mathbf{S}_i = (1/2) \sum_{\alpha,\beta} c^{\dagger}_{i,\alpha}\boldsymbol{\sigma}_{\alpha,\beta}c_{i,\beta}$, and the sum runs over nearest neighbors only. This model corresponds to the limiting case of the Hubbard model in the limit $U \gg t$. In two dimensions, on a square lattice, this model is believed to capture the physics of high temperature superconductors. Other low-dimensional cuprates forming chains, or n-leg ladders (i.e. n coupled chains) with $n = 2$ and 3, were recently synthesized [8], with the same local chemical units, such that they can also be described by the n.n. t-J model.

Since analytical approaches can deal only with few special cases ($J/t = 2$ [9], $J/t \to 0$ [10]) and do not provide definitive insight, numerical methods constitute then a fundamental tool to study the n.n. t-J model as shown by exact diagonalization [11], or recent advances in density matrix renormalization group (DMRG) [12], and quantum Monte Carlo (QMC) simulations [13]. We apply a newly developed QMC algorithm, the *hybrid-loop* QMC, that is able to deliver accurate results for static and dynamical observables for the n.n. t-J model. Although there are a number of algorithms dealing with the n.n. t-J model, like the ones deriving from the Green Function Monte Carlo [14, 15, 16], ours is the only one able to calculate dynamical correlation functions.

The starting point in order to formulate the *hybrid-loop*-algorithm is a canonical transformation [17, 18] of the model (1) that leads to

$$\tilde{H}_{tJ} = \sum_{<i,j>} t_{ij}\, P_{ij} f_i^\dagger f_j + \frac{1}{2} \sum_{<i,j>} J_{ij} \Delta_{ij} (P_{ij} - 1), \tag{2}$$

where $P_{ij} = (1 + \boldsymbol{\sigma}_i \cdot \boldsymbol{\sigma}_j)/2$, $\Delta_{ij} = (1 - n_i - n_j)$ and $n_i = f_i^\dagger f_i$. In this mapping, one uses the following identities for the standard creation ($c_{i,\sigma}^\dagger$) and annihilation ($c_{i,\sigma}$) operators $c_{i\uparrow}^\dagger = \gamma_{i,+} f_i - \gamma_{i,-} f_i^\dagger$, $c_{i\downarrow}^\dagger = \sigma_{i,-}(f_i + f_i^\dagger)$, where $\gamma_{i,\pm} = (1 \pm \sigma_{i,z})/2$ and $\sigma_{i,\pm} = (\sigma_{i,x} \pm i\sigma_{i,y})/2$. The spinless fermion operators fulfill the canonical anti-commutation relations $\{f_i^\dagger, f_j\} = \delta_{i,j}$, and $\sigma_{i,a}$, $a = x, y$, or z are the Pauli matrices. The constraint to avoid doubly occupied states transforms to the conserved and holonomic constraint $\sum_i \gamma_{i,-} f_i^\dagger f_i = 0$. The Hamiltonian (2) describes free fermions interacting with quantum mechanical spins.

In the case of finite doping, and for $T = 0$, the ground-state is projected out of a trial wave function $|\Psi_T\rangle$ as usually:

$$\frac{\langle \Psi_0 | O | \Psi_0 \rangle}{\langle \Psi_0 | \Psi_0 \rangle} = \lim_{\Theta \to \infty} \frac{\langle \Psi_T | e^{-\Theta \tilde{H}_{tJ}} O e^{-\Theta \tilde{H}_{tJ}} | \Psi_T \rangle}{\langle \Psi_T | e^{-2\Theta \tilde{H}_{tJ}} | \Psi_T \rangle} \tag{3}$$

The above equation is valid provided that $\langle \Psi_T | \Psi_0 \rangle \neq 0$ and O denotes an arbitrary observable. By using a Trotter decomposition

$$\mathrm{e}^{-\Theta \tilde{H}_{tJ}} = \left(\mathrm{e}^{-\Delta\tau \tilde{H}_{tJ}} \right)^{L/2} , \tag{4}$$

with $\Delta\tau = 2\Theta/L$, and introducing a complete set of spin-states at each time-slice, the weight of a configuration is given by

$$W(\boldsymbol{\sigma}) = \langle \psi_T \mid B_L \cdots B_1 \mid \psi_T \rangle \,. \tag{5}$$

The trial wavefunction for a given number of holes in the system is a Slater determinant, since after the canonical transformation, $\tilde{H}_{tJ}$ is bilinear in the fermionic operators. The evolution of the holes from one time-slice to the next is given by

$$B_\ell = \langle \sigma_{\ell+1} \mid e^{-\Delta\tau \tilde{H}_{tJ}} \mid \sigma_\ell \rangle \ , \tag{6}$$

where $\mid \sigma_\ell\rangle$ is a configuration of the quantum spin system at the ℓ-th (imaginary) time-slice. A new configuration is proposed according to loops for the spin-background (loop-algorithm) [19], that are built up in the same way as in the pure Heisenberg model, but the acceptance of the new configuration of the spin-background is determined by eq. (5). Equation (5) is formally the same as in the determinantal algorithm for the Hubbard model using a discrete Hubbard-Stratonovich transformation [20]. The crucial difference is that in our case, we are dealing with quantum mechanical spins instead of an auxiliary classical field. Since the formal structure is the same as in the determinantal method, observables related to the charge degrees of freedom can be calculated in the same way. The name *hybrid-loop*-algorithm comes from the need to merge two different well known QMC algorithms to study the n.n. t-J model: the loop-algorithm for the spin degrees of freedom, and the determinantal algorithm for the fermionic degrees of freedom.

This algorithm has several advantages. For a given spin realization, fermions are evolved exactly in a quantum mechanical way. Due to the determinantal part it is possible to measure static and in particular dynamical observables that are not accessible with other techniques. The loop-algorithm with its global update for the spin lead to short autocorrelation-times and avoids metastability problems compared to algorithms based on local updates. Dynamical data is obtained from the imaginary time Green's function and analytically continued using stochastic analytic continuation [21].

The implementation of this new *hybrid-loop*-algorithm is being presently carried out on the one-dimensional case, where a rich phase diagram with a Luttinger liquid, a superconducting region, a region with a spin-gap, and a phase-separated region is expected [22, 23].

One of the most striking properties of the Luttinger liquid phase is that the low energy elementary excitations do not act like a coherent electron but they split into two independent elementary excitations one of pure spin (*spinon*) and one of pure charge (*holon*), i.e. what is called *spin-charge separation* takes place.

To study the elementary excitations of the 1D n.n t-J model we have studied some dynamical properties like the photoemission ($A^-(k,\omega)$, Fig. 1), the inverse photoemission ($A^+(k,\omega)$, Fig. 1), the dynamical charge ($N(k,\omega)$, Fig. 2) and spin ($S(k,\omega)$, Fig. 3) correlation functions.

We have already with success realized that it is possible to have a better understanding of the n.n. model via a comparison with another type of t-J model in which both the hopping and the interaction term scale like $1/r^2$ and which, thanks to its high symmetry, is at $J/t = 2$ mostly analytically solvable [13]. In the $1/r^2$ t-J model it has been proved that only a small

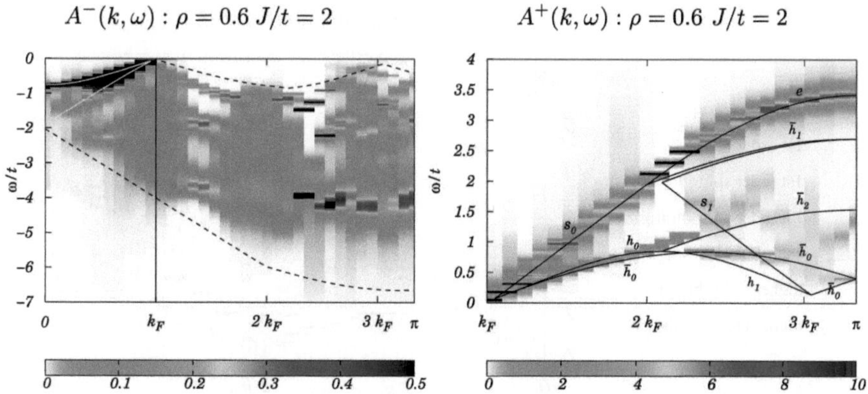

Fig. 1. One-particle spectral function: $A^-(k,\omega)$ photoemission and $A^+(k,\omega)$ inverse photoemission at $J/t = 2$ for $N = 70$ sites and density $\rho = 0.6$ from hybrid-loop algorithm calculation. The Fermi energy is set to zero. *Solid lines*: dispersion curves obtained from Bethe-Ansatz equations. Identification of elementary excitations: *s spinons*, *h holons*, $\bar{h}$ *antiholons* and *e electron*

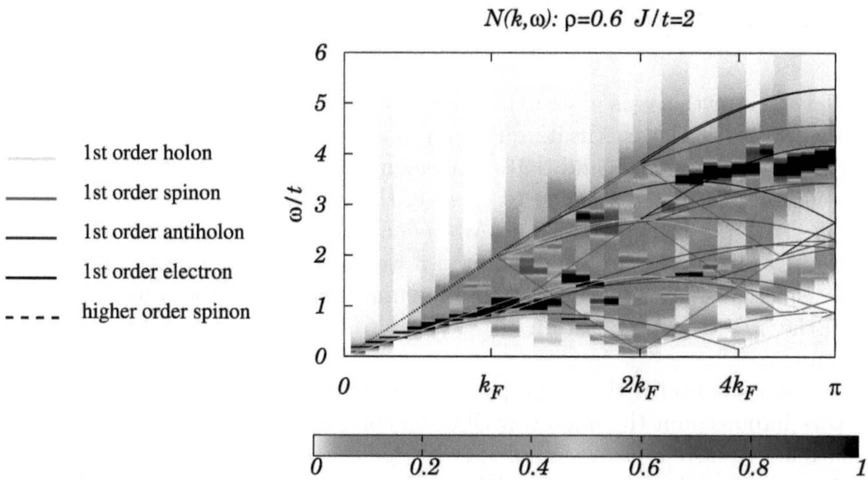

Fig. 2. Two-particle spectral functions: dynamical charge correlation function $N(k,\omega)$ at $J/t = 2$ for $N = 70$ sites and density $\rho = 0.6$ from hybrid-loop algorithm calculation. *Lines*: dispersion curves obtained from Bethe-Ansatz equations for the identification of the elementary excitations

number of elementary excitations contribute to the spectral functions and that *spin-charge separation* takes place at all energies and in the form of three free excitations: a spinon with charge $Q = 0$ and spin $S = 1/2$, a holon with charge $Q = -e$ and spin $S = 0$, and an antiholon with charge $Q = 2e$ and spin $S = 0$. Here e is the charge of the electron. Now we use that the n.n. and the $1/r^2$ t-J model are two different singular limits of a more general integrable

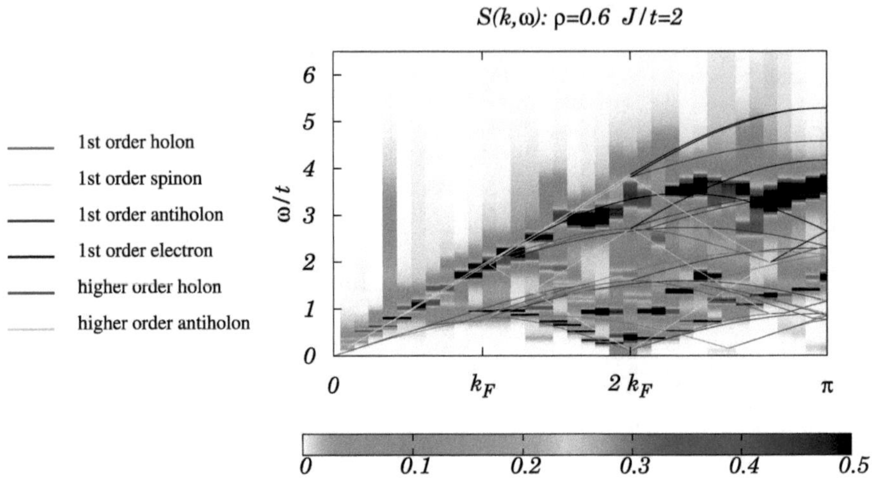

Fig. 3. Two-particle spectral functions: dynamical spin correlation function $S(k,\omega)$ at $J/t = 2$ for $N = 70$ sites and density $\rho = 0.6$ from hybrid-loop algorithm calculation. *Lines*: dispersion curves obtained from Bethe-Ansatz equations for the identification of the elementary excitations

model [24]. For this reason we expect that excited states expressed by the same configuration of Bethe-Ansatz quantum numbers are adiabatically connected between the two models and that the elementary excitations of the $1/r^2$ model provide the main contribution to those of the n.n. model.

We have performed the Bethe-Ansatz calculation for the n.n. model using these ingredients and in Figs. 1, 2 and 3 we show that this procedure makes sense since all major features of the dynamical spectral functions calculated with the *hybrid-loop*-algorithm are well described by the Bethe-Ansatz calculation [25].

For what concerns the single-particle spectral function (Fig. 1) it is enough to take into account the main contribution coming from the Bethe-Ansatz calculation. We have a clear sign of *spin-charge separation* in the photoemission sector $(A^-(k,\omega))$ of the spectral function $(\omega < 0, 0 \le k \le k_F)$. At $k = 0$ we see a single δ peak as expected [26]. The main spectral weight is on the lines coming from the *one-spinon, one-holon* contribution of the $1/r^2$ model. The rest of the $A^-(k,\omega)$ spectrum is incoherent. In the inverse photoemission sector $(A^+(k,\omega))$ of the spectral function $(\omega > 0, k_F \le k \le \pi)$ the compact support for the $1/r^2$ model contains all the spectral weight and we are able to associate to each branch a meaning in terms of all three elementary excitations of the $1/r^2$ model. This means that there are signals of a new type of *spin-charge separation* for strongly correlated systems at finite energy where *holons*, *spinons* and *antiholons* are the elementary excitations.

For what concerns the two-particle spectral functions the main contribution coming from the Bethe-Ansatz calculation still explains the main spectral weight, but to complete the compact support even some higher order contributions must be taken into account.

For the dynamic charge correlation function (Fig. 2) the main contribution comes from the *two-holon, one-antiholon* and the *two-spinon, two-holon, one-antiholon* contributions. To complete the compact support is needed a second order contribution given by *two-spinon* in the small momentum region. It is possible to identify clearly the expected singularity at $2k_F$, at $4k_F$ the expected singularity is less pronounced.

For the spin dynamics (Fig. 3) the main contribution comes from the *two-spinon* and the *two-spinon, two-holon, one-antiholon* contributions. The second order contribution given by *two-holon, one-antiholon* in the small momentum region is needed to complete the compact support. The expected singularity at $2k_F$ is clearly present and there is no weight at $4k_F$ as expected from bosonization.

3 Ultra-cold Atom Gases in Optical Lattices

Since the first realizations of BEC in magnetically trapped dilute alkali vapors [27, 28, 29], the study of ultra-cold atomic gases (of temperatures down to fractions of microkelvins) has become an active research areas of physics. After these first experiments with weakly interacting bosons, among many other achievements the creation of spinor [30] and dipolar [31] condensates has extended the range of observed phenomena. Furthermore, quantum degeneracy was observed in the fermionic case [32], and first steps towards strongly correlated systems have been made [4, 5]. Confining the atomic cloud to an optical lattice formed by interfering laser beams leads to physical situations similar to the one encountered in solid state physics [33]. A gas of bosonic atoms under such conditions is described by the Hamiltonian of the Bose-Hubbard model [34],

$$H = -t \sum_{\langle i,j \rangle} \left(b_i^\dagger b_j + h.c. \right) + \frac{U}{2} \sum_i n_i(n_i - 1) + \sum_i V_i n_i . \tag{7}$$

Here, t denotes the nearest neighbor hopping amplitude, and U an on site repulsion between the bosons. Furthermore, b_i ($b_i^\dagger$) denote annilation (creation) operators for bosons on lattice site i, and $n_i = b_i^\dagger b_i$ the local density. The ratio t/U is tuneable by varying the depth of the optical lattice potential [33], which in particular allows to access the regime $U \gg t$ of strongly correlated bosons. V_i denotes a local potential due to the presence of an (usually harmonic) external trapping potential, which confines the atomic gas. In the uniform case ($V_i = 0$), this model has a superfluid phase for low values of U/t and Mott-insulator regions of commensurate densities for stronger interactions [34]. The transition from a superfluid BEC to a Mott-insulator has been achieved for atoms in both one- and three-dimensional optical lattices upon increasing the optical lattice depth [5, 35]. Quantum Monte Carlo simulations allow for a qualitative analysis of these experiments.

Apart from the QMC studies discussed in the following sections, a determinantal QMC algorithm was implemented to investigate the formation of a Mott-insulator with fermionic atoms [36, 37, 38]. Also novel exact numerical methods, which allow for the study of both equilibrium and non-equilibrium properties of ultra-cold atoms in optical lattices, were developed in our group [39, 40, 41, 42, 43, 44, 45]. Here we found a quasi-condensate emerging in the free expansion of an atomic cloud out of a Fock state, where the coherent matter wave forms an atom laser [41, 44]. On the other hand, the expansion of a quasi-condensate of hard-core bosons leads to a bosonic cloud, where the momentum distribution function presents a Fermi-edge [42].

4 Stochastic Series Expansion Quantum Monte Carlo

We use the stochastic series expansion (SSE) quantum Monte Carlo technique [46, 47], which is based on a high temperature series expansion of the partition function Z of the quantum lattice model in Eq. (1) in the inverse temperature $\beta = 1/k_B T$:

$$Z = \mathrm{Tr} \exp(-\beta H) = \sum_{n=0}^{\infty} \frac{\beta^n}{n!} \sum_{\{i_1,\ldots,i_n\}} \sum_{\{b_1,\ldots,b_n\}} \langle i_1| - H_{b_1} |i_2\rangle \cdots \langle i_n| - H_{b_n} |i_1\rangle . \tag{8}$$

The Hamiltonian H is decomposed into a sum of single-bond terms $H = \sum_b H_b$, and we inserted complete sets of basis states. For a bosonic system with a positive hopping amplitude $t > 0$ all terms contributing to Eq. (8) have a positive weight, and thus a Monte Carlo importance sampling of Z can be performed efficiently. Each Monte Carlo step consists of two consecutively applied update schemes: First, in a local, diagonal update, the expansion order n changes by adding/removing diagonal single-bond terms, while keeping the intermediate states and offdiagonal terms fixed. Then in a second, nonlocal update scheme, the offdiagonal terms and intermediate states are modified using a directed loop update scheme [48, 49], which allows efficient simulations at low temperatures and quantum phase transitions. The results presented below were obtained using an highly optimized C++ implementation of the algorithm based on the ALPS library [50] with native checkpointing and MPI inter-node communication.

5 The Superfluid to Mott-insulator Transition

The presence of a magnetic confinement potential in the experiments on bosonic atoms in optical lattices [5, 35] leads to spatial confinement and an inhomogeneous density distribution of the atoms inside the trap [33]. The local density of the atoms can however not be measured in current experiments. Instead, absorption images are taken during free expansion of the atomic cloud, which reveal the initial momentum distribution $n(\mathbf{k})$ of the atoms. The gradual loss of interference patterns in such images upon increasing the optical lattice depth gave first indications for the passage from a coherent superfluid BEC to the coexistence of large incoherent Mott-insulator and small superfluid regions [5]. Using quantum Monte Carlo simulations, the corresponding changes in the density distribution of the confined Bose gas inside the optical lattice can be analyzed [51, 52, 53]. As an example, in Fig. 4 density distributions are shown for the case of bosons confined to a two-dimensional lattice in (a) the superfluid and (b) the coexistence regime. In the latter case, the strong interactions lead to the formation of a Mott-insulating region with integer density (here $n_i = 1$) at the trap center, surrounded by a superfluid shell. We confirmed the coexistence of superfluid and Mott-insulating regions can be confirmed by analyzing the local compressibility κ in these inhomo-

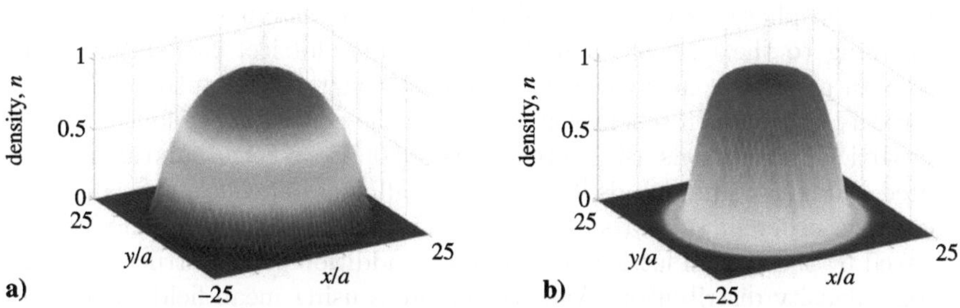

Fig. 4. Local density distribution of two-dimensional confined bosonic atoms, (**a**) in the superfluid phase for $U/t = 6.7$, and (**b**) in the coexistence regime for $U/t = 25$

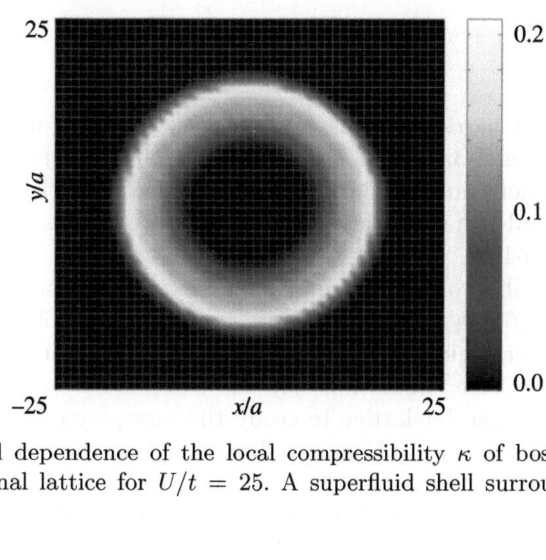

Fig. 5. Spatial dependence of the local compressibility κ of bosons confined to a two-dimensional lattice for $U/t = 25$. A superfluid shell surrounds the central Mott-insulator

geneous systems [51, 52]. As an example, the spatial dependence of κ for the case of Fig. 4 b) is shown in Fig. 5, clearly resolving a compressible superfluid ring surrounding the cental incompressible Mott-insulator. In the following, we consider possible extensions of the experimental setup, by including novel lattice geometries and the effects of disorder in our numerical simulations.

6 Supersolid Lattice Bosons

Recently, evidence was reported for a possible supersolid phase of ^{4}He, derived from a non-classical momenta of inertia in torsional oscillator experiments [54]. Such a state of matter is characterized by the simultaneous presence of both diagonal and off-diagonal long range order in form of a superfluid with periodic density modulations, breaking both U(1) and translational symmetry [55, 56]. Whether the recent observations on ^{4}He are indeed due to the presence of a supersolid state, is still under debate [57, 58, 59], and thus the possibility of supersolid phases in translational invariant systems remains unsettled.

Turning to the case of an underlying regular lattice, various proposals have been presented, how to realize a supersolid by loading ultra-cold atoms in optical lattices: such schemes are based on the generation of longer ranged interparticle interactions using dipolar gases [60], Bose-Fermi mixtures [61] or excited states in higher bands [62]. The crystalline order relevant for diagonal long range order in such a supersolid is not the trivial density modulation enforced by the optical lattice but implies an additional superstructure in the bosonic density distribution. Analytical studies using mean-field theory and renormalization group methods indeed found stable supersolid phases in many models, such as the extended Bose-Hubbard model on the square lattice,

$$H = -t \sum_{\langle i,j \rangle} \left(b_i^\dagger b_j + h.c. \right) + V \sum_{\langle i,j \rangle} n_i n_j + \frac{U}{2} \sum_i n_i (n_i - 1) - \mu \sum_i n_i, \quad (9)$$

in particular in the hard-core limit, $U/V \to \infty$ close to half-filling. Here, V denotes a nearest-neighbor repulsion and μ the chemical potential of the bosons. However, subsequent numerical calculations showed, that the supersolid state is unstable towards phase separation for $U > 4V$, i.e. for dominant on-site interactions [63, 64].

Since it is possible to generate optical lattices which depart from the square lattice geometry [65], the question arises, if stable supersolid phases exist in realistic parameter regimes using different lattice structures. We performed quantum Monte Carlo simulations for the extended Bose-Hubbard model, Eq. (9), on the triangular lattice to study the interplay of supersolidity and geometric frustration [66]. In Fig. 6 we show the phase diagram obtained from our simulations in the hard-core limit.

In addition to the superfluid phase at large values of t/V, the system shows two solid phases for low values of $t/V < 0.2$, with densities $\rho = 1/3$ and

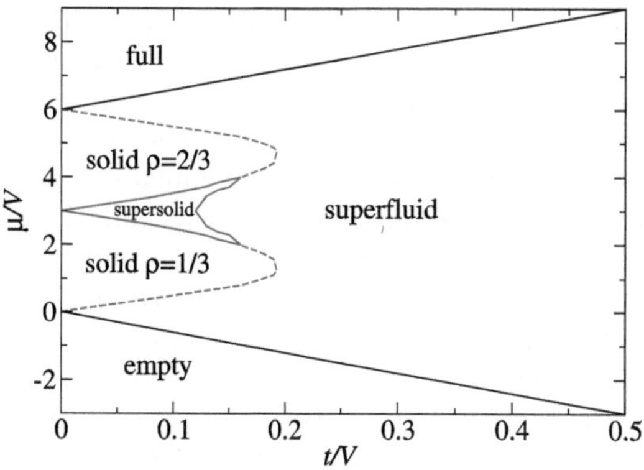

Fig. 6. Ground state phase diagram of hard-core bosons on the triangular lattice, obtained from quantum Monte Carlo simulations. *Solid lines* denote continuous quantum phase transition lines, whereas *dashed lines* denote first-order transitions. The system is half-filled for $\mu/V = 3$

$\rho = 2/3$, respectively. We found that upon doping these solid phases towards half-filling, $\rho = 1/2$, two supersolid phases emerge, with a first order transition line at $\rho = 1/2$, separating the low- and high-density supersolids [66]. Supersolidity in this model emerges by an order-by-disorder effect [67] out of a hugely degenerate state of the frustrated classical model at $t = 0$ [68], driven by quantum fluctuations [66]. Doping the $\rho = 2/3$ solid with additional bosons (or the $\rho = 1/3$ solid with holes), a possible super-solid is unstable towards phase separation due to the proliferation of domain-walls, giving rise to a first-order transitions to the superfluid [66, 64]. Our results are in qualitative agreement with analytical findings [69], which however overestimated the extends of the solid and supersolid phases. While an earlier numerical study [70] did not find a supersolid phase at half-filling, recent studies confirm our calculations [71, 72].

Our preliminary results for the case of hard-core bosons on the Kagomé-lattice, for which a supersolid phase was obtained in spin-wave approximation [69], indicate that the increased quantum fluctuations destroy supersolidity. Compared to the case of the square lattice, the triangular lattice thus offers the experimentally easiest possibility for realizing order-by-disorder phenomena and supersolid phases of ultra-cold atoms on optical lattices.

7 Bosons with Random Interaction Strength

Another means of realizing novel quantum phases of bosons in optical lattices is randomness, produced by e.g. additional incommensurable lattices [73], or by laser speckles [74]. They can lead to Anderson localization [75] and Bose-glass phases [34].

We proposed a novel means of realizing randomness for bosons in optical lattices, by employing the extreme sensitivity of the bosonic scattering potential at the verge of a Feshbach resonance [76, 77], leading to a random interaction strength U in the Bose-Hubbard model [78]. In our scenario, bosons on an atom chip [79] are considered close to an electric wire, producing a spatially random magnetic field [79]. This will induce random variations in the local interaction strength, if the bosons are set near the Feshbach resonance by the overall off-set field [78]. We studied the phase diagram of the one-dimensional random-U Bose-Hubbard model using both a strong coupling expansion (SCE) [80] and SSE quantum Monte Carlo simulations [78], and contrasted our model to the case of randomness in the chemical potential [34]. The resulting zero-temperature phase diagram for a uniformly distributed interaction strength, $U(1-\epsilon) \leq U_i \leq U(1+\epsilon)$, is shown in Fig. 7 for $\epsilon = 0.25$. Similar to the case of a random chemical potential [34], the disordered system exhibits a Bose-glass regime, identified as an insulating, but compressible phase. However, in the random-U case, the disorder selectively destroys all Mott-insulating regions above an ϵ-dependent filling factor ($n \geq 3$ for $\epsilon = 0.25$). Furthermore, we find that the Bose-glass phase does not extend into the low-density region of the phase diagram, $\mu < 0$, giving rise to a tricritical point along the lower boundary of the $n = 1$ Mott-lobe. Estimates of the relevant length scales indicate that our scenario can indeed be realized using currently available experimental techniques [78].

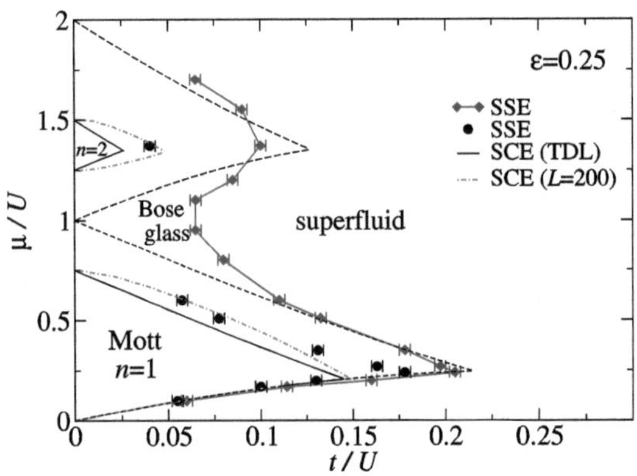

Fig. 7. Zero-temperature phase diagram of bosons on a one-dimensional optical lattice with random interaction strength of $\epsilon = 0.25$, obtained from SSE simulations of 200 sites and a third-order SCE for the thermodynamic limit (TDL). The extent of the Mott-lobes in the pure case ($\epsilon = 0$) from SCE is indicated by the *dashed line*, whereas the *dot-dashed line* shows the SCE results for a finite system of $L = 200$ sites

8 Conclusions

We used a newly developed quantum Monte Carlo (QMC) algorithm, the *hybrid-loop* QMC to study the single-particle and the two-particles (dynamical spin and charge correlation functions) spectral functions of the 1D n.n. t-J model at finite doping. We have shown that the major features and the excitation content of the n.n. t-J model can be well understood via comparison with Bethe-Ansatz solution, where the Bethe-Ansatz equations for the n.n. model are solved using the knowledge of the excitation content of the $1/r^2$ t-J model.

We furthermore used state-of-the-art quantum Monte Carlo methods to analyze prospective novel phases of strongly correlated matter in ultra-cold atom gases. An emergent supersolid state of matter was found on the triangular lattice, which is accessible experimentally. For the future, we plan to extend our analysis by considering finite temperature properties and the effects of longer ranged interactions, which are relevant for Bose-Einstein condensates of dipolar atom gases such as Chromium.

Acknowledgements

We thank M. Arikawa for collaborations. We wish to thank HLRS-Stuttgart (Project CorrSys) and NIC-Jülich for the allocation of computer time.

References

1. P. Fulde, J. Keller and G. Zwicknagl, Solid State Physics - Advances in Research and Applications **41**, 1 (1988).
2. For a recent review see e.g. P.A. Lee, N. Nagaosa, and X.-G. Wen, Rev. Mod Phys **78**, 269 (2006).
3. U. Schollwöck, J. Richter, D. J. J. Farnell and R. F. Bishop (Eds.), *Quantum Magnetism*, Lecture Notes in Physics 645, Springer, Berlin (2004).
4. B. Paredes *et al.*, Nature **429**, 277 (2004).
5. M. Greiner *et al.*, Nature **415**, 39 (2002).
6. A. Foussats, A. Greco, M. Bejas, and A. Muramatsu, Phys. Rev. B **72**, 020504(R) (2005).
7. S.R. Manmana, A. Muramatsu, and R.M. Noack, AIP Conf. Proc. **789**, 269 (2005); *ibid* **816**, 198 (2006).
8. E. Dagotto and T. M. Rice, Science **271**, 618 (1996).
9. P. A. Bares and G. Blatter and M. Ogata, Phys. Rev. B **44**, 130 (1991).
10. K. Penc, K. Hallberg, F. Mila, and H. Shiba, Phys. Rev. Lett. **77**, 1390 (1996).
11. E. Dagotto, Rev. Mod. Phys. **66**, 763 (1994).
12. H. Benthien, F. Gebhard and E. Jeckelmann, Phy. Rev. Lett. **92**, 256401 (2004).
13. C. Lavalle, M. Arikawa, S. Capponi, F. Assaad, and A. Muramatsu, Phys. Rev. Lett. **90**, 216401 (2003).
14. H. J. M. van Bemmel, *et al.*, Phys. Rev. Lett. **72**, 2442 (1994).
15. D. F. B. ten Haaf, H. J. M. van Bemmel, J. M. J. van Leeuwen, W. van Saarloos and D. M. Ceperley Phys. Rev. B **51**, 13039 (1995).
16. S. Sorella and L. Capriotti, Phys. Rev. B. **61**, 2599 (1999).
17. G. Khaliullin, JETP Lett. **52**, 389 (1990).
18. A. Angelucci, Phys. Rev. B **51**, 11580 (1995).
19. H. G. Evertz, Adv. Phys. **52**, 1 (2003).

20. A. Muramatsu, in *Quantum Monte Carlo Methods in Physics and Chemistry*, edited by M. Nightingale and C. Umrigar (NATO Science Series, Kluwer, Dordrecht, 1999).
21. K. S. D. Beach, cond-mat/0403055.
22. M. Ogata, *et al.*, Phys. Rev. Lett **66**, 2388 (1991).
23. M. Nakamura, K. Nomura and A. Kitazawa, Phys. Rev. Lett. **79**, 3214 (1997).
24. V. Inozemtsev, J. Stat. Phys. **59**, 1143 (1990).
25. C. Lavalle, M. Arikawa and A. Muramatsu, in preparation.
26. S. Sorella and A. Parola, Phys. Rev. B **57**, 6444 (1998).
27. M. H. Anderson *et al.*, Science **269**, 198 (1995).
28. C. C. Bardley *et al.*, Phys. Rev. Lett. **75**, 1687 (1995).
29. K. B. Davis *et al.*, Phys. Rev. Lett. **75**, 3969 (1995).
30. J. Stenger *et al.*, Nature **396**, 345 (1999).
31. A. Griesmaier *et al.*, Phys. Rev. Lett. **94**, 160401 (2005).
32. B. DeMarco and D. S. Jin, Science **285**, 1703 (1999).
33. J. Jaksch *et al.*, Phys. Rev. Lett. **81**, 3108 (1998).
34. M. P. A. Fisher *et al.*, Phys. Rev. B **40**, 546 (1989).
35. T. Stöferle *et al.*, Phys. Rev. Lett. **91** 130403 (2004).
36. M. Rigol A. Muramatsu, G.G. Batrouni, and R.T. Scalettar, Phys. Rev. Lett. **91**, 130403 (2003).
37. M. Rigol, A. Muramatsu, Phys. Rev. A **69**, 053612 (2004).
38. M. Rigol and A. Muramatsu, Opt. Commun. **243**, 33 (2004).
39. M. Rigol and A. Muramatsu, Phys. Rev. A **70**, 031603(R) (2004).
40. M. Rigol and A. Muramatsu, Phys. Rev. A **70**, 043627 (2004).
41. M. Rigol and A. Muramatsu, Phys. Rev. Lett. **93**, 230404 (2004).
42. M. Rigol and A. Muramatsu, Phys. Rev. Lett. **94**, 240403 (2005).
43. M. Rigol and A. Muramatsu, Phys. Rev. A **72**, 013604 (2005).
44. M. Rigol and A. Muramatsu, Mod. Phys. Lett. B bf 19, 861 (2005).
45. M. Rigol *et al.*, Phys. Rev. Lett. **95**, 218901 (2005).
46. A. W. Sandvik and J. Kurkijärvi, Phys. Rev. B **43**, 5950 (1991).
47. A. W. Sandvik, Phys. Rev. B **59**, R14157 (1999).
48. O. F. Syljuåsen and A. W. Sandvik, Phys. Rev. E **66**, 046701 (2002).
49. F. Alet, S. Wessel, and M. Troyer, Phys. Rev. E **71**, 036706 (2005).
50. F. Alet *et al.*, J. Phys. Soc. Jpn. Suppl. **74**, 30 (2005); source codes available at `http://alps.comp-phys.org/`.
51. S. Wessel, *et al.*, Adv. Solid State Phys. **44**, 265 (2004).
52. S. Wessel, *et al.*, Phys. Rev. A **70**, 053615 (2004).
53. S. Wessel, *et al.*, J. Phys. Soc. Jpn. Suppl. **74**, 10 (2005).
54. E. Kim and M. H. W. Chan, Nature **427**, 225 (2004); Science **305**, 1941 (2004).
55. O. Penrose and L. Onsager, Phys. Rev. **104**, 576 (1956).
56. A. J. Leggett, Phys. Rev. Lett. **25**, 1543 (1970).
57. A. Leggett, Science **305**, 1921 (2004).
58. N. Prokof'ev and B. Svistunov, Phys. Rev. Lett. **94**, 155302 (2005).
59. E. Burovski *et al.*, Phys. Rev. Lett. **94**, 165301 (2005).
60. K. Góral, L. Santos and M. Lewenstein, Phys. Rev. Lett. **88**, 170406 (2002).
61. H. P. Büchler and G. Blatter, Phys. Rev. Lett. **91**, 130404 (2003).
62. V. W. Scarola and S. Das Sarma, Phys. Rev. Lett. **95**, 03303 (2005).
63. G. G. Batrouni and R. T. Scalettar, Phys. Rev. Lett. **84**, 1599 (2000).
64. P. Sengupta *et al.*, Phys. Rev. Lett. **94**, 207202 (2005).

65. L. Santos *et al.*, Phys. Rev. Lett. **93** 030601 (2004).
66. S. Wessel and M. Troyer, Phys. Rev. Lett. **95**, 127205 (2005).
67. J. Villain *et al.*, J. Phys. **41**, 1263 (1980).
68. G. H. Wannier, Phys. Rev. **79**, 357 (1950).
69. G. Murthy, D. Arovas and A. Auerbach, Phys. Rev. B **55**, 3104 (1997).
70. M. Boninsegni, J. Low Temp. Phys. **132**, 39 (2003).
71. D. Heidarian and K. Damle, Phys. Rev. Lett. **95**, 127206 (2005).
72. R. Melko *et al.*, Phys. Rev. Lett. **95**, 127207 (2005).
73. R. B. Dimer *et al.*, Phys. Rev. A **64**, 033416 (2001).
74. P. Horak, J.-Y. Courtois and G. Grynberg, Phys. Rev. A **58**, 3953 (2000).
75. P. W. Anderson, Phys. Rev. B **109**, 5 (1958).
76. E. Tiesinga *et al.*, Phys. Rev. A **47**, 4114 (1993).
77. S. Inouye *et al.*, Nature **392**, 151 (1998).
78. H. Gimperlein, S. Wessel, J. Schmiedmayer, and L. Santos, Phys. Rev. Lett. **95**, 170401 (2005).
79. S. Wildermuth *et al.*, Nature **435**, 440 (2005).
80. J. K. Freericks and H. Monien, Phys. Rev. B **53**, 2691 (1996).

Chemistry

Christoph van Wüllen

Institut für Chemie, Sekr. C3, Technische Universität Berlin,
Straße des 17. Juni 135, D-10623 Berlin, Germany

The ever increasing power of computational resources allows that computational modeling in chemistry is applied to more and more complex systems. Problems of nearly arbitrary complexity can arise if phenomena at (non-ideal) surfaces or in large biomolecules are modeled. In these fields we witness applications that would not be possible without the availability of powerful supercomputers.

Surfaces are of particular interest because much of "real life" chemistry actually happens at surfaces. This includes processes that one wants (like heterogeneous catalysis) and those which one does not want so much (like corrosion). Ideal surfaces are those with complete regularity. They are certainly quite aesthetic but also boring, and have not much to do with real chemistry which usually takes place at defects like steps, pits or kinks. This complicates the theoretical modeling considerably but it is also an experimental problem, since one has to find and investigate those defects. One of the most widely used experimental technique to do so is scanning tunneling microscopy (STM), but the question arises what one actually sees when one looks at the pictures generated by these devices. This problem is addressed in a contribution by Kovacik, Meyer and Marx, who model the surface free-energy differences of different defects on zinc oxide and compute the STM images these defects would produce. To model the STM measurment requires to include the measurement apparatus (in this case, the STM tip) in the computational model. These calculations then allow for a direct comparison of computed and measured data and can help experimentalists to analyze the STM images they obtain.

Everyday experience tells us that real life gets more complicated if things happen, and this is also the case on surfaces. Atoms and defects move around, and it is by now means clear how exactly an atom moves from A to B, and how much activation energy is required to do so. These questions stimulated the investigation by Schmickler, Pötting, and Jacob. Their computations simulate the diffusion of gold atoms on a gold surface, which is an important model problem. It turns out that diffusion of gold atoms on an ideal surface is quite different from diffusion in the vicinity of defects. For example, a step in the surface serves as a handrail to the moving atom along which it can wander much more easily.

What is the essence of life? The most basic function of a cell is to have a border which separates inside from outside. But is not enough, cells need to be smarter than soap bubbles. A cell can live because there are selective channels through the border (cell membrane) through which only specific molecules can pass, channels that can be opened and closed according to the needs of the cell. One class of channels for small, uncharged molecules such as water are the aquaporins. Biological functions of this kind usually do *not* involve breaking or making of covalent chemical bonds. Instead, hydrogen bonding and van der Waals forces determine structure and dynamics of such systems. This is the reason why empirical force field methods are much used in biophysical investigations. Dynowski and Ludewig, in their report, describe the application of such force field methods to model water transport through aquaporin channels at the atomic level. Typical questions addressed by these simulations are why molecules of one sort are accepted (transported) by the channel while others cannot pass.

Progress in scientific computing has many fathers. The theory has to be improved and refined, and the resulting formulas have to be implemented to form a highly efficient computer code. These efforts would not be made, and these developments simply would not take place, if supercomputers were not available on which, at the end of the day, the codes run and produce results. The following three reports give a feeling what can be done today to model our world at the atomic level.

Characterization of Catalyst Surfaces by STM Image Calculations

Roman Kovacik, Bernd Meyer, and Dominik Marx

Lehrstuhl für Theoretische Chemie, Ruhr–Universität Bochum, D-44780 Bochum, Germany
roman.kovacik@theochem.ruhr-uni-bochum.de

Summary. ZnO and Cu/ZnO are important industrial catalysts for many hydrogenation reactions, for example, the methanol synthesis from synthesis gas. The identification of the dominant surface structures and surface defects under reaction conditions is essential for a microscopic understanding of the activity of a catalyst.

Using density functional theory (DFT) we have calculated the formation energy for the most important atomic defects on the ZnO(10$\bar{1}$0) surface as a function of the redox properties of a surrounding gas phase. To give guidelines on how these defects may appear in a scanning tunneling microscopy (STM) experiment, STM images have been calculated using our recent implementation of Bardeen's tunneling formula into the Car–Parrinello Molecular Dynamics (CPMD) code. We find significant differences in the tunneling properties between the ideal surface and O, Zn, and ZnO vacancies, which may allow to identify these defects in STM measurements.

As a first step to study the morphological changes of the Cu particles in the binary Cu/ZnO catalyst when exposed to an oxidizing environment we have studied the stability of various copper oxide surfaces by combining DFT calculations with a thermodynamic formalism. STM images of the most stable surface structures were calculated to be compared with experimental results.

1 Introduction

The methanol synthesis on ZnO and Cu/ZnO based catalysts has become an intensively studied process over the last years due to its importance in chemical industry and the possible usage of methanol as a fuel. In order to understand the catalytic reaction mechanisms on a microscopic level, the properties of atomic surface defects as active sites and the interplay between copper, the zinc oxide surfaces and the surrounding gas phase are some of the important issues which have to be addressed.

The catalytic activity of pure oxide surfaces is usually attributed to the presence of atomic defects as active sites, particularly to F–centers. Therefore, as a first step, the nature of the dominating atomic defects under realistic reaction conditions together with the adsorbate structures involved in the catalytic process have to be identified.

Using the DFT–based Car–Parrinello Molecular Dynamics (CPMD) program package [1, 2] we have calculated the relaxed atomic structure and energy of ideal, defective and adsorbate-covered ZnO surfaces. In order to determine which surface defects and surface structures can be expected at certain experimental conditions, we introduce appropriate chemical potentials and apply a thermodynamic formalism to extend our DFT results to relevant temperatures, pressures and different compositions of a gas phase environment [3, 4, 5].

Scanning tunneling microscopy (STM) as a real space method is an ideal tool for identifying surface defects in experiment. However, the interpretation of STM experiments in terms of atomic structure and chemical identity is often difficult since STM does not provide a direct image of the underlying atomic structure, but the tunneling current depends on the combined electronic structure of the tip and the substrate. In many cases it is not possible to connect measured STM images with structural concepts without additional assumptions – on the contrary, it may happen that wrong interpretations are made based on observed symmetries or structure elements in the STM images. Therefore, the simulation of STM data using known atomic structures is an essential feedback for experiments and indispensable for a correct interpretation of characteristic features in the STM images.

Different methods have been developed to calculate the tunneling current in STM experiments from the electronic structure of the substrate and the tip as a function of the tip position [6]. Theoretical methods treating the surface and the tip as one interacting system and describing the electron tunneling as a true electron transport process are computationally very expensive and today still not feasible on the DFT level. Therefore, certain approximations have to be made. The two most commonly used methods are Bardeen's perturbation approach [7] and the Tersoff–Hamann approximation [8] – both computationally cheap enough to allow the study of large surface unit cells, but still providing a good qualitative or even quantitative picture.

The atomic structure of the tips used in STM experiments is rarely known, and there are not many reports about it. However, the experimental and theoretical investigation of tip models is important to understand their influence on the STM data. Variations in the crystallographic orientation of the tip or adsorbed atoms at the apex may yield substantially different results. Recent studies using field ion microscopy [9, 10, 11] give some hints about realistic tip structures which are present in experiments. Relying on this information, we decided to build systematically a library of various tip models and study their influence on the appearance of the STM images.

For the binary Cu/ZnO catalysts, several experimental investigations have given evidence that the ZnO substrate and the surrounding gas phase have a strong influence on the properties and the morphology of the supported Cu particles ("strong metal–support interaction effect") [12]. For example, Topsøe and coworkers [13] showed with in–situ transmission electron microscopy (TEM) measurements that the geometric form of ZnO supported Cu particles reversibly changes depending on the redox potential of the gas phase environment. While the Cu clusters show a rather spherical form under oxidizing conditions, a wetting of the ZnO substrate is observed when the environment is switched to be reducing. In–situ EXAFS measurements [14] furthermore showed that the local chemical environment of the Cu atoms changes at the same time, what is considered to be a consequence of reduced ZnO_x species being incorporated into the Cu particles.

Changes in the chemical composition of Cu particles deposited on ZnO surfaces were also observed by Dulub et al. [15] in a combined STM/LEED study. In an oxidizing environment the formation of a $(\sqrt{3} \times \sqrt{3})\,R30^\circ$ superstructure on top of the Cu islands was found which was attributed to the incorporation of O atoms and the start of Cu_2O formation [16]. As a first step to attempt to understand the formation of such structures, the relative stability of pure copper oxide surfaces with different structures and compositions in thermodynamic equilibrium with a surrounding gas phase was studied by combining DFT calculations with a thermodynamic formalism. These investigations are currently extended to the full copper/zinc oxide interface.

2 Computational Methods

2.1 *Ab-initio* Calculations

The atomic relaxations of the different surface structures were done using the DFT–based CPMD program package [1, 2]. All surfaces were represented by periodically repeated slabs, and the tips were modeled by pyramids supported on a slab. An appropriate vacuum region between the the slabs was introduced to avoid self–interactions of the slabs with its repeated images.

The slab structures were constructed by using an even number of atomic layers. The atoms in the bottom half of slabs were kept fixed at the bulk positions to mimic the underlying bulk structure. The coordinates of atoms in the top half of the slabs were relaxed until the atomic forces were smaller than $4 \cdot 10^{-4}$ Ha/bohr, and the residuum of the wave function was converged to be smaller than $1 \cdot 10^{-6}$. Due to the large supercells employed in the calculations only the Γ–point was used for the Brillouin–zone integrations. The exchange–correlation part of Hamiltonian was treated in the generalized-gradient approximation using the functional of Perdew, Burke and Ernzerhof (PBE) [17].

Vanderbilt ultrasoft pseudopotentials [18] and a 25 Ry plane wave cut-off energy were used for the relaxation of the atomic structures. For the generation of the orbitals for the STM calculations the ultrasoft potentials were replaced by Gödecker normconserving pseudopotentials [19] and a large cut-off energy of 120 Ry was used to obtain smooth wave function tails far from the surface.

2.2 Thermodynamics

In order to determine the stability of a particular surface as a function of environmental properties, a thermodynamic formalism was used to extend the zero temperature and zero pressure results from the DFT calculations to realistic environmental conditions [3, 4, 5]. This method has been applied successfully in many previous studies (see Refs. [4, 5] and references therein). The surface free energy γ of a surface in thermal equilibrium with particle reservoirs at temperature T and pressure p is given by

$$\gamma(T,p) = \frac{1}{A}\left(G(T,p,N_i) - \sum_i N_i\,\mu_i(T,p)\right) \quad , \tag{1}$$

where $G\,(T,p,N_i)$ is the Gibbs free energy of the system with the studied surface, A is the surface area, and μ_i, N_i are the chemical potential and the number of particles of the present species, respectively.

The most stable surface termination for a given set of chemical potentials is the one with the lowest free energy. The chemical potentials can be related to temperature and partial pressures conditions via experimental thermochemical data, or, in the case of simple gas phases, via the ideal gas equation.

Usually we compare the surface free energy of a particular surface with certain atoms removed or added with the ideal surface termination. The relative stability of the two systems is given by the difference of their respective surface free energies

$$\Delta\gamma(T,p) = \frac{1}{A}\left(G_{\text{slab}}(T,p,\Delta N_i) - G_{\text{slab}}^{\text{ideal}}(T,p) - \sum_i \Delta N_i\,\mu_i(T,p)\right) \quad , \tag{2}$$

where G_{slab} and $G_{\text{slab}}^{\text{ideal}}$ are the energy of the modified and the ideal reference surface, respectively, and ΔN_i are the differences in atomic numbers between the two surfaces for the present species.

In principle, the Gibbs free energy of the ideal and the defective systems would have to be calculated including the entropy and volume terms. However, since we are only interested in relative stabilities, only the difference terms enter Eq. (2). It has been shown in Ref. [4] that for similar systems as we study, the entropy and volume contributions cancel to a large extent so that we can neglect them and simply replace the Gibbs free energies by the total energies from our DFT calculation.

If we assume that the surfaces are in thermodynamic equilibrium with the bulk, the chemical potentials of the species are related via

$$\sum_i x_i\,\mu_i = E^{\text{bulk}} \quad , \tag{3}$$

where x_i are the stoichiometric coefficients of the compound and E^{bulk} is the energy of one stoichiometric compound bulk unit cell. This reduces the number of independent variables in Eq. (2). In addition, we have to set appropriate ranges for the chemical potentials. The chemical potential for each species has to be lower than the total energy of its most stable elemental phase. Otherwise the Gibbs free surface energy could be lowered by simply forming precipitates of the elements on the surfaces. In our case, the chemical potential of the metals M have to be lower than the energy of a metal bulk unit cell $E_{\mathrm{M}}^{\mathrm{bulk}}$ and the O chemical potential has to be lower than half the energy of an oxygen molecule $E_{\mathrm{EO}_2}^{\mathrm{mol}}$

$$\mu_{\mathrm{M}} \leq E_{\mathrm{M}}^{\mathrm{bulk}} \quad , \qquad \mu_{\mathrm{O}} \leq \frac{1}{2} E_{\mathrm{O}_2}^{\mathrm{mol}} \quad . \tag{4}$$

2.3 Bardeen's Perturbation Approach (BPA)

In Ref. [7] Bardeen showed that if a STM tip is placed far enough above a surface so that direct interactions between the tip and the surface can be neglected, the tunneling current I between the surface and the tip is given in first order perturbation theory by

$$I = \frac{2\pi e}{\hbar} \sum_{\mu,\nu} \left(f\left(E_\mu - E_F\right) - f\left(E_\nu - E_F\right) \right) \left| M_{\mu\nu} \right|^2 \delta\left(E_\mu + eV - E_\nu\right) \tag{5}$$

$$M_{\mu\nu} = \frac{\hbar^2}{2m} \int d\mathbf{S} \cdot \left(\psi_\nu^* \nabla \psi_\mu - \psi_\mu \nabla \psi_\nu^* \right) \quad , \tag{6}$$

where V is the bias voltage between surface and tip, $M_{\mu\nu}$ is the tunneling matrix element of the current density operator between a surface wave function ψ_μ and a tip wave function ψ_ν, E_F is the Fermi energy, f is the Fermi distribution function giving the occupation of the states μ and ν, and S is a plane in the vacuum region separating surface and tip. The Fermi energy E_F is chosen to be aligned for the surface and tip when $V = 0$.

In combination with DFT calculations the BPA allows a parameter–free determination of STM data and has been shown to give very reliable and realistic results in comparison with experiment [6]. However, the BPA is computationally very demanding.

2.4 Tersoff–Hamann Approximation (THA)

To derive a more efficient computational scheme for the calculation of STM data, Tersoff and Hamann proposed a very simple approximation for evaluating Bardeens tunneling formula [8]. The tip is approximated by a sphere with fixed radius and distance $\mathbf{r}_0$ between the center of the sphere and the substrate. The electronic structure of this spherical tip is represented by a single wave function which is assumed to be a damped spherical s–wave. Other details of electronic structure of the tip are neglected. The tunneling current then turns out to be proportional to the electron density of the substrate at the position of the sphere center, where only substrate wave functions with an energy close to the Fermi energy are contributing

$$I \propto \sum_{\mu} |\psi_{\mu}(\mathbf{r}_0)|^2 \delta(E_{\mu} - E_F) \quad . \tag{7}$$

Despite these rough approximations and their simplicity, the THA method often provides the correct interpretation of STM images of simple surfaces and adsorbate structures. Especially, it makes transparent that STM experiments do not image the atomic structure directly, but rather the electronic structure of a surface.

3 Results

3.1 Tungsten Tip Models

Two models of a tungsten tip have been constructed to test the dependence of the STM images on the specific tip geometry. Tungsten forms a bcc lattice with an experimental lattice constant of 3.16 Å which was used for our models. The first tip model consists of a slab with three (100) planes (3×3 surface unit cells) and a 5 atom pyramid with a square base. The second model consists of a slab with two (110) planes (4×3 surface unit cells) and a 13 atom pyramid with a rhombohedral base. Only the atoms of the pyramids were allowed to relax in the geometry optimization, while the slab atoms were kept at their bulk positions. For the relaxed structures we measured the opening angles between the pyramid edges and planes with the apex atom as a center to determine the sharpness of the tips. The sharper tip is expected to provide better resolved STM images. For the (100) model we find an edge angle of 113° and a planar angle of 96°, whereas the (110) model gives edge angles of 92° and 81° in the x and y direction, respectively, and a planar angle of 71°.

3.2 Defects on the ZnO(10$\bar{1}$0) Surface

We turn now to the characterization and identification of atomic defects on the nonpolar ZnO(10$\bar{1}$0) surface. The most simple atomic defects on this surface are (i) oxygen vacancies (O–v), (ii) zinc vacancies (Zn–v), and (iii) missing zinc–oxygen dimers (ZnO–v). If we assume that the surface is in thermodynamic equilibrium with the bulk, the chemical potentials μ_{Zn} and μ_{O} of Zn and O are related via Eq. (3)

$$\mu_{\mathrm{Zn}} + \mu_{\mathrm{O}} = E_{\mathrm{ZnO}}^{\mathrm{bulk}} \quad , \tag{8}$$

where $E_{\mathrm{ZnO}}^{\mathrm{bulk}}$ is the energy of one ZnO bulk unit cell. The upper limits for the chemical potentials according to Eq. (4) are given by

$$\mu_{\mathrm{Zn}} \leq E_{\mathrm{Zn}}^{\mathrm{bulk}} \quad , \qquad \mu_{\mathrm{O}} \leq \tfrac{1}{2} E_{\mathrm{O}_2}^{\mathrm{mol}} \quad . \tag{9}$$

Eq. (8) allows us to eliminate μ_{Zn}, which simultaneously introduces a lower bound for μ_{O}. If we define the formation energy $E^{\text{form}}_{\text{ZnO}}$ of bulk ZnO from metallic Zn and molecular oxygen

$$E^{\text{form}}_{\text{ZnO}} = E^{\text{bulk}}_{\text{ZnO}} - E^{\text{bulk}}_{\text{Zn}} - \tfrac{1}{2}E^{\text{mol}}_{\text{O}_2} \quad , \tag{10}$$

and if we use the upper bound for μ_{O} as a new zero point of energy

$$\Delta\mu_{\text{O}} = \mu_{\text{O}} - \tfrac{1}{2}E^{\text{mol}}_{\text{O}_2} \quad , \tag{11}$$

we can rewrite the allowed range for $\Delta\mu_{\text{O}}$ as

$$E^{\text{form}}_{\text{ZnO}} \leq \Delta\mu_{\text{O}} \leq 0 \quad . \tag{12}$$

The formation energies of the different defects can now be written as simple functions of the calculated DFT energies and one free parameter $\Delta\mu_{\text{O}}$ which describes the dependence of the defect formation energies on the environmental conditions:

$$\begin{aligned} E^{\text{vac}}_{\text{O}} &= E^{\text{O-v}}_{\text{slab}} + \tfrac{1}{2}E^{\text{mol}}_{\text{O}_2} - E^{\text{ideal}}_{\text{slab}} + \Delta\mu_{\text{O}} \\ E^{\text{vac}}_{\text{Zn}} &= E^{\text{Zn-v}}_{\text{slab}} + E^{\text{bulk}}_{\text{ZnO}} - \tfrac{1}{2}E^{\text{mol}}_{\text{O}_2} - E^{\text{ideal}}_{\text{slab}} - \Delta\mu_{\text{O}} \\ E^{\text{vac}}_{\text{ZnO}} &= E^{\text{ZnO-v}}_{\text{slab}} + E^{\text{bulk}}_{\text{ZnO}} - E^{\text{ideal}}_{\text{slab}} \end{aligned} \tag{13}$$

where $E^{\text{X-v}}_{\text{slab}}$ is the total energy of a slab calculation with a X vacancy, and $E^{\text{ideal}}_{\text{slab}}$ is the reference energy of the ideal, defect–free surface.

To express the formation energies of the defects as a function of the temperature T and pressure p_{O_2} of a surrounding oxygen gas phase environment we used the standard formula for the chemical potential of the ideal gas

$$\Delta\mu_{\text{O}}(T, p_{\text{O}_2}) = \tfrac{1}{2}\left(\tilde{\mu}_{\text{O}_2}(T, p^\circ) + kT\ln\left(p_{\text{O}_2}/p^\circ\right)\right) \quad . \tag{14}$$

$\tilde{\mu}_{\text{O}_2}(T, p^\circ)$ is obtained from tabulated values of standard enthalpies and entropies as a function of the temperature T at the standard pressure $p^\circ = 1$ atm.

ZnO crystallizes in the hexagonal wurtzite structure. The cell parameters are calculated to be $a = 3.29$ Å and $c/a = 1.615$ in our DFT–PBE setup. The surface structures were modeled by periodically repeated slabs using an orthorhombic box with the (x, y, z) axes aligned to the crystallographic $[1\bar{2}10]$, $[0001]$, and $[10\bar{1}0]$ directions, respectively (see Fig. 1). To converge the formation energies with respect to the supercell size, the defect structures were relaxed for different slab thicknesses of 4, 6, 8 and 10 atomic layers and for lateral extensions of (3×2), (4×2), and (5×3) surface unit cells.

The results for the defect formation energies are listed in Table 1. The formation energies are found to be converged with respect to the supercell size within the error of our DFT method for a slab thickness of 8 atomic

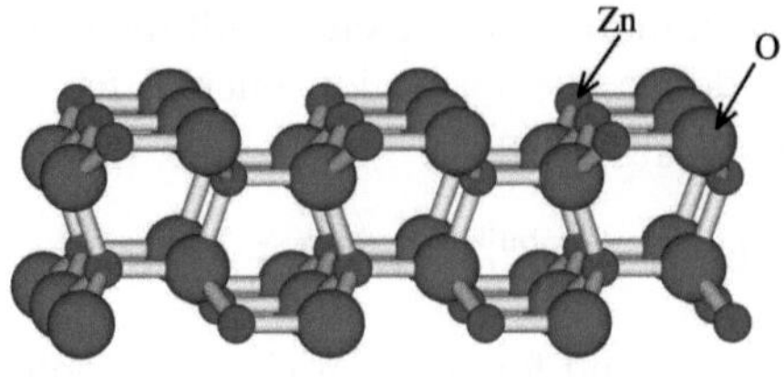

Fig. 1. Atomic structure of the nonpolar $ZnO(10\bar{1}0)$ surface

Table 1. Formation energy (in eV) of simple atomic defects (O, Zn and ZnO dimer vacancy) as a function slab thickness N_L and size of the surface unit cell

	N_L	N_{at}	E_O^{vac}	E_{Zn}^{vac}	E_{ZnO}^{vac}
3×2	4	48	2.89	–	1.52
	6	72	2.97	0.52	1.16
	8	96	2.86	0.46	0.94
	10	120	3.01	–	1.04
4×2	6	96	3.05	0.50	1.07
	8	128	3.05	0.53	0.97
	10	160	3.05	0.47	0.95
5×3	8	240	3.17	0.51	1.03

layers and a lateral extension of (4 × 2) surface unit cells. In summary, the defect formation energies are given by

$$E_O^{vac} = 3.1\,\text{eV} + \Delta\mu_O \qquad E_{Zn}^{vac} = 0.5\,\text{eV} - \Delta\mu_O \qquad E_{ZnO}^{vac} = 1.0\,\text{eV} \quad . \tag{15}$$

To determine the most abundant type of atomic defect which will form on the $ZnO(10\bar{1}0)$ surface depending on temperature and pressure, we have constructed a pT phase diagram indicating for each atomic defect the region where it is the one with the lowest formation energy. The boundaries between these regions are given as curves of constant O chemical potential as determined by Eq. (15). The O chemical potential at the transition between the O–v/ZnO–v and the ZnO–v/Zn–v region is $\Delta\mu_O^{O/ZnO} = -2.11\,\text{eV}$ and $\Delta\mu_O^{Zn/ZnO} = -0.44\,\text{eV}$, respectively. Two other important boundaries are given by the limits of the O chemical potential $\Delta\mu_O^{UP} = 0.0\,\text{eV}$ and $\Delta\mu_O^{LOW} = -3.5\,\text{eV}$ according to Eq. (12). Beyond these boundaries the Gibbs free energy would be lowered by decomposing the ZnO crystal and forming the elemental phases. Using Eq. (14) we plotted the function

$$log_{10}\left(\frac{p}{p^{\circ}}\right) = \frac{2\Delta\mu_O - \tilde{\mu}_{O_2}(T, p^{\circ})}{ln\,(10)\,kT} \tag{16}$$

for $\Delta\mu_O = \{\Delta\mu_O^{UP}, \Delta\mu_O^{Zn/ZnO}, \Delta\mu_O^{O/ZnO}, \Delta\mu_O^{LOW}\}$. The resulting phase diagram is shown in Fig. 2.

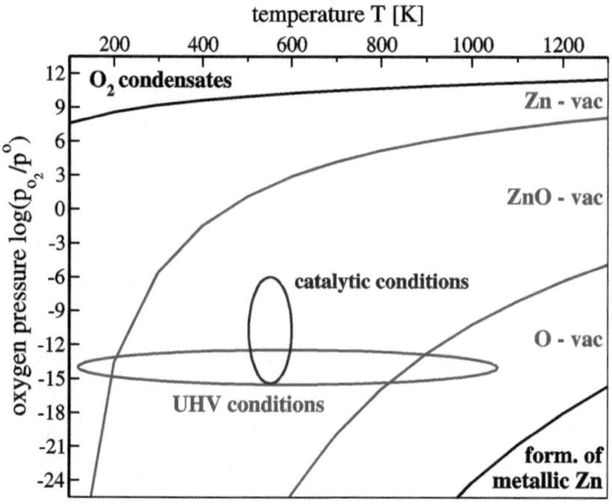

Fig. 2. Phase diagram of the most abundant atomic defect on the ZnO(10$\bar{1}$0) surface depending on temperature and pressure

Most surface science experiments are done at ultrahigh vacuum conditions at base pressures below 10^{-13} atm. During sample preparation typical annealing temperatures of about 1000 K are applied. As can be seen in Fig. 2, a certain fraction of O vacancies may be formed at these conditions, but during the slow cooling after the annealing process we expect that most of O vacancies either convert into ZnO dimer vacancies or are removed due to traces of oxygen in the gas phase. At low temperatures, Zn vacancies become the most favorable type of atomic defect. However, their formation might be suppressed if the activation barriers for the diffusion of atoms are too high to be overcome at those low temperatures. So overall we expect to observe almost exclusively ZnO dimer vacancies at the ZnO(10$\bar{1}$0) surface.

The methanol synthesis under industrial conditions is performed at about 10 atm pressure and a temperature of 550 K. For the synthesis gas, a mixture of H_2, CO and CO_2, only a rough estimate for the O chemical potential can be given. A typical range is marked in Fig. 2. At these conditions also the ZnO dimer vacancies will be the dominating atomic defects and single O vacancies will only play a minor role. In a recent joint theoretical/experimental study it was argued that probably O vacancies (so called F–centers) are responsible for the catalytic activity of ZnO [20]. The suppression of O vacancies on the ZnO(10$\bar{1}$0) surface might therefore be the explanation why it has been observed that this surface is less active for the methanol synthesis than the two polar ZnO surface terminations [21].

3.3 STM Images of the ZnO(10$\bar{1}$0) Surface

For an experimental verification of our conclusions derived from the analysis of the *pT* phase diagram we have calculated STM images to provide guidelines on how the different atomic defects might be identified in STM measurements. To obtain realistic images, the STM calculations were done for large (5×3) surface unit cells. However, to reduce the computational cost, only the upper

4 layers of the relaxed slabs were taken into account for the determination of the orbitals which enter the THA and BPA formulas. As tests have shown, 4 atomic layers are enough to provide well converged STM data. STM images were calculated for a set of different bias voltages. Negative and posititve voltages correspond to "filled state images" (tunneling from the surface) and "empty state images" (tunneling to the surface), respectively. In the THA calculations the tunneling current was converted into an electron density for the density isosurface according to an estimate given in Ref. [6]. BPA calculations were done for both tungsten tip models to evaluate their influence.

To point out characteristic features which may be useful to identify the different atomic defect structures in experiments, we present series of selected STM images. Bias voltages of 1 – 3 V and tunneling currents of 1 – 10 nA were considered, which are typical experimental values for imaging oxide surfaces.

Ideal surface (Fig. 3): At negative bias voltages the pronounced peaks in the STM images match the position of the surface O atoms. At low positive voltages (up to 1 V), the whole ZnO dimers are imaged, causing elongated spots in the [0001]–direction, whereas at higher positive voltages only the Zn atoms are visible. The spots of the Zn atoms get more and more smeared in the [1$\bar{2}$10]–direction leading to a row-like appearance of the Zn atoms.

O vacancy (Fig. 4): For low negative bias the O vacancy appears as a pronounced peak due to the localized defect state. If the voltage is further decreased the peak at the position of the O vacancy becomes less and less pronounced and blends in with the peaks of the surrounding O atoms. At low

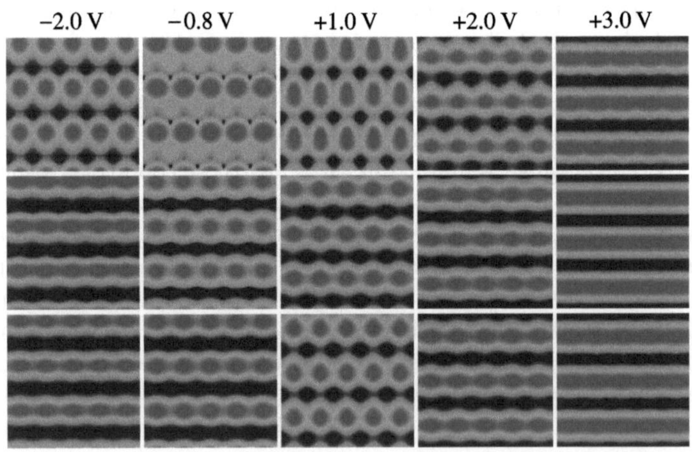

Fig. 3. STM images of the ideal, defect-free ZnO(10$\bar{1}$0) surface displaying (5 × 3) surface unit cells. The bias voltage is given on top. All images are calculated for 1 nA constant current mode. Three STM setups were used: THA (*first row*), BPA with W(100) tip model (*second row*), BPA with W(110) tip model (*third row*). The height of the tip above the surface is *color–coded* with *red–green–blue* corresponding large–medium–small distances between tip and surface

positive voltages the spot at the O vacancy is separated by a deep pit from the other dimer peaks which again gets more and more suppressed when the bias voltage is increased.

Zn vacancy (Fig. 5): The O atom next to the removed Zn atom strongly relaxes in the [0001]–direction by 1.04 Å, opening a gap which is visible for either bias polarity. At low negative and positive voltages this O atom produces a high peak in the STM images due to an additional small vertical relaxation of 0.15 Å. For low negative voltages also peaks of second layer O atoms (which were previously bounded to the removed Zn atom) are visible. At positive bias the Zn vacancy is marked by a rectangular shaped hole, independent of the voltage.

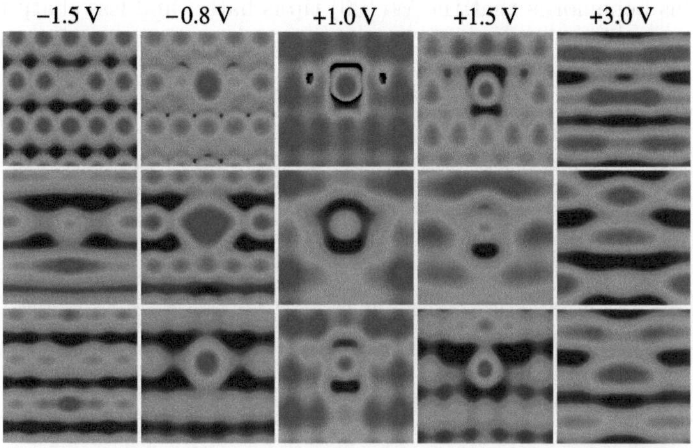

Fig. 4. STM images of the O vacancy. The same setup as in Fig. 3 is used

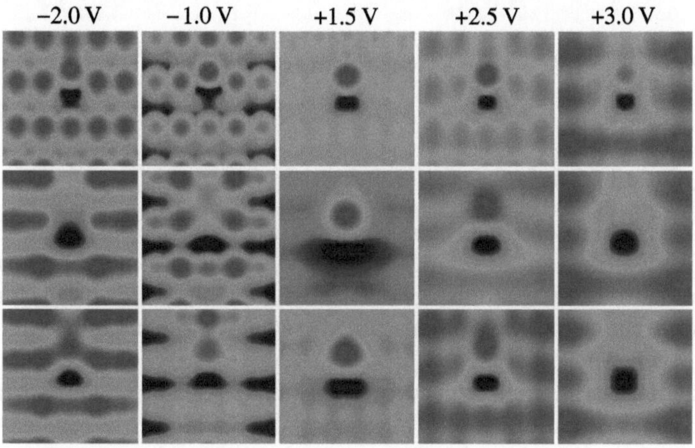

Fig. 5. STM images of the Zn vacancy. The same setup as in Fig. 3 is used

Zn+O vacancy (Fig. 6): The missing O and Zn atom of the ZnO dimer are imaged as a pronounced holes for negative bias and positive bias, respectively. The O vacancy hole is slightly elongated in the [0001] direction, whereas the Zn vacancy hole is triangular shaped.

In general, far away from the vacancies the main characteristics of the ideal surface (peaks at the position of the O atoms at negative voltages, in the [0001]–direction elongated dimer peaks for low positive voltages, and in the [1$\bar{2}$10]–direction smeared Zn rows for higher positive bias) are well recovered for each of the defective surfaces. The THA method yields STM images which correspond well with the BPA calculations for both tungsten tip models. Compared to the THA results, the W(100) tip causes larger deviations and a stronger smearing of structural features than the W(110) tip. The reason for this behavior is that the W(100) tip is flatter and less sharp than the W(110) tip so that the tip wave functions, which probe the surface, are more delocalized in space.

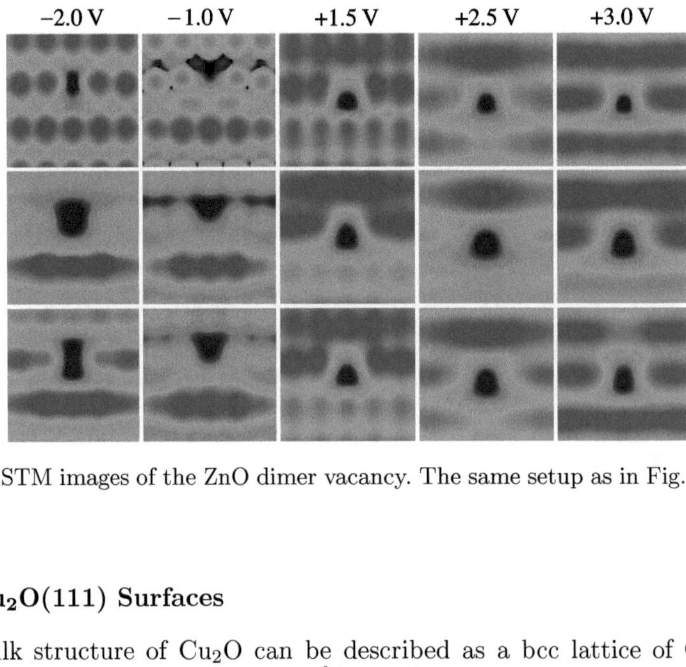

Fig. 6. STM images of the ZnO dimer vacancy. The same setup as in Fig. 3 is used

3.4 $Cu_2O(111)$ Surfaces

The bulk structure of Cu_2O can be described as a bcc lattice of O atoms (with a lattice constant of $a = 4.30$ Å), in which all O atoms are surrounded by a tetrahedron of Cu atoms. The Cu atoms, on the other hand, form a fcc lattice with the same lattice constant, in which the Cu atoms are linearly linked to two O neighbors.

Altogether 29 different structures of the hexagonal $Cu_2O(111)$ surface with various concentration of Cu and O vacancies and adatoms have been studied. The $Cu_2O(111)$ surface can be characterized by a stacking sequence of hexagonal Cu layers which are sandwiched between an upper and a lower O layer –

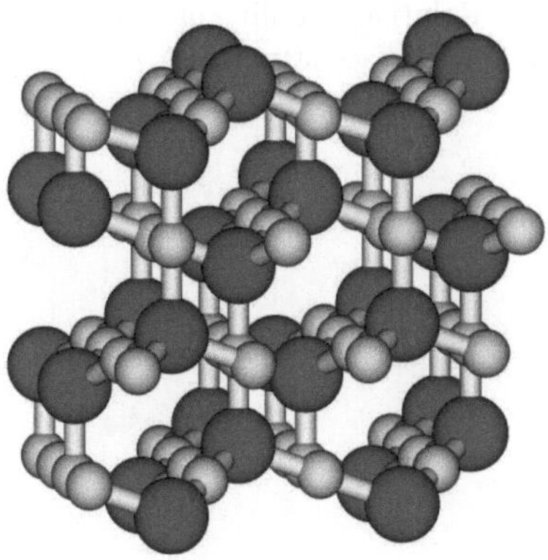

Fig. 7. Structure of Cu_2O. Cu atoms are depicted in *yellow*, O atoms in *blue*

in the following denoted by $\overline{\mathrm{O}}$ and $\underline{\mathrm{O}}$. One out of four Cu atoms in the Cu layer (in the following denoted by Cu^1) has no bonds to O atoms in those two neighboring O layers but only to O atoms in the underlying second layer.

We apply the same thermodynamic formalism as in the previous section. For the chemical potentials μ_{Cu} and μ_{O} of Cu and O we get

$$2\mu_{\mathrm{Cu}} + \mu_{\mathrm{O}} = E^{\mathrm{bulk}}_{\mathrm{Cu_2O}} \quad , \qquad \mu_{\mathrm{Cu}} \leq E^{\mathrm{bulk}}_{\mathrm{Cu}} \quad , \qquad \mu_{\mathrm{O}} \leq \frac{1}{2} E^{\mathrm{mol}}_{\mathrm{O_2}} \quad , \tag{17}$$

where $E^{\mathrm{bulk}}_{\mathrm{Cu_2O}}$ and $E^{\mathrm{bulk}}_{\mathrm{Cu}}$ are the energy of one Cu_2O and Cu bulk unit cell, respectively, and $E^{\mathrm{mol}}_{\mathrm{O_2}}$ is the energy of an oxygen molecule. Again, we relate the O chemical potential to its upper bound, and we introduce the Cu_2O bulk formation energy $E^{\mathrm{form}}_{\mathrm{Cu_2O}}$

$$\Delta\mu_{\mathrm{O}} = \mu_{\mathrm{O}} - \tfrac{1}{2} E^{\mathrm{mol}}_{\mathrm{O_2}} \quad , \qquad E^{\mathrm{form}}_{\mathrm{Cu_2O}} = E^{\mathrm{bulk}}_{\mathrm{Cu_2O}} - 2E^{\mathrm{bulk}}_{\mathrm{Cu}} - \tfrac{1}{2} E^{\mathrm{mol}}_{\mathrm{O_2}} \tag{18}$$

to get the allowed range for $\Delta\mu_{\mathrm{O}}$

$$E^{\mathrm{form}}_{\mathrm{Cu_2O}} \leq \Delta\mu_{\mathrm{O}} \leq 0 \quad . \tag{19}$$

From Eq. (2), (17) and (18) we calculate the stability of the different surface structures relative to the ideal surface termination. We find only two surface structures to appear as thermodynamically most stable surface terminations over the entire $\Delta\mu_{\mathrm{O}}$ range. These two structures are: (i) the ideal surface structure with an additional adsorbed O_2 molecule per surface unit cell (O–ads) at high O chemical potential (oxidizing conditions), resulting in a CuO surface stoichiometry, and (ii) the ideal surface structure with all Cu^1 atoms removed (Cu–vac) at low O chemical potential (reducing conditions), which corresponds to a Cu_3O_2 surface stoichiometry (see energy diagram in Fig. 8).

For the calculation of the STM images the same computational setup as in the previous Sect. 3.2 was used. Up to now only the THA results are available. The ideal surface termination (which is thermodynamically unstable according

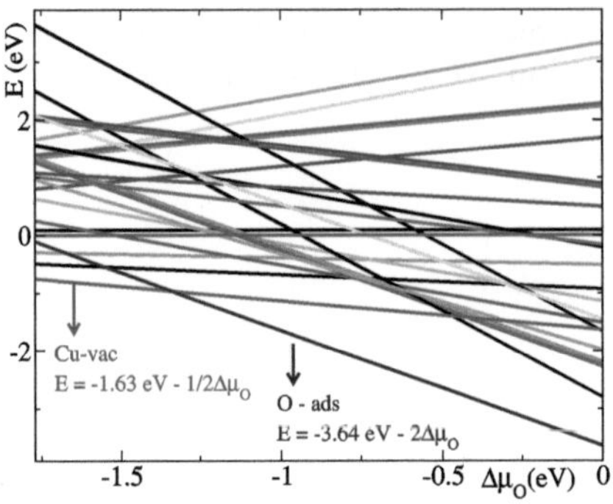

Fig. 8. Relative stability of different $Cu_2O(111)$ surface structures

to the energy diagram) yields for negative voltages peaks at the position of the Cu^1 and $\overline{O}$ atoms which are three–pointed star shaped due to the surrounding Cu atoms. With increasing bias the Cu^1 peaks become suppressed. For positive bias only the Cu^1 atoms are visible.

The adsorption of additional O atoms break the hexagonal symmetry of the surface. The result is a row–like appearance of the STM images of the O–ads surface structure with neighboring rows shifted by half of the surface unit cell. For negative voltages, the highest peaks are connected to the adsorbed O atoms, followed by slightly lower $\overline{O}$ peaks. At low negative bias also the Cu^1 atoms become visible. In the case of positive voltages, the dominant $\overline{O}$ peaks at low bias are replaced by triangular shaped peaks of the adsorbed O atoms at higher bias.

On the Cu–vac surface, the Cu^1 vacancies are imaged as large holes, independently of the bias voltage. The $\overline{O}$ atoms appear as pronounced three-pointed star shaped peaks (similar to the ideal surface termination), which form a honeycomb structure at high voltages.

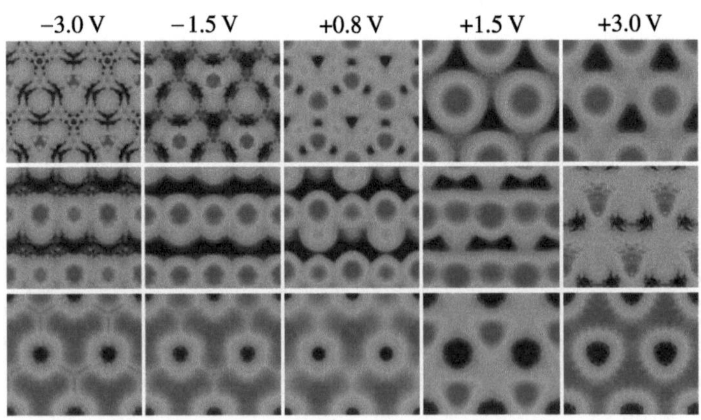

Fig. 9. STM images of the Cu_2O surface: ideal surface (*first row*), O–ads (*second row*), Cu–vac (*third row*). The same setup as in Fig. 3 is used

4 Conclusions

Ideal, defective and adsorbate-covered surfaces of ZnO and Cu/ZnO catalysts have been studied using DFT calculations in combination with a thermodynamic formalism. The stability of various surface structures as function of environmental conditions was examined, and for the most stable structures STM images were calculated using the Tersoff–Hamann approximation and Bardeens perturbation approach, employing two models of a tungsten tip with (100) and (110) orientation.

The analysis of the formation energy of the O, Zn and ZnO dimer vacancy as function of the O chemical potential showed that the dominating defect type on the ZnO($10\bar{1}0$) surface is the ZnO dimer vacancy for both, UHV conditions and for typical temperatures and pressures applied in chemical reactions. The O vacancy (F–center), which is assumed to be the important active center in ZnO catalysts, is predicted to be present in much lower concentrations, which might explain the reduced catalytic activity of the ZnO($10\bar{1}0$) surface compared to its polar counterparts. STM images for the ideal and defective surfaces were calculated. Characteristic features in the STM images are pointed out which allow the identification of the different defects in STM experiments.

An extensive search for the most stable structure and composition of the $Cu_2O(111)$ surface was performed. Two structures, depending on the environmental conditions, were found to be the thermodynamically the most stable terminations. In an oxidizing atmosphere, the chemisorption of oxygen leads to a surface layer with a CuO stoichiometry. On the other hand, in a reducing environment, the removal of one fourth of the surface Cu atoms (leading to a Cu_3O_2 stoichiometry of the surface layer) yields the lowest surface energy. The STM images of the two structures are characterized by a break of the hexagonal symmetry and a honeycomb structure, respectively.

References

1. D. Marx and J. Hutter, *Ab Initio Molecular Dynamics: Theory and Implementation*, in: *Modern Methods and Algorithms of Quantum Chemistry*, pp. 301–449, Editor: J. Grotendorst, NIC, FZ Jülich (2000); Electronic Version: `www.theochem.ruhr-uni-bochum.de/go/cprev.html`.
2. J. Hutter *et al.*, see: `www.cpmd.org`.
3. E. Kaxiras, K.C. Pandey, Y. Bar-Yam, and J.D. Joannopoulos, Phys. Rev. Lett. **56**, 2819 (1986); Phys. Rev. B **35**, 9625 (1987); G.-X. Qian, R.M. Martin, and D.J. Chadi, Phys. Rev. B **38**, 7649 (1988).
4. K. Reuter and M. Scheffler, Phys. Rev. B **65**, 035406 (2001).
5. B. Meyer, Phys. Rev. B **69**, 045416 (2004).
6. W.A. Hofer, A.S. Foster, and A.L. Shluger, Rev. Mod. Phys. **75**, 1287 (2003).
7. J. Bardeen, Phys. Rev. Lett. **6**, 57 (1961).
8. J. Tersoff and D.R. Hamann, Phys. Rev. B **31**, 805 (1985).

9. G. Antczak, R. Blaszczyszyn, and T.E. Madey, Prog. Surf. Sci. **74**, 81 (2003).
10. Y.C. Kim and D.N. Seidman, Met. Mater. Int. **10**, 97 (2004).
11. P.V.M. Rao, C.P. Jensen, and R.M. Silverc, J. Vac. Sci. Technol. B **22**, 636 (2004).
12. S.J. Tauster, Acc. Chem. Res. **20**, 389 (1987).
13. P.L. Hansen, J.B. Wagner, S. Helveg, J.R. Rostrup-Nielsen, B.S. Clausen, H. Topsøe, Science **295**, 2053 (2002).
14. J.-D. Grunwaldt, A.M. Molenbroek, N.-Y. Topsøe, H. Topsøe, and B.S. Clausen, J. Catal. **194**, 452 (2000).
15. O. Dulub, M. Batzill, and U. Diebold, Topics in Catalysis **36**, 65 (2005).
16. S.V. Didziulis, K.D. Butcher, S.L. Cohen, and E.I. Solomon, J. Am. Chem. Soc. **111**, 7110 (1989).
17. J.P. Perdew, K. Burke, and M. Ernzerhof, Phys. Rev. Lett. **77**, 3865 (1996).
18. D. Vanderbilt, Phys. Rev. B **41**, 7892 (1990).
19. S. Goedecker, M. Teter, and J. Hutter, Phys. Rev. B **54**, 1703 (1996); C. Hartwigsen, S. Goedecker, and J. Hutter, Phys. Rev. B **58**, 3641 (1998); M. Krack, Theor. Chem. Acc. **114**, 145 (2005).
20. M. Kurtz, J. Strunk, O. Hinrichsen, M. Muhler, K. Fink, B. Meyer, and Ch. Wöll, Angew. Chem. Int. Ed. **44**, 2790 (2005).
21. H. Wilmer, M. Kurtz, K.V. Klementiev, O.P. Tkachenko, W. Grünert, O. Hinrichsen, A. Birkner, S. Rabe, K. Merz, M. Driess, Ch. Wöll, and M. Muhler, Phys. Chem. Chem. Phys. **5**, 4736 (2003).

Theoretical Investigation of the Self-Diffusion on Au(100)

K. Pötting[1], T. Jacob[2], and W. Schmickler[1]

[1] Department of Theoretical Chemistry, University of Ulm, 89081 Ulm, Germany
kay.poetting@uni-ulm.de, wolfgang.schmickler@uni-ulm.de

[2] Fritz-Haber-Institut der Max-Planck-Gesellschaft,
Faradayweg 4–6, 14195 Berlin-Dahlem, Germany
jacob@fhi-berlin.mpg.de

1 Introduction

Adatom and vacancy diffusion is most important for mass transport processes on metal surfaces, e.g. adsorption, desorption, epitaxial growth or coarsening. Surface diffusion may occur by *hopping* of an adatom between minima of the potential energy surface, i.e. between stable or metastable adsorption sites. Alternatively, the adatom can replace an underlying surface atom, which moves up to an adjacent surface position. This so-called *exchange diffusion* was first found for adatom diffusion on Pt(011) by Bassett and Weber [1]. While on most surfaces self-diffusion is expected to occur by hopping events, calculations by Feibelman indicated that on Al(100) self-diffusion preferentially proceeds via atom exchange [2]. In addition, the latter process was also found to be dominating in case of Au(100) [4].

While improvements in experimental techniques led to a considerable progress in the investigation of surface diffusion [3], there is still a lack on experimental data for self-diffusion on Au(100). Difficulties arise from the fact that experimentally activation energies for diffusion ΔE_{act} are obtained by measuring diffusion constants D for several temperatures, which are then used with the following equation to deduce activation barriers:

$$\ln D = \ln D_0 - \Delta E_{\mathrm{act}}/RT. \tag{1}$$

Self-diffusion on the defect-free Au(100) surface is the simplest diffusion process in the Au/Au(100) system. Several theoretical investigations on this system have been performed using *ab initio* density-functional theory (DFT). For instance, Yu and Scheffler [4] reported that for Au(100) the surface diffusion is predicted to proceed by atom exchange. Similarly, Chang and Wei [5] studied the self-diffusion of single adatoms and dimers on the (100)-surfaces of different fcc-crystals. They also found that for Au the exchange mechanism is favored compared to surface hopping.

Compared to the rather simple diffusion on terraces, there are much less theoretical investigations on non-perfect surfaces, whose morphologies might be influenced by defects or the presence of islands. On these more rough surfaces, which are more realistic than idealized (perfect) terraces, one of the most active sites is certainly the kink position. It plays an important role for epitaxial growth and island formation, since removing or adding an atom from or to this position does not cause a structural change.

Using *ab initio* density-functional theory, the focus of the present work is on the self-diffusion of adatoms on Au(100) terraces as well as on surface defects such as step edges, kinks, or vacancies. In addition to the changes in the energetic barriers, the rates of the different diffusion processes, which are important to study the dynamics of growth or morphological changes, can be deduced by knowing the activation energies and the vibrational frequencies. While in most cases the difference between the potential energies from initial and final state strongly influence the diffusion, for the migration on a perfect Au(100) terrace both states have the same energy that is why in this case temperature or entropy determine the inital step of diffusion. Besides these environmental influences, especially in the field of electrochemistry the presence of an electrode potential can also modify the self-diffusion. For instance on Au(100) Giesen *et al.* [6] found that the mobility of point defects as mass carriers depends of the electrode potential. Adatoms and defects have an associated dipole moment, which depends on the surface position and interacts with the electric field resulting from the electric double-layer. Thus, changes in the electric field induced by tuning the electrode potential correspondingly, may cause different diffusion processes to be more favorable. Since this should have an effect on the final surface structure (due to growth or island formation), the electrode potential might offer a way to actively control surface nanostructuring.

In this paper, we present detailed investigations on various single-atom diffusion processes on plane and rough Au(100) surfaces, which play an important role in crystal growth or island formation, e.g. Ostwald ripening. In addition to the energetics, for each process we deduced the diffusion barriers as well as the vibrational frequencies of the adsorbates at the initial and final positions, which later will allow us to study the influence of the electrode potential on the growth behavior.

2 Method

2.1 Density-functional Theory

For the calculations presented in this paper we used SeqQuest [7, 8], a periodic DFT program with the PBE Generalized Gradient Approximation (GGA) [9] exchange-correlation functional. The 68 core electrons of each Au atom were replaced by a (standard) norm-conserving pseudopotential [10], leaving the

11 5d and 6s electrons to be treated explicitly by contracted Gaussian functions on the "double-zeta plus polarization"-level. This basis set was optimized for the Au atom and different bulk structures (fcc, bcc, hcp, a15, diamond, sc). Moreover a converged Brillouin zone (BZ) sampling with 12×12 **k**-points corresponding to the (1×1)-surface unit cell (SUC) was used.

If not specified differently, in all calculations the (semi-)infinite Au(100) surface was modeled by a five-layer slab, where the bottom two layers were fixed at the calculated bulk crystal Au–Au distance of $a_0 = 4.164$ Å (exp.: 4.08 Å) [11] and the remaining layers plus the adsorbate atom were allowed to fully optimize (to < 0.02 eV/Å). To estimate the influence of the surface relaxation for the different systems we also performed calculations allowing only the diffusing Au atom to relax, while fixing all other Au atoms at their bulk-crystal positions.

Because of its importance for understanding the diffusion kinetics we also calculated vibrational frequencies along the diffusion coordinate, obtained by a harmonic fit to the potential energy curve around the minimum. Thus, starting from the same position each diffusion pathway might result in a different vibrational frequency.

2.2 Surface Unit Cells

While adsorption and diffusion on the perfect Au(100) surface can be studied with relatively small unit cells, the reduced symmetry introduced by step edges, kinks, or vacancies requires extensive investigations to determine adequate unit cells for each diffusion process. For each case the size should be sufficiently large to minimize the lateral interactions between the adsorbate atoms and the periodically repeated images of surface defects.

For all systems except terrace diffusion a five-layer slab was used, giving converged energies for adatom adsorption at various sites. In contrast the horizontal extension of the unit cells strongly depends on the symmetry of the surface and the particular defect. After an extensive phase of convergence tests, we found the following SUCs to be sufficient in size:

- *Terrace*: In order to model diffusion of a single Au adatom on the terrace a (3×3) surface unit cell was used. Although for a single adsorbate this leads to an effective coverage of 0.11 ML, the lateral adsorbate interactions are expected to be rather small (see [4]). The perfect terrace was modeled with a six-layer slab. This additional layer allowed us to study the surface diffusion by vertical atom exchange more accurately.
- *Step edge*: Due to the reduced symmetry caused by the step edge this system requires more extended SUCs. The diffusion along the edge was modeled with a (3×4) SUC, while the diffusion away from the step edge required a (3×6) SUC. We caution here of using a too small unit cell for the latter process, since already with a (3×5) SUC the adatom binding at the final position after diffusion (hollow-site) is influenced by the periodically repeated images of the step edge by 0.08 eV.

- *α-kink*: This system was studied using a (5×5) unit cell. While this should be extended enough to calculate the diffusion along the step, one might consider this SUC to be too small for modeling diffusion from the step to the four-fold hollow site accurately. However, corresponding tests were performed and will be discussed in Sect. 3.2.
- *β-kink*: Due to the similarity to the α-kink, here the same (5 × 5) SUC was used.

3 Results and Discussion

To investigate self-diffusion of Au on Au(100), we calculate the most relevant migration pathways with different surface morphologies. This includes diffusion on terrace sites, near step-edges, step-vacancies as well as various kink sites (see Figs. 1, 3, 6, 9). Below we will discuss the results in the following order of systems: defect-free Au(100) surface, the immediate vicinity of a step edge, and various kink sites. Binding energies, activation barriers and frequencies are presented both for the relaxed systems and the non-relaxed systems (see Tables 1 to 8). Binding energies for the relaxed, respectively non-relaxed systems were calculated using

$$E_{\text{bind}}^{(\text{non}-)\text{relaxed}} = E_{\text{system}}^{(\text{non}-)\text{relaxed}} - E_{\text{slab}}^{(\text{non}-)\text{relaxed}} - E_{\text{Au}-\text{atom}}^{\text{vac}} \tag{2}$$

While in the relaxed system the adatom (except along the diffusion coordinate) and the surface were allowed to freely relax, in the non-relaxed system this holds only true for the adsorbate, leaving the entire surface to be kept fixed. If not mentioned explicitly, in the following all discussed energies refer to the relaxed cases. Here we would also like to mention that due to the reduced number of degrees of freedom each diffusion process was simulated by successively moving the adatom along the diffusion coordinate rather than taking any transition state finding procedure (e.g. NEB–Nudged Elastic Band).

3.1 Terrace Diffusion

Adsorbing a single Au atom on the Au(100) surface is most strongly on the four-fold hollow position ($E_{\text{bind}} = -3.46$ eV). Therefore, diffusion along the defect-free surface may proceed by hopping over bridge or top positions, or by a vertical exchange mechanism. The calculated binding energies, activation barriers and vibrational frequencies along the corresponding diffusion coordinates are given in Tables 1 and 2.

While the highest activation barrier is for hopping diffusion over an underlying atom (see Fig. 1, A → C), the diffusion barrier for diffusion over the bridge position (see Fig. 1, A → B) is about 0.69 eV smaller. However, the calculated energy needed to migrate from position A to C by exchange is 0.55 eV, which is a decrease of about 20 % compared to the bridge-diffusion and about

Fig. 1. Adsorption sites on the clean Au(100) surface

Table 1. Binding energies and vibrational frequencies for Au diffusion on the clean Au(100) surface (see Fig. 1). For each adsorption site values with and without relaxation of the surface atoms are given. The subscripts for position A indicate the direction for which the vibration was evaluated.

System	**Position**	**non-relaxed**		**relaxed**	
		E_{bind} [eV]	ν [cm^{-1}]	E_{bind} [eV]	ν [cm^{-1}]
terrace	$A_{\to B}$	–3.39	356	–3.46	307
	$A_{\to C}$	–3.39	414	–3.46	275
	A_{exchange}	–	–	–3.46	287
	E (bridge)	–2.76	–	–2.76	–
	On top	–2.17	–	–2.07	–

Table 2. Activation energies for Au diffusion along different pathways on the clean Au(100) surface (see Fig. 1). For each pathway the activation barrier for the non-relaxed and relaxed system is given.

System	**Pathway**	E_{act} [eV]	
		non-relaxed	relaxed
terrace	A ⇌ B (bridge)	0.63	0.70
	A ⇌ C (top)	1.22	1.39
	A ⇌ C (exchange)	–	0.55

60 % compared to top-diffusion. As shown in Table 2, the activation barrier is smaller for the non-relaxed system. This effect could be explained in terms of relaxation effects occurring in the second layer of the slab during geometry optimization. Comparison of the systems with and without surface relaxation shows that in the relaxed case the binding energy of the adatom in the four-fold hollow position A decreases about 0.07 eV (stronger bond). Adsorption at the bridge position gives almost the same binding energies, while on top binding even leads to a 0.1 eV weaker bond on the relaxed surface. Taking into account that the difference in total energies between the adsorbate-free relaxed and the non-relaxed surface is only 6 meV, both surfaces refer almost to the same level. Therefore, the adatom induces a bigger change in the geometry of the underlying atoms on the four-fold hollow site as on the bridge site. This leads to a negative shift in the binding energy at the four-fold hol-

low position and consequently to a higher activation barrier compared to the non-relaxed system. In the case of hopping diffusion from A to C over the top position, it is more difficult to model the migration pathway of the adatom, because one has to define two geometric constraints for the adsorbate (in x- and y-direction). This may not lead to the right minimum energy pathway. However, the diffusion barrier over top was calculated as $E_{\mathrm{act}} = 1.39$ eV, which is two times higher as the barrier for bridge diffusion.

3.2 Step Diffusion

To estimate the diffusion barriers in the presence of a step edge, we calculate the migration along and away from the step edge (see Tables 3 and 4). As expected, the activation barrier for diffusion along the step edge (A $\rightarrow$ B, $E_{\mathrm{act}} = 0.38$ eV) is much smaller than the barrier for detaching the adsorbate from the step edge (A $\rightarrow$ C, $E_{\mathrm{act}} = 0.84$ eV). Therefore, one might expect that during the process of island growth a particle finds its optimum shape faster than being disturbed by additional atom attachment. As shown in Table 4 and Fig. 5 there is an intermediate state for the step edge diffusion, a fact which is important to understand the decrease of the activation barrier of about 57 % compared to the pathway A $\rightarrow$ C. During the surface relaxation, the distance between the adsorbate and the step edge atoms T1 and T2 (see Fig. 5) decreases to 2.86 Å in the transition state, about 97 % of the ideal bulk atom distance. Since two bonds are formed with the step edge atoms T1 and T2, a repulsive force acts on the underlying atom T3, leading to a displacement

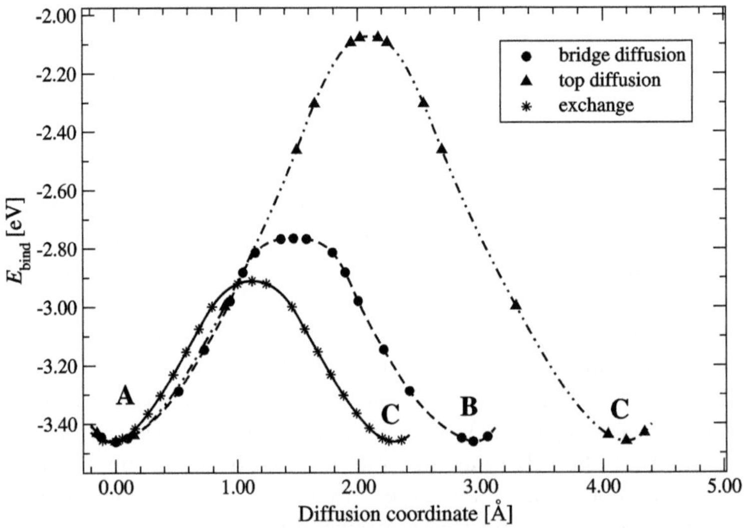

Fig. 2. Migration pathways of an adatom on the clean Au(100) surface (relaxed system). The adsorption sites correspond to the positions in Fig. 1

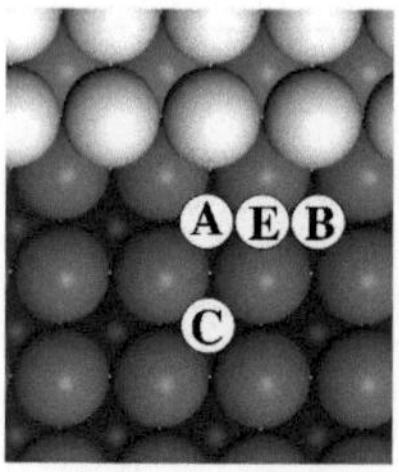

Fig. 3. Adsorption sites on the stepped Au(100) surface

Table 3. Binding energies and vibrational frequencies for Au diffusion on the stepped Au(100) surface (see Fig. 3). For each adsorption site values with and without relaxation of the surface atoms are given. The subscripts indicate the direction for which the vibration was evaluated.

System	Position	non-relaxed		relaxed	
		E_{bind} [eV]	ν [cm^{-1}]	E_{bind} [eV]	ν [cm^{-1}]
step edge	$A_{\rightarrow B}$	-3.71	320	-3.66	231
	$A_{\rightarrow C}$	-3.71	418	-3.66	313
	$C_{\rightarrow A}$	-3.43	352	-3.48	233
	$E_{\rightarrow A/B}$	-3.28	174	-3.24	72

Table 4. Activation energies of Au diffusion along different pathways on the stepped Au(100) surface (see Fig. 2). For each pathway the activation barrier for the non-relaxed and relaxed system is given.

System	Pathway	E_{act} [eV] non-relaxed	relaxed
step edge	A/B → E	0.45	0.38
	E → A/B	0.02	0.01
	A → C	0.93	0.84
	C → A	0.65	0.56

of $\Delta r = 0.223$ Å into the bulk with respect to its perfect lattice position. The distance between the adsorbate and T3 is 2.70 Å, while the distance between the adsorbate and T4 is 2.86 Å. As can be seen in Fig. 5 two bonds to T1 and T2 are formed, which lead to the double-well shaped diffusion pathway shown in Fig. 4. Consequently this stabilizes the adsorbate and decreases the diffusion barrier to 0.38 eV. Moving from the intermediate state E to the final state B, the adatom has to overcome an energy barrier of 0.01 eV, which is very small and should be negligible already at very low temperatures. However, interesting for the overall diffusion from A to B is the broadening of the energy curve, which has an influence on the kinetics. The binding energy at position A is –3.66 eV, which is 0.20 eV more negative than the binding energy on the defect-free Au(100) surface ($E_{\mathrm{bind}}^{\mathrm{terrace}} = -3.46$ eV). Comparing the binding energies at position C calculated with a (3×5) and a (3×6) SUC

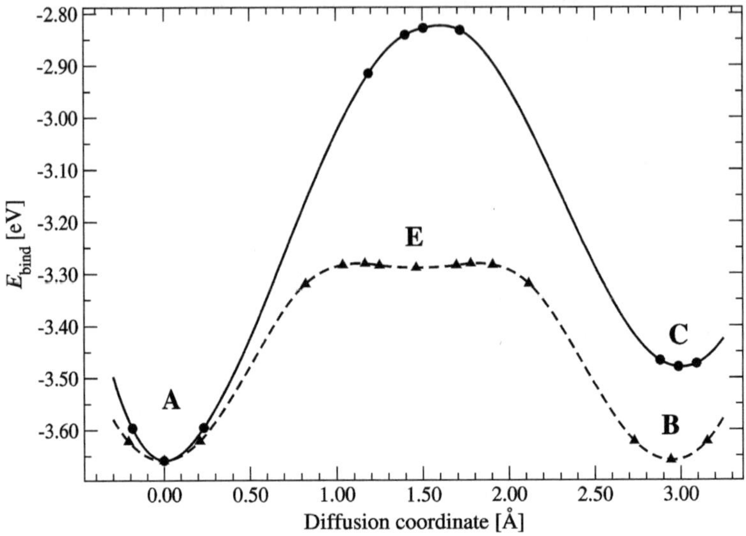

Fig. 4. Migration pathways of an adatom on the stepped Au(100) surface. The adsorption sites correspond to the positions in Fig. 3

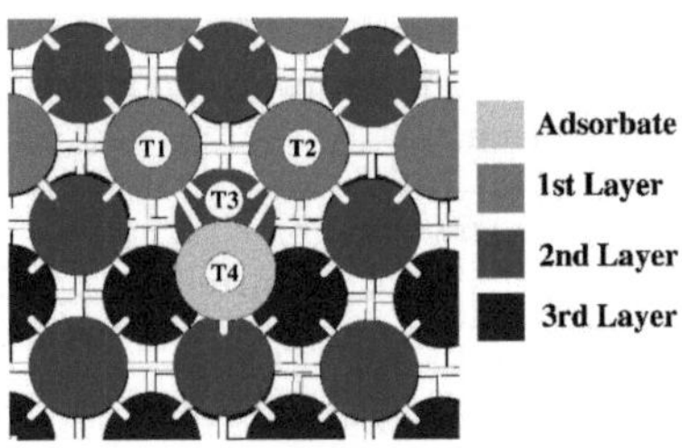

Fig. 5. Detailed view of the structure at the transition state E, see Fig. 3, A → B

results in a binding energy of –3.40 eV for the smaller SUC and –3.48 eV with the (3 × 6) unit cell. The adsorbate-adsorbate distances at position C in each case are $d^{(3\times5)}_{ads-ads} = 2{\cdot}a_0$ and $d^{(3\times6)}_{ads-ads} = 3{\cdot}a_0$, where a_0 is the calculated next-neighbor distance. Hence the interaction of the adsorbates with the image of the step causes a difference in the adsorption energy of 0.08 eV. While in the (3 × 6) cell the underlying atoms of the adatom at site C are able to relax, the same atoms are much less mobil in the (3 × 5) cell, because their mobility is reduced through the repeated step edge. Compared to the adsorption energy at position A on the terrace (see Fig. 1), the difference in the adsorption energy at the four-fold hollow site (position C) is 0.02 eV.

3.3 Kink Diffusion

3.3.1 α-Kink

Adatom migration from or to a kink site is most interesting, since it involves a variety of possible diffusion proccesses. For diffusion from site A, which in the following we assume to be the starting configuration, to site D (see Fig. 4) different pathways have to be distinguished. Considering the case of removing an atom from the kink position A, the adatom might move from position A either directly to position B, or first diffuse along the step-edge (position C) and afterwards desorb to position D. Besides these basic pathways two exchange processes might also occur. The first of them includes a place exchange between the adatom adsorbed at site A and the second-layer atom at position T1, moving to position D (see Fig. 8). During the second process the adatom replaces the atom at position T2, which thus moves to site T3. In the following we first focus on the basic diffusion processes, which do not include place exchange. For these mechanisms Fig. 7 shows the corresponding behavior of the adatom binding energy. Moving from position A to B, the adatom has to overcome an energy barrier of 0.87 eV. Although the presence of a nearby kink-atom causes certain constraints to the relaxation of the underlying surface, this barrier is comparable to adatom desorption from a perfect step edge to an adjacent four-fold position (see Fig. 3, A→C). Usually one would expect that the reduction of surface symmetry should increase the diffusion barrier. In order to estimate this increase, we compare the following bridge-diffusion processes:

(1) terrace, $E_{\text{act}} = 0.70$ eV
(2) step edge, $E_{\text{act}} = 0.84$ eV
(3) α-kink, $E_{\text{act}} = 0.87$ eV
(4) step vacancy, $E_{\text{act}} = 1.06$ eV

While on the terrace the adatom can freely migrate in four directions (considering bridge-diffusion), there are only three directions at the step edge. This results in a 0.14 eV higher activation energy. Moving out of a step edge, where the adatom can only hop in a single direction, the diffusion barrier is 0.36 eV higher than on the terrace. This is about three times the value of 0.14 eV mentioned before.

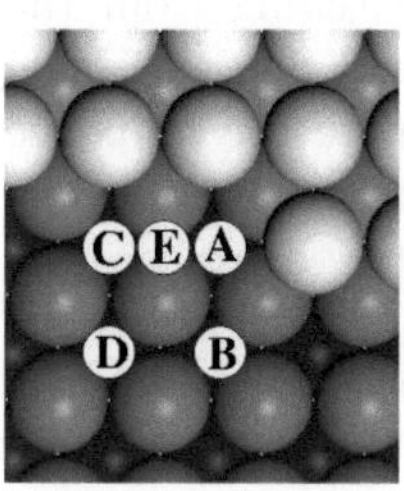

Fig. 6. Adsorption sites around a kink site on a Au(100) surface. Position A is called the α-kink position

Table 5. Binding energies and vibrational frequencies for Au diffusion on Au(100) in the vicinity of a kink (see Fig. 6). For each adsorption site values with and without relaxation of the surface atoms are given. The subscripts indicate the direction for which the vibration was evaluated.

System	Position	non-relaxed		relaxed	
		E_{bind} [eV]	ν [cm^{-1}]	E_{bind} [eV]	ν [cm^{-1}]
α-kink	$A_{\rightarrow B}$	–3.95	396	–3.87	272
	$A_{\rightarrow C}$	–3.95	395	–3.87	277
	$B_{\rightarrow A}$	–3.46	352	–3.47	230
	$C_{\rightarrow A}$	–3.70	– [1]	–3.61	233
	$C_{\rightarrow D}$	–3.70	– [1]	–3.61	268
	$D_{\rightarrow C}$	–3.41	354	–3.43	228
	$E_{\rightarrow A/C}$	–3.26	123	–3.27	83

[1] Calculations are in progress

Table 6. Activation energies of Au diffusion along different pathways in presence of a kink (see Fig. 2). For each pathway the activation barrier for the non-relaxed and relaxed system is given.

System	Pathway	E_{act} [eV] non-relaxed	relaxed
α-kink	A → B	0.97	0.87
	B → A	0.49	0.46
	A → E	0.69	0.61
	E → A	0.003	0.003
	E → C	0.015	0.008
	C → E	– [1]	0.351
	C → A	– [1]	0.346
	A → C	– [1]	0.615
	C → D	0.95	0.81
	D → C	0.66	0.62

[1] Calculations are in progress

Therefore, it can be assumed that each reduction in the degrees of freedom increases the barrier by about 0.12 – 0.14 eV. Consequently, at the α-kink one would expect an activation energy in the range of 0.93 to 0.97 eV. Instead, we find only a small increase of the energy (0.03 eV) switching from the step edge to the α-kink system, resulting in E_{act} = 0.87 eV. Since a single bond is formed moving from A to B (see Fig. 8), we estimate the influence of this bond on the activation barrier by comparison between the bridge-diffusion on the terrace and along the perfect step edge. In the latter case the adatom forms two bonds with step edge atoms (in the transition state, see Fig. 5). There, the presence of the perfect step edge reduces the barrier by 0.32 eV. Since at the α-kink there is only a single bond between A and T2

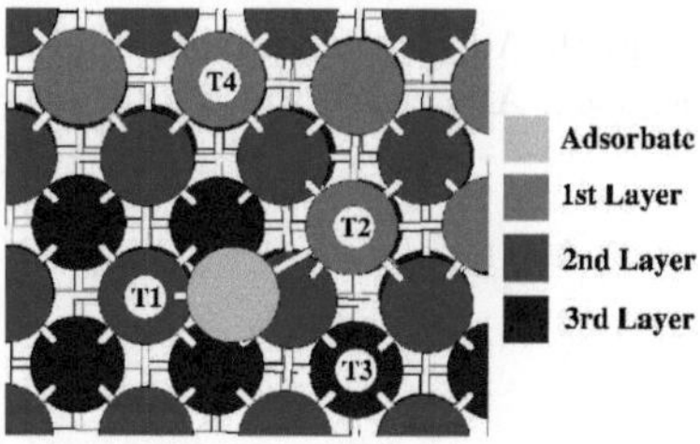

Fig. 7. Detailed view of the structure at the transition state, corresponding to the diffusion path A → B, see Fig. 5

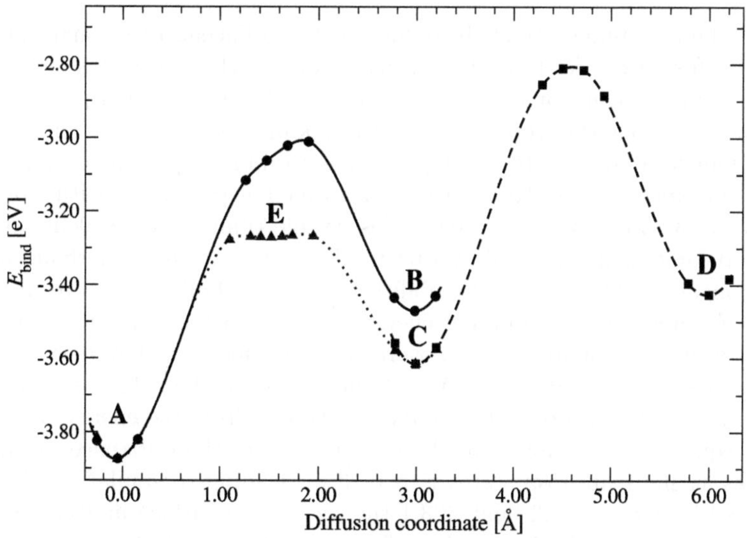

Fig. 8. Migration pathways for the α-kink system. Position A to E correspond to the adsorption sites labeled in Fig. 6

at the transition state d(A–T2)=3.04 Å, we expect half the reduction of the barrier (0.16 eV). This reduction leads to an activation barrier of about 0.77 to 0.81 eV, which is close to the investigated value of 0.87 eV. Compared to the step edge system (see Fig. 3, C) one would expect a binding energy at positions B and D of about −3.56 to − 3.58 eV. As first approximation, since the mobility of the underlying atoms is reduced by the repeated cell image, adding 0.08 eV (corresponding to the effect explained in Sec. 3.3) leads to an adsorption energy of –3.55 eV at site B and –3.51 eV at site D. This is in the range of the expected value mentioned before, because there is an interaction with the atom T2.

Fig. 9. Adsorption sites around a kink site on a Au(100) surface. A is called the β-kink position

3.3.2 β-Kink

In the β-kink system we treat the influence of an adjacent orthogonal step edge on the diffusion barrier (see Fig. 9). In contrast to the α-kink system (except exchange mechanisms or vacancy migration) only two diffusion pathways are possible, one along the step away from the β-kink, another away from the step to the four-fold hollow site (see Fig. 9). The results are presented in Tables 7 and 8. As expected, the strongest bond is formed at position A with a binding energy of –3.83 eV, which is 0.04 eV less stable than the equivalent position at the α-kink. This is certainly due to the fact that the underlying atoms are stronger inhibited in their mobility as in the α-kink system, resulting in a 0.04 eV more positive binding energy. Moving to the adjacent position B along the step the adatom has to overcome a barrier of 0.59 eV (see Table 8). This is close to the barrier for A $\rightarrow$ C migration in the α-kink system. The adsorption energy at position B differs by 0.03 eV from the energy calculated at the equivalent position C in the α-kink system, the difference in binding energies at the transition states E is 0.03 eV leading to similar barriers. As discussed in Sec. 2.2, 3.2, and 3.3.1 the size of the surface unit cell is quite important, because the interaction of the adsorbate with its image can lead to non-realistic binding energies. Cause to this effect a surface unit cell size of

Table 7. Binding energies and vibrational frequencies for Au diffusion on Au(100) in the vicinity of the β-kink (see Fig. 9). For each adsorption site values with and without relaxation of the surface atoms are given. The subscripts indicate the direction for which the vibration was evaluated.

System	Position	non-relaxed		relaxed	
		E_{bind} [eV]	ν [cm^{-1}]	E_{bind} [eV]	ν [cm^{-1}]
β-kink	$A_{\rightarrow E}$	–	–	–3.83	278
	$B_{\rightarrow E}$	–	–	–3.58	268
	$B_{\rightarrow C}$	–	–	–3.58	229
	$C_{\rightarrow B}$	–	–	–3.28	246
	$E_{\rightarrow A/B}$	–	–	–3.24	124

Table 8. Activation energies of Au diffusion along different pathways in presence of a β-kink (see Fig. 9). For each pathway the activation barrier for the non-relaxed and relaxed system is given.

System	Pathway	E_{act} [eV] non-relaxed	relaxed
β-kink	A → E	–	0.59
	E → A	–	< 0.001
	E → B	–	0.01
	B → E	–	0.34
	A → B	–	0.60
	B → C	–	0.93
	C → B	–	0.63

(6×6) may lead to more realistic binding energies[1]. Anyway, for a qualitative analysis of both systems (α- and β-kink), this should be sufficient. As shown in Figs. 7 and 10 , the diffusion process from the kink site to the four-fold hollow site (A → D in Fig. 6, A → C in Fig. 9) is energetically favored. So the process consists of two parts: first moving away from the kink along the step edge followed by detachment from the step to the terrace. In experiments the migration of an adatom to a step edge is described by the sticking energy [12]:

$$E_{\text{stick}} = E_{\text{act}}^{\text{step edge}} - E_{\text{act}}^{\text{terrace}} \tag{3}$$

The sticking energy E_{stick} is the additional energy barrier for *sticking* an adatom to an island edge. Calculations of the sticking energies for each system give the following values

(1) step edge: $E_{\text{stick}} = 0.14$ eV
(2) α-kink $E_{\text{stick}} = 0.11$ eV
(3) β-kink: $E_{\text{stick}} = 0.23$ eV

The sticking energies should be of the same magnitude, since the surface geometry close to the sticking position is similar. For the β-kink the sticking energy is about the double, which could be explained again with the dimension of the surface unit cell.

4 Conclusions and Perspectives

We have studied the self-diffusion behavior of adatoms on Au(100) surfaces including surface defects using *ab initio* density-functional theory. For self-diffusion of an adatom on the clean (100)-surface we find that exchange diffusion is favored, followed by bridge diffusion. If a perfect step edge is present

[1] Calculations with a (6×6) SUC both for the α-kink and the β-kink system are in progress.

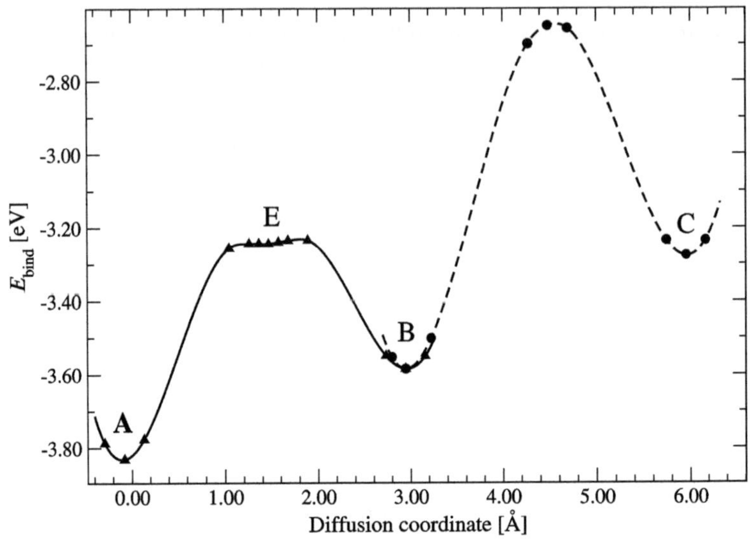

Fig. 10. Migration pathway for the β-kink system. Starting from position A, the adsorbate first diffuses along the step, followed by detachment from the step (B $\rightarrow$ C)

we find the smallest activation barrier for the diffusion pathway along the step edge. The decrease of the diffusion barrier along the step edge results from the formation of two bonds between the adatom and two step edge atoms. Afterwards we studied diffusion for two kink systems. In the α-kink system the diffusion pathway to a free four-fold hollow site can be realized by two different hopping processes. The desorption from the kink site to the step edge is favorable since the activation barrier is rather small compared to the pathway directly to a four-fold hollow site.

In addition, we calculated diffusion pathways to form dimers, migration of a step atom moving out of a step-edge (step vacancy) and diffusion around a corner. Since the final goal of these studies is to perform Kinetic Monte Carlo (KMC) simulations on the self-diffusion on Au(100), further DFT-investigations related to exchange mechanisms and collective adsorbate movements are required. Finally, knowing energies and frequencies of all important diffusion processes we will be able to perform large-scale simulations on the growth, stability, and structure of Au(100).

5 Computational Resources

Each structure along the diffusion pathways was calculated within 2 to 5 days depending on the size of the studied system, using two CPUs for small (3×3), (3×5) or (4×4) SUCs, and four CPUs for larger SUCs. The parallel calculations were performed with a MPI-parallelized DFT-code. Postprocessing calculations on charge distributions and charge localizations were usually performed in serial mode. The memory requirement varied between 500 MB and 2 GB per calculation.

Acknowledgements

K. P. and W. S. greatly acknowledge financial support by the "Landesstiftung Baden-Württemberg" and by the "Deutsche Forschungsgemeinschaft"

T. J. greatly acknowledges support by the "Fonds der Chemischen Industrie" (VCI).

References

[1] D.W. Basset and P.R. Webber, Surf. Sci. **70**, 520 (1978)
[2] P.J. Feibelman, Phys. Rev. Lett. **64**, 3143 (1990)
[3] Surface Diffusion: Atomistic and Collective Processes, edited by M.C. Tringides, Plenum Press, New York (1997)
[4] B.D. Yu and M. Scheffler, Phys. Rev. B **56**, R15569 (1997)
[5] C.M. Chang and C.M. Wei, Chi. J. Phys. **43**, 169 (2005)
[6] M. Giesen, G. Beltramo, S. Dieluweit, J.E. Mueller, H. Ibach and W. Schmickler, Surf. Sci. **595**, 127 (2005)
[7] P.A. Schultz, unpublished; A description of the method is in: P.J. Feibelmann, Phys. Rev. B **35**, 2626 (1987)
[8] C. Verdozzi, P.A. Schulze, R. Wu, A.H. Edwards, N. Kioussis, Phys. Rev. B **66**, 125 408 (2002)
[9] J.P. Perdew *et al.*, Phys. Rev. B. **46**, 6671 (1992)
[10] D.R. Hamann, Phys. Rev. B **40**, 2980 (1989)
[11] C. Kittel, *Einführung in die Festkörperphysik*; Oldenbourg Verlag: München (1991)
[12] M. Giesen, Prog. Surf. Sci. **68**, 1–153 (2001)

TrpAQP: Computer Simulations to Determine the Selectivity of Aquaporins

M. Dynowski, U. Ludewig

Zentrum für Molekularbiologie der Pflanzen (ZMBP), Pflanzenphysiologie, Universität Tübingen, Auf der Morgenstelle 1, 72076 Tübingen, Germany
uwe.ludewig@zmbp.uni-tuebingen.de

1 Summary

Membrane channels of the **M**ajor **I**ntrinsic **P**rotein (MIP) family are pores for water and solutes. Such channels are found in almost all organisms and are especially abundant in plants. Experimental evidence suggests that some MIPs are aquaporins and are therefore relatively selective for water. Others, in contrast, conduct small solutes such as glycerol, urea or boric acid, and are therefore called aquaglyceroporins. Some MIPs conduct solutes that in isolation would be gases, such as NH_3. The structure of many aquaporins, including one from plants, has been resolved on the molecular level and all have a similar fold. To identify the molecular determinants of the selectivity of MIPs from plants and to identify the physiologically relevant transported solutes, a strategy based on computer modeling, combined with experimental verification has been set up. MIP homologs from plants were homology modeled and various simulations have been applied to measure their conductance for specific solutes, such as ammonia and/or urea. Urea is quantitatively the most important nitrogen fertilizer used worldwide, and its transport in plants is therefore of high interest. The molecular dynamics simulations suggest that specific MIPs have preferences for urea and specific solutes such as urea and/or ammonia are transported. Although the size of the solute and the pore diameter are important for transport, these do not exclusively determine if a solute is transported.

1.1 Diversity and Structure of Major Intrinsic Proteins (MIPs)

In the plant *Arabidopsis thaliana*, 35 genes encode MIPs, which can be grouped by sequence similarity into four subfamlilies, the **P**lasmamembrane **I**ntrinsic **P**roteins (PIPs), **T**onoplast **I**ntrinsic **P**roteins (TIPs), **N**OD-26-like **I**ntrinsic **P**roteins (NIPs) and **S**mall **I**ntrinsic **P**roteins (SIPs). In unicellular organisms, such as bacteria, much less MIPs are identified. Often the genome of such simpler organisms encodes only two MIP genes, a water-specific and a glycerol-specific channel. In humans, eleven MIPs have been identified, most of these can transport water at high rate, but some can also transport other solutes such as glycerol. In plants, all analyzed members of the PIP group are plasma membrane localized, while those of the TIP group are localized to an intracellular membrane, the tonoplast. At least some NIPs are localized to the plasma membrane, while SIPs are located at the membrane of the endoplasmic reticulum. After our initial observation that some MIP homologs from *Arabidopsis* transport ammonia (Loqué et al., 2005) and/or urea (Liu et al., 2003) in addition to water, we sought to identify

1. if small solute transport is a general property of plant MIPs
2. what are the molecular determinants of small solute transport
3. if computer simulations be used to predict novel and possibly physiological substrates of MIPs.

Computer simulations of water and glycerol transport in MIPs has been reported in numerous publications; these simulations more or less reproduce the observed water transport rates and give an idea about the selectivity of aquaporins (Ash et al., 2004). Computer simulations of water transport in aquaporin channels are a meaningful test for molecular dynamics (MD) simulations, since the transport rates in water channels are in the range of 10^8 per second. These fast rates allow resolving individual transport events even with current computer power, even though such calculations involve the simultaneous calculation of the movement and interaction of about 106'000 atoms. In addition, many MIP channels have been crystallized, and already 18 MIP structures are available at atomic resolution in the PDB database. Thus MIPs are an outstanding protein family to check and verify programs used for homology modeling of membrane proteins. In addition, homology models can be made with outstanding confidence as they are based on many different templates. All MIP homologs have a similar overall structure with four identical pores arranged in a tetrameric assembly. Each pore forms an identical pathway for water in its center. If the central pore within the tetramer can also conduct any solute, is currently unclear but unlikely; mutations in the individual pores of the monomers have been shown to alter conductance for several solutes. These experiments and MD calculations show that the pores in the center of each monomer are the major pathways for water and glycerol.

The pore within individual monomers has specific molecular features: first, close to the external pore mouth, the pore diameter is minimal and the residues surrounding the narrowest pore part are called selectivity filter residues. Second, a pair of Asn-Pro-Ala residues face into the center of each monomer; these residues are highly conserved between MIPs and are important for proton exclusion in these channels.

Only recently, the structures of the first plant aquaporin have been deposited in the PDB database (Törnroth-Horsefield et al., 2006). Structures of an open and a closed form of the spinach SoPIP2;1 aquaporin were resolved;

these structures show that plant aquaporins are gated channels and that the gating involves movement of a cytoplasmic loop that occludes the pore. The overall pore size and diameter at the narrowest region of the pore are unchanged in the open and closed conformation. By gating, a cytoplasmic loop plugs the internal pore mouth to close the channel. The overall pore architecture is unchanged by that gating also the pore diameter at the "selectivity filter" was not affected.

1.2 Modeling the Structure of Plant Major Intrinsic Proteins (MIPs)

Based on the homology to three-dimensional structures of several MIPs solved by X-ray crystallography, homology models of MIPs from *Arabidopsis* were constructed using the program modeller (Fiser and Sali, 2003), which has been compared with other homology prediction programs and is among the best performing programs to date (Wallner and Elofsson, 2005). Initial models were mostly based on the high-resolution structures of bovine and bacterial aquaporins. The homology structure of AtPIP2;1 was analyzed in most detail, but other homologs were analyzed as well. AtPIP2;1 was chosen because it is highly expressed in plants, can be functionally expressed in heterologous systems, such as oocytes and yeast, and the functional tests have revealed that it is highly selective to water. The analysis of the AtPIP2;1 homology model is shown in Fig. 1. The monomeric structures of a homology model were analyzed for bad stereochemistry and unrealistic bonds. The best models had only few residues with "bad" stereochemistry, but these were generally

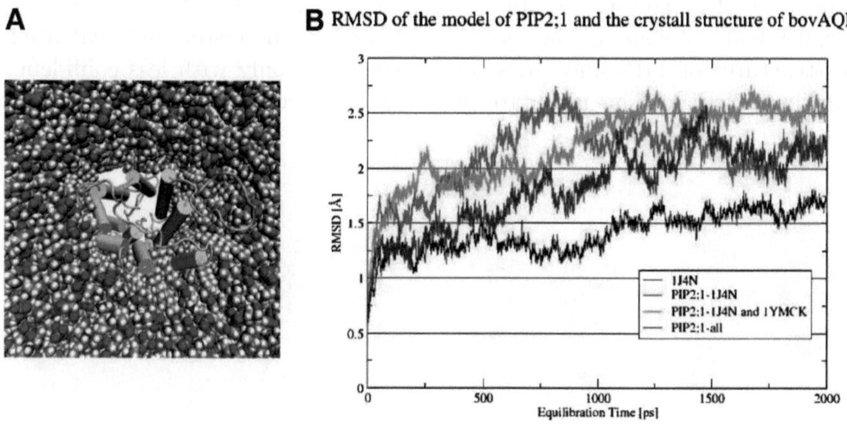

Fig. 1. Verification of homology model monomers by equilibration in a lipid POPE membrane. (**A**) Top view of the monomer in *yellow* (**B**) Time dependence of the RMSD of the original structure (*black*), or models of PIP2;1 based on bovAQP1 (*red*), based on bovAQP1 and AQPZ (*green*) or based on all available MIP structures (*blue*)

located outside of the pore structure. Further evaluation was done by measureing the stability of the different models. After a few steps of minimization to eliminate "clashes", the monomeric structres were equlibrated for 2ns. The change of the **R**oot **M**ean **S**quare **D**eviation (RMSD) of the $C\alpha$ atoms over the time was recorded and the obtained values from the models were compared with the values from the crystal structure of bovAQP1 (PDB accession code: 1J4N). In Fig. 1 it is shown, that, irrespective of which template was used for modeling AtPIP2;1, a slightly worse stability of the models was encountered. However, the models showed reasonable stability over time, suggesting that they provide reasonable models for further simulations. Similar calculations were done with tetrameric MIPs embedded in a patch of palmitoyloleoyl phosphatidylethanolamine (POPE) bilayer ; such models did not give evidence for an influence of neighboring subunits on the structure of a monomer. Many Simulations have now been shown that at least in short simulations, individual pores in a tetramer behave identical. If neighboring pores affect the gating of individual pores cannot be tested by such short simulations; a single example for such inter-subunit communication has been reported in the literature.

In the meantime, the availability of the PDB structure of the first plant aquaporin from spinach allowed to predict even superior homology models: SoPIP2;1 has 78% protein identity with AtPIP2;1 compared to only 32% identity with mammalian aquaporins. Most of the divergence is in extrahelical loops which are not involved in determining the selectivity. In the core structure of the pore, however, excluding the extrahelical external and cytoplasmic loops, the predicted homology structures and pore diameters were highly similar, independent of the template used. Homology models of several MIPs from *Arabidopsis* were similarly based on the closed form of SoPIP2;1 (which has the highest resolution) and are shown in Fig. 2. Although the overall fidelity of some of the models still needs to be tested and evaluated, the structures of TIPs and NIPs can be predicted only with less confidence than those of PIPs. This is due to the lower sequence similarity, but as a rule,

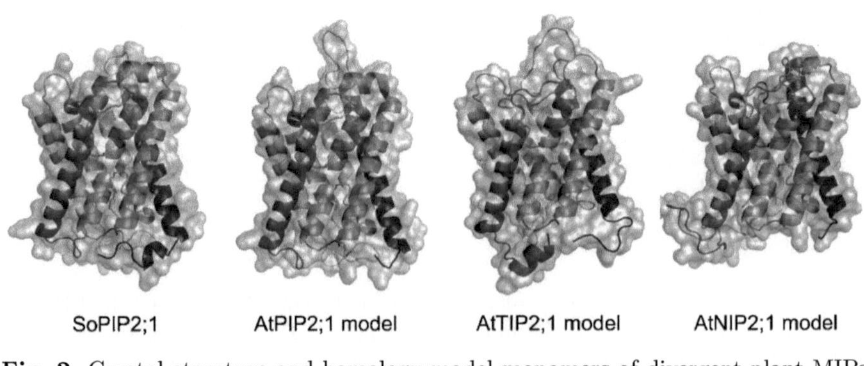

Fig. 2. Crystal structure and homology model monomers of divergent plant MIPs based on SoPIP2;1 (closed conformation, PDB accession code: 1Z98)

the narrowest and central part of the pore, which is critically involved in the selectivity, appears to be always predicted with relative high confidence.

Due to the exceptional high similarity of AtPIP2;1 and SoPIP2;1, we also modeled pore mutants of AtPIP2;1. Such mutants have the protein backbone of AtPIP2;1, but are mutated in only three residues. The mutated residues are in the center of the pore and surround the most constricted region of the pore. Thus, these residues constitute the selectivity filter of the pore. As an example, the pore mutant AtPIP2;1-K5 is explicitly shown: in that mutant the selectivity filter residues are exchanged into the respective residues that are identified in AtNIP5;1: Phe57 is exchanged to Ala, His186 is exchanged to Ile and Thr195 is exchanged to Gly. The position of the residues and a pore analysis of the homology model are shown in Fig. 3. It is evident that the exchange of these residues leads to a wider pore. As a result, it is expected that larger solutes may conduct the pore. Indeed, heterologous expression of

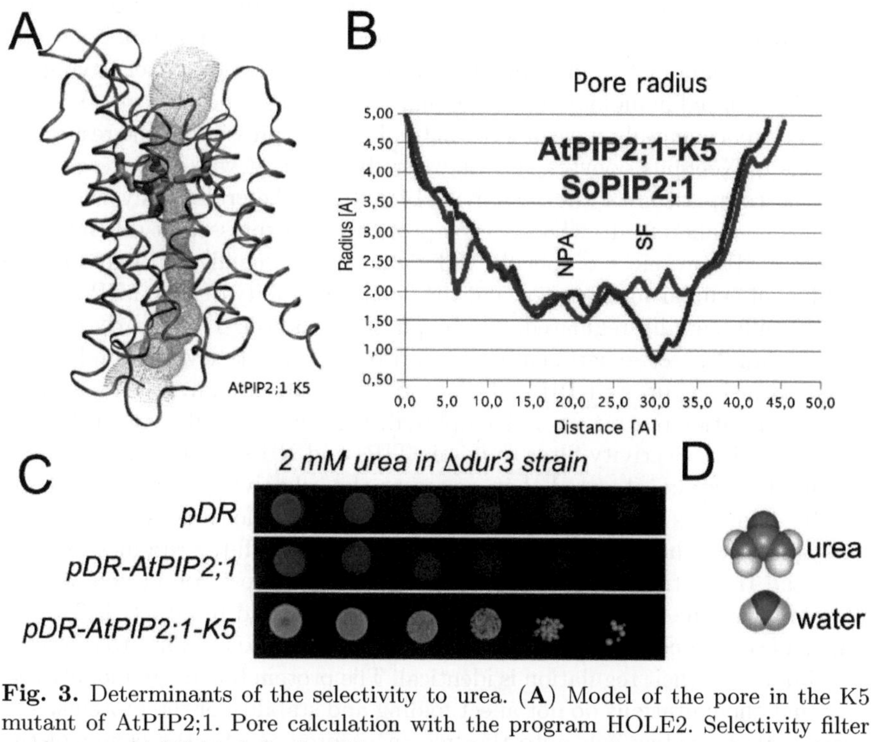

Fig. 3. Determinants of the selectivity to urea. (**A**) Model of the pore in the K5 mutant of AtPIP2;1. Pore calculation with the program HOLE2. Selectivity filter residues are highlighted. In the K5 mutant, three residues from the selectivity filter are exchanged: F57A,H186I,T195G. In that mutant, the selectivity filter corresponds to that of AtNIP5;1. (**B**) Pore sizes in SoPIP2;1 and the model of AtPIP2;1-K5. (**C**) Experimental verification of urea transport by yeast transformed with different constructs in a growth assay with urea as sole nitrogen source. pDR: vector control: pDR-AtPIP2;1 yeast expressing the water selective AtPIP2;1. (**D**) size comparison of urea and water

the pore mutant AtPIP2;1-K5 in yeast shows evidence for urea transport of that mutant. In contrast, the original AtPIP2;1 water channel does not show detectable urea transport (Fig. 3). AtNIP5;1, which has the same selectivity filter as AtPIP2;1-K5, also transports urea (data not shown).

AtNIP5;1 also transports boron, and it has been shown that the boron transport is physiologically relevant (Takano et al., 2006). Interestingly, while transporting boric acid at high rate, AtNIP5;1 only marginally transports water, which is much smaller than boric acid. This shows that solutes are not exclusively selected by size. The reason for the low water permeability can therefore not be deduced from the pore diameter and the transported substrates cannot be identified by simply analyzing the pore size.

1.3 Independent Experimental Analysis of the Selectivity of MIPs

The quality of computer simulations (and the prediction of transported solutes) requires independent experimental validation and verification. While MIP monomer models are equally accessible in computer simulations and molecular dynamics, the proteins themselves are clearly not. After having set up and evaluated homology models of plant MIPs, the selectivity of MIPs was directly experimentally measured by heterologous expression of the respective proteins in Xenopus oocytes and/or yeast. However, it was early recognized that the comparison of the selectivity of different heterologously expressed proteins has inherent drawbacks: each protein may expressed at different levels, each MIP protein has specific, individual target signals and may localize to different cellular membranes, and the open probability of each MIP protein may be differentially regulated, e.g. by phosphorylation. This clearly impairs a simple and direct experimental comparison of the selectivity of individual heterologously expressed PIPs, TIPs and NIPs. To circumvent these problems, a mutational strategy was employed: the water specific AtPIP2;1 was mutated in its selectivity filter to mimic TIPs and NIPs. Using that strategy, 8 mutants were constructed that represent the selectivity filter of 32 of the 35 MIP homologs in *Arabidopsis*. Only the 3 SIPs are not represented in that analysis. Since they are more distant, the selectivity filter residues are less clear and their structure can only be predicted with less confidence. As the protein backbone is identical in AtPIP2;1 wild type and in all AtPIP2;1 mutants, all constructs have the same target signals and thus localize to the same membrane and their regulation is identical. The protein homology models suggest that the mutations do not affect folding and stability of the proteins, and the mutated residues fold into the free space provided by the center of the pore. Importantly, the homology models of all constructs with the AtPIP2;1 backbone can all be made with the same confidence. The derived models are indeed identical, with the exception in the selectivity filter region. Expression of the proteins in selective yeast strains revealed that several, but not all, mutants conduct urea and/or ammonia. Comparison with the respective TIP or NIP channels with identical selectivity residues showed for several examples (we have not cloned all TIPs and NIPs) that the respective, "original" TIP or NIP channels conduct the urea and/or ammonia similar as the mutant constructs (in AtPIP2;1, in which only the selectivity filter is changed TIP- or NIP-like). These data experimentally verify additionally that the "selectivity filter" residues indeed determine the selectivity in the MIPs.

Basic molecular dynamics simulations on the selectivity of MIPs Initial computer simulations were performed using a very simplified setup: Only a monomer was used, the outer atoms of protein were fixed, allowing no overall movements of the protein, but movements of the residues along the pore. The use of a monomer, but not the full aquaporin tetramers in simulations of water/solute transport has been reported and found reasonable (Grayson et al., 2003). This is probably in part due to the fact that the selectivity filter is deeply burried within the central pore. Simulations with homology model monomers appear to be a reasonable abstraction of the real situation and have been published (Law and Sansom, 2004). The substrate of interest was placed close to the pore entrance and pushed using a small constant force through the pore. Although the Brownian motion is "accelerated" and "steered" in such simulations by the force applied to the substrate, such "steered" MD simulations have been shown to give reasonable results on aquaporin conduction (Jensen et al., 2002). In these first attempts, the interaction of the solute with water in the pore was neglected as the pore was not hydrated. Thus, competition between water and substrate molecules was excluded, an unrealistic situation. Despite these simplifications and the simple setup, initial results suggested that urea is not conducted by AtPIP2;1, but is conducted by AtTIP2;1. These results from the initial computer simulations reflect the real situation in MIPs: Expression in selective yeast strains showed that AtPIP2;1 does not conduct urea, while AtTIP2;1 conducts urea. During the analysis with urea as the solute of interest, it became evident that AtPIP2;1 mutants with the selectivity filter as in AtTIP2;1 did not conduct urea. This was surprising, as AtTIP2;1 had been shown to conduct urea. This was analyzed in more detail. Simulations with models of the AtPIP2;1 mutant with the selectivity filter as in AtTIP2;1 (AtPIP2;1-K1) showed, that the exchange of F87 to His blocks the pore in such mutants. In AtPIP2;1, a neighbouring residue to F87 is T55. The side chain of that threonine points into the direction of F87. In contrast, the corresponding residue in AtTIP2;1 is 87H in combination with 55G. As the glycine at the corresponding position has no side chain, the His in AtTIP2;1 can be positioned and coordinated well; if that glycine is exchanged to threonine, the His points into the center of the pore and blocks the pore (Fig. 4). The expression of the AtPIP2;1-K1 construct with the additional mutation T55G in yeast identified urea conductance for this modified AtPIP2;1-K1 construct and verified the predictions made from the modeling. The single exchange T55G did not affect transport in simulations and yeast experiments. These results show that even residues in the "second line", behind the primary selectivity filter residues, can affect the solute conductance in combination with other residues of the selectivity filter in plant aquaporins.

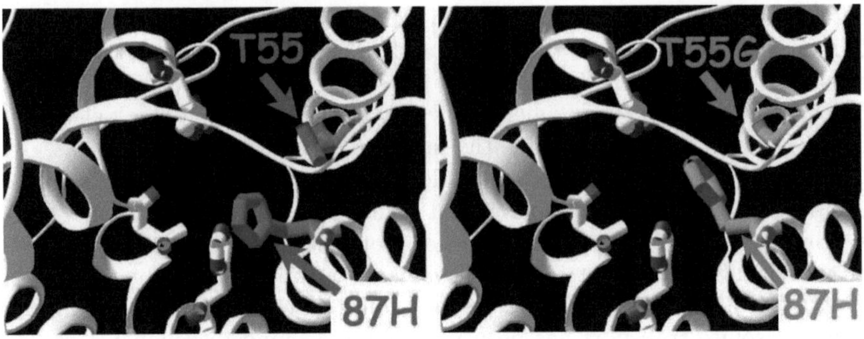

Fig. 4. Selectivity filter residues affected by nearby residues. Top views of a mutant aquaporin with selectivity filter as in TIP2;1. Note that residue 55H is bent into the pore center and partially blocks it because of nearby residue T55 (*left*). A further mutation, T55G, allows enough space for residue 55H to be correctly positioned in the mutant (*right*)

Further simulations were done with a hydrated pore. Such simulations were tested for several solutes and different crystal structures and homology models. As protein residues of the MIP were fixed, conformational changes in such simulations are excluded; however this may not be too artificial, as conformational changes occur on a slower time scale and thus should not influence the observed transport events. A major problem, however, may occur in the way that the substrate is forced into a wrong and artificial direction. To circumvent such problems, further simulations were set up.

A more realistic situation is now reflected in the more sophisticated models, which likely much better reflect the realistic situation of solute conductance: the monomer is embedded in lipid, minimized and equilibrated, a water shell surrounds the system and a constant hydrostatic pressure is applied to more distant water. The setup is shown in Fig. 5. Such pressure stimulated water transport in aquaporins has been analyzed by another group (Zhu et al., 2002). In such simulations, periodic boundary conditions are used and the solute is not forced into the channel by a directly applied force. As a result, the probability of entering the pore is much reduced, and longer calculation times are necessary to identify transport events. The homology modeled MIP monomer of interest can be easily exchanged in these models and statistical analysis is needed to evaluate the quality of such simulations. The comparison with the experimental results from expression of the respective proteins in yeast will identify if the simulations can give valuable results for determination of the selectivity. A final analysis needs to clarify which computational effort is minimally needed to give a qualitative and quantitative estimate of the transport or exclusion of a solute by a specific MIP.

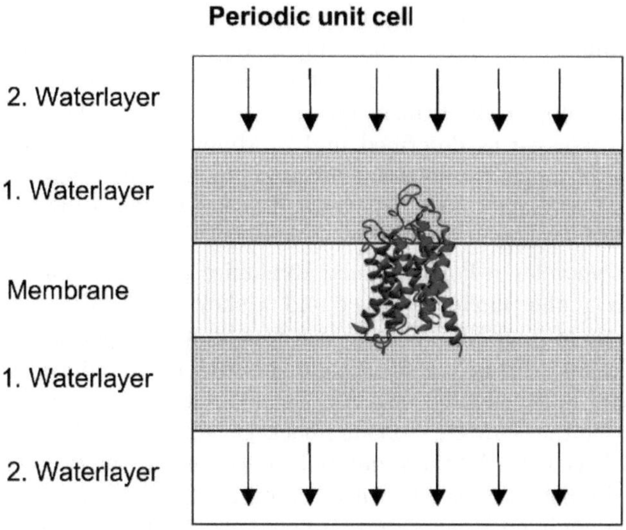

Fig. 5. Model of the conditions used to test the selectivity. Periodic boundary conditions are applied, and an equilibrated monomer in a lipid membrane is surrounded by water. A small force that mimics hydrostatic pressure is applied to the water in more distant layers, but not to the solutes, which are only placed into the water layer close to the membrane. The substrate of interest is placed close to the pore entrance to ensure a high probability to hit the pore entrance

1.4 Outlook

After the basic setup for the computational analysis for MIP selectivity has been set up, several runs need to be conducted to get some statistics of the transport events. Nearly identical simulations with several MIP homology models need to be compared to give a reasonable estimate how the computer simulations reflect the real situation of transport in plant MIPs. After initial tests with many different MIPs, the simulations have revealed that probably only some MIPs of the NIP subfamily are of potential physiological importance in ammonia and urea transport. We now concentrate our efforts on understanding the transport selectivity of those MIPs.

1.5 Computational Methods

Homology modeling – The primary sequence of plant aquaporin homologs were aligned with several MIPs of known crystal structure using ClustalW, followed by manual inspection of functionally important residues. Structural fitting was done using MODELLER[1] (Version 7v7, later 8v2), a well-established 3D structure homology modelling program (Wallner and Elofsson, 2005). The homol-

[1] http://salilab.org/modeller/

ogy models were evaluated using the programs WHATIF[2] and Procheck[3]. The obtained Ramachandran plots and comparsion with crystal strutures showed, that, generally, residues with "bad" stereochemistry were located outside of the core structure of the transmembrane pore. Further the stability of the models was assessed as described in 1.2. Minor attention was given to reconstruction of loops, as these are expected to minimally influence the core structure and selectivity filter. The pore diameter was measured with the program HOLE2[4] (Smart et al., 1993).

Molecular Dynamics – For molecular dynamics (MD), a monomer was embedded into a phospholipid bilayer and a shell of TIP3P water using the LEAP module of AMBER7. The system was limited to phospholipid and water molecules at a distance of 18 Å from the protein, to reduce the number of elements used in the simulations. All MD simulations presented in this work were done with the NAMD2[5] software package (Phillips et al., 2005), using the AMBER98 force field. All the simulations were carried out at 300 K. The time step of the simulations was 2 fs, with a cutoff at 12 Å for the nonbonded interactions. The nonbonded pairs were updated every 20 steps. All atoms within 18 Å of the protein monomer were allowed to move. Atoms 18Å far away from the center of the pore were fixed to prevent the moving of the protein during simulation. Prior to MD simulations, a series of minimizations were carried out with each complex model. A convergence criterion for the energy gradient of 0.5 kcal mol^{-1} $Å^{-1}$ was attained in all cases. After minimization, the lipid embedded protein was stepwise heated to 300 K and equilibrated for at least 200 ps at the final temperature. Data collection was then carried out for 0.5–3 ns. The substrate molecules were random positioned near the pore entrance and serveral simulation were done. To force the substrate molecule through the pore a constant force vector of 69-483fN (0.1-0.7 kcal mol^{-1} $Å^{-1}$) was applied to the center-of-mass (c.o.m.) of the substrate molecule or to each atom of the molecule. The trajectory of the solute on the same axis of the applied force was measured by the time of residency of the solute at a given position on each simulation. The ammount of force which was needed to "push" the molecule through the pore was used to assess the selectivity of the channel. Details of further simulations are detailed in the text.

References

W.L. Ash, M.R. Zlomislic, E.O. Oloo, and D.P. Tieleman. Computer simulations of membrane proteins. *Biochim Biophys Acta*, 1666 (1-2):158–189, Nov 2004. doi: 10.1016/j.bbamem.2004.04.012. URL `http://dx.doi.org/10.1016/j.bbamem.2004.04.012`.

[2] http://biotech.ebi.ac.uk:8400/
[3] http://www.biochem.ucl.ac.uk/ roman/procheck/procheck.html
[4] http://d2o.biop.ox.ac.uk:38080//, (Release 2.002)
[5] http://www.ks.uiuc.edu/Research/namd/

A. Fiser and A. Sali. Modeller: generation and refinement of homology-based protein structure models. *Methods Enzymol*, 374:461–491, 2003. doi: 10.1016/S0076-6879(03)74020-8. URL http://dx.doi.org/10.1016/S0076-6879(03)74020-8.

P. Grayson, E. Tajkhorshid, and K. Schulten. Mechanisms of selectivity in channels and enzymes studied with interactive molecular dynamics. *Biophys J*, 85(1):36–48, Jul 2003.

M. . Jensen, S. Park, E. Tajkhorshid, and K. Schulten. Energetics of glycerol conduction through aquaglyceroporin GlpF. *Proc Natl Acad Sci U S A*, 99(10):6731–6736, May 2002. doi: 10.1073/pnas.102649299. URL http://dx.doi.org/10.1073/pnas.102649299.

R.J. Law and M.S.P. Sansom. Homology modelling and molecular dynamics simulations: comparative studies of human aquaporin-1. *Eur Biophys J*, 33(6):477–489, Oct 2004. doi: 10.1007/s00249-004-0398-z. URL http://dx.doi.org/10.1007/s00249-004-0398-z.

L.-H. Liu, U. Ludewig, B. Gassert, W.B. Frommer, and N. von Wirén. Urea transport by nitrogen-regulated tonoplast intrinsic proteins in Arabidopsis. *Plant Physiol*, 133(3):1220–1228, Nov 2003. doi: 10.1104/pp.103.027409. URL http://dx.doi.org/10.1104/pp.103.027409.

D. Loqué, U. Ludewig, L. Yuan, and N. von Wirén. Tonoplast intrinsic proteins AtTIP2;1 and AtTIP2;3 facilitate NH3 transport into the vacuole. *Plant Physiol*, 137(2):671–680, Feb 2005. doi: 10.1104/pp.104.051268. URL http://dx.doi.org/10.1104/pp.104.051268.

J.C. Phillips, R. Braun, W. Wang, J. Gumbart, E. Tajkhorshid, E. Villa, C. Chipot, R.D. Skeel, L. Kal, and K. Schulten. Scalable molecular dynamics with NAMD. *J Comput Chem*, 26(16):1781–1802, Dec 2005. doi: 10.1002/jcc.20289. URL http://dx.doi.org/10.1002/jcc.20289.

O.S. Smart, J.M. Goodfellow, and B.A. Wallace. The pore dimensions of gramicidin A. *Biophys J*, 65(6):2455–2460, Dec 1993.

J. Takano, M. Wada, U. Ludewig, G. Schaaf, N. von Wirn, and T. Fujiwara. The Arabidopsis major intrinsic protein NIP5;1 is essential for efficient boron uptake and plant development under boron limitation. *Plant Cell*, 18(6):1498–1509, Jun 2006. doi: 10.1105/tpc.106.041640. URL http://dx.doi.org/10.1105/tpc.106.041640.

S. Törnroth-Horsefield, Y. Wang, K. Hedfalk, U. Johanson, M. Karlsson, E. Tajkhorshid, R. Neutze, and P. Kjellbom. Structural mechanism of plant aquaporin gating. *Nature*, 439(7077):688–694, Feb 2006. doi: 10.1038/nature04316. URL http://dx.doi.org/10.1038/nature04316.

B. Wallner and A. Elofsson. All are not equal: a benchmark of different homology modeling programs. *Protein Sci*, 14(5):1315–1327, May 2005. doi: 10.1110/ps.041253405. URL http://dx.doi.org/10.1110/ps.041253405.

F. Zhu, E. Tajkhorshid, and K. Schulten. Pressure-induced water transport in membrane channels studied by molecular dynamics. *Biophys J*, 83(1): 154–160, Jul 2002.

Computational Fluid Dynamics

Prof. Dr.-Ing. Siegfried Wagner

Institut für Aerodynamik und Gasdynamik, Universität Stuttgart,
Pfaffenwaldring 21, 70550 Stuttgart

The following paragraph represents the selection of papers submitted to HLRS that revealed a very high scientific standard and demonstrated also the unalterable usage of high performance computers (HPC) for the solution of the problem. CFD in combination with HPC allows the scientist many new insights into the mechanisms of flow and will continue to do so in the future when the performance and storage capacity of the computers and the capability of numerical methods will still grow. In this respect the availability of the NEC SX-8 high performance computer demonstrated clearly the big step forward compared to the situation at HLRS before.

One of the great challenges in fluid mechanics is still the proper simulation of turbulence. The most accurate way is direct numerical simulation (DNS) where turbulence is calculated and not simulated by a model. However, the extremely high computational efforts restrict application of DNS to simple flow problems and to small Reynolds numbers, orders of magnitude smaller than in engineering application. Despite of this restriction, DNS allowed already a large number of new insights into the mechanisms of turbulence. In order to reduce the computational effort, large eddy simulation (LES) is applied where only the large eddies are calculated whereas the small ones are represented by a model. Thus, two thirds of the papers in this paragraph deal with DNS and LES.

The first paper by Hetsch and Rist deals with DNS and the analysis of the stages of laminar-turbulent transition in the flow field around a swept laminar separation bubble. The high performance of the NEC SX-5 (around 5 GFlop/s per CPU) and the generous amount of memory (3730×545 grid points in x and y direction, respectively, and 31 Fourier modes in spanwise direction) allowed a new insight into the complex flow structures and stages of transition. Although a high speed and a vectorisation rate of 99.3% were reached a simulation time between 110 and 145 hours per run was required (220 to 2610 CPU hours).

Sander and Weigand investigate primary break-up phenomena in liquid sheets by DNS. Only through the availability of HPC and of sophisticated numerical models DNS has become the proper tool to study these phenomena, especially the influence of inflow conditions. The studies were performed on the CRAY-Opteron Cluster with 4 CPUS at HLRS using a grid resolution of 589,824 cells. One simulation required approximately 1500 CPUh. In accordance with experimental results the authors found that the break-up phenomena depend for instance primarily on the character of the mean axial velocity profile inside a turbulent nozzle flow.

Denev, Fröhlich and Bockhorn study the mixing and chemical reactions in a round jet impinging into a cross flow with the help of DNS. The simulations were performed on the HP XC-6000 Cluster of SSC Karlsruhe using 31 or 32 processors. The computational domain contained 22.3 Mio cells and was partitioned into 219 numerical blocks. The authors calculated so far 81.4 dimensionless time units. Each one consisted of 1220 time steps.

Marxen and Henningson investigate the bursting of a laminar separation bubble using DNS and show that a super computer is truly essential for this kind of studies. They used combined SMP and MPI parallelisation and 3 nodes (times 8 CPUs) as well as 33.36 GB memory on the NEC SX-8. The performance of the NEC SX-8 was 4.4 GFlop/s per CPU or 105.6 GFlop/s in total. Despite this high performance the bursting case used 5196 CPUh with an average vector length of 241. If the physical time of a bubble bursting has to be simulated by DNS it would take 30.000 CPU hours on the NEC SX-8 as estimated by the authors.

The next six papers are devoted to LES. Hauser addresses LES simulation with the software package UG with the goal to accelerate simulation processes, in the present case by grid adaptation and parallel computation. The fluid mechanical problems investigated are a 2-D jet in cross flow and a turbulent flow in a static mixer. In the first case a sequential computation would have needed more than half a year. On the HP XC-6000 Cluster using 8 processors on top level the complete computation could be carried out within several weeks. In the second case 32 nodes with 2 processors and a RAM allocation of 3.5 GB (2 million elements) were used requiring 25 CPU days.

Raufeisen, Breuer, Kumar, Botsch and Durst computed the flow and heat transfer in an idealized cylindrical Czochralski configuration using LES. This device is used to produce large crystals for the electronic industry. The simulation helps to produce crystals of high quality. The largest computational mesh consisted of about 8×10^6 nodes. An excellent average performance of 7.9 GFlop/s per CPU (or 63.2 GFlop/s total) on 8 processors with a vectorization ratio of 99.6% was achieved .

Hickel and Adams provide with the adaptive local deconvolution method (ALDM) a systematic framework for the implicit large-eddy simulation (ILES) of incompressible turbulent flows. With ALDM the unmodified conservation law is discretized. They also present a simplification of the algorithm and call it simplified adaptive local deconvolution (SALD) that is almost twice as fast as ALDM. They demonstrate the capability of ALDM and SALD by applying them to decaying grid-generated turbulence and turbulent channel flow. They reached a performance on a single NEC SX-8 CPU for the deconvolution operator and for the numerical flux function of the order 12–13 GFlop/s and 8–10 GFlop/s, respectively. They used open MP and MPI. An LES of a so-called periodic-hill flow required about 700 CPU hours using 8 SX-8 CPUs.

Alkistriwi, Meinke and Schröder performed LES of tendish flows. The goal was to develop a procedure to investigate flows where compressible and nearly incompressible flow regimes occur simultaneously. The simulations were conducted on the NEC SX-8 at HLRS using 28 CPUs and 15 GB memory. 7 GFlop/s were achieved on a single SX-8 processor. For a statistically converged solution of tendish flow they needed 250 CPU hours on 28 SX-8 CPUs with a total performance of 196 GFlop/s.

Large Eddy Simulations (LES) of open-channel flow over spheres were performed by Stösser and Rodi and were compared with recent laboratory experiments. The goal was to examine the complex flow structures, like low and high speed streaks as well as vortex growth. The simulations were performed on the HP XC-6000 Cluster of SSC Karlsruhe using up to 2,4 million grid points. Typically 11 seconds of CPD time were needed for one of the 600.000 time steps.

Magagnato, Pritz, Buchner and Gabi investigated resonance characteristics on Helmholtz resonator-type combustion chambers on the basis of LES. High pressure oscillations can lead to higher emissions and structural damage of the combustion chamber. The simulations agree well with experimental data and an analytical model. A typical simulation on the HP XC-6000 of SSC Karlsruhe using 32 processors ran for about 300 hours (elapsed time).

Zeiser studied the flow and species transport in packed beds, e.g. fixed bed reactors, by lattice Boltzmann (LBM) simulations and compared the results with CFX-5 simulations. Furthermore, a random walk particle tracking algorithm was used to complement the LBM method in order to widen the range of Peclet numbers. Finally, preliminary performance results of a new 1-D list based LBM implementation are shown that will soon replace the present full array based code and that has an advantage on vector computers as well as on cache based parallel architectures.

Harting and Giusepponi studied rheological properties of binary and ternary amphiphilic fluid mixtures, e.g. blood polymers, using the Lattice Boltzmann method. The simulation code LB-3D is written in FORTRAN 90 and uses MPI. It was run on several computers, NEC SX 8, IBM-p-Series, SGI Altrix and Origin, Cray T3E, Compaq Alpha clusters, as well as low cost 32- and 64-bit Linux clusters. Unfortunately, the authors give no information on the performance of these different platforms. They just mention that a 128^3 lattice would require around 2.2 GB of memory. Furthermore, they remark as a conservative rule of thumb, that the code would run over 10^4 lattice site updates per second per CPU on a "fairly recent machine" and would roughly linearly scale up to order 10^3 computer nodes. Since a 128^3 simulation contains around 2.1×10^6 lattice sites, running it for 1000 time steps would require about an hour of real time split across 64 CPUs.

C.F. Dietz, Henze, Neumann, von Wolfersdorf and Weigand investigated the effects of vortex generator arrays on heat transfer and flow field. They used a full differential Reynolds stress model and explicit algebraic models to simulate turbulence effects and heat transfer, respectively. The computations were performed on the CRAY Opteron Cluster of HLRS using 4 CPUs. The grids contained between 4,8 and 5,6 million cells and required a memory of about 7 GB. One simulation required usually 90 hours real-time on 4 CPUs.

Garcia-Villalba, Fröhlich and Rodi investigated the influence of the inlet geometry on the flow in a swirl burner, especially the flow oscillations and the noise generated. They got a good agreement between experiments and their simulations. They found different coherent structures depending on the inlet duct and the position of the inner jet. The computations were carried out on 32 processors on the HP XC-6000. Each simulation required a total number of 300,000 time steps on average. The total CPU time for one run was of the order of 47.000 hours. It was possible to get 32 processors for approximately 24 hours a day; the elapsed time was in the case 1495 hours, i.e. roughly 62 days.

Krause and Ballmann perform numerical investigations and simulations of transition effects in hypersonic intake flows of scramjets. Two different 2 D intake geometries were studied using different turbulence models. They showed that the influence of the wall temperature on flow separation and on the boundary layer thickness is very important. They use the DLR FLOWer Code. Three parallel computer systems were used, the SUN cluster of RWTH Aachen, the Jump cluster of the Research Centre Jülich and the NEC SX-8 of HLRS. A comparison of the computing times on these platforms for one simulation is presented. The next step should be parallel computing on NEC SX-8.

Dietz, Kessler and Krämer deliver a contribution to aeroelastic simulations of isolated rotors using weak fluid-structure especially when additional higher harmonic control flaps are attached to the trailing-edge of the rotor blade. The goal is an improvement of the reproduction of blade vortex interaction (BVI) induced oscillations in coupled CFD/CSD (coupled structural dynamics) computations. In order to detect the occurring higher harmonic oscillations of the blade high performance computing was mandatory. Up to 20 million grid cells were necessary for the resolution of the oscillations. 24 CPUs of the NEC SX-8 were used reaching a performance of 67.6 GFlop/s with a vector operation of 98.7%. Wall clock time per rotor revolution was 6,5 hours. The memory required was approximately 2 KB/cell.

Reimer, Braun und Ballmann performed a computational study of the aeroelastic equilibrium configuration of a swept wing model in subsonic flow of a wind tunnel. They applied three-dimensional Reynolds-averaged Navier-Stokes (RANS) equations using structured and unstructured grids. The goal of this project was to get information about the exchange of energy via the aerodynamic surface and about the change of the surface shape in order to properly predict the aeroelastic behaviour of a wing. The results of the simulation were validated by comparison with measurements. With FLOWer code (95% of computer time is necessary for CFD) they reached a code performance of 1300 MFlop/s and a vector operation ratio not less than 95% on a single NEC SX-8 processor. A simulation of one run took less than 1.5 hours and about 1.2 GB memory on 4 NEC SX-8 processors for a grid of about one million grid points.

Direct Numerical Simulation and Analysis of the Flow Field Around a Swept Laminar Separation Bubble

Tilman Hetsch[1] and Ulrich Rist[2]

[1] Aeronautics and Astronautics, School of Engineering Sciences, University of Southampton, Highfield, Southampton, SO17 1BJ, United Kingdom
hetsch@soton.ac.uk
[2] Institut für Aerodynamik und Gasdynamik, Universität Stuttgart, Pfaffenwaldring 21, 70550 Stuttgart, Germany.
rist@iag.uni-stuttgart.de

Abstract. The transition process around a short leading-edge separation bubble subjected to a sweep angle of 30° is studied in detail by means of direct numerical simulation, spatial linear stability theory and solutions of the parabolised stability equations. The combined analysis of the averaged flow field, instantaneous flow visualisations and postprocessing data as amplification curves leads to the distinction of four succeeding stages qualitatively comparable to the unswept case. It is shown that the saturation of background disturbances is the key event, after which a rapid breakdown of transitional structures occurs. The mechanism of the final breakdown of this swept scenario of fundamental resonance is best described as an "*oblique K-type transition*". Great care is taken to isolate and describe of typical structures within each stage as a foundation for the analysis of complex transition scenarios.

1 Introduction

Separation bubbles are observed when laminar boundary layers encounter strong adverse pressure gradients, as on high-lift devices of commercial aircrafts or turbine blades. For instance, a swept separation bubble was measured at a Mach number of 0.245 by Greff [3] on the slat of an Airbus A310-300 in landing configuration. Although most modern passenger airplanes exhibit a sweep angle, research efforts so far have been focused almost exclusively on the easier unswept case. The largest body of experimental data devoted to swept separation bubbles known to the authors was published by Young & Horton [17] and Horton [8] in the late 1960s. Apart from Horton's benchmark data, literature on swept separation bubbles is still extremely rare. Davis, Carter & Reshotko [2] successfully validated a boundary layer code against Horton's data and thus confirmed the experiments. More recently, Kaltenbach & Janke [10] published the direct numerical simulation *(DNS)* of a separation bubble in the flow behind a swept, rearward-facing step. For the lack of literature little is known about the structure, behaviour and the transition mechanisms of swept separation bubbles and especially no systematic picture is established how they can be related to their by now well-investigated unswept counterparts. Note that separation bubbles can be subclassified into laminar or transitional separation bubbles depending on a laminar or turbulent reattachment. This paper is organised in the following way: After a description of the investigated series of swept laminar separation bubbles in Sect. 2 one disturbance scenario is chosen for a detailed analysis in Sect. 3. The aim is two-fold: Firstly it is determined how typical transitional structures express themselves in flow visualisations. This enables us to distinguish a succession of transitional stages in the flow field around a swept laminar separation bubble. Both topics will be addressed in the main Sect. 3.4. Before, the disturbance contents of the flow is analysed in Sect. 3.1, properties of the averaged flow field are studied in 3.2 and the onset of turbulent flow is discussed in Sect. 3.3. Section 4 then describes the *DNS*-codes and the necessary computational resources followed by the conclusions in Sect. 5.

2 Description of the Flow Field

The unswept prototype of the present leading edge bubble was extensively studied by Rist [13] by means of *DNS* and linear stability theory *(LST)*. Its extension to swept flows, a verification and validation, as well as the effect of sweep on the base-flow and a *LST* analysis were published in Hetsch & Rist [5]. The reliability and accuracy of *LST* and the parabolised stability equations *(PSE)* in swept laminar separation bubbles were subject of a quantitative investigation in Hetsch & Rist [7]. All *PSE*-results are obtained by the linear version of the code 'nolot' of the DLR-Göttingen. It is described by Hein [4], who used the unswept version of the present base flow to prove the applicability of *PSE* to laminar separation bubbles for the first time. Finally, first results about the effect of an increasing sweep angle on the disturbance development in this configuration were reported in Hetsch & Rist [6]. All simulations are split in a *DNS* of the steady laminar base flow Q and a succeeding unsteady *DNS* of the disturbance propagation q' within this base flow. So for any flow quantity $q \in \{u, v, w\}$ the solution takes the form of the decomposition $q(t, x, y, z) = Q(x, y) + q'(t, x, y, z)$.

Only important parameters of the base flow already described in [5] are repeated here: All quantities in the paper are non-dimensionalised by the reference length $\overline{L} = 0.05\,\mathrm{m}$ and the chordwise free-stream velocity $\overline{U}_\infty = 30\,\frac{\mathrm{m}}{\mathrm{s}}$, which is held constant for all cases. The x- y- and z-direction are taken normal to the leading edge, wall-normal, and parallel to the leading edge with the associated base flow velocity components U, V and W, respectively. Period-

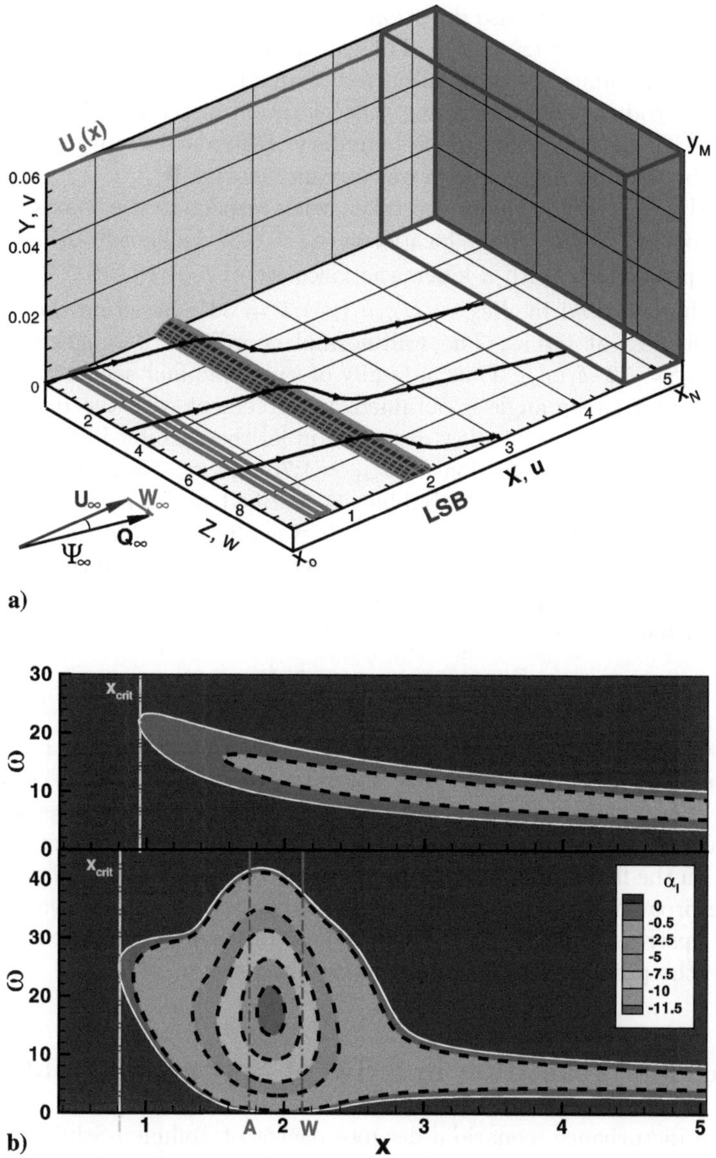

Fig. 1. Overview and properties of the present 30°-base flow. (**a**) Computational domain with 30°-separation bubble and streamlines. *Inflow*: Sweep angle Ψ_∞, freestream velocity Q_∞ with components U_∞, W_∞. *Inside*: disturbance strip, dividing streamline $\boldsymbol{\Psi}_0$ of the bubble. *Outflow*: schematic sketch of damping zone. Upper boundary condition U_e: potential flow deceleration. (**b**) Comparison of *LST*-amplification rates for waves with spanwise wave number $\gamma = 0$ of present flow (*bottom*) with 30°-'Blasius' (*top*), which results *without* potential flow deceleration. Upstream shift of x_{crit} (point of first disturbance amplification). *A*: separation, *W*: reattachment

icity is assumed in spanwise direction only, resulting in a quasi-2D base flow with ($\frac{\partial}{\partial z} \equiv 0$), but $W(x, y) \neq 0$. The calculation domain shown in Fig. 1(a) consists of an infinite flat plate subjected to an adverse pressure gradient. The latter is introduced by prescribing a deceleration of the chordwise potential flow velocity $U_e(x)$ at the upper boundary. Different sweep angles Ψ_∞ are realised by varying the spanwise free stream velocity $W_\infty = U_\infty \tan(\Psi)$ and setting $W_e(x) \equiv W_\infty$. Angles are taken with respect to the x-axis throughout the paper. At the inflow located at $x_o = 0.37$ Falkner-Scan-Cooke profiles are prescribed. With a kinematic viscosity of $\overline{\nu} = 15 \cdot 10^{-6} \frac{\mathrm{m}^2}{\mathrm{s}}$ the flow can be characterised by $Re_{\delta_1} = \overline{U}_\infty \overline{\delta}_1(x_o)/\overline{\nu} = 331$, based on the displacement thickness at inflow. The wall-normal coordinate y ranges from 0 to $y_M = 0.238 = 72 \cdot \delta_1(x_o)$. Thus, a family of swept laminar separation bubbles with arbitrary sweep angle is obtained. In agreement with the *independence principle* of incompressible flow discussed in [5] they exhibit identical separation and reattachment positions at $x_{\text{sep}} = 1.75$ and $x_{\text{reat}} = 2.13$, respectively. The steady calculation of the bubbles is justified by its small size and experience with the unswept case in [13]. It was already shown in [7] that the linear stability theory is very accurate in predicting the streamwise wave number α_r. Therefore, the propagation direction Ψ, wavelength λ or phase speed c_r of a disturbance wave

$$\Psi := \arctan(\gamma/\alpha_r), \quad \lambda := 2\,\pi/\sqrt{\alpha_r^2 + \gamma^2}, \quad c_r := \omega/(\alpha_r^2 + \gamma^2) \qquad (1)$$

are based on *LST* throughout this paper, if not stated otherwise. Figure 1(b) displays an overview over the linear stability properties of the 30°-separation bubble in comparison the same flow field *without* the adverse pressure gradient. The presence of the small leading-edge bubble obviously has a remarkable impact on the flow stability: The amplification rates are up to 16 times higher, a much broader frequency-spectrum of disturbances is amplified and the elliptical upstream influence of the separation bubble may be noticed by a shift of x_{crit}, the point where the base flow first becomes unstable.

3 Stages of Transition in a Swept Separation Bubble

For each disturbance scenario a discrete packet of Tollmien-Schlichting (TS) waves is generated by means of suction and blowing through a disturbance strip at $x \in [0.5; 0.64]$. One selected "*primary disturbance*" (*PD*) is excited with an initial amplitude 5 orders of magnitude larger than all other modes. Additionally, 10 low-amplitude "*background disturbances*" (*BD*) with systematically varying spanwise wave numbers $\gamma \in [-50, -40, \ldots, 50]$ are introduced as partners for non-linear interactions. As we are interested in scenarios of *fundamental resonance* all waves share the (angular) frequency $\omega = 2\,\pi\,(\bar{L}/\bar{U}_\infty)\,f$ of the primary disturbance. After a initial transient phase steady boundary conditions and periodic wave excitation lead to a quasi-periodic state in time.

Together with periodic boundary conditions in the spanwise direction this allows for a double Fourier analysis in time and span in the postprocessing. It provides a decomposition of any disturbance quantity q' into Fourier modes (ω/γ) with amplitudes $\widehat{A}^q_{(\omega/\gamma)}$, from which amplification curves $q'_{(\omega,\gamma)}(x) = \max_y(\widehat{A}^q_{\omega,\gamma}(x,y))$ can be obtained. As modern passenger planes typically exhibit sweep angles of about $\Psi_\infty = 30°$ the disturbance scenario 30°-(20/20) was selected for a detailed analysis of the stages of transition in the flow field around a swept laminar separation bubble. Its primary disturbance showed the greatest disturbance amplification for the sweep angle $\Psi_\infty = 30°$ according to linear stability theory.

3.1 Non-Linear Wave Generation in the Disturbance Spectrum

From the initial disturbance spectrum further disturbances will develop by non-linear mechanisms. Mathematically non-linear wave generation and interaction have their origin in the non-linear convective terms of the Navier-Stokes equations. By studying the multiplication of two Fourier modes it can be shown that any non-linear interaction '⊕' results in the generation

$$(\omega_1/\gamma_1) \oplus (\omega_2/\gamma_2) = \underbrace{(2\,\omega_1/2\,\gamma_1) + (2\,\omega_2/2\,\gamma_2) + (0/0)}_{\text{autointeraction: } \mathcal{O}(amp^2)} + \underbrace{(\omega_1 \pm \omega_2/\gamma_1 \pm \gamma_2)}_{\text{interaction: } \mathcal{O}(amp_1 \cdot amp_2)},$$

plus its complex conjugate. Each time step every disturbance mode present in the spectrum generates its first higher harmonic and a contribution to the mean flow deformation (0/0). The amplitude of the higher harmonic will be approximately the square-amplitude of its generator. Furthermore, it interacts with every other disturbance by generating the new modes $(\omega_1 + \omega_2/\gamma_1 + \gamma_2)$ and $(\omega_1 - \omega_2/\gamma_1 - \gamma_2)$ with an initial amplitude of approximately the product of those of its generators. Because of the low amplitudes of the background disturbances only *direct* interactions with the primary disturbance will be large enough to contribute to the overall flow development:

$$\underbrace{(20/20)}_{PD} \oplus \underbrace{(20/\gamma)}_{BD} = \underbrace{(40/40)}_{1.HH} + (0/0) + \underbrace{(40/20+\gamma)}_{TS} + \underbrace{(0/20-\gamma)}_{CF}. \qquad (2)$$

Any background disturbance in the initial spectrum generates therefore a Tollmien-Schlichting wave and a steady crossflow wave (*CF*) with the primary disturbance. With the exception of further higher harmonics of the primary disturbance itself any subsequent interactions of these new modes will again be too small to be of importance. All modes predicted in (2) can be detected in the postprocessing as shown in the amplification curves of Fig. 2.

3.2 Analysis of the Averaged Flow Field

The direct comparison of the time- and spanwise averaged spanwise vorticity field $[\omega_z] := \Omega_z + (0/0)_{\omega_z}$ with the undisturbed base flow in Fig. 2 allows

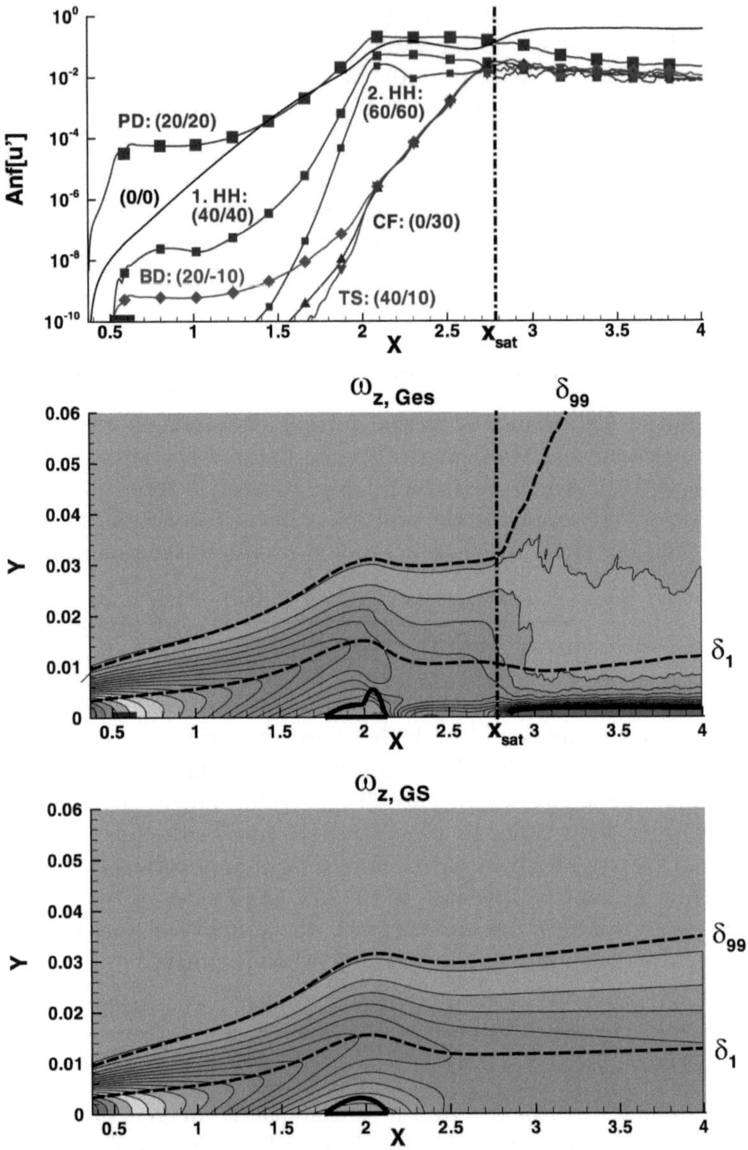

Fig. 2. 30°-scenario with primary disturbance (20/20): Comparison of amplification curves (shown are only the *PD*, the *BD* (20/ − 10) and modes directly generated by them, *top*) with contour plots of the time- and spanwise averaged total flow $[\omega_z'] = \Omega_z + (0/0)_{\omega_z}, [\omega_z'] \in [-16.9; 277.5]$: 23 iso-levels $[-45; -35; \ldots; 175]$, *middle*) and the undisturbed base flow ($\Omega_z \in [-5.4; 172.6]$: 23 iso-levels $[-45; -35; \ldots; 175]$, *bottom*). Additionally: The boundary layer thicknesses δ_{99} calculated from $[u_{pseu}']$ and U_{pseu} respectively and the momentum thickness δ_1, as well as the saturation position of the background disturbances $x_{\text{sat}} = 2.78$, the dividing streamline $\boldsymbol{\Psi}_0$ of the separation bubble and the disturbance strip at $x \in \{0.50; 0.63\}$

the immediate distinction of three zones. However, the deceleration of the freestream prevents the use of the classical formulas for the boundary layer parameters, which would become misleading and domain height dependant. Following Spalart & Strelitz [15] and Marxen, Lang, Rist & Wagner [12] all boundary layer parameter were determined by the so-called *pseudo velocity* $[u_{pseu}]$:

$$[u_{pseu}](x,y) := \int_o^y [\omega_z](x,\tilde{y})\, d\tilde{y} \implies \delta_1(x) = \int_o^\infty \left(1 - \frac{[u_{pseu}](x,\tilde{y})}{[u_{e,pseu}](x)}\right) d\tilde{y}.$$

A preliminary classification of the flow field can now be established as follows: Until $x \approx 1.90$, where the mean flow deformation $(0/0)_u$ reaches about $1.5\%\, U_e$, no difference to the undisturbed flow appears. As shown in Fig. 2 the disturbances are still too small to generate a sufficient mean flow deformation to visibly influence the base flow terms of order $\mathcal{O}(1)$. The next zone still resembles the base flow, but distinct differences appear especially in the near-wall region. There, a noticeable rise in the averaged spanwise vorticity $[\omega_z]$ and therefore the wall friction is indicated by an accumulation of isolines and confirmed by Fig. 3(a). Under the influence of the disturbances the rear part of the separation bubble changes towards a roughly triangle-shaped outline observable in experiments as shown in Fig. 1.5 in [13]. It also develops the typical pressure plateau displayed in Fig. 3(a). Qualitatively it compares well with a measured pressure distributions of an unswept transitional separation bubble reported by Lang in Fig. 3 in [12]. But with the simultaneous saturation of all background disturbances at $x_{sat} = 2.78$ any similarity with the base flow abruptly ends. The hitherto layered structure of the flow switches over to a more chaotic development of the isolines. This is accompanied by a steep rise in the wall friction and the boundary layer thickness δ_{99}. Other indicators also show a shift towards turbulent flow: In the regions of constant $[u_{e,pseu}]$ before and after the separation bubble the shape parameter H_{12} of Fig. 3(b) can be compared with classical results from Schlichting's book [14] for a two-dimensional flat plate without pressure gradient. It follows from the independence principle that the sweep angle Ψ_∞ will not have a major impact on the comparison. At about x_{sat} H_{12} drops quickly approaching the typical turbulent value of $H_{12} = 1.29$. From the classical theory one can also expect a ratio of $\delta_{99} : \delta_1 = 2.9 : 1$ for a Blasius flow, which is nearly exactly satisfied in the laminar inflow region yielding $2.86 : 1$. The ratio of $8 : 1$ for a fully developed turbulent boundary layer is approached, but not yet reached, towards the end of the domain, where a ratio of $6.8 : 1$ is found at $x = 4.0$. Finally, the velocity profiles in this region are noticeably fuller as compared to the laminar profiles in the first region. Thus after the saturation of the background disturbances a turbulent flow field emerges which quickly approaches the criteria of a fully developed turbulent boundary layer.

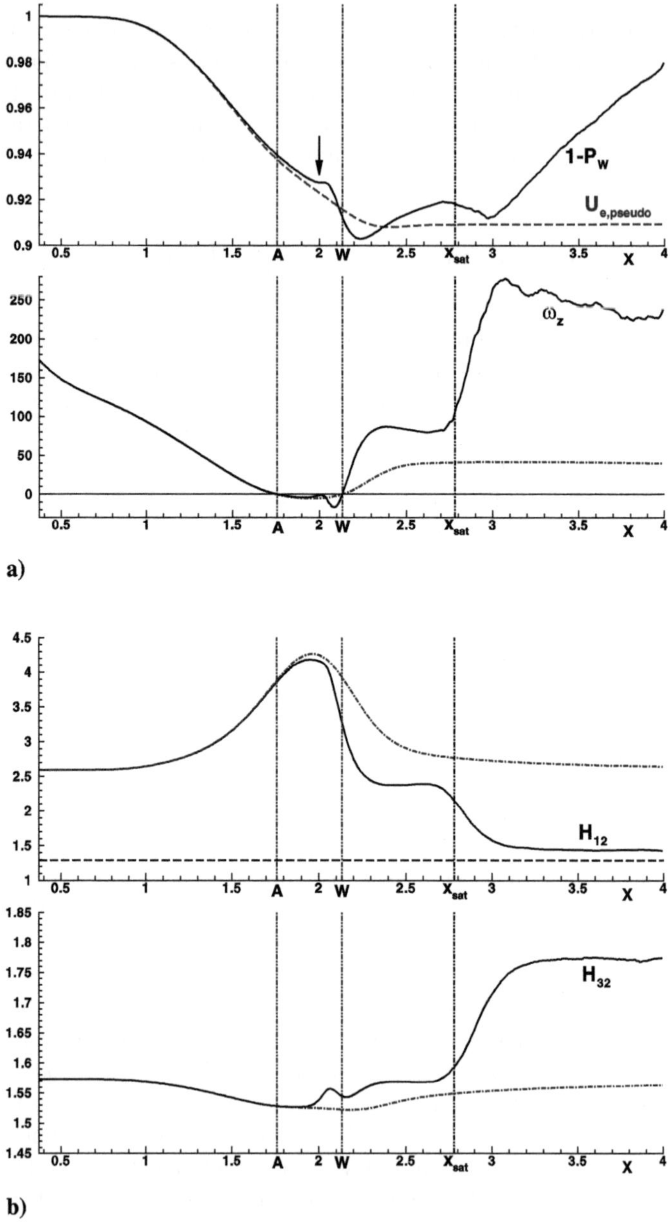

Fig. 3. Properties of the total flow $[q'] = Q + (0/0)_q$ averaged in time and span. The shape parameters were evaluated using u_{pseu}. *Vertical lines*: A $= x_{\mathrm{sep}}$, W $= x_{\mathrm{reat}}$, and position of the saturation of background disturbances x_{sat}. (**a**) Potential flow velocity $u_{e,pseu}$ at upper boundary, pressure plateau (*arrow*) of $1 - p_w/Re$ of the wall pressure and spanwise vorticity component ω_z there. (**b**) Shape parameters H_{12} and H_{32} compared to base flow (*dash-dotted*). Dashed: $H_{12} \equiv 1.29$, typical for fully developed turbulent boundary layer

3.3 On the Onset of Turbulence

Although laminar-turbulent transition is a process developing over a certain downstream region in stages, it is sometimes necessary – like in *RANS* calculations – to decide after which point the flow should be regarded as turbulent. According to Sect. 3.2 clearly the position of the saturation of the background disturbances x_{sat} should be considered for the present flow. To derive a criterion which yields an exact x-position without interpretation and can thus be programmed for automatic detection, the intersection of the amplification curves of the mean flow deformation $(0/0)_u$ and the primary disturbance was chosen. Physically, this marks the point where the background disturbances have reached a sufficiently high amplitude, such that their combined contribution to the mean flow deformation makes it surpass the primary disturbance as the dominating mode. Just as in Fig. 2 this point agrees remarkably well with the x_{sat}-position which one would choose on an intuitive level in all investigated scenarios. Furthermore, in disturbance scenarios *without* transition no such intersection occurs. But it should be noted that this intersection is not necessarily unique as also observable in Fig. 2. In further investigations not presented here the most effective primary disturbance for a given sweep angle was determined by comparison of the 'earliness' of the onset of turbulent flow. To this end the above mentioned criterion was deployed. The criterion also seems to work in totally different scenarios like a crossflow-vortex-induced transition published by Wassermann & Kloker, see Fig. 15 in [16]. Additionally, in the right picture of Fig. 14 of that publication the background disturbances where switched off. Consequently no transition occurred and the mean flow distortion mainly generated by the dominating vortex could never surpass its generator.

3.4 Characterisation and Properties of the Transition Stages

Figure 4 gives a complete overview over the transition scenario 30°-(20/20) at the end of the 60*th* disturbance period: Shown are alternating iso-surfaces of the disturbance-component of the spanwise vorticity $\omega_z' = \pm 0.0001$ up to $x = 1.45$ followed by iso-surfaces $\lambda_2 = -1$ and $\lambda_2 = -200$ of the λ_2-criterion of Jeong & Hussain [9]. Additionally, the boundary layer thickness δ_{99} and the two-dimensional dividing streamline $\boldsymbol{\Psi}_0$ of the bubble were calculated from data averaged in time and span. The latter is used to visualise the extent of the separation bubble with $\boldsymbol{\Psi}_0$ defined as the iso-surface $\boldsymbol{\Psi} = 0$ of the stream function $\boldsymbol{\Psi}(x, y) := \int_0^{ymax} [u](x, \widetilde{y})\, d\widetilde{y}$. Note that the figure shows twice the spanwise extent, but only the lower half of the actual calculation domain which was moreover cut at the beginning of the damping zone at $x = 4.0$. In order to decompose the overall picture into the different stages of transition, the top view is directly compared with postprocessing data in Fig. 5:

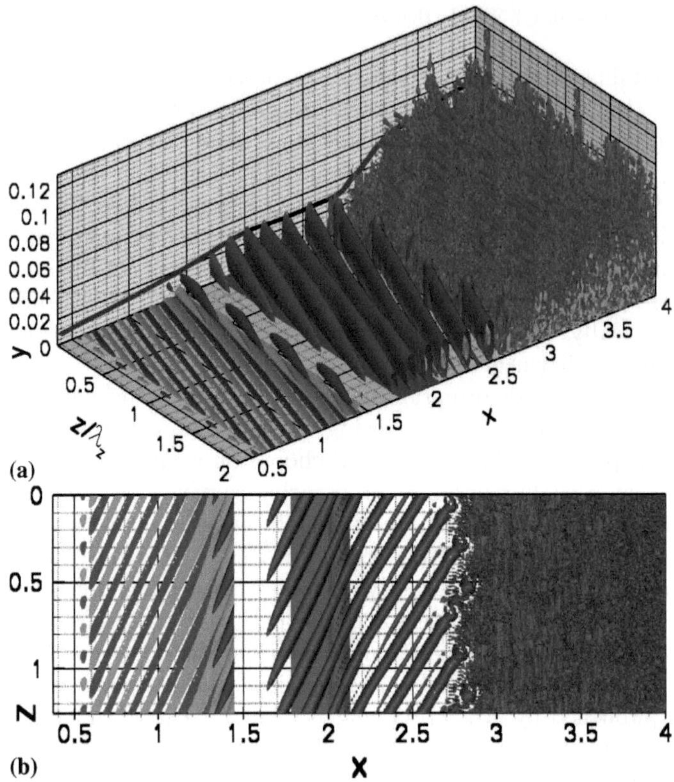

Fig. 4. 3D-visualisation of instantaneous data of the swept transition scenario 30°-(20/20) at the end of the 60*th* disturbance period: Restricted to $x \in [0.37; 1.45]$: ω_z'-iso-surfaces with $\omega_z' \equiv -0.0001$ (*red*) and $\omega_z' \equiv 0.0001$ (*orange*). Afterwards: λ_2-iso-surfaces with $\lambda_2 \equiv -1$ (*blue*), $\lambda_2 \equiv -200$ (*vortex axes, green*). From the averaged total flow solution: The dividing streamline $\boldsymbol{\Psi}_0$ of the separation bubble and the boundary layer thickness δ_{99} (*dark green stripe* in x, y-plane). (**a**) *Three-dimensional view*: As below twice the spanwise extent is displayed. (**b**) *Top view*: The x- and z-axis are to scale

Stage *(I)*: Linear Disturbance Amplification

One wavelength after the disturbance strip at the latest the boundary layer has filtered out additional disturbance waves of neighbouring wave length which are necessarily co-excited in the process of disturbance generation. Therefore $x = 0.8$ marks the beginning of the *linear domain*, where the growth of any mode can be very accurately described by *LST* and *PSE* as demonstrated in Fig. 5. At $x_{\text{LinEnd}} \approx 1.91$ the primary disturbances reach an amplitude of $3\% \, [u_{e,pseu}]$ and the amplification curves depart from the *PSE*-solution. Note that the *LST*-solutions depart earlier at 1.85, so that an investigation with *LST* only would yield a slightly inaccurate smaller linear domain ending with a primary disturbance of $1 - 2\% \, [u_{e,pseu}]$. As separation occurs at $x = 1.75$,

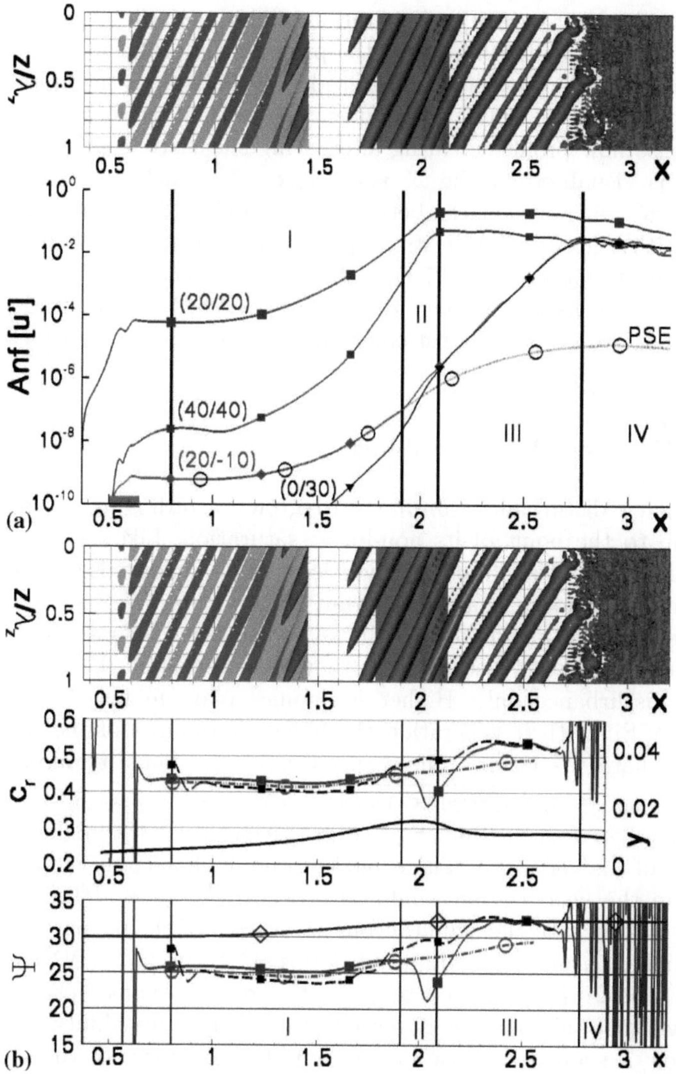

Fig. 5. Comparison of the visualisation 4(b) with the most important amplification curves from Fig. 2 (*top*) and parameters of the primary disturbance (20/20) (*bottom*). *Bottom*: *Circle*: *LST*, *squares*: *DNS* (*large*: close to wall, *small*: freestream), *diamonds*: Ψ_e. *Line without symbol*: δ_1. (**a**) Direct comparison with selected amplification curves from Fig. 2. (**b**) Direct comparison with phase speed c_r and propagation direction Ψ

the flow in the front part of the separation bubble can *still* be predicted by linear theories. Throughout stage *(I)* the primary disturbance dominates other disturbances by $2-5$ orders of magnitude. Visualisations therefore show this *TS*-wave in its pure form: Its oscillations periodically accelerate and decelerate the base flow profiles creating alternating shear stress at the wall, which in terms is visualised by the ω'_z iso-surfaces. The inclination of their wave fronts of 26° and wavelengths between 0.141 and 0.136 taken from the visualisation 4(b) agree well with its propagation angle $\Psi_{(20/20)}$ in Fig. 5(b) and its wavelength $\lambda_{(20/20)} = 0.137$ ($\bar{\lambda} = 6.9\,\text{mm}$) calculated from linear stability theory. As soon as the primary disturbance reaches an amplitude of about $0.1\%\,[u_{e,pseu}]$ at $x = 1.6$ the fist emergence of vortices can be detected by the λ_2-criterion.

Stage *(II)*: Secondary Instability

The primary disturbance *continues* to grow according to linear stability theory up to the point of its non-linear saturation. Likewise the dominant structures of the visualisation – vortices with a clockwise sense of rotation – show a smooth changeover from stage (I) to (II). As confirmed by the same angle of inclination as the ω'_z iso-surfaces before and by the absence of other relevant disturbances, they are induced by the high-amplitude primary disturbance only. Higher harmonics play *no* major part in their emergence: Since their generation the higher harmonics of the primary disturbance share the latter's phase speed $c_{r,(20/20)}$. Thus it follows from (1) that $\alpha_{r,(40/40)} = \sqrt{2\,\omega_{(20/20)}/c_{r,(20/20)} - 4\,\gamma_{(20/20)}} \approx 2\,\alpha_{r,20/20)}$ and therefore $\Psi_{(40/40)} \approx \Psi_{(20/20)}$, but $\lambda_{(40/40)} \approx \frac{1}{2}\lambda_{(20/20)}$. If the higher harmonics were part of the vortices, a noticeable shortening of the intervals between two vortices in the visualisation would have occurred. Stage *(II)* naturally ends at $x_{\text{PDsat}} = 2.08$ with the simultaneous saturation of the primary disturbance and its higher harmonics at amplitude levels of $22\%\,[u_{e,pseu}]$ (PD), $5\%\,[u_{e,pseu}]$ ((40/40)) and $2\%\,[u_{e,pseu}]$ ((60/60)). For the respective unswept case Rist [13] has shown that this phase is governed by secondary instability theory. Both cases show a sudden increase of the amplification rates of the background disturbances as a result of their resonance with the primary disturbance.

Stage *(III)*: Coherent Structure of Saturated *TS*-Waves

Sharing a common speed and direction the higher harmonics of the primary disturbance travel together with their generator, which remains unchanged after their simultaneous saturation. These saturated *TS*-waves form now a new entity – a *coherent structure* – which massively influences the background disturbances. A description of this coherent structure for unswept separation bubbles can be found in Rist [13]. The presence of the strong vortices formed by the coherent structure forces all background disturbances into a common dependency indicated by an identical growth rate in Fig. 2, which is slightly damped compared to the steep rise in stage *(II)*. The primary vortices in stage *(III)* are accompanied by weak secondary vortices at their rearward side close to the wall. Such secondary vortices are frequently observed whenever a strong vortex interacts with a wall, reported e.g. at the updraft side of a crossflow vortex in [16]. Particular to the present swept case is a sudden rise in the propagation direction $\Psi_{(20/20)}$ until it *exactly* matches the local freestream direction Ψ_e. Once Ψ_e is reached, $\Psi_{(20/20)}$ stays constant in Fig. 5(b) up to the final breakup of the coherent structure. According to (1) this rise must correspond to a drop in the streamwise wave number α_r which in term leads to an increased wavelength $\bar{\lambda}(x = 2.5) = 8.4mm$ and a sudden increase of the phase speed c_r by 20%. Both events are visible in the flow visualisations in forms of a larger spacing of the vortices in stage *(III)* and a slight bend in their axes at $x = 2.15$. The presence of the laminar separation bubble complicates this process by stretching the coherent structure: The phase speed development $c_r(PD)$ in Fig. 5b after the separation indicates that the near-wall parts of the vortices are retarded by the presence of back flow while parts above the bubble are locally accelerated due to its displacement. After leaving the bubble both parts share a common rise in c_r and Ψ again. The readjustment of the vortex axes until they are normal to the free stream is therefore an inherent property of the structure, independent of the separation bubble. Stage *(III)* ends abruptly after the saturation of the background disturbances at $x_{\text{sat}} = 2.78$ with the breakdown of the coherent structure. The mechanisms of the breakdown can be clarified by the vortex core lines in Fig. 6: Viewed in the potential flow direction they display a striking similarity to the classical K-type breakdown of a flat plate boundary layer as described by Bake *et al.* [1]. 'Spanwise' modulations along the vortex axes increase downstream until they break the wave fronts apart. The pieces form Λ-vortices which propagate in an aligned fashion exactly in freestream direction. First modulations are observable around $x = 2.6$ where the background disturbances reach amplitudes

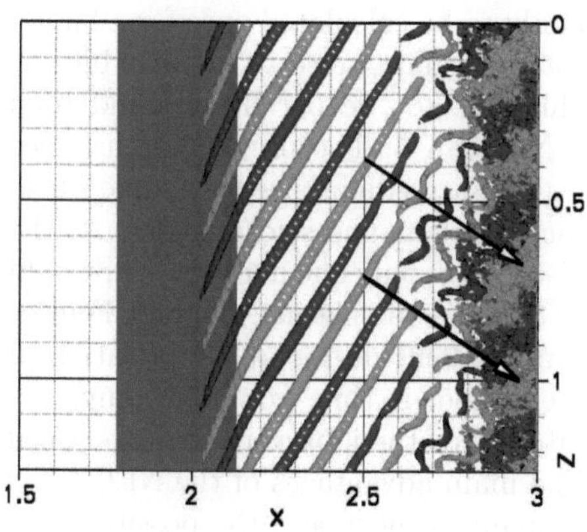

Fig. 6. Close-up on the vortex axes of picture 4(b): Snapshots of λ_2-iso-surfaces for $\lambda_2 \equiv -200$ at two different instants of time after 59.5 (*red*) and 60 disturbance periods (*green*). *Arrows*: direction of potential flow $\Psi_e(x = 2.5) = 32.4°$

of 1%. Most likely, this rich spectrum of background disturbances provides the missing 'oblique' partners for an *oblique K-type transition.* Moreover, in the present disturbance scenario based on so-called "fundamental resonance" such a transition process can be expected.

Stage *(IV)*: Turbulent Flow

After the breakdown of the coherent structures turbulent flow develops as described in Sect. 3.2. Figure 5(a) demonstrates again how well the saturation of the background disturbances coincides with the onset of turbulence.

4 Computational Aspects

4.1 The Algorithms for the Direct Numerical Simulations

Our *DNS*-code solves the three-dimensional Navier-Stokes equations for unsteady, incompressible flow in vorticity-velocity formulation. The quasi-two-dimensional *base flow* equations are discretised with central finite differences of $4th$ order accuracy. The steady state is reached by the help of a dissipative, semi-implicit, pseudo-temporal ADI approach for the vorticity transport equations. A vectorisable stripe-pattern LSOR technique is employed to solve the Poisson equations for the velocity components. For the *disturbance simulations* a complex Fourier spectral ansatz is used to decompose the flow field in z. Compact finite differences of mostly $6th$ order accuracy guarantee a highly accurate spatial wave transport and a $4th$-order Runge-Kutta scheme is used for the time-stepping. For an in-depth description of the *DNS* algorithms see Wassermann & Kloker [16], from which the present code version differs only in minor details.

4.2 Performance and Computational Resources

For each sweep angle the base flow had only to be calculated once for a highly resolved case with 2786×1537 grid points in x and y to serve for an arbitrary number of disturbance simulations. This can not be parallelised and took $37h$ user time on a single CPU on the NEC SX-5. The present disturbance scenario – the middle one in Table 1 – was a medium sized example of an extensive series with different disturbance contents and sweep angles. As the code was designed for the NEC SX-4/SX-5/SX-6, only one node per run would be requested on the NEC SX-8 with its 8 CPUs calculating different Fourier modes in parallel as micro-tasks with an excellent degree of vectorisation of 99%. In a single run as many disturbance periods in time would be completed as possible within the run-time limit. Afterwards the jobs were restarted until the quasi periodic state was reached. The main advantages of the NEC SX-8 were its speed, the generous amount of main memory and the possibility to work with $3-4$ different scenarios simultaneously by allocating a node each.

Table 1. Performance data of a small, medium & large run with *'Perio'* disturbance periods per single run out of final number, *'time/run'* user time per single run, so whole scenario finished after *'CPU'* CPU-hours. Further: Grid points in x & y, number of Fourier modes in z, giga-flops/CPU, main memory, vectorisation rate

Name	$X \times Y$	K_{max}	Gflops	GByte	Vect	Perio	time/run	CPU
30°-sweep	2402×273	15	4.85	9	99.0%	24/48	$110h$	$220h$
30°-visu	2402×273	31	4.89	18	99.1%	4/60	$38h$	$570h$
CF-TS-int.	3730×545	31	4.95	55	99.3%	2/48	$145h$	$2610h$

5 Conclusions

The stages of laminar-turbulent transition in the flow field around a swept laminar separation bubble have been analysed and visualised in detail. The development of a discrete spectrum of oblique Tollmien-Schlichting waves dominated by a "primary disturbance" of higher amplitude was qualitatively similar to the unswept case: *(I)* Linear disturbance amplification until the primary disturbance reaches a sufficient amplitude of 3% of the local free-stream velocity. Towards the end of this phase vortices induced by the primary disturbance emerge, after its amplitude surpasses 0.1%. *(II)* A stage of secondary flow instability with strong resonance of the background disturbances with the high-amplitude vortices of the primary disturbance, which still grows according to *LST* until saturation at an amplitude level of 22%. Because of the strong amplification rates inside the bubble this phase is quite short, so that secondary instability represents a comparatively unimportant mechanism in separation bubbles in general. *(III)* A coherent structure is formed by the simultaneously saturating primary disturbance and its higher harmonics, which forces any background disturbance into a common dependency indicated by identical amplification rates. The vortices change their orientation until everything evolves exactly in freestream direction. The stage ends with a rapid breakdown immediately after the saturation of the background disturbances. The event is triggered by the breakup of the vortex cores into aligned Λ-vortices, which resemble an oblique K-type transition. *(IV)* Emergence of turbulent flow which quickly approaches criteria of a fully turbulent boundary layer. Within all scenarios of fundamental resonance for a sweep angle of 30° the chosen case exhibits the greatest linear amplification of the primary disturbance. Furthermore, it could be shown that the intersection point between the amplification curves of the mean flow deformation and the dominant disturbance marks the saturation of the background disturbances. It is therefore a good indicator for the onset of turbulent flow in different transition scenarios with one dominating disturbance. The interpretation of flow visualisation by direct comparison with postprocessing data within the same figure has been highly fruitful. The knowledge, how typical flow structures express themselves in the visualisations and how such structures develop and interact, provide the "building blocks" for the analysis of more complex disturbance scenarios. To utilise them, a computer aided decomposition of a complex flow field into its elements will be necessary as proposed by Linnick & Rist in [11].

Acknowledgements

The authors would like to thank the DLR-Göttingen and especially Dr. Hein for the possibility to use the linear 'nolot'-PSE-code. The financial support by the Deutsche Forschungsgemeinschaft (DFG) under contract number RI 680/12 is gratefully acknowledged. All CPU-time for the simulations was provided by the HLRS in Stuttgart.

References

1. S. Bake, D.G.W. Meyer, U. Rist (2002): *Turbulence mechanisms in Klebanoff transition: a quantitative comparison of experiment and direct numerical simulation.* J. Fluid Mech., **459**, pp. 217-243.
2. R. Davis, J. Carter, E. Reshotko: *Analysis of Transitional Separation Bubbles on Infinite Swept Wings.* AIAA Vol. **25**, No. 3, (1987), pp. 421–428.
3. E. Greff: *In-flight Measurements of Static Pressures and Boundary-Layer State with Integrated Sensors*, J. Aircraft, **28**, No. 5, (1991), pp. 289–299.
4. S. Hein: *Linear and nonlinear nonlocal instability analyses for two-dimensional laminar separation bubbles.* Proc. IUTAM Symp. on Laminar-Turbulent Transition, Sedona, USA, Sep. 1999, Springer, pp. 681–686.
5. T. Hetsch, U. Rist (2004): *On the Structure and Stability of Three-Dimensional Laminar Separation Bubbles on a Swept Plate.* In: Breitsamter, Laschka et al. (Eds.): New results in numerical and experimental fluid mechanics IV: Proceedings of the 13th DGLR/STAB-Symposium, Munich 2002. NNFM, **87**, Springer, pp. 302–310.
6. T. Hetsch, U. Rist (2006): *The Effect of Sweep on Laminar Separation Bubbles.* In: R. Govindarajan (Ed.): Sixth IUTAM Symposium on Laminar-Turbulent Transition, Proc. IUTAM-Symp. Bangalore, India, 2004, Fluid Mechanics and its Applications, Vol. 78, Springer-Verlag, pp. 395–400, 2006.
7. T. Hetsch, U. Rist (2006): *Applicability and Quality of Linear Stability Theory and Linear PSE in Swept Laminar Separation Bubbles.* Proc. of the 14.th DGLR/STAB Symposium. Accepted for publication. To appear 2006.
8. H. Horton (1968): *Laminar Separation Bubbles in Two and Three Dimensional Incompressible Flow.* PhD thesis, Department of Aeronautical Engineering, Queen Mary College, University of London.
9. J. Jeong, F. Hussain (1995): *On the identification of a vortex.* J. Fluid Mech., **285**, pp. 69–94.
10. H.-J. Kaltenbach, G. Janke (2000): *Direct numerical simulation of a flow separation behind a swept, rearward-facing step at* $Re_{\mathrm{H}} = 3000$. Physics of Fluids, **12**, No. 9, pp. 2320–2337.
11. M. Linnick, U. Rist: *Vortex Identification and Extraction in a Boundary-Layer Flow.* In: G. Greiner, J. Hornegger, H. Niemann, M. Stamminger (Eds.): Vision, Modelling, and Visualization 2005 Proc. November 16-18, 2005. Erlangen, Germany, Akad. Verlagsges. Aka, Berlin, pp. 9–16.

12. O. Marxen, M. Lang, U. Rist, S. Wagner: *A Combined Experimental/Numerical Study of Unsteady Phenomena in a Laminar Separation Bubble.* Flow, Tub. and Comb. **71**, Kluwer Acad. Pub., 2003, pp. 133–146.
13. U. Rist: *Zur Instabilität und Transition in laminaren Ablöseblasen.* Habilitation, Universität Stuttgart, Shaker 1999.
14. H. Schlichting (1982): *Grenzschicht-Theorie*, 8. Auflage, Verlag G. Braun.
15. P. Spalart, M. Strelets (2000): *Mechanisms of Transition and Heat Transfer in a Separation Bubble.* JFM, **403**, pp. 329–349.
16. P. Wassermann, M. Kloker: *Mechanisms and passive control of crossflow-vortex-induced transition in a three-dimensional boundary layer.* J. Fluid Mech. **456**, Camb. Univ. Press 2002, pp. 49–84.
17. A.D. Young, H.P. Horton (1966): *Some Results of Investigations of Separation Bubbles.* AGARD CP No. 4, pp. 780–811.

Direct Numerical Simulation of Primary Breakup Phenomena in Liquid Sheets

Wolfgang Sander and Bernhard Weigand

Institute of Aerospace Thermodynamics, University of Stuttgart,
Pfaffenwaldring 31, 70569 Stuttgart, Germany
wolfgang.sander@itlr.uni-stuttgart.de

Abstract. Starting from the first experimental and analytical studies on primary breakup phenomena, many interesting results have been published in the past. It is known that in addition to the typical dimensionless groups (Reynolds and Weber number), inflow conditions can drastically influence primary breakup phenomena. Now that high computational resources are available, direct numerical simulation (DNS) has become a powerful tool in order to study primary breakup phenomena. Nevertheless only a few DNS studies concerning breakup phenomena and the influence of inflow conditions are available. This might be due to the fact that besides high demands of computational resources, sophisticated numerical models are also required in order to prescribe realistic inflow conditions and capture all length scales in the flow. This paper mainly focuses on the influence of different inflow conditions, such as the integral length scale or the fluctuation level inside the turbulent nozzle flow. For this, the breakup phenomena of water sheets at moderate Reynolds numbers injected into an quiescent air environment are considered. Since this study is performed as an numerical experiment by varying the character of the inflow velocity data, it was found that not only the mean axial velocity profile but also the integral length scale and the fluctuation level can have an influence on breakup phenomena.

1 Introduction

Growing demands concerning combustion efficiency and severe emission limits in the past led to increasing research activities in the field of injection technology. Therefore atomization of liquid jets is a major research domain, for example, in combination with combustion processes for gas turbines or piston engines. The physical mechanism leading to disintegration of a liquid sheet or jet injected into an gaseous environment is mainly caused by the aerodynamic interaction between the two fluids. From experimental and theoretical investigations it is known that unstable wave-growth and turbulence inside the nozzle flow leads to disintegration of liquid jets. In this context, primary breakup phenomena occurring adjacent to the nozzle exit are particularly crucial. Therefore many studies in this field focus mainly on the influence of instability growth based on linear stability analysis ([1], [2] or [3]). Apart from this, more recent studies focus on the influence of the nozzle flow on the breakup phenomena. From [4] it was experimentally demonstrated what influence different nozzle geometries and inflow conditions can have at the same nondimensional parameters. Similar investigations were also carried out numerically by [5] and [6].

It is evident that technical designs for injection systems at this time work at the mechanical and physical limit of each component. In particular, high pressure injection systems suffer from high mechanical stresses due to heavy loads such as caused by high pressure and pressure fluctuations or cavitation. Apart from this, a large quantity of power is needed in order to generate these pressure levels. Because this concept is mainly based on the influence of unstable wave-growth at very high nondimensional parameters, it might be reasonable from an engineering point of view to modify the nozzle design and accordingly the nozzle flow in order to achieve a higher turbulence level. This would basically result in strong instabilities of the injected jet but cause lower mechanical stresses due to the additionally enforced instability mechanism caused by the higher turbulence intensity of the nozzle flow. Of course this idea is justified and evident but the main question is – how can this be achieved? It would be easy to change the shape of the nozzle hole, deflect the flow inside the nozzle, or generate additional distortions inside the nozzle (see [7]). Many designs based on this idea have been developed in the past, but, in most car engines for example, very simple nozzle designs are still installed. This is one of the aspects which leads to the conclusion that basic mechanisms causing the instabilities of liquid sheets or jets are yet not well understood.

2 Numerical Method

The in-house 3D CFD program FS3D [8] has been developed to compute the Navier-Stokes equations for incompressible flows with free surfaces. Based on the idea of direct numerical simulation (DNS), turbulent fluctuations are captured by an extremely fine spatial resolution, hence a turbulence model is not necessary. The flow is governed by the conservation equations for mass and momentum

$$\nabla \cdot \mathbf{u} = 0 \tag{1}$$

$$\frac{\partial (\rho \mathbf{u})}{\partial t} + \nabla \cdot (\rho \mathbf{u}) \otimes \mathbf{u} = -\nabla p + \nabla \cdot \mu \left[\left(\nabla \mathbf{u} + (\nabla \mathbf{u})^T \right) \right] + \rho \mathbf{k} + \mathbf{f}_\gamma \tag{2}$$

where $\mathbf{u}$ denotes the velocity vector, t the time, ρ the density, μ the dynamic viscosity and p the pressure. Furthermore, the capillary stress tensor $\mathbf{f}_\gamma$ and an external body force denoted by $\mathbf{k}$ are added to the momentum equation, whereas $\mathbf{f}_\gamma$ is only non-zero in the interface region.

The presence of a liquid and gaseous phase is considered based on the Volume-of-Fluid (VOF) method [9]. Consequently an additional transport equation

$$\frac{\partial f}{\partial t} + \nabla \cdot (\mathbf{u} f) = 0 \tag{3}$$

is defined in order to describe the temporal and spatial evolution of the two phase flow. The variable f, called the VOF-variable, represents the volume of (liquid phase) fluid fraction.

$$f(\mathbf{x}, t) = \begin{cases} 0 & \text{outside the liquid phase} \\ 0 < f < 1 & \text{at the interface} \\ 1 & \text{inside the liquid phase} \end{cases} \tag{4}$$

Since fluid properties are required every time step in order to solve the Navier-Stokes equation the following relations for density ρ and viscosity μ are evaluated

$$\rho(\mathbf{x}, t) = \rho_g + (\rho_l - \rho_g)\, f(\mathbf{x}, t) \tag{5}$$

$$\mu(\mathbf{x}, t) = \mu_g + (\mu_l - \mu_g)\, f(\mathbf{x}, t) \tag{6}$$

where the subscript g indicates the gas phase and l the liquid phase.

For the numerical solution of the relevant transport equations, a spatial discretization of the computational region is realized by a finite volume scheme on a structured staggered grid. In each phase, the discretization is second-order accurate. To ensure a sharp interface of all flow discontinuities and to suppress numerical dissipation of the liquid phase during each time step, the interface is reconstructed by the PLIC (piecewise linear interface calculation) method proposed by [10]. After the reconstruction, the liquid phase is transported on the basis of its reconstructed distribution. Due to the high gradients across the interface, a limiter is used to prevent the generation of oscillations and spurious solutions.

The formulation for the fully conservative momentum advection and volume fraction transport, the momentum diffusion, and surface tension is treated explicitly. The pressure is determined implicitly by a Poisson equation during a projection step in order to ensure the incompressibility constraint $\nabla \cdot \mathbf{u} = 0$. Due to discontinuous coefficients resulting from the discontinuous density field, a robust multigrid solver based on a Galerkin coarse grid approximation is used for solving the Poisson equation for the pressure. In order to accomplish all possible computational requirements the program is parallelized by the domain decomposition technique using the communication library *MPI*, hence it can either be used on single-processor or on multi-processor systems.

Additionally, inflow velocity data required during runtime as inflow conditions for the inflow region are generated during the initialization stage of the program. This procedure is split into two different parts, where, in a first step, the mean velocity profile can be prescribed as such:

- top-hat velocity profile
- mean velocity profile of a fully developed channel flow calculated by means of turbulent mixing length

In a second step, velocity fluctuations based on a digital filtering technique developed by [11] are added to the mean velocity profile. In order to prescribe different inflow conditions such as the before mentioned fluctuation level or length scale it is possible to prescribe a constant fluctuation level and length scale in each direction, but also to use calculated results for the Reynolds stress tensor from [12]. Further details regarding the numerical setup are given in the following chapter.

3 Numerical Setup

Discretization

The results presented here were obtained from 2D-simulations of a liquid water film injected in air. For all simulations, the Reynolds number was 3000 or 5000 based on the inflow width of the nozzle D and the bulk velocity U_0 at the inflow. These simulations were performed on a rectangular geometry with an uniform mesh as presented in Fig. 1.

The extension of the computational domain based on the inflow width D, was 24D × 6D in the axial (x) and vertical (y) directions. The computational domain was discretized with 1536 × 384 grid points, resulting in a minimum resolution of 60 grid points across the inflow region. In order to ensure the typical mass entrainment effect occurring during the spreading of the jet, the vertical and outflow boundaries are set to continuous boundary conditions where the pressure is set to zero. At the inflow side, the vertical and normal velocity component is set to zero, except in the inflow region where velocity profiles from the inflow-generation procedure are used. In Fig. 1, the domain

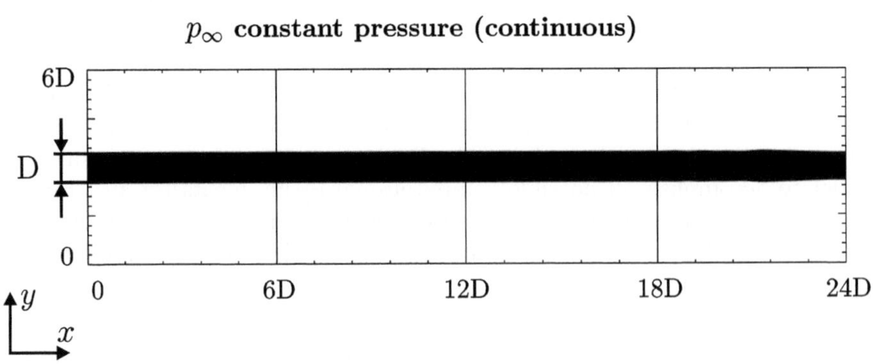

Fig. 1. Computational domain and grid

decomposition is also presented. This figure demonstrates the idea of dividing the complete domain into rectangular sub domains of the size 6D×6D. This results in the setup of a grid size of 384×384 grid points on each processor, which represents an optimum for the coarsening technique of the multigrid solver.

Inflow Conditions

The main focus of this study is dedicated to the influence of different inflow conditions on primary breakup phenomena. Therefore variations in the mean velocity profile are presented in the following, as well as the fluctuation level and the integral length scale of the generated inflow velocity data which were used for the simulations.

The procedure used for the generation of inflow velocity data u_i was based on the following relation

$$u_i = \overline{u_i} + a_{ij}\mathcal{U}_j \tag{7}$$

where $\overline{u_i}$ denotes the mean axial velocity value at a certain grid point across the inflow region and the coefficient a_{ij} is related to the Reynolds stress tensor according to [13] while $\mathcal{U}_j$ represents the velocity fluctuation.

In order to prescribe the mean axial velocity profile, two different approaches were taken. According to Fig. 2, a top-hat profile can be used in this context, nearly representing the mean axial velocity profile for a conical contraction nozzle (A). Alternatively, the mean axial velocity profile of a fully developed channel flow is also considered since it nearly represents the velocity field inside a long converging nozzle (B). Both of these nozzle designs are widely used and therefore represent an interesting parameter due to the different resulting mean axial velocity fields.

The velocity fluctuation $\mathcal{U}_j$ is generated by means of digital filtering of random data [11]. It basically approximates a prescribed autocorrelation function

$$R_{uu}(r) = exp\left(-\frac{\pi r^2}{4L^2}\right) \qquad \left(\text{where } L = L(t) = \sqrt{2\pi\nu(t-t_0)}\right) \tag{8}$$

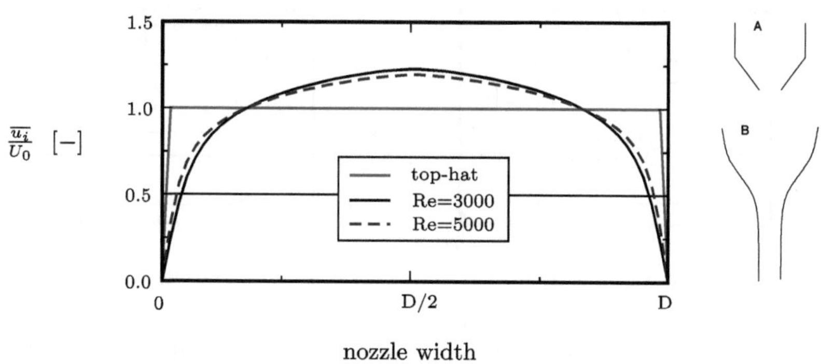

Fig. 2. Mean axial velocity profile depending on the nozzle design

for homogeneous turbulence in a late stage given by [14]. In Eqn. 8 $r = |\mathbf{r}|$ denotes the distance vector whereas L refers to the integral length scale which should be prescribed. In the procedure used, the length scale at the inflow plane is specified in the vertical direction as $L_y = n_y \Delta y$ (a multiple of the grid spacing) and a time scale $L_x = n_x \Delta t$ in the axial direction which can be transformed into a length scale by Taylors hypothesis. Additionally, the fluctuation level must be defined in order to fully prescribe the characteristics of the inflow velocity profile. Based on the bulk inflow velocity, the standard deviation of the fluctuations $\sqrt{\overline{u'u'}}/U_0$ and $\sqrt{\overline{v'v'}}/U_0$ were set equal in both directions to a constant level of 2.5% or 5%, whereas the shear stress was set to zero.

According to Fig. 3, the influence of the velocity data depending on the prescribed length scale can be clearly recognized. The velocity fluctuations around the mean velocity $\overline{u}$ are all at a constant level whereas the variation in frequency around $\overline{u}$ increases due to the decreasing length scale from $L_y = D/10$ to the limit of random fluctuations. This fluctuation pattern is typical and already indicates the different length scales of a flow which were prescribed in this case according to Fig. 3 (right). In order to demonstrate the accuracy of the procedure, a spatial correlation of the generated velocity data was also performed for $L_y = D/20$.

The ability to prescribe the fluctuation level and additionally the length scale of the velocity data represents a simplified but powerful feature in this context. This is due to the fact that the prescription of real inflow conditions, which generally requires a full discretization and simulation of the nozzle flow, far exceeds computational resources compared to the procedure used. In contrast, the generation of velocity data based on randomly generated fluctuations is not a realistic approach to the aimed intention because random fluctuations basically represent purely random white noise which is not characteristic of turbulent flows. This relationship can clearly be recognized from the power spectral density distribution in Fig. 4. It can also be recognized from Fig. 3 that randomly fluctuating velocity data immediately falls to an uncor-

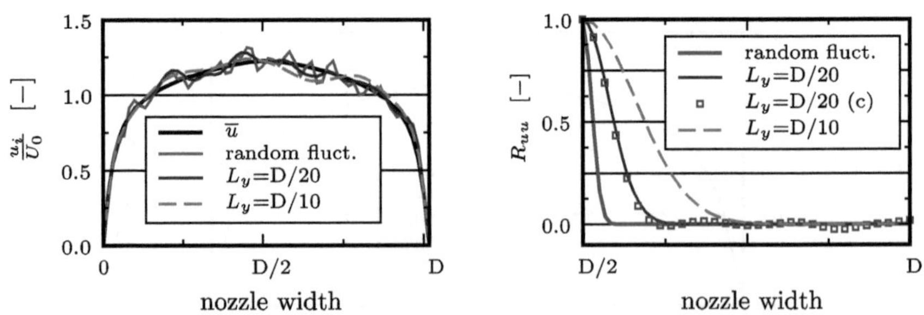

Fig. 3. Axial velocity profile with different length scales at a certain time point (*left*) – Prescribed and calculated spatial correlation of the inflow velocity data in vertical direction (*right*)

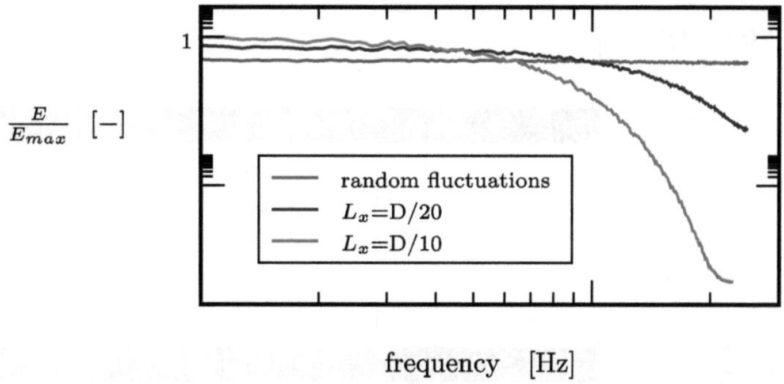

Fig. 4. Power spectral density of the prescribed inflow velocity data

related level and additionally the power spectral density is constant within the frequency range.

4 Results

In order to demonstrate the strong influence of inflow conditions and the difference between the various possible parameters, numerous numerical experiments were performed. Since the mean axial velocity profile and therefore the Reynolds and Weber number can be varied, and additionally the length scale and fluctuation level, the enormous number of different options cannot be reviewed in one study. Therefore selected combinations of the different parameters are compared in order to examine the most crucial differences and accordingly the major effects on breakup phenomena. For all the following simulation results, a constant mass flow rate and turbulent kinetic energy level is ensured, independent of the mean axial velocity character.

4.1 Influence of the Mean Axial Velocity

Starting with a completely undisturbed velocity profile as presented in Fig. 5, the result is a completely undisturbed surface of the liquid sheet. Within the considered range of moderate Reynolds-numbers, no distortions or breakup phenomena can be observed. Therefore a more realistic approach must be taken in order to generate distortions at the inflow, hence obtaining the aimed breakup phenomena. Considering the two above mentioned mean axial velocity profiles, it appears evident to add fluctuations on the undisturbed mean axial velocity profile. The result of this approach is, as presented in Fig. 5, strongly dependent on the character of the mean axial velocity profile. It can directly be seen that even high fluctuation levels of 10% achieve only very weak distortions of the liquid film surface in case of the top-hat velocity profile, whereas stronger distortions are obtained by using a channel flow velocity profile.

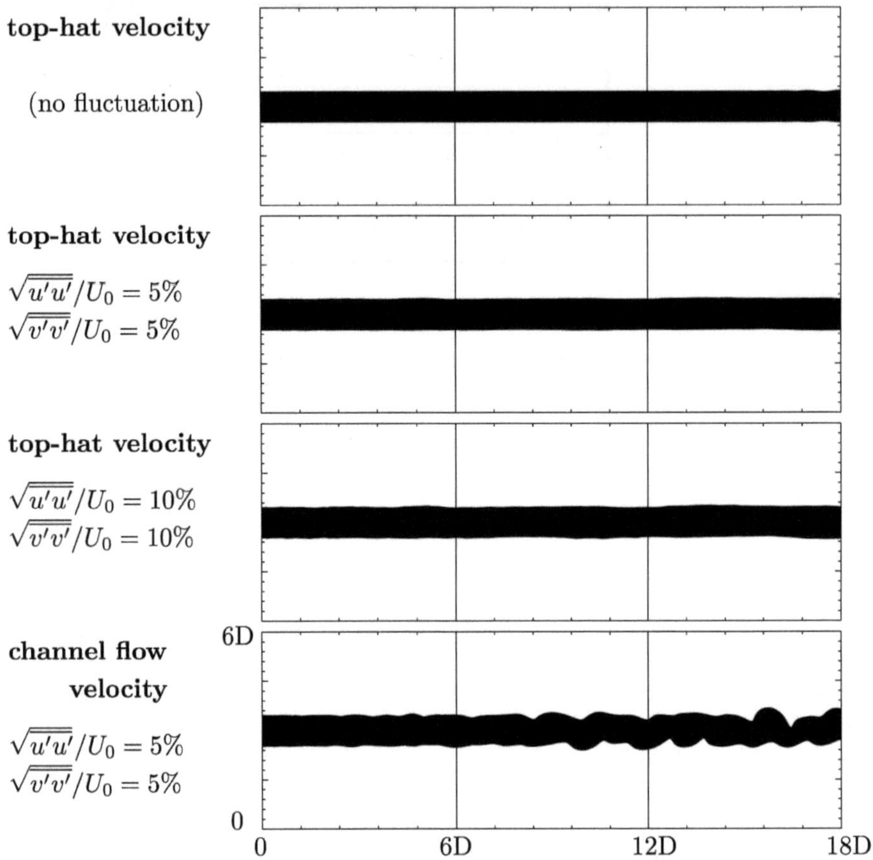

Fig. 5. Different instability phenomena of the liquid film due to random fluctuations superposed on different mean axial velocity profiles (Re = 3000/We = 190)

Regarding these first results, it appears interesting to look at the development of the distortions inside the liquid film in more detail. In particular, the difference concerning the development and propagation of the introduced disturbances between the two cases might be interesting due to the strong influence of the mean axial velocity characteristics. By comparing the velocity field and the vorticity of the two cases, these differences appear directly. According to Fig. 6, a clear difference between the two liquid films can also be observed by looking at the velocity magnitude. As a result of the flat top-hat velocity profile, a homogeneous velocity field is present across the liquid film. Regarding the lower contour plot, the parabolic character of the velocity distribution across the liquid film can be seen. Also the fluctuations that were introduced can be recognized in both images represented by the irregular local distribution of the velocity magnitude, especially at the inflow region.

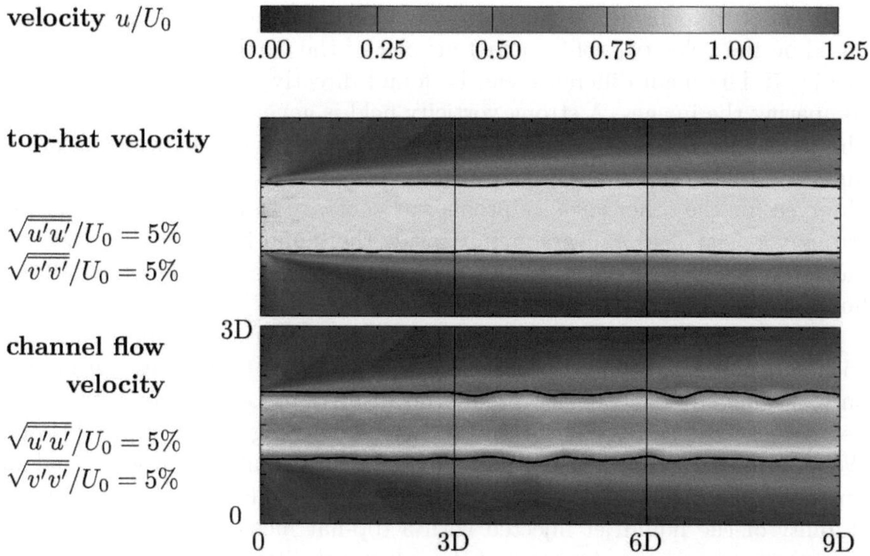

Fig. 6. Contour plot of the velocity magnitude for two different mean axial velocity profiles at an identical fluctuation level of the superposed random fluctuations (Re = 3000/We = 190)

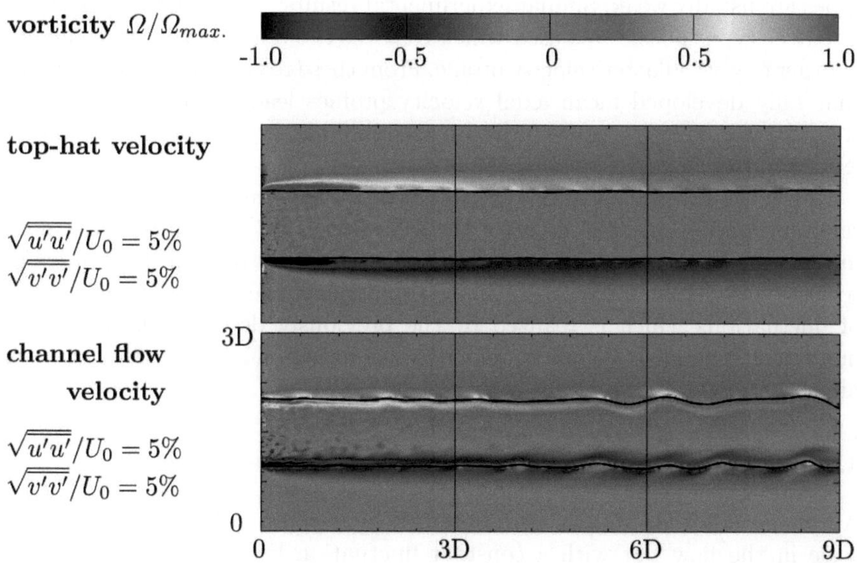

Fig. 7. Contour plot of the vorticity field in the flow for two different mean axial velocity profiles at an identical fluctuation level of the superposed random fluctuations (Re = 3000/We = 190)

In order to focus on the question about the different spreading of the film based on the velocity profile, a contour plot of the vorticity field is presented in Fig. 7. The main difference can be found directly at the inflow region by comparing the images. A strong vorticity field is generated outside the liquid film in the gas flow in the case of the top-hat velocity profile, while all introduced perturbations are damped further downstream. A similar behavior is observed for the other velocity profile but contrary to the previous case, the strong vorticity field appears mainly inside the liquid film and obviously immediately leads to an effective distortion of the surface due to the introduced fluctuations.

It should be pointed out that this difference was only found in simulations with additional introduced perturbations, whereas in both cases considering only the mean axial velocity profile, no distortions of the liquid film have been observed (see Fig. 5) within the considered range of the Reynolds and Weber number. Regarding the fluctuation level in combination with random fluctuations, it was found that even higher fluctuation levels did not affect the stability of the liquid jet injected with a top-hat mean axial velocity. From results concerning the character of the inflow velocity presented in [4], it can be assumed that fluctuations are reduced to a minor level due to acceleration and relaminarization of the flow at the nozzle exit. This clearly explains why film instabilities of plane liquid jets injected from conical contraction nozzle types are usually weak. Similar experimental results were found by [15], where a "cutter-type" nozzle was used which removes the boundary layer of the flow in order to give a flatter velocity profile. From these results, it can be concluded that fully developed mean axial velocity profiles lead to higher instabilities than top-hat velocity profiles due to the different liquid vorticity of the jet exit condition.

Since the previous results were all obtained from inlet conditions based on random fluctuations, the focus of the following is on a more realistic reproduction of the spectrum of turbulent fluctuations via defined integral length scales. This demand requires a more sophisticated and effective generation of fluctuations which is realized by the previously described procedure for numerical generation of inflow velocity data based on a digital filtering technique.

4.2 Influence of the Integral Length Scale

All of the following results were obtained by variation of the integral length scale in the flow but with a constant fluctuation level. In order to demonstrate the influence of the integral length scale, numerous simulations have been performed with different setups for the integral length and time scale. According to Fig. 8, the spreading of the liquid film at a certain fluctuation level can strongly be influenced by the prescribed length scale. Since random fluctuations are not representative of turbulent flows, the advantage of the inflow procedure used can be recognized by comparing the different contour

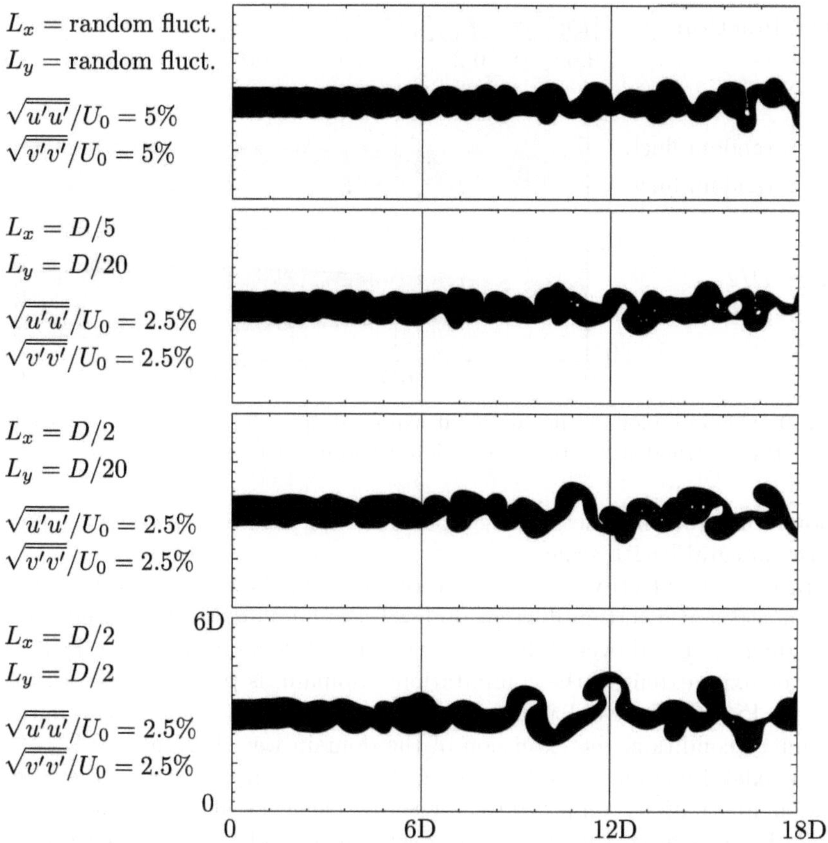

Fig. 8. Different breakup of the liquid film due to prescribed length scales at a fluctuation level of 2.5% in both directions (Re = 5000/We = 540), compared with random fluctuations at 5%

plots. Based on the volume concentration, the location of first distortions and their level varies due to the prescribed length scale.

Considering the presented results it appears evident that fluctuations generated basically on pure random white noise are not a realistic approach for turbulent flows. Since the amount of introduced turbulent kinetic energy at a certain frequency can mainly be varied by prescribing a certain length scale (refer to Fig. 4) this quantity represents a crucial parameter in this context, in addition to the fluctuation level. As fluctuations at high frequencies are continuously dissipated in the flow, it is reasonable to shift the level of turbulent kinetic energy to a lower frequency range, hence a more effective perturbation can be introduced. From this point of view, it is reasonable that the spreading of a jet can be increased by augmenting the fluctuation level, however, a similar behavior can also be achieved by increasing the integral length scale of the inflow velocity field. This is basically related to the fact that a specific

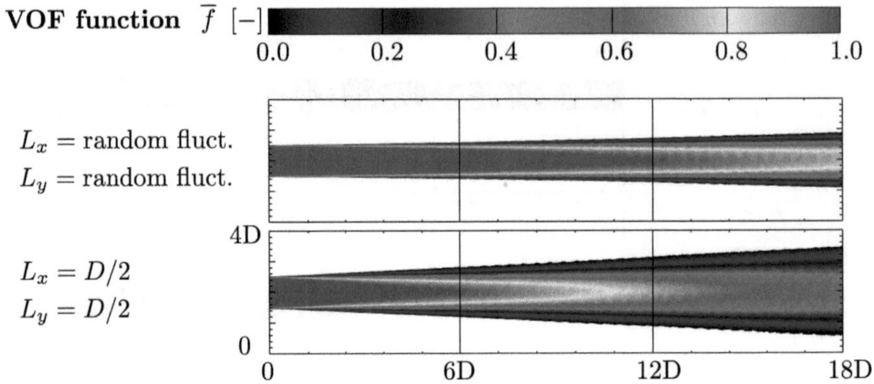

Fig. 9. Contour plot of the temporal averaged spreading based on the volume concentration (random fluctuations with a constant fluctuation level of 5%)

amount of turbulent kinetic energy distributed at a certain frequency depends on the integral length scale.

In order to identify the influence of turbulent flows containing different length scales at a certain fluctuation level, the liquid distribution and the jet spreading, averaged over 12 flow through times based on the bulk velocity U_0 and the axial extent of the computational domain, is presented as a contour plot in Fig. 9. To avoid influences on the solution caused by the outflow boundary condition, the extension of the domain was shortened to L = 18D in the axial direction. From the averaged volume concentration in the case of random fluctuations compared to a prescribed length scale of $L_x = L_y = D/2$, it can be recognized that the liquid distribution and the spreading rate can essentially be influenced. Furthermore it is possible to change these quantities of breakup behavior with a variation of the length scale in each direction.

In order to demonstrate the influence of different length scales in a flow, several simulation results are compared in Fig. 10. As a result of the variation

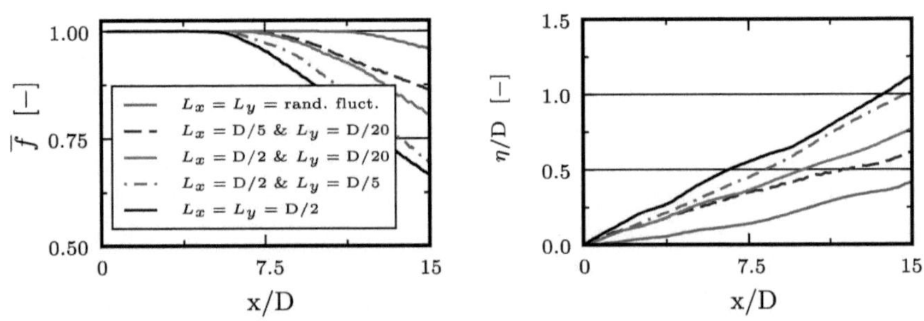

Fig. 10. Influence of different length scales on the film breakup at a constant fluctuation level of 2.5% (random fluctuations with a constant fluctuation level of 5%) – averaged liquid distribution along the center line (*left*) and max. spreading based on the averaged volume concentration (*right*)

of different inflow velocity data due to the prescribed length scale, the averaged liquid distribution $\overline{f}$, and the spreading rate η/D, varies depending on the inflow condition. According to this, a large length scale such as $L_x = L_y = D/2$ and the combination of a large length scale in the axial direction and a small length scale in the vertical direction ($L_x = D/2$ & $L_y = D/20$) cause strong distortions and a higher spreading rate compared to smaller length scales or random fluctuations.

5 Computational Resources

All results presented in this paper have been computed on the CRAY-Opteron Cluster at the HLRS. The computational time for the presented 2D-simulations of the breakup phenomena mainly depends on the extent of the domain and the grid size used. The setup presented with a grid resolution of 1536×384 (589,824 grid points) required approximately 1500 CPUh for 10–15 flow-through times. The number of required CPUs is usually connected to the constraint of the domain decomposition but also to a balance between the needed computational time in order to obtain all results within a adequate period. In this case 4 CPUs were used for all simulations.

6 Concluding Remarks

The results presented in this study were obtained by various numerical experiments performed in order to study the influence of numerically generated inflow conditions on the liquid film instability. The main focus of this study is to identify the major parameters which lead to the disintegration of the liquid films studied.

It was found that the character of the mean axial velocity depending on the nozzle type can strongly influence film instabilities. Since even high fluctuation levels combined with a top-hat mean axial velocity profile did not lead to any noticeable distortions of the liquid film, it can be concluded in accordance with several experimental results that breakup phenomena are primarily dependent on the character of the mean axial velocity profile. In particular, the results found by using the mean axial velocity profile of a fully developed channel flow confirm this result. Additionally, it could be demonstrated how far disintegration of the liquid film can be controlled by different fluctuation levels or integral length scales of the flow. Although all presented results show good qualitative agreement with experimental data, there was no serious effort to prescribe realistic inflow conditions from an experiment. The simplifications introduced by using a constant fluctuation level across the inflow region and the simplified 2D setup must be considered in this context.

Further studies will focus on these aspects, whereas the non-constant fluctuation level and forced breakup phenomena are of major interest. Additionally, 3D simulations accompanied by experiments will be performed once the major influencing parameters are identified within the simplified 2D simulations.

Acknowledgements

The authors greatly appreciate the *High Performance Computing Center Stuttgart* (HLRS) for support and supply of computational time on the high performance computers.

References

1. F.R.S. Rayleigh and J.S.W. Lord. On the instability of jets. *Proc. London Math. Soc.*, Vol. 10, pp. 4–13, 1878.
2. C. Weber. Zum Zerfall eines Flüssigkeitsstrahles. *Zeitsch. für angew. Math. und Mech.(ZAMM)*, Vol. 11(2), pp. 136–154, 1931.
3. S.P. Lin. *Breakup of Liquid Sheets and Jets.* Cambridge University Press, 2003.
4. K. Heukelbach. *Untersuchung zum Einfluss der Düseninnenströmung auf die Stabilität von flächigen Flüssigkeitsstrahlen.* PhD thesis, Technische Universität Darmstadt, 2003.
5. M. Klein. *Direkte numerische Simulation des primären Strahlzerfalls in Einstoffzerstäuberdüsen.* PhD thesis, Technische Universität Darmstadt, 2002.
6. Y. Pan and K. Suga. Simulation of liquid jet breakup by the level set method. In *In Proc. 9th Int. Conf. on Liquid Atomization and Spray Systems*, Aichi 480–1192, Japan, 2003.
7. H. Hiroyasu. Spray breakup mechanism from the hole-type nozzle and its applications. *Atomization and Sprays*, Vol. 10, pp. 511–527, 2000.
8. M. Rieber. *Numerische Modellierung der Dynamik freier Grenzflächen in Zweiphasenströmungen.* PhD thesis, Universität Stuttgart, 2004.
9. C.W. Hirt and B.D. Nichols. Volume of fluid (VOF) method for the dynamics of free boundaries. *Journal of Computational Physics*, Vol. 39, pp. 201–225, 1981.
10. W.J. Rider and D.B. Kothe. Reconstructing volume tracking. *Journal of Computational Physics*, Vol. 141, pp. 112–152, 1998.
11. M. Klein, A. Sadiki, and J. Janicka. A digital filter based generation of inflow data for spatially developing direct numerical or large eddy simulations. *Journal of Computational Physics*, Vol. 186, pp. 652–665, 2003.
12. R.D. Moser, J. Kim, and N.N. Mansour. Direct numerical simulation of turbulent channel flow up to $\mathrm{Re}_{\mathrm{tau}} = 590$. *Physics of Fluids*, Vol. 11, pp. 943–945, 1999.
13. T. Lund, X. Wu, and D. Squires. Generation of turbulent inflow data for spatially-developing boundary layer simulations. *Journal of computational physics*, Vol. 140, pp. 233–258, 1998.
14. G.K. Batchelor. *The theory of homogeneous turbulence.* Cambridge University Press, Cambridge, 1953.
15. P. Wu, R. Miranda, and Faeth G. Effect of initial inflow conditions on primary break-up of nonturbulent and turbulent jets. *Atomization and Sprays*, Vol. 5, pp. 175–196, 1995.

Direct Numerical Simulation of Mixing and Chemical Reactions in a Round Jet into a Crossflow – a Benchmark

J.A. Denev, J. Fröhlich, and H. Bockhorn

Institute for Technical Chemistry and Polymer Chemistry
{denev,froehlich,bockhorn}@ict.uni-karlsruhe.de

Abstract. A benchmark simulation of the jet in crossflow (JICF) configuration is presented in detail. A Direct Numerical Simulation (DNS) was carried out with a low Reynolds number equal to 275 and a jet-to-crossflow velocity ratio equal to 2.4. The benchmark is carefully selected to provide reference data concerning the following phenomena: the flowfield, the mixing process of passive scalars and three chemical reactions. The data presented concern both instantaneous and time-averaged values as well as the corresponding fluctuations. To facilitate the quantitative comparison with the data from the present work various one-dimensional plots are presented. To allow easy repetition of the present numerical benchmark, both the jet and the crossflow are supplied at laminar flow conditions. As a result of this a transition zone occurs which in turn constitutes a severe test for any simulation methodology.

1 Introduction

The present investigation is part of the Priority Programme SPP-1141 "Analysis, modeling and computation of flow mixing apparatus with and without chemical reactions" of the German Research Foundation (DFG). The overall goal of this programme is to develop analytical and numerical methods for the reliable prediction of flow and concentration fields in mixings devices, including chemical reactions. To achieve this, a detailed understanding of the flow physics is necessary. With this purpose a Direct Numerical Simulation (DNS) of a jet in crossflow is carried out in the present investigation. The careful selection of both the flow and the numerical parameters targets the development of a suitable benchmark configuration for the JICF.

The specification of a benchmark configuration is a delicate issue. On one hand, it has to represent a physically meaningful and challenging test for the models to be scrutinized. On the other hand, it must be simple enough to allow unambiguous specification of all conditions.

Since the specification of turbulent inflow conditions for DNS and Large Eddy Simulations (LES) is a delicate task which often inhibits close comparison between different simulations, it was decided to devise a case with entirely laminar inflow conditions. Thus the use of a precursor simulation for generation of unsteady turbulent inflow data is avoided and only the time-average shape of the inflow profile has to be specified in the simulation. In accordance to the previous experience of the authors [6], [4], up to 25% of the CPU-Time for LES and DNS could be saved when such precursor simulations are dropped. At the same time, the laminar inflow conditions lead to the presence of a transition zone in the flow, thus presenting a hard test for statistical turbulence models for RANS (Reynolds Averaged Navier Stokes) simulations or subgrid models for LES.

DNS benchmarks exist for many other flow types, but the complex geometry prevents most of the spectral-based codes from simulating the complete JICF configuration. Just recently increasing computer power allowed finite volume codes to meet the challenges of DNS simulations of the JICF. In their DNS Hahn and Choi [8] make use of a Carthesian grid and simulate only the crossflow. The effect of the round pipe has been accounted for by specifying a parabolic profile at the bottom wall of the domain. In their study the jet-to-crossflow velocity ratio $R = U_{jet}/U_{crossflow}$ was set to 0.5 to examine a film cooling configuration. The Reynolds number throughout this paper is defined with the jet-diameter and the velocity of the crossflow, $Re = U_\infty D/\nu$, where ν is the kinematic viscosity.

Sharma and Acharya [15] simulate a configuration with $R = 0.25$ and a rectangular jet relevant to film cooling of gas turbine blades and combustor walls. Muldoon [10], [1] presents DNS results of pulsating jets to study the effect of the Strouhal number and the waveforms on the flow structures and the mixing of a JICF. The investigations of Muppidi and Mahesh [13, 11, 14] follow the experiments of Su and Mungal [16] and use a relative high velocity ratio $R = 5.7$. In order to specify the boundary conditions in the pipe, a separate simulation of a fully turbulent pipe flow has been conducted. The data from a plane of this separate simulation have been stored and interpolated at the pipe inflow (pipe length equal to two diameters) thus making any repetition of the simulation by others quite difficult. In [12] the same authors present results only for the jet trajectories. However, in this study they use a different approach for the boundary conditions. They specify a "parabolic" and a "mean-turbulent" profile at the entrance of the pipe (10 diameters long) and do not report using any turbulence-generating approach for the different flow configurations with pipe-Reynolds numbers $Re = 1500$, $R = 1.5$ and $Re = 5000$, $R = 5.7$.

Unlike previous numerical investigations, the present work uses a set of well-specified laminar boundary conditions for both, the pipe and the crossflow which are easy to impose in any numerical code. To support further the target of benchmarking, quantitative results are presented for a wide range of phenomena: fluid flow and its coherent structures, mixing of passive scalars and simple chemical reactions. The jet-to-crossflow velocity ratio $R = 2.42$ used in the present study specifies a jet trajectory which is away from any adjacent channel wall.

2 The Flow Configuration and the Boundary Conditions

In a companion project at the same institute experimental work is currently performed for the same configuration [17]. These measurements have not yet been terminated, but the geometry of the flow and some of the boundary conditions are selected according to the experimental setup to support future data validation. The flow configuration is shown in Fig. 1 together with the employed coordinate system. All lengths are made dimensionless with the reference value being the jet diameter ($D = 8\ mm$ in the experiment): $L_x = 20$, $L_y = L_z = 13.5$, $l_x = 3$ and $l_z = 2$. The reference velocity $U = U_\infty = U_{crossflow} = U_{max}$ is the velocity of the crossflow at the middle of the main channel which appears to be the maximum velocity (and not the bulk velocity) of the channel. The most important physical parameters for this flow are the Reynolds number and the velocity ratio between jet and crossflow. The latter is defined here with the bulk velocity of the jet ($U_{b,jet}$), i.e. $R = U_{b,jet}/U$. Its value in the present case is $R = 2.42$ and has been selected to obtain a location of the jet remote from the channel walls. The Reynolds number based on the bulk velocity in the pipe hence is $Re_{jet} = U_{b,jet}D/\nu = 666$ so that both, crossflow and pipe flow appear laminar.

In the simulation a parabolic velocity profile:

$$w(r)/U_{b,jet} = 2[1 - (r/(D/2)^2] \tag{1}$$

was imposed at the origin of the pipe. Here, r is the radial coordinate and w the vertical velocity component. With the kinematic viscosity of air $\nu = 1.406 \times$

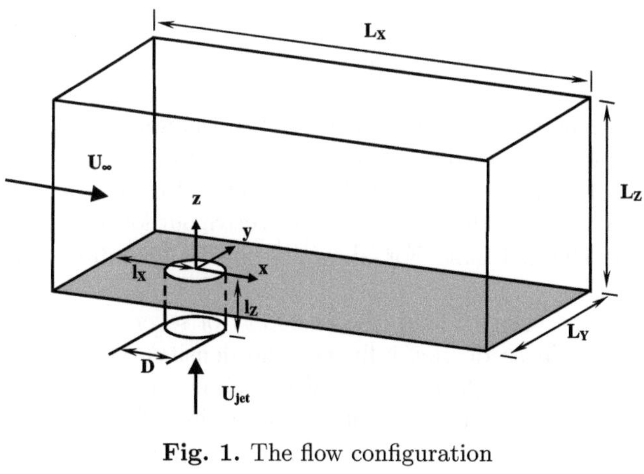

Fig. 1. The flow configuration

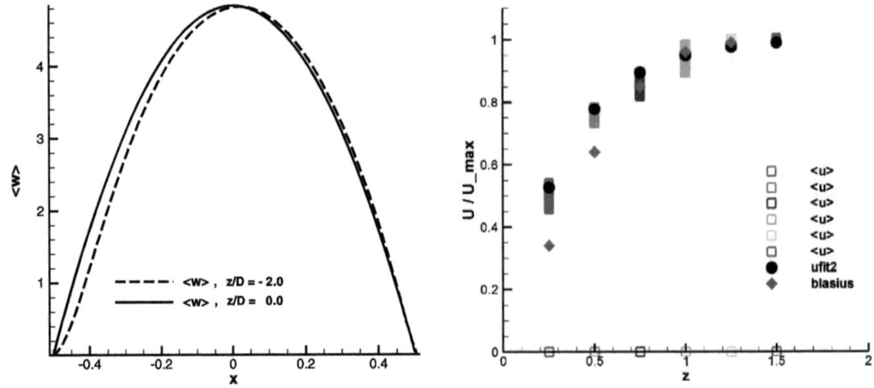

Fig. 2. *Left*: The profile in the pipe at $z/D = -2.0$ used as boundary condition and at the jet entrance, $z/D = 0.0$. *Right*: The experimentally measured velocities, the Blasius profile and the curve fit used in the simulation

10^{-05} m^2/s the dimensional bulk velocity of the jet corresponds to $U_{b,jet} = 1.16$ m/s. This boundary condition is shown together with the resulted profile at the end of the pipe (at $y/D = 0$) in Fig. 2. The profile at $z/D = -1.0$ (not shown) is practically the same as the boundary condition, i.e. unaffected from the crossflow.

The low Reynolds number creates a relatively large boundary layer along the walls of the duct. The shape of the boundary layer along the bottom wall is given in Fig. 2. The squares represent the measured data (PIV without the jet flow, planes $z = const$) at different distances from the bottom wall. The circles show the fit to these data used in the simulation, yielding a boundary layer thickness $\delta_{99\%} = 1.54D$. A Blasius profile with the same thickness is also shown for comparison. The boundary layer was not explicitly measured at the other walls, but for reasons of symmetry the same boundary layer shape was assumed in the simulation so that the u-velocity component at the inflow boundary has been defined by:

$$u(d_n) = 1.0 - \exp(-3.0\ d_n) \quad , \tag{2}$$

where d_n is the normal distance to the closest channel wall. The two other velocity components (v, w) are set to zero at the inflow boundary. Equation (2) yields $U_{max} = 1.0$ in the middle of the channel and $U_b/U_{max} = 0.905$, where U_b is the bulk velocity in the channel. A convective outflow boundary condition is used at the channel exit. For all walls no-slip boundary conditions are applied.

Apart from the velocity field, the transport of seven scalar variables have been calculated. These do not influence the density and hence are passive scalars for the flow field. The transport of all scalars can therefore be computed in the same simulation without mutual interaction. The first scalar is non-reactive and introduced with the jet where its value is set to 1 while being 0

Table 1. Schmidt and Damköhler numbers used in the present study

	A	B	Da
No reaction:	$Sc_1 = 0.96$		
Reaction 1:	$Sc_2 = 1.0$	$Sc_3 = 1.0$	$Da_{2,3} = 1.0$
Reaction 2:	$Sc_4 = 1.0$	$Sc_5 = 1.0$	$Da_{4,5} = 0.5$
Reaction 3:	$Sc_6 = 2.0$	$Sc_7 = 1.0$	$Da_{6,7} = 1.0$

in the crossflow. Its Schmidt number $Sc_1 = 0.96$ has been selected to match the diffusion of NO_2 in air which is the non-reacting tracer in the experiment. The remaining six scalars are reactive. Variations of Schmidt and Damköhler number have been performed as detailed in Table 1 in order to investigate their respective influence. Three model reactions of type $A + B \rightarrow P$ are considered in the study with reaction rate $\omega_{A,B} = Da_{A,B} Y_A Y_B$ where Y_A and Y_B are the mass concentration of the species A and B, respectively. Since Y_A, Y_B and Y_P sum up to unity, the concentration of the product Y_P needs not be computed. The scalars in column A of Table 1 are introduced with the jet while those of column B have concentration 1 in the crossflow and 0 in the jet. The Damköhler number Da characterizes the ratio of the characteristic flow time to the characteristic chemical reaction time. Hence, large values correspond to fast combustion occurring in very thin layers requiring a correspondingly fine numerical grid to resolve these reaction fronts. The values of Da have been selected here such that the cost of the simulation remains within the available limits.

3 Numerical Method

The Finite Volume code LESOCC2 (Large Eddy Simulation On Curvilinear Coordinates, version 2) [9] has been used in the present simulations. It solves the incompressible Navier-Stokes equations using second order central differences for the convective and the viscous fluxes together with the Rhie-Chow momentum interpolation on collocated grids. A three-step low-storage Runge-Kutta method is used for time-advancement. A Poisson-type equation is solved for the pressure-correction using the Strongly Implicit Procedure Procedure (SIP). The code uses body-fitted curvilinear block-structured grids and information between processors is exchanged via MPI.

If block-structured grids allow only matching cells at block boundaries, the overall number of control volumes increases quickly when the mesh is refined in any area of the computational domain. To overcome this, a local grid refinement technique has been implemented recently in the code. It allows local refinement of blocks by integer fractions compared to neighbouring block. As shown below, this can be used to improve the distribution of control volumes within the computational domain. The local refinement technique has been tested prior to this investigation on a Taylor-vortex and channel flows and data from the latter have been reported in [7].

4 Details of the Numerical Grid and Parallelization

The present grid containing 22.3 Mio cells has been generated with ICEM-CFD v4.3.3. It is partitioned into 219 numerical blocks. Out of these, 42 are locally refined in all three spatial directions by a factor of 3 as displayed in Fig. 3. The refined blocks are located close to the jet exit where gradients are large and transition takes place. The location of refinement was chosen after a preliminary simulation on a coarser grid. Despite the fact, that the refined blocks only cover a small portion of the computational domain, they contain 89% of the cells. Load balancing was obtained by an evolutionary algorithm based on the number of cells per processor. An optimal distribution was obtained with 31 or 32 processors. With 31 processors of the HP XC6000 Cluster at the Computer Centre Karlsruhe this yielded an average "user time" of 90.97% while the communication required was 9.03% (Overhead: 8.46%; Blocking: 0.57%).

An O-grid topology was used in the pipe containing 144 almost equally distributed points along the pipe diameter. In vertical direction the pipe contains 60 gridpoints. Their size Δz is equally decreasing toward the pipe exit, where it reaches the value $\Delta z = 0.026D$. Thirty-three points are located in the $\delta_{99\%}$ boundary layer of the bottom wall of the crossflow channel. The resolution capacity of the grid was assessed in physical terms by evaluating an LES subgrid-scale model during one time step of the developed solution. For a true DNS its contribution should be very small. In the present case,

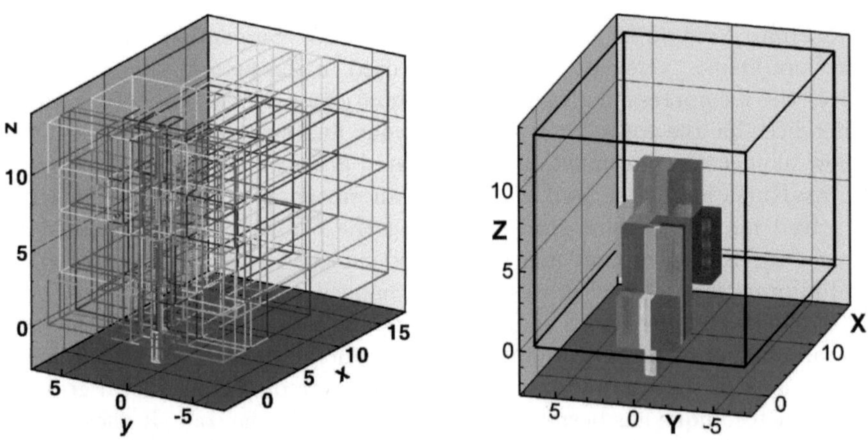

Fig. 3. Block boundaries of all 219 blocks of the grid (*left*) and overview of the refined blocks near the jet-exit (*right*)

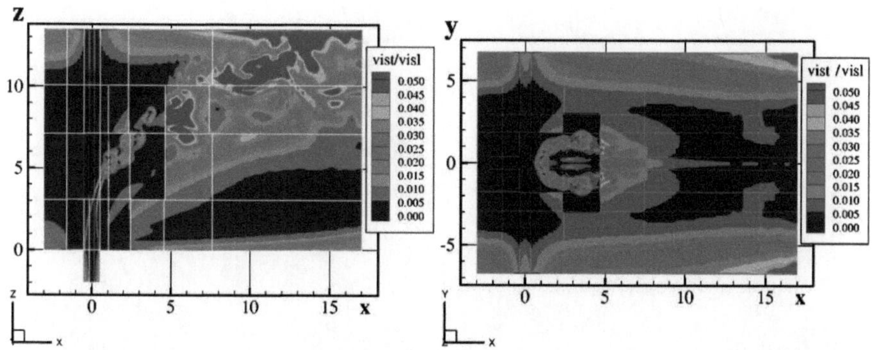

Fig. 4. The ratio of turbulent viscosity to molecular viscosity obtained with the Smagorinsky model for one iteration, planes $y = 0$ and $z/D = 5$

the Smagorinsky model with the standard choice of $C_s = 0.1$ for the constant involved in this model yielded an eddy viscosity below 3% of the molecular viscosity in the refined blocks as illustrated in Fig. 4. Bearing in mind that this model yields a turbulent viscosity even in the laminar case, this shows the resolution to be adequate for a DNS with the present grid.

5 Results

All figures below, including representations of the grid, have been obtained by consideration of every second cell in the full computational grid.

5.1 Statistical Data

The unsteady flow was computed over an initial period of 60 time units D/U before statistical averaging was started. The averaging has been performed over 50 time units.

The CFL number has to be restricted due to the explicit time integration scheme and for physical reasons, to resolve the flow in time. Its value for the present computation is CFL = 0.96 yielding a dimensionless time step of 0.00082. The timestep was set constant in view of the Fourier and wavelet analyses for the timesignals to be made at a later stage.

5.2 Instantaneous Data and Flow Structures

A first evaluation of the flow structures is based on the three dimensional pressure field. An iso-surface of the instantaneous pressure fluctuation which allows the visualization of the large vortex structures is plotted in Fig. 5. A companion visualization by the Q-criterion [5] is provided as well. These pictures show three types of structures: the boundary of the jet close to the

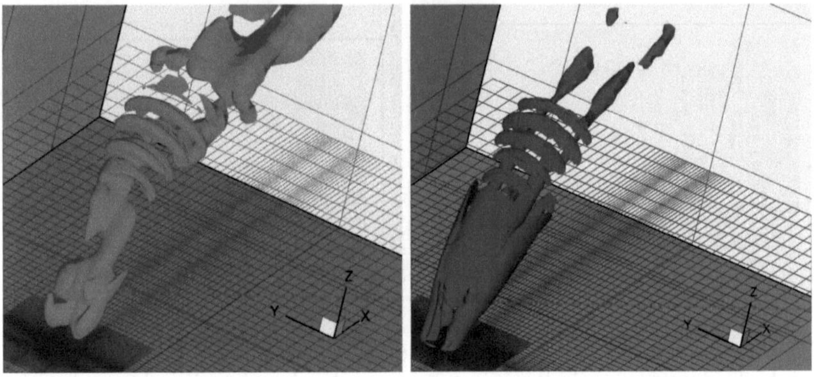

Fig. 5. *Left*: Coherent structures of the flow visualized by the isosurface $p - p_{average} = -0.035$. *Right*: The same structures visualized by the Q-criterion, $Q = 1.6$

outlet, spanwise rollers on the upper surface of the jet and further downstream two vortices aligned with the axis of the jet, the so-called counter-rotating vortex pair (CVP). The vortices shown are typical for the jet in crossflow configuration. Unlike previous simulations of the authors [6, 4] performed at higher *Re* with a turbulent jet-pipe flow, the present flow structures are clear cut and easier to identify due to the low Reynolds number.

Instantaneous plots of the non-reactive scalar S_1 and the reactive scalar S_2 originating in the jet (see Table 1) are presented in Fig. 6 for the centerplane $y = 0.0$. Both plots are taken at the same time at which the jet trajectory is also shown. Subsequent plots of the concentration field reveal that the position of the transition point changes with time. The position of the transition point has been found to vary between $x/D = 2.0$ and $x/D = 3.0$ for the different times plotted, which corresponds to a path of $s = 4.8D$ to $s = 6.0D$ along the jet trajectory, respectively. Those values, defined here as the position of the first waves, appearing near the jet-trajectory in the level-plots of scalar S_1 (see Fig. 6), correspond well with the the experimental data of [2].

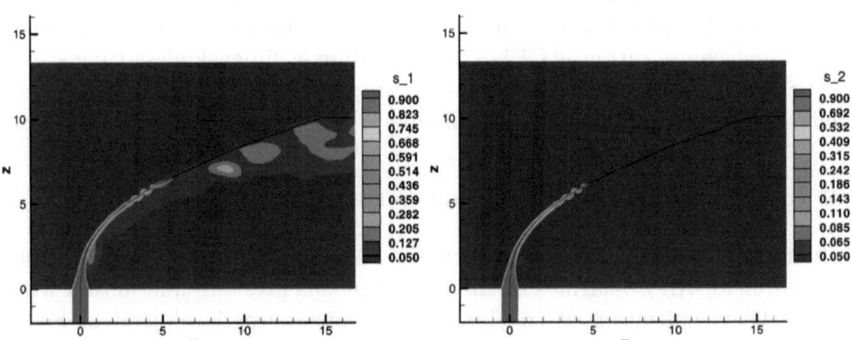

Fig. 6. The concentration of S_1 (*left*) and S_2 (*right*) in the symmetry plane

5.3 Statistical Data in the Centerplane

Figure 7 shows the jet trajectory in the centerplane, determined by the streamtrace of the time-averaged velocity field, which originates at the middle of the pipe-exit ($x, y, z = 0$). Furthermore, selected streamlines are visualized which demonstrate the complicated flow patterns behind the jet with locally reverse and locally descending flow. The saddle point in this graph, where all streamtraces originate, is located at $(x/D, z/D) = (1.16, 0.55)$. The near-bottom streamtrace at the left hand side of the jet shows that, similar to the case with $R = 5.7$ of [12], there exists no hovering vortex near the pipe-end, which is due to the relative strong jet momentum investigated.

The contour plot on the right hand side of Fig. 7 shows the scalar concentration at $x/D = 4.0$ together with the streamtraces in that plane. The counter-rotating vortices are clearly seen together with the upward flow of ambient fluid in the middle of the jet. This upward flow is in accordance with the streamtraces on the left picture of Fig. 7. It is the main reason for the good mixing properties of the JICF configuration as shown also by the mixing indices studied in [3].

Figures 8–12 show quantitative data for means and fluctuations along vertical cuts in the centerplane. The axial positions have been normalized by RD and correspond to $x/D = 0.0$, $x/D = 1.45$, $x/D = 2.90$ and $x/D = 4.35$, respectively. In Fig. 8 the negative $\langle u \rangle$-component is clearly apparent together with negative values of $\langle w \rangle$ near the bottom. The position of the jet is related to narrow peaks in both components. The corresponding fluctuations in Fig. 9 also exhibit peaks at these positions while their level remote from the jet is very low. An exception is w_{RMS} at $x/RD = 0.6$.

Figure 10 displays the four scalars introduced with the jet, S_1 being the non-reactive, while the other three are described in Table 1. It is apparent that the reactive scalars are consumed and hence attain a lower level downstream.

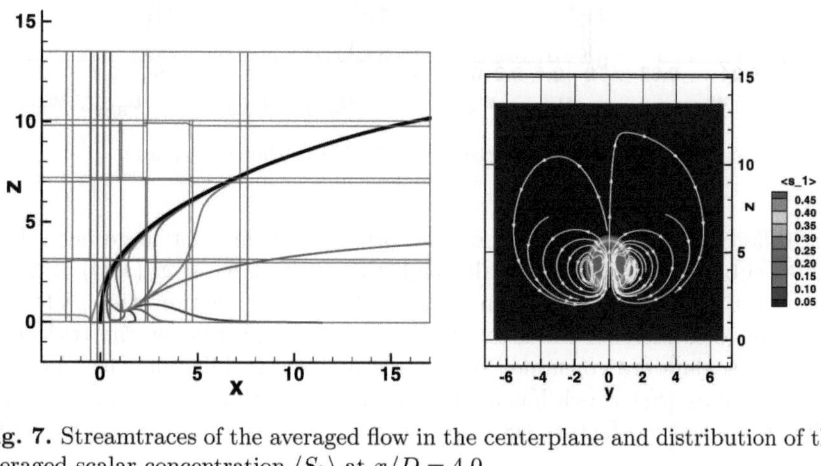

Fig. 7. Streamtraces of the averaged flow in the centerplane and distribution of the averaged scalar concentration $\langle S_1 \rangle$ at $x/D = 4.0$

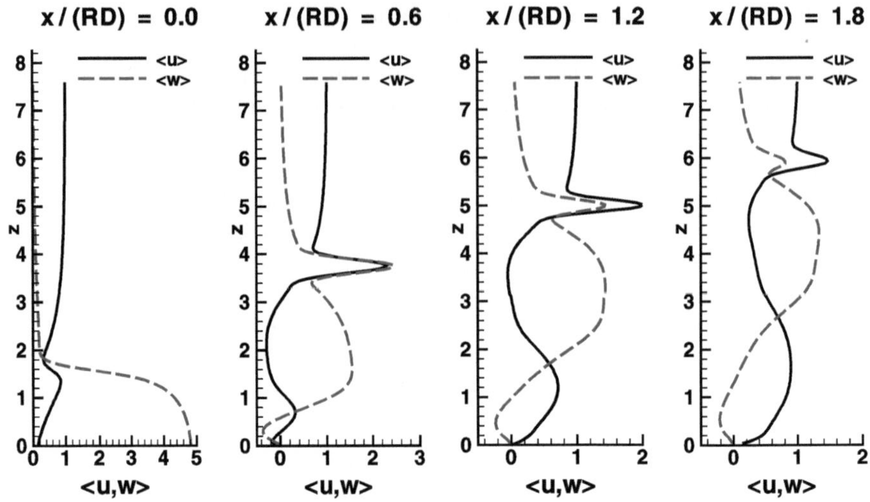

Fig. 8. Mean velocity $\langle u \rangle$ and $\langle w \rangle$ in the centerplane

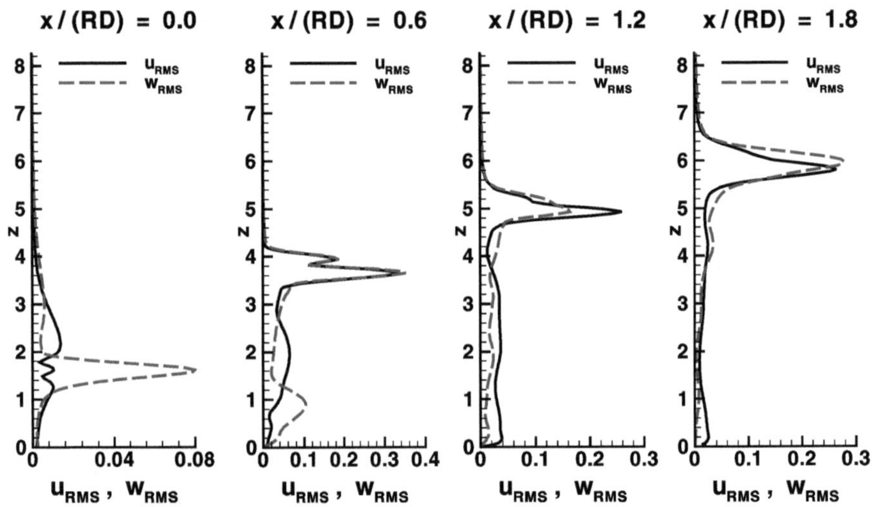

Fig. 9. Velocity fluctuations (RMS) in the centerplane

The differences in Da and Sc, on the other hand, are not large enough to induce substantial differences, although the peak of S_2 at $x/(RD) = 1.8$ is only two thirds of the maximum of $\langle S_4 \rangle$. Due to the reaction, the CVP does not manage to increase the level of $\langle S_2 \rangle$, $\langle S_4 \rangle$ and $\langle S_6 \rangle$ below the trajectory since the scalars are consumed on their way. This is different with the non-reactive scalar $\langle S_1 \rangle$ which has a value of about one thirds the peak level just below the jet at $x/(RD) = 1.8$.

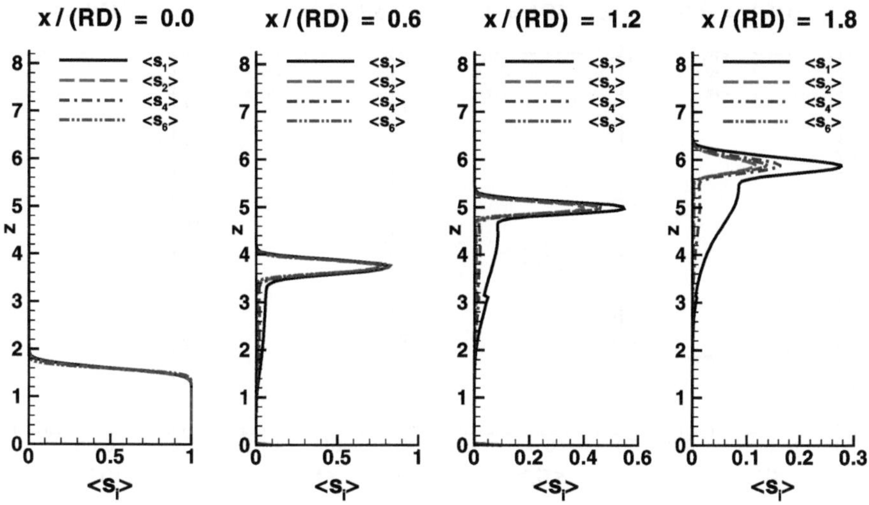

Fig. 10. Mean scalar concentrations in the centerplane

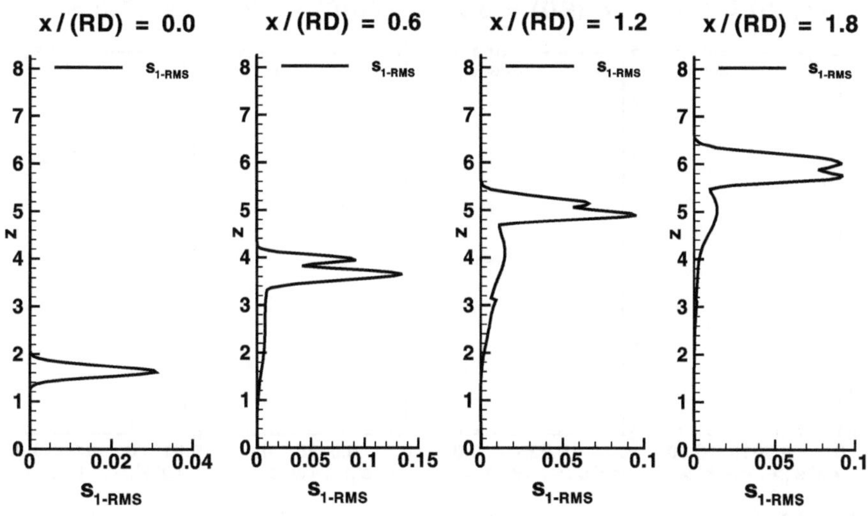

Fig. 11. RMS-fluctuations of the non-reactive scalar S_1 in the centerplane

The RMS-values of S_1 are given in Fig. 11. They exhibit a double peak in the jet and small values below the jet, for the reasons explained above. Figure 12 shows the three reaction rates and illustrates the consumption in and below the jet. Figure 13 finally provides the turbulent fluxes of S_1 in streamwise and vertical direction. Apart from $x = 0$ where the jet is vertical, they both have about the same level. A particularity is the negative peak of $\langle w' s_1' \rangle$ at the right-most position.

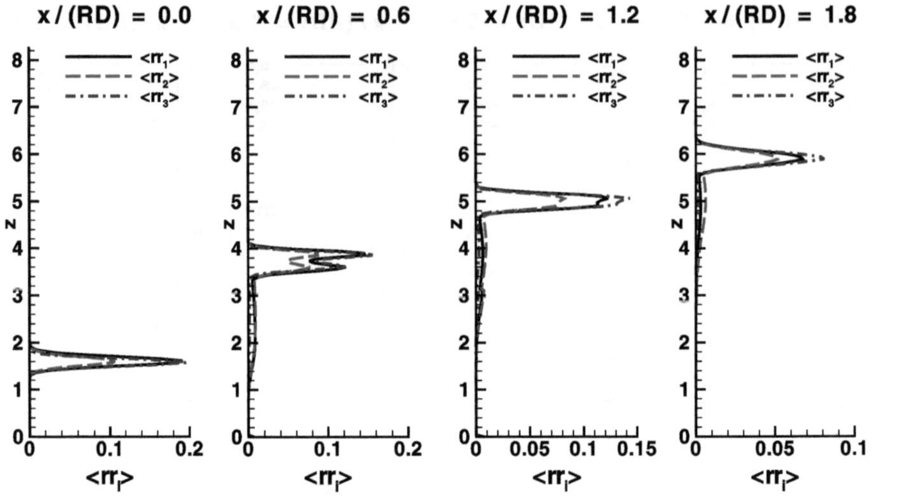

Fig. 12. Reaction rate of reactions 1–3 in Table 1

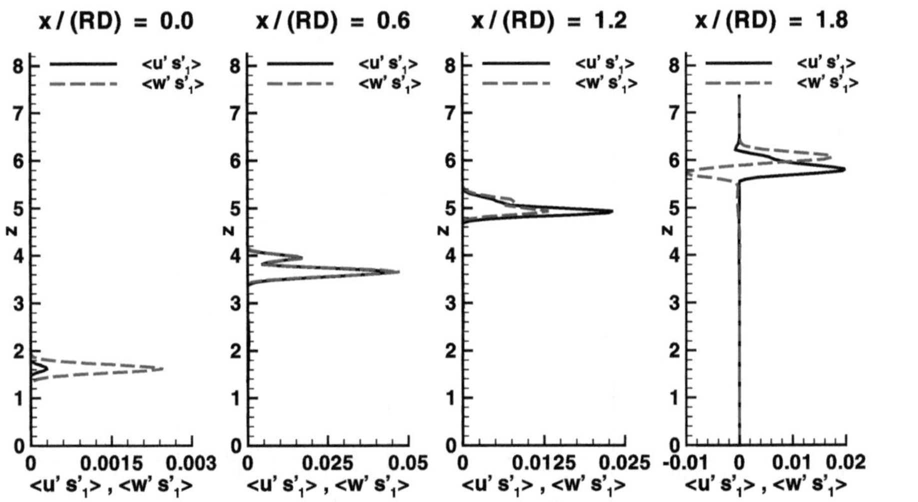

Fig. 13. Turbulent fluxes of the non-reactive scalar in streamwise and vertical direction, $\langle u's_1'\rangle$ and $\langle w's_1'\rangle$, respectively

5.4 Computational Resources

The computation was performed on 31 processors of the HP XC6000 Cluster of the Computer Centre at the University of Karlsruhe. In total 110 dimensionless time units D/U were calculated. Each dimensionless time unit consists of 1220 time steps and requires 21.3h CPU-Time (on 31 processors). The computation of the seven scalar variables with the three chemical reactions takes approximately 54% of the total CPU-time, 46% are spent for the velocity field, mainly for the pressure-correction which results from the solution of a Poisson equation. The required RAM is below 1 Gb per processor. The disk space to store the compressed (zipped) binary data containing both instantaneous and averaged values is 4,6 GB. Another 10 GB are required for postprocessing these data.

6 Conclusions

Direct numerical simulations are exempt from any turbulence model and can hence provide reference data for the calibration and validation of LES (Large Eddy Simulation) subbgrid models or turbulent models for RANS (Reynolds Averaged Navier Stokes equations) simulations. The present computations were devised to serve as a benchmark of the jet in crossflow configuration at low Reynolds numbers. For this purpose the physical and numerical parameters have been adjusted to yield a compromise between diverging requirements:

- simplifying the numerical efforts and decreasing the numerical uncertainty by imposing laminar boundary conditions in both the jet and the crossflow;
- specifying values for the Reynolds number and the velocity ratio R which yield a jet trajectory away of the channel walls;
- specifying values for the Reynolds number and the velocity ratio R to obtain a possibly short transition length of the jet – in order to decrease the region with fine grid and to keep numerical resources bounded.

The last requirement was not fully met in the present investigation and a transition has been found to occur approximately 5–6 diameters downstream the jet trajectory, varying with the time. However, the fact that a large area of the computational domain is still laminar, allows the present data to be compared also with coarse-grid DNS.

Detailed description of the present boundary conditions aims at easy comparison of the present data with the results of other numerical groups. This target is supported also by the various one-dimensional plots of averaged and fluctuating quantities presented.

A further feature of the present benchmark is the presentation of data regarding also mixing of passive scalars and simple chemical reactions with different values of the Schmidt and Damköhler numbers.

After publishing the results, the data from the one-dimensional plots will be made available for comparison at:
http://www.ict.uni-karlsruhe.de/index.pl/themen/dns/index.html.

Acknowledgements

The authors gratefully acknowledge funding by the DFG through SPP 1141 and provision of computing time at the Karlsruhe Supercomputer Center. The authors would like to thank their colleagues Priv. Doz. Dr. habil. Rainer Suntz and Dipl.-Ing. Camilo Cardenas for delivering the data for the boundary layer of the crossflow.

References

1. *url = www.navo.hpc.mil/Navigator/sp04_Feature3.html.*
2. R. Camussi, G. Guj, and A. Stella. Experimental study of a jet in a crossflow at very low reynolds number. *J. Fluid Mech.*, 454:113–144, 2002.
3. J.A. Denev, J. Fröhlich, and H. Bockhorn. Evaluation of mixing and chemical reactions within a jet in crossflow by means of LES. In *Proc. European Combustion Meeting, April 3–6, Louvain, Belgium, 2005*, pages CD–ROM, Louvain-la-Neuve, Belgium, 2005. Université Catholique de Louvain.
4. J.A. Denev, J. Fröhlich, and H. Bockhorn. Structure and mixing of a swirling transverse jet into a crossflow. In J.A.C. Humphrey, T.B. Gatski, J.K. Eaton, R. Friedrich, N. Kasagi, and M.A. Leschziner, editors, *Proceedings of 4th Int. Symp. on Turbulence and Shear Flow Phenomena, June 27–29 2005, Williamsburg, Virginia*, pages 1255–1260, 2005.
5. Y. Dubief and F. Delcayre. On coherent-vortex identification in turbulence. *J. Turbulence*, 1(11), 2000.
6. J. Fröhlich, J.A. Denev, and H. Bockhorn. Large eddy simulation of a jet in crossflow. In *Proceedings of ECCOMAS 2004, Jyväskylä, Finland, July 24–28, ISBN 951-39-1868-8, CD-ROM*, Barcelona, 2004. CIMNE.
7. J. Fröhlich, J.A. Denev, C. Hinterberger, and H. Bockhorn. On the impact of tangential grid refinement on subgrid-scale modelling in large eddy simulations. In *Sixth International Conference on Numerical Methods and Applications – NM&A'06 August 20–24, 2006, Borovets, Bulgaria*, 2006.
8. S. Hahn and H. Choi. Unsteady simulation of jets in a cross flow. *J. Comput. Phys.*, 134:342–356, 1997.
9. C. Hinterberger. *Dreidimensionale und tiefengemittelte Large–Eddy–Simulation von Flachwasserströmungen.* PhD thesis, Institute for Hydromechanics, University of Karlsruhe, http://www.uvka.de/univerlag/volltexte/2004/25/, 2004.
10. F. Muldoon. *Numerical Methods for the Unsteady Incompressible Navier-Stokes Equations and Their Application to the Direct Numerical Simulation of Turbulent Flows.* PhD thesis, Louisiana State University, 2004.
11. S. Muppidi and K. Mahesh. Direct numerical simulation of turbulent jets in crossflow. In *43 rd AIAA Aerospace Sciences Meeting and Exibit*, Reno, Nevada, Jan 10–13 2005. AIAA Paper 2005–1115.
12. S. Muppidi and K. Mahesh. Study of trajectories of jets in crossflow using direct numerical simulations. *J. Fluid Mech.*, 530:81–100, 2005.
13. S. Muppidi and K. Mahesh. Velocity field of a round turbulent transverse jet. In J.A.C. Humphrey, T.B. Gatski, J.K. Eaton, R. Friedrich, N. Kasagi, and M.A. Leschziner, editors, *Fourth International Symposium. on Turbulence and Shear Flow Phenomena*, Paper TSFP4–197, pages 829–833, Williamsburg, Virginia, 2005.
14. S. Muppidi and K. Mahesh. Passive scalar mixing in jets in crossflow. In *44th AIAA Aerospace Sciences Meeting and Exhibit*, Reno, Nevada, Jan 9–12 2006. AIAA Paper 2006–1098.

15. Ch. Sharma and S. Acharya. Direct numerical simulation of a coolant jet in a periodic crossflow. Technical report, National Aeronautics and Space Adminisatration, October 1998.
16. L.K. Su and M.G. Mungal. Simultaneous measurements of scalar and velocity field evolution in turbulent crossflowing jets. *J. Fluid Mech.*, 513:1–45, 2004.
17. R. Suntz and C. Cardenas. Analysis, modeling and computation of flow mixing aparatus with and wihtout chemical reactions. Technical report, DFG-Report, Priority Programme SPP-1141, June 2006.

Numerical Simulation of the Bursting of a Laminar Separation Bubble

Olaf Marxen and Dan Henningson

Department of Mechanics, Royal Institute of Technology (KTH)
Stockholm, Sweden
olaf.marxen@mech.kth.se

Abstract. Numerical simulations of laminar separation bubbles are carried out to investigate the so-called bubble bursting, i.e. the changeover from a short to a long bubble by means of very small variation of one governing parameter. A laminar separation bubble is formed if a laminar boundary layer separates in a region of adverse pressure gradient on a flat plate and undergoes transition, leading to a reattached turbulent boundary layer. Bubble bursting denotes a phenomenon, in which a local, in average closed region of reverse flow (the short separation bubble) suddenly becomes considerably longer as a result of only small changes in the conditions of the surrounding flow. Here, this condition is the disturbance input upstream of separation.

Both, long laminar separation bubbles and bubble bursting, are not yet well understood on a fundamental level, but it is commonly accepted that the transition process plays an important role. Simulations in which transition is or is not explicitly triggered are carried out. Depending on this triggering, either a short laminar separation bubble develops or the bursting process is initiated and the flow develops towards a long-bubble state. If the flow is tripped to turbulence prior to the adverse pressure gradient, the boundary layer remains attached. Performance data on a NEC SX-8 super computer are given for two different resolutions.

1 Introduction

Transition to turbulence in a separated boundary layer often leads to reattachment of the turbulent boundary layer and the formation of a laminar separation bubble (LSB). In environments with a low level of disturbances fluctuating in time, the transition process is governed by strong amplification of these disturbances. Such a scenario is typical for a pressure-induced LSB, e.g. found on a (glider) wing in free flight, or for an experiment where the region of pressure rise is preceded by a favorable pressure gradient that damps out unsteady perturbations (Watmuff, 1999). In the region of an adverse pressure gradient, disturbance waves are subject to strong amplification, and their saturation marks the location of transition to turbulence.

LSBs can occur on the surface of slender bodies, e.g. wing profiles, and may strongly affect their performance, such as lift and drag, as well as their noise production. These phenomena are observable on different aerial vehicles, on gas and wind turbines as well as on or behind obstacles in the flow in general.

Laminar-turbulent transition in laminar separation bubbles has been subject of numerous studies in the past. Watmuff (1999); Häggmark (2000); Lang *et al.* (2004) carried out experimental studies of LSBs, while Spalart & Strelets (2000), Alam & Sandham (2000), Häggmark *et al.* (2001), Marxen *et al.* (2003), and Wissink & Rodi (2004) tackled the flow by means of direct numerical simulations (DNS). Most of the cited studies conclude that some type of linear instability (Tollmien-Schlichting or Kelvin-Helmholtz instability) is the cause for transition.

Owen & Klanfer (1953) were the first to distinguish between *short* and *long separation bubbles*. This initial classification was based on the bubble length in comparison to the chord of an airfoil. Tani (1964) introduced a classification that depends on the (upstream) influence of the separation bubble, which is supposed to be local for short and global for long bubbles. In particular, short bubbles on airfoils shall have a small impact on the overall pressure distribution, while a strong impact arises for long bubbles.

Under certain conditions, for slight changes in the flow conditions, e.g. by increasing the angle of attack of the airfoil, short bubbles can break-up into long ones, the so-called *bubble bursting*. In some cases (especially on airfoils), the flow might not reattach at all, leading to stall. On a semi-infinite flat plate, the flow will finally reattach in any case. Usually, bubble bursting refers to the phenomenon of sudden change in final (mean) state with only minor changes of the governing parameters and not to the dynamical process of the actual bursting of a short to a long bubble.

Gaster (1966) carried out a detailed investigation to settle characteristics of short and long bubbles, and to establish parameters that shall allow prediction under which (external) circumstances short or long separation bubbles appear, i.e. develop a bursting criterion. However, no criteria have found general acceptance so far (Diwan *et al.*, 2006), and a physical explanation for bursting is still lacking.

Only short LSBs have been successfully simulated by means of full 3-d DNS. Pauley *et al.* (1990) claim to have simulated long LSBs using 2-d simulations, yet this seems questionable in view of the current knowledge of (short) separation bubbles underlining the necessity of full 3-d simulations. Time-accurate three-dimensional numerical simulations of long LSBs and bubble bursting appear now within reach using today's supercomputers. The numerical simulations reported here can be viewed as a first step towards DNS of such flows. Furthermore, they aim at offering a physical explanation for the occurrence of bubble bursting by demonstrating the importance of the transition process for this phenomenon.

2 Numerical Method for Direct Numerical Simulations

Numerical simulation of the three-dimensional unsteady incompressible Navier-Stokes equations using the code *n3d* serve to compute the flow field with a separation bubble.

The code *n3d* is a solver for highly accurate computation of transitional and turbulent flows, developed at the Institut für Aerodynamik und Gasdynamik (IAG), Universität Stuttgart. It is based on the vorticity-velocity formulation of the Navier-Stokes equations for an incompressible fluid. The present method is based on a discretization suggested and carefully analyzed by Kloker (1998), and thereafter slightly refined, extended, and optimized by many others (see also Marxen & Rist, 2005).

The method uses finite differences of fourth/sixth-order accuracy on a Cartesian grid for downstream (NMAX) and wall-normal (MMAX) discretization. Grid stretching in wall-normal direction allows to arbitrarily distribute grid points. In spanwise direction, a spectral ansatz is applied (KMAX + 1 modes). An explicit fourth-order Runge-Kutta scheme is used for time integration (LPER time steps per period of the fundamental disturbance). Upstream of the outflow boundary a buffer zone in the interval $[x_{st,oBZ}, x_{en,oBZ}]$ smoothly returns the flow to a steady laminar state. The solution of a Poisson equation for the wall-normal velocity is obtained by a direct method based on a Fourier expansion. Further details of the code *n3d*, i.e. the numerical method and the implementation, can be found in Meyer *et al.* (2003). The code *n3d* applies a hybrid parallelization approach employing two fundamentally different techniques at the same time: a directive-based shared-memory intra-nodal parallelization (similar to OpenMP) is combined with an explicitly coded distributed-memory MPI inter-nodal parallelization. The code was provided by the project partner of this HLRS project *long_lsb*, U. Rist of the Institut für Aerodynamik und Gasdynamik (IAG), Universität Stuttgart. Extensions to the code made within this project, e.g. implementation of the volume forcing and adaption of the Poisson solver to a total-flow formulation, are available to the Stuttgart group.

3 Description of the Configuration and Parameters

The present investigation is restricted to pressure-induced laminar separation on a flat plate. Following a strategy persued in many experiments (e.g. Gaster (1966); Watmuff (1999); Lang *et al.* (2004)) and some numerical simulations (e.g. Marxen *et al.* (2003)), the streamwise pressure gradient is designed in a way that we have a favorable pressure gradient followed by an adverse one. A detailed description of the potential flow governing the pressure distribution will be given in the following section.

3.1 Slip Flow

At a distance of y_{max}, a distribution of the streamwise velocity is set according to the following formula (assuming non-dimensional quantities, see Sect. 3.2):

$$u_{slip}(y_{max}) = 1 + A_1 \cdot exp(-B_1 x^2) + A_2 x^2 \cdot exp(-B_2 x^2) \quad (1)$$

with the choice $A_1 = 10.0$, $A_2 = -41.0$, $B_1 = 200.0$, and $B_2 = 12.0$. The coordinate system is chosen such that the maximum of this distribution is located at $x = 0$. Prescribing such a distribution at $y_{max} = 160$, placing a (slip) wall, i.e. no-through-flow condition, at $y = 0$, and prescribing a constant velocity at $x_{ifl} = -266.\overline{6}$ and $x_{ofl} = 1040$ would allow to compute the potential-flow solution by solving Laplace's equation numerically.

However, for simplicity we assume periodicity in x, which is a good approximation to constant u (and vanishing v) at the inflow and outflow. In this case, the maximum streamwise wave length is $\lambda_x = 1306.\overline{6}$, giving a fundamental wave number $\alpha_0 = 2\pi/\lambda_x \approx 4.8 \cdot 10^{-3}$. We can thus compute the slip-flow field u_{slip}, v_{slip} after transforming u_{slip} resulting from eq. (1) to Fourier space (quantities in Fourier spaces are denoted by $\hat{}$: $u_{slip} = \sum \hat{u}^{(\mathsf{g})}_{slip}\, exp(i\mathsf{g}\alpha_0 x) + c.c.$ with a wave-number coefficient $\mathsf{g} = 0 \ldots \mathsf{G}_{max}$):

$$\hat{u}^{(\mathsf{g})}_{slip}(y) = \hat{u}^{(\mathsf{g})}_{slip}(y_{max}) \frac{\cosh(\mathsf{g}\alpha_0 y)}{\cosh(\mathsf{g}\alpha_0 y_{max})}, \quad \forall \mathsf{g}, \quad (2a)$$

$$\hat{v}^{(\mathsf{g})}_{slip}(y) = -i\, \hat{u}^{(\mathsf{g})}_{slip}(y_{max}) \frac{\sinh(\mathsf{g}\alpha_0 y)}{\cosh(\mathsf{g}\alpha_0 y_{max})}, \quad \forall \mathsf{g}. \quad (2b)$$

Note that a flow field according to eqs. (2) does not obey the no-slip condition at the wall. In case of the real flow, it will thus be valid only (far) outside the boundary/shear layer that develops close to the wall. Furthermore, it might be altered if the latter has a (strong) displacement effect on the potential-flow field (so-called viscous-inviscid interaction).

The velocities $u_{slip}(x, y), v_{slip}(x, y)$ at any wall-distance $y \in [0, y_{max}]$ can be computed from the given distribution $u_{slip}(x, y_{max})$ by transforming the latter to Fourier space, evaluating eqs. (2) at the desired y and transforming $\hat{u}^{(\mathsf{g})}_{slip}(y), \hat{v}^{(\mathsf{g})}_{slip}(y)$ back to physical space. Considering also complex conjugates yields real quantities as a final result. The number of streamwise harmonics G_{max} was chosen to be 140 in the present case.

By wall-normal integration of the streamwise velocity, the streamfunction can be obtained (Fig. 1). The potential-flow field resembles one created by a cylinder (i.e. a dipole) above a wall. However, the present distribution possesses the advantage that it results in almost constant streamwise (and vanishing wall-normal) velocity at the inflow and outflow boundary for the chosen streamwise domain size. This would not be the case for a flow field created by a single dipole at ($x = 0$, y_{max}) that possesses a strength which gives a comparable pressure minimum at the wall.

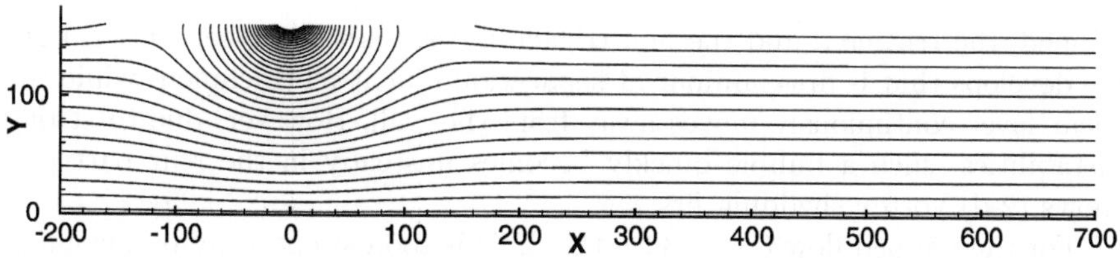

Fig. 1. Contours of the stream function Ψ for the potential flow used to generate the pressure distribution at the wall

3.2 Computational Parameters and Initial Condition

All quantities are non-dimensionalized using the boundary-layer displacement thickness δ^* and the free-stream velocity U_∞ at the inflow. The Reynolds number based on δ^* at the inflow amounts to $Re_{\delta^*} = 600$.

At the inflow boundary $x_{ifl} = -266.\overline{6}$, a Blasius similarity solution is prescribed. The outflow boundary is located at $x_{ofl} = 333.0612$ with a preceding damping region starting at $x_{st,oBZ} = 293.8776$. The useful region of the integration domain extends up to $x \approx 270$. The height y_{max} was chosen to be 160 and the spanwise extent is given by $\lambda_z = 2\pi/\gamma_0$ with $\gamma_0 = 0.12$.

Three cases were run: a flow tripped to turbulence using volume forcing, a laminar flow where transition was triggered by small disturbance input with a period of $T_0 = 2\pi/\beta_0$ with $\beta_0 = 0.3$, and an essentially untriggered case. Main parameters of the simulations are summarized in Table 1.

Simulations with trip-forcing used a lower resolution (half the resolution in x and y, KMAX = 23) to reduce computational effort due to a lower required accuracy. Runs with complete absence of forcing or triggering were initially carried out with low resolution as well, but are currently repeated with increased resolution. While being different in the details, both, low and high resolution simulations exhibit a development towards a long bubble state (see Sect. 5.2). In all low resolution calculations, the outflow boundary was located considerably further downstream at $x_{ofl} = 847.8912$ with a preceding damping region starting at $x_{st,oBZ} = 798.2585$.

As an initial condition a Blasius boundary-layer profile is prescribed throughout the integration domain. The profile is then multiplied with the local streamwise slip velocity at the wall. Finally, both boundary-layer and potential solution are summed up and the streamwise slip velocity at the wall is subtracted. Starting the calculation with such a field, in all cases (the

Table 1. Simulation parameters for numerical simulations

$Re_{\delta^*_{ifl}} = 600$	$y_{max} = 160.0$	$x_{ifl} = -266.\overline{6}$	$x_{st,oBZ} = 293.8776$
$\beta_0 = 0.3$	$\gamma_0 = 0.12$	$x_{ofl} = 333.0612$	$x_{en,oBZ} = 328.7075$
NMAX = 6001	MMAX = 1537	KMAX = 2×135	LPER = 2000

tripped, the triggered and the non-triggered case) initially a separation bubble develops that is first laminar at separation *and* at reattachment while the shear layer continuously moves away from the wall. However, due to strong instabilities, such a bubble quickly becomes unsteady in the rear part and shows (2-d) vortex shedding first.

For the tripped flow (Sect. 3.3), this LSB is soon swept away by incoming fluctuations. For the triggered LSB (Sect. 4), the forced oblique waves finally play a dominant role in the transition process, leading to a short LSB that converges towards a statistically stationary (and almost periodic) state. The untriggered simulations approaches a long LSB and will be subject of Sect. 5. Simulations were initiated on a coarse grid and only after some time the resolution was increased in order to save computer time.

3.3 Flow Tripped to Turbulence

To get an estimate of the pressure distribution imposed on the boundary layer by the potential flow, Gaster (1966) tripped the flow to turbulence for the same configuration he used to investigate the laminar separated flow. In all cases, he found that the boundary layer remained attached if tripped. Even though the slip-flow pressure distribution is known here, tripping the flow to turbulence appears still interesting to see if the flow remains attached also for the present configuration. The computation for tripped flow assumed spanwise symmetry and a coarser resolution was applied, namely NMAX = 5121, MMAX = 769, KMAX = 23, LPER = 900.

To trip the flow to turbulence, a volume force is applied according to a procedure by Herbst & Henningson (2006). Here, the normalized amplitude of the forcing amounts to 0.2% of the freestream velocity and it is centered at $x = -236.\overline{6}$. A new random force distribution (among the first 8 spanwise modes) was generated every $1/5 \cdot T_0$.

Indeed, with such a trip forcing, the boundary layer remains attached within the whole integration domain. The presssure distribution at the wall matches the one of the slip flow as expected (Fig. 2, left). The acceleration

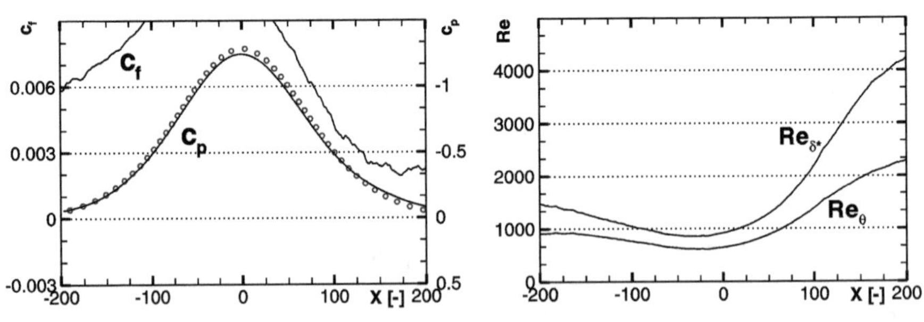

Fig. 2. Coefficients for surface pressure c_p, skin friction c_f (*left*), and Reynolds numbers Re_{δ^*} and Re_θ (*right*). Comparison of numerical results for tripped flow (*solid lines*) and slip-flow results $c_{p,slip}$ (*symbols*)

($x < 0$) is sufficiently strong to cause a decrease in quantities related to boundary-layer thicknesses, while in the deceleration region ($x > 0$) a strong increase in these quantities is observable (Fig. 2, right).

4 Short Laminar Separation Bubble

Simulations are performed with deterministic disturbance input. A pair of oblique time-harmonic perturbations is triggered via blowing and suction at the wall through a disturbance strip. The streamwise location of the strip is $x \in [1.5238, 15.4558]$ while the amplitude of the disturbance input amounts to $A_v = 10^{-4}$ for the wall-normal velocity. The fundamental frequency is $\beta_0 = 0.3$ and the fundamental spanwise wavenumber $\gamma_0 = 0.12$. The disturbance strip is placed downstream of the c_p-minimum, since the accelerated boundary layer is very stable and the forced perturbation would otherwise decay quickly.

The resulting flow field is visualized in Fig. 3 by means of contours of the instantaneous spanwise vorticity component for the spanwise harmonic $\mathsf{k} = 0$, together with time-averaged streamlines. Very close to the wall, the mean dividing streamline $\bar{\psi} = 0$ visualizes the short LSB. Its start is marked by the point of separation S ($x_S \approx 34$). The transition location as judged from the appearance of small-scale structures is roughly situated at $x \approx 66$. Note that in the figure, the wall-normal axis is *not* stretched compared to the streamwise one to give a more realistic impression on the shape and size of the LSB.

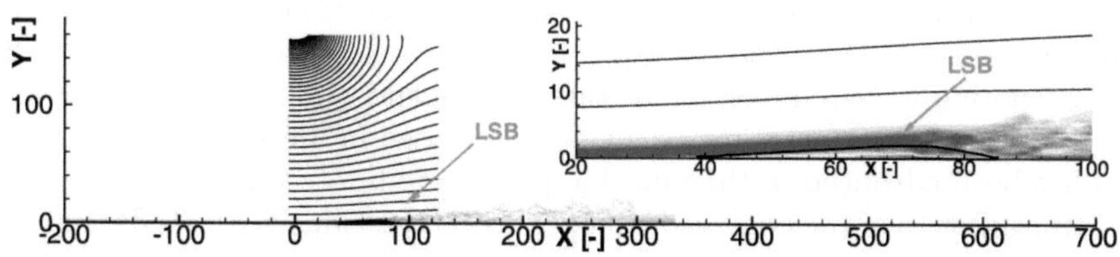

Fig. 3. Contours of instantaneous spanwise-averaged spanwise vorticity ω_z from simulations of a short LSB together with mean streamlines (insert: enlargement)

4.1 Mean Flow and Boundary-Layer Parameters

Triggering the flow is not sufficient to suppress separation altogether, but transition to turbulence of the initially laminar shear layer results in mean reattachment which follows almost immediately after the transition process has been completed. This can be seen from the typical drop in skin-friction at the transition location (see curve c_f in Fig. 4, left) quickly fol-

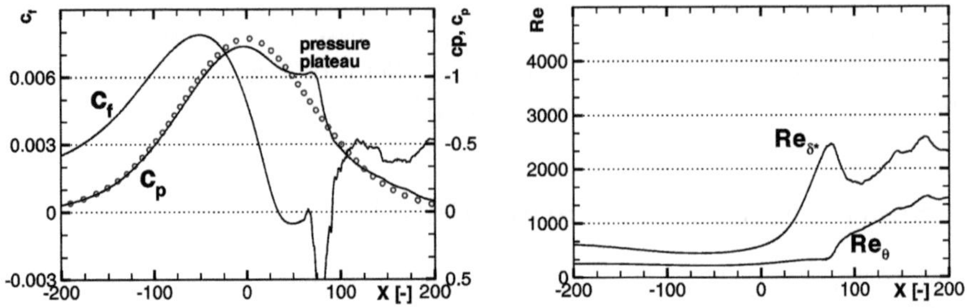

Fig. 4. Coefficients for surface pressure c_p, skin friction c_f (*left*), and Reynolds numbers Re_{δ^*} and Re_θ (*right*). Comparison of numerical results for a short LSB (*solid lines*) and slip-flow results $c_{p,slip}$ (*symbols*)

lowed by a strong rise to positive values. A deviation of the pressure coefficient at the wall c_p from the case of slip flow is observable where the LSB is located. In particular the formation of a pressure plateau is clearly seen.

Quantities related to boundary-layer thicknesses are computed using a pseudo velocity that is obtained by integrating the spanwise vorticity in wall-normal direction (see Spalart & Strelets (2000)). Again, first a drop and then a strong increase in the quantities related to the displacement of the boundary layer (such as the displacement thickness δ^* or Re_{δ^*}) is visible along the integration domain.

4.2 Transition

A double Fourier transform in time and spanwise direction yields disturbance amplitudes and phases. The notation (h, k) is used to specify modes, with h and k denoting wave-number coefficients in time and span, respectively. Four periods of the fundamental frequency β_0 where used in the analysis. The flow field has been advanced in time until a quasi-periodic state is reached.

Strong growth of the triggered oblique waves $(1, \pm 1)$ inside the LSB (Fig. 5, left) is a result of a linear, essentially inviscid instability of the separated shear-layer (Kelvin-Helmholtz instability). Growth of this disturbance favorably compares with streamwise integrated eigenvalues obtained from a local solution of the Orr-Sommerfeld equation.

The frequency of the triggered perturbation was chosen such that it corresponds roughly to the integrally most amplified disturbance, as it can be deduced from the stability diagram of the mean flow (Fig. 5, right). Finally, disturbances saturate and fill up the spectrum in time and span, i.e. cause transition to turbulence. The appearance of steady 3-d disturbances that are observable in the front part of the LSB and their role remains to be clarified. Possible candidate mechanism seem to be transient growth (Schmid & Henningson, 2001; Marxen *et al.*, 2005) and global instabilities.

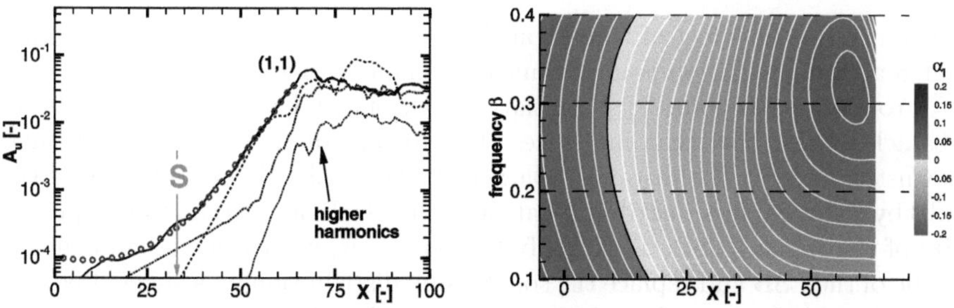

Fig. 5. *Left*: Amplification curves for the maximum (in y) streamwise velocity fluctuation $|\hat{u}'^{(h,k)}_{max}|$. Results from numerical simulations with $A_v = 10^{-4}$ (*lines*) and from linear stability theory based on the Orr-Sommerfeld (OS) equation (*symbols*). *Right*: Stability diagram for a spanwise wave number $\gamma = 0.12$ (imaginary part α_i of the eigenvalues of the OS equation), with neutral-stability curve (*black line*)

5 Bubble Bursting

As pointed out in the introduction, the expression bubble bursting is often not meant in the sense of a dynamic process but rather refers to the fact that with a small change of parameter(s), a strong difference in the mean flow is observed. On the other hand, the bursting can as well be regarded as a process, i.e. as the actual switch-over from short to long bubbles. This viewpoint shall be taken here and such a process will be illustrated by means of a numerical simulation based on the same configuration as before.

5.1 Deficiencies of Known Bursting Criteria

All hitherto known bursting criteria have deficiencies as pointed out by Diwan *et al.* (2006) and do not suffice for a clear distinction between, or classification of, long and short bubbles. In particular, they do not offer an explanation as to *why* bursting should occur. While a classification seems interesting from an engineering point of view, this project is more concerned with understanding the physical mechanism behind bubble bursting. Such an understanding is believed to be a necessary condition for prediction of bursting – with bursting prediction being of major importance for technical devices involving laminar separated flows. Two main features of laminar separation bubbles play a central role: the influence of the LSB on the surrounding potential flow and vice versa (so-called viscous-inviscid interaction) and transition to turbulence in the separated shear layer together with its effect on the mean flow.

As an example of the deficiency to explain why bursting should occur, the criterion by Tani (1964) is examined more closely. This is based on the influence of the LSB on the pressure distribution, which should be local for short and global for long bubbles. His criterion therefore considers viscous-inviscid interaction as the important feature, but does not appreciate possible differences in the effect of transition. It shall now be argued that considering differences in transition is important to explain bubble bursting.

From the viewpoint of potential-flow theory, a LSB can be seen as an obstacle in the flow that influences the pressure distribution both up- and downstream of it. For instance, Marxen (2005) modeled this influence of the LSB by a source/sink distribution at the wall, which depends on the shape and size of the (mean) LSB. To simplify matters a bit, we assume a half-cylinder shape of the LSB and replace the source/sink distribution by a single dipole at the wall in the center of the LSB, leaving us with the dipole strength (which corresponds to the LSB's size) as the only parameter. A short LSB would give a small dipole strength while a long LSB is expected to give a larger dipole strength. Hence, in both cases, the upstream impact on the velocities (and with it on the pressure distribution) is roughly proportional to $1/x_{ud}^2$, with x_{ud} being the upstream distance from its center.

For a short LSB it is known that transition to turbulence is usually governed by a Kelvin-Helmholtz mechanism, i.e. amplification and saturation of the linearly most amplified wave[1], and that transition is responsible for reattachment of the flow in the mean. In fact, Augustin *et al.* (2004) demonstrated that the length of the bubble (and with it the dipole strength in our simple model) is roughly proportional to $|\log A_v|$ for small A_v: the decreasing impact upstream – with increasing amplitude – is a gradual process.

It it is not unlikely to assume that also long LSBs can be shortened by explicitly triggering transition to turbulence if only the amplitude were sufficiently large, since Gaster (1966) showed that even long bubbles can be removed completely by tripping the flow to turbulence prior to the APG: obviously, increasing the mixing *early enough* in the APG region is sufficient. This suggests that even for parameters were long bubbles exist, short bubbles can be obtained by triggering the flow with a sufficiently large amplitude.

If we now start with such a short bubble – triggered by a certain amplitude – and then lower this amplitude, the enlargement of the LSB should be a gradual process again, *under the assumption that transition to turbulence is always responsible for reattachment.* However, since at one point a long-bubble state has to be reached, i.e. bubble bursting – a fairly sharp switch-over between short and long bubbles – must have occurred, this assumption concerning the transition process is probably wrong. Rather, the *transition process is expected to be different for short and long bubbles and might not be able to reattach the flow*, and this might be the fundamental difference between the two cases. This is in line with the observation that almost all studies of bubble bursting conducted in the past indicate more or less that the transition process plays a central role in the appearance of bursting.

Existence of turbulent separation bubbles suggests that the increased mixing of turbulent flow is not always sufficient to cause reattachment. The

---
[1] Note that only the amplification rate can be computed by linear theory, and not saturation location or saturation amplitude.

same could apply for transition: the ability of laminar-turbulent transition to achieve (roughly immediate) reattachment could mark the border-line between short and long bubbles. Keeping in mind that triggering transition with increasing amplitudes results in smaller and smaller (short) LSBs, a *decrease* in amplitude could lead to bursting for a sufficiently strong pressure rise.

5.2 Numerical Simulation of the Development of a Long LSB

Since the experiments of Gaster (1966) are not documented well enough, in particularly with respect to the disturbance content, it was not found desirable to exactly trying to reproduce one of his cases. Rather, his settings served as a guide line in the design of the pressure gradient and the size of the integration domain. To initiate bubble bursting, the simulation of the short bubble was continued, however the disturbance input was switched off at some time. As mentioned before, both, coarse- (using the same resolution as applied in the tripped case) as well as fine-grid simulations exhibit a continuously growing separation bubble following the switch-off of the disturbance input. Due to a considerably lower computational cost, the coarser simulations could be advanced further in time and corresponding results shall be reported here.

Time-averaged (over the last $36T_0$ of a total of $360T_0$ computed periods) streamlines are shown in Fig. 6. It can be seen that in fact a huge LSB has developed, which becomes particularly clear if it is compared to the short LSB as given in Fig. 3. Even though transition to turbulence takes place still at around the same streamwise location (in fact, the transition location has moved a bit upstream compared to the short LSB case), saturating perturbations are no longer able to reattach the flow.

As pointed out before, bubble bursting shall here be interpreted as a dynamical process. This process is visualised in Fig. 7, where contours of the spanwise vorticity at the wall give a hint on the continuously growing LSB. It shows that the bubble is on its way towards a long LSB state. It is unclear why this process is so slow, but the formation of low-frequency large-scale turbulent vortexes (visible in Fig. 6, $x > 500$) with very low frequency ($\approx 1/50 \cdot T_0$) might offer an explanation: only if a considerable number of such shedding cycles has taken place will the flow probably be in an equilibrium.

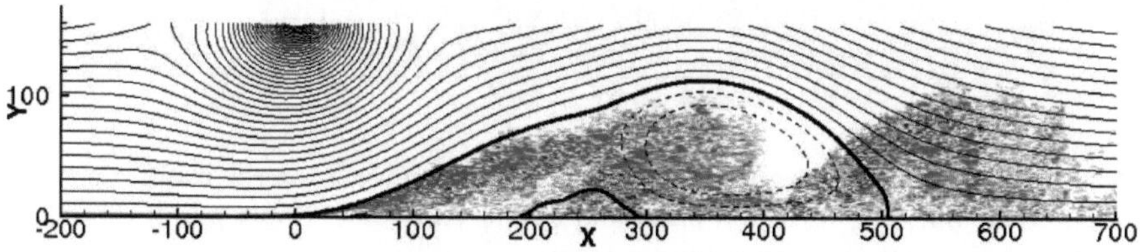

Fig. 6. Contours of instantaneous spanwise-averaged spanwise vorticity ω_z from (the coarser) simulation of the bursting LSB together with mean streamlines

Unfortunately, the simulation of the bursting bubble had to be stopped since rotational perturbations finally hit the upper boundary in violation of the zero vorticity condition used there. Still, the present simulations clearly show how big a separation bubble develops even though it is not yet fully converged to a statistically steady state.

6 Computational Aspects

The present investigation can be viewed as a feasibility study of the proposed bubble bursting mechanism due to the coarse grid used, but high-resolution simulations are currently underway to confirm the present findings. This confirmation is essential but very expensive, since the bursting does not take place immediately as it was expected, but instead seems to be a rather slow process that involves a considerably longer computing time compared to a short LSB. This becomes clear when looking at Fig. 7: while the short LSB is quickly established and can be assumed statistically steady from $t/T_0 \approx 50$ onwards, the long-bubble state has not yet converged even at $t/T_0 = 360$.

The simulations here can be viewed as large-scale simulations, since the total number of grid points exceeds by far the one typically used in studies of short LSBs (Marxen & Rist, 2005). Therefore, for the present study a super computer is truly essential, in particular if considering the long times that have to be computed. For both resolutions (see Table 1 and Sect. 3.3), and the applied, SX-specific optimized version of the code, the memory consumption and the performance observed on NEC SX-8 using combined SMP and MPI parallelization is given in Table 2. The lower resolved bursting case used 5196 $CPU\,h$ on the NEC SX-8. The average vector length was 237–241.

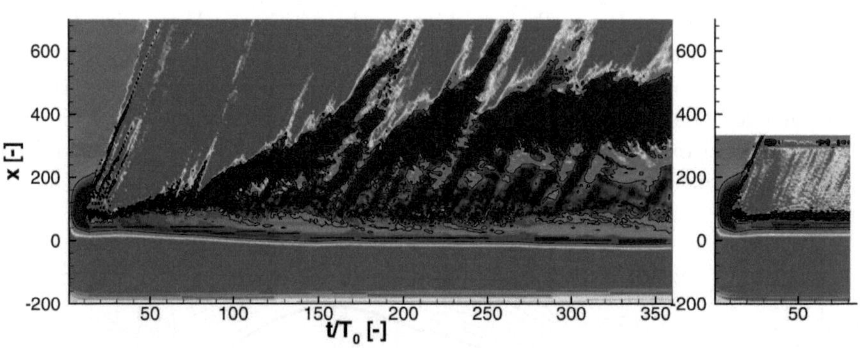

Fig. 7. Contours of spanwise-averaged spanwise vorticity ω_z at the wall from simulation of the bursting LSB (*left part*) and for the short LSB (*right part*). The black line marks the contour of vanishing vorticity ($\omega_z = 0$)

Table 2. Performance of *n3d* in the present case for a typical run

Machine	CPU×Nodes	GFLOPS/CPU	Memory	CPU Time/Period T_0
NEC SX-8, MPI	8×3	4.4	33.36 GB	$14.43\,h$
NEC SX-8, MPI	8×17	4.4	618.82 GB	$712.47\,h$

7 Conclusions and Outlook

For the first time, time-dependent 3-d computations of the bursting process of a laminar separation bubble have been carried out. To demonstrate the difference between short and long bubbles, results from three different numerical simulations that were all based on the same (slip-flow) configuration are reported. A main goal of these simulations was to establish a suitable parameter setting for DNS of long laminar separation bubbles and bubble bursting. The bursting process could be initiated by switching off the explicit forcing of perturbations upstream of separation that otherwise causes a short LSB. Yet, transition to 3-d turbulence occurs inside the separation region in both cases.

In contrast to what is known from short LSBs, in the bursting case the saturated disturbances are not able to reattach the flow soon after transition. Instead, reattachment occurs only considerably downstream of the transition location. This underlines the important role of the transition process. Furthermore, it strongly indicates that the disturbance content in the flow (turbulence level) is an important parameter to consider in investigations of bubble bursting.

It was found that the wall-normal extent was particularly crucial and had to be moved further away compared to some earlier computations. Furthermore, the bursting process was considerably slower than expected, which unfortunately increases the required computational time a lot.

Computations of bubble bursting are currently repeated with higher resolution in all spatial dimensions to meet the requirements for a well-resolved DNS. Furthermore, the transition mechanism in both, short and long separation bubble deserves more attention. In case of a short bubble, the occurrence of steady 3-d disturbances that showed up but were not explicitly triggered requires a profound investigation. In case of the bursted bubble, the transition mechanism remains unclear altogether as for now, however there is some indication (e.g the amount of reverse flow) that an absolute instability could be in operation. Finally, possible mechanisms leading to bubble bursting processes and to reattachment of the long LSB have to be studied in more detail – particularly the relation of the observed large-scale turbulent vortex shedding.

Acknowledgements

OM acknowledges gratefully financial support by the Deutsche Forschungsgemeinschaft DFG under grant Ma 3916/1-1 and computing time on NEC SX-8 granted by the Höchstleistungsrechenzentrum Stuttgart (HLRS) within the project *long_lsb*. Furthermore, he thanks Ulrich Rist and Markus Kloker, IAG, Uni Stuttgart for providing the DNS code *n3d*.

References

ALAM, M. & SANDHAM, N.D. 2000 Direct Numerical Simulation of 'Short' Laminar Separation Bubbles with Turbulent Reattachment. *J. Fluid Mech.* **410**, 1–28.

AUGUSTIN, K., RIST, U. & WAGNER, S. 2004 Control of laminar separation bubbles by small-amplitude 2D and 3D boundary-layer disturbances. In *RTO Specialists Meeting on "Enhancememnt of NATO Military Flight Vehicle Performance by Management of Interacting Boundary Layer Transition and Separation"*, pp. 6-1–6-14. Prague, Cz, 4–8 October 2004.

DIWAN, S.S., CHETAN, S.J. & RAMESH, O.N. 2006 On the bursting criterion for laminar separation bubbles. In *Laminar-Turbulent Transition* (ed. R. Govindarajan), pp. 401–407. 6th IUTAM Symposium, Bangalore, India, 2004, Springer, Berlin, New York.

GASTER, M. 1966 The structure and behaviour of laminar separation bubbles. *AGARD* CP–4, pp. 813–854.

HÄGGMARK, C. 2000 Investigations of disturbances developing in a laminar separation bubble flow. PhD thesis, Royal Institute of Technology, Stockholm, TRITA-MEK 2000:3.

HÄGGMARK, C.P., HILDINGS, C. & HENNINGSON, D.S. 2001 A numerical and experimental study of a transitional separation bubble. *Aerosp. Sci. Technol.* **5** (5), 317–328.

HERBST, A.H. & HENNINGSON, D.S. 2006 The Influence of Periodic Excitation on a Turbulent Separation Bubble. *Flow, Turbulence and Combustion* **76** (1), 1–21.

KLOKER, M. 1998 A Robust High-Resolution Split-Type Compact FD Scheme for Spatial Direct Numerical Simulation of Boundary-Layer Transition. *Appl. Sci. Res.* **59**, 353–377.

LANG, M., RIST, U. & WAGNER, S. 2004 Investigations on controlled transition development in a laminar separation bubble by means of LDA and PIV. *Experiments in Fluids* **36**, 43–52.

MARXEN, O. 2005 Numerical Studies of Physical Effects Related to the Controlled Transition Process in Laminar Separation Bubbles. Dissertation, Universität Stuttgart.

MARXEN, O., LANG, M., RIST, U. & WAGNER, S. 2003 A Combined Experimental/Numerical Study of Unsteady Phenomena in a Laminar Separation Bubble. *Flow, Turbulence and Combustion* **71**, 133–146.

Marxen, O. & Rist, U. 2005 Direct Numerical Simulation of Non-Linear Transitional Stages in an Experimentally Investigated Laminar Separation Bubble. In *High Performance Computing in Science and Engineering '05* (ed. W. E. Nagel, W. Jäger & M. Resch), pp. 103–117. Transactions of the HLRS 2005, Springer.

Marxen, O., Rist, U. & Henningson, D.S. 2005 Steady three-dimensional Streaks and their Optimal Growth in a Laminar Separation Bubble. In *Contributions to the 14th STAB/DGLR Symposium, Nov. 16–18, 2004, Bremen, Germany*. Accepted for publication, Springer.

Meyer, D., Rist, U. & Kloker, M. 2003 Investigation of the flow randomization process in a transitional boundary layer. In *High Performance Computing in Science and Engineering '03* (ed. E. Krause & W. Jäger), pp. 239–253. Transactions of the HLRS 2003, Springer.

Owen, P.R. & Klanfer, L. 1953 On the laminar boundary layer separation from the leading edge of a thin airfoil. *Tech. Rep.* No. Aero 2508. Royal Aircraft Establishment, UK, In: A.R.C. Technical Report C.P. No. 220, 1955.

Pauley, L.L., Moin, P. & Reynolds, W.C. 1990 The structure of two-dimensional separation. *J. Fluid Mech.* **220**, 397–411.

Schmid, P.J. & Henningson, D.S. 2001 *Stability and Transiton in Shear Flows*, 1st edn. Springer, Berlin, New York.

Spalart, P.R. & Strelets, M.K. 2000 Mechanisms of transition and heat transfer in a separation bubble. *J. Fluid Mech.* **403**, 329–349.

Tani, I. 1964 Low-speed flows involving bubble separations. *Prog. Aerosp. Sci.* **5**, 70–103.

Watmuff, J.H. 1999 Evolution of a wave packet into vortex loops in a laminar separation bubble. *J. Fluid Mech.* **397**, 119–169.

Wissink, J. & Rodi, W. 2004 DNS of a laminar separation bubble affected by free-stream disturbances. In *Direct and Large-Eddy Simulation V* (ed. R. Friedrich, B. Geurts & O. Métais), *ERCOFTAC Series*, vol. 9, pp. 213–220. Proc. 5th internat. ERCOFTAC Workshop, Munich, Germany, Aug. 27–29, 2003, Kluwer Academic Publishers, Dordrecht, Boston, London.

Parallel Large Eddy Simulation with UG

Andreas Hauser[1] and Gabriel Wittum[2]

[1] Simulation in Technology, IWR, University of Heidelberg
Andreas.Hauser@iwr.uni-heidelberg.de
[2] Simulation in Technology, IWR, University of Heidelberg
wittum@techsim.org

1 Introduction

Large Eddy Simulation (LES) is a popular technique for simulating turbulent flows. Kolmogorov's theory of self similarity [11] assumes small eddies to be universal and, hence, independent of the geometry. This feature now suggests resolving the large eddies and modeling the small eddies using subgrid-scale (sgs) models. Although the number of scales are reduced tremendously compared with a Direct Numerical Simulation (DNS), where all length scales are resolved, a LES is still much more complex than computations using statistical methods [13]. Further, when applying LES to complex geometries in $\mathbb{R}^3$, the computations must be carried out in parallel in order to solve the problem within a reasonable time. For the simulations carried out here, the software package *UG* [3] is applied, which has proven good scaling properties even on 256 processors and more.

As an example of the powerful simulation tools available in *UG*, that have been developed partly on parallel clusters like the XC6000 in Karlsruhe, a LES of the fluid flow in a static mixer is presented. Static mixers are heavily used in chemical process industry in order to mix substances and fluids respectively. For the purpose of optimization, detailed information of the flow characteristics within the static mixer is needed. The numerical results are then validated with experimental data. These experiments were carried out exclusively for the specific mixer used for these simulations.

The contribution is now structured as follows: First, the geometry and grid of the mixer is presented, followed by the mathematical model, the boundary conditions, the numerical schemes and solving strategy. Finally, the numerical results are shown and compared with experimental data.

1.1 Grid and Grid Hierarchy

The modelled CAD geometry is used for the fabrication of the mixing device for the experiments as well as for the import into *UG* via the CAD-

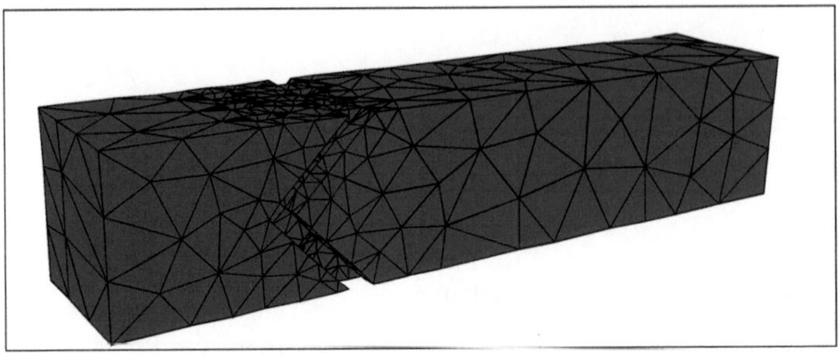

Fig. 1. Coarse triangulation of the static mixer with 3913 Tetrahedra

interface [1]. The geometry is triangulated and also imported into *UG*. As we make use of fast geometric multigrid solvers, the coarse grid, shown in Fig. 1, is uniformly refined during the solving process. This leads to the grid hierarchy given in Table 1. The number of unknowns of 1.4e+7 and

Table 1. Grid hierarchy through uniform refinement

level i	#Tetrahedra	#Nodes	#Unknowns	min h_i	max h_i
0	3913	1086	4344	2.73	68.62
1	31304	6975	27900	1.84	34.31
2	2.5e+5	48766	1.95e+5	0.83	17.15
3	2.0e+6	3.62e+5	1.45e+6	0.41	8.58
4	1.6e+7	2.78e+6	1.40e+7	0.20	4.29

the fact, that transient turbulent computations generally cover a large time interval using several thousands of time steps, underline the need for parallel resources.

2 Mathematical Model and Numerics

Starting point for every LES are the Navier-Stokes equations. In order to separate now the large from the small eddies, a quantity is filtered as follows:

$$\bar{\phi}(\xi) = \frac{1}{\Delta}\int_{-\infty}^{+\infty} G(\xi-\eta)\phi(\eta)\,d\eta, \qquad -\frac{1}{2}\Delta \leq \eta \leq \frac{1}{2}\Delta, \tag{1}$$

where $G(\boldsymbol{\xi})$ represents the kernel of the convolution integral and the bar denotes a filtered quantity, whereas Δ is the filter width. Because filtering nonlinear quantities does not commute, the application of the filter operator (1) to the Navier-Stokes-equations leads to an unclosed system of PDEs. Similar to statistical methods, a stress-tensor is introduced that has to be modelled. The filtered incompressible Navier-Stokes-equations with the velocities u_i, pressure p, time t and the kinematic viscosity ν now reads:

$$\frac{\partial \bar{u}_i}{\partial t} + \frac{\partial}{\partial x_j}(\bar{u}_i\bar{u}_j) + \frac{\partial \bar{p}}{\partial x_i} - \nu \frac{\partial^2 \bar{u}_i}{\partial x_j \partial x_i} + \frac{\partial \tau_{ij}}{\partial x_j} = 0$$
$$\frac{\partial \bar{u}_j}{\partial x_j} = 0 \tag{2}$$

The stress-tensor τ_{ij} in the LES context is called sgs-tensor and can be modelled in several ways. Here we use the well established dynamic model [7], where the usage of an additional filter with larger filter size $\bar{\Delta}$ is applied in order to get automatically the space- and time-dependent parameter $C = C(x_i, t)$ in

$$\tau_{ij} = \overline{u_i u_j} - \bar{u}_i\bar{u}_j = 2C\Delta^2|\bar{S}|\bar{S}_{ij}, \tag{3}$$

with the deformation tensor

$$\bar{S}_{ij} = \frac{1}{2}\left(\frac{\partial \bar{u}_i}{\partial x_j} + \frac{\partial \bar{u}_i}{\partial x_i}\right)$$

and

$$|\bar{S}| = \sqrt{2\bar{S}_{ij}\bar{S}_{ij}}.$$

With $C = const$, (3) represents the oldest but rather dissipative Smagorinsky model.

Because the dynamically determined parameter might vary heavily in time causing numerical instabilities and artifacts, the parameter is low-pass-filtered in the following sense [4]:

$$C^{n=1}_{filtered} = (1-\epsilon)C^n + \epsilon C^{n+1}, \qquad \epsilon = 10^{-3}. \tag{4}$$

2.1 Boundary Conditions

The experiments [3] show a slightly distorted profile at the inflow of the mixer due to the redirection of the fluid flow within the experimental setting. Therefore, the measured velocities at the inflow are used as the inlet boundary condition for the numerical simulation. The raw data show high frequency-wiggles due to experimental uncertainties (see Fig. 2 left). As these disturbances might cause numerical problems, the raw data were low-pass-filtered leading to a smooth profile given in Fig. 2 (right). With the viscosity of water the Reynolds number counts $Re = 562$. For the outlet boundary condition, a simple convection-diffusion convection is solved for the velocities, whereas Dirichlet boundary condition for the pressure with $p = 0$ is prescribed. At the walls no-slip conditions for the velocities and no condition at all for the pressure are used.

[3] The experiments were carried out by S. Leschka and Prof. Dr. Thévenin at the LSS, University of Magdeburg.

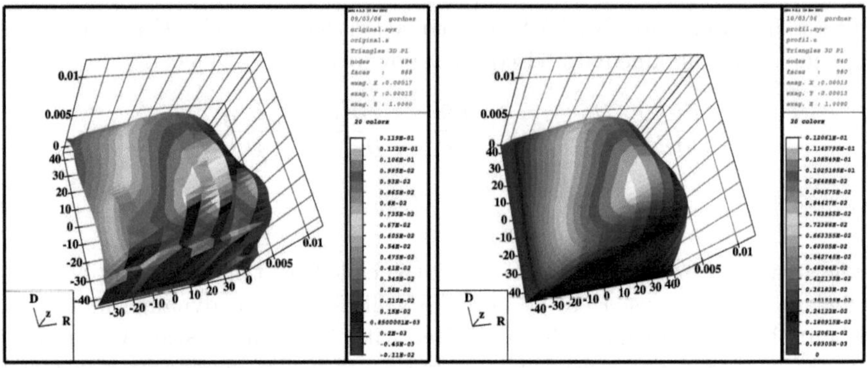

Fig. 2. Raw (*left*) and smoothed (*right*) inflow profile

2.2 Spatial Discretization

The discretization in space is carried out with a vertex-centered finite volume scheme with colocated variables using continuous, piecewise linear trial functions. The LBB-stabilization is carried out using a local momentum balance equation on element level inducing a streamline-diffusion-type stabilization for the convective term. This idea stems from [16] and was realized in *UG* by [15, 12]. The discretization is locally mass-conserving and second-order consistent.

With the solution vector

$$\mathbf{x} = [\mathbf{u}, p]^T, \qquad \mathbf{x} \in \mathbb{R}^n,$$

the continuous system (2) can be written in semi-discrete form

$$M\dot{\mathbf{x}} + kA(\mathbf{x})\mathbf{x} = kb,$$

with $b \in \mathbb{R}^n$, the scaled time step k, and with $A \in \mathbb{R}^{n\times n}$ being a nonlinear function of $\mathbf{x}$. The matrix $M \in \mathbb{R}^{n\times n}$ denotes the singular mass matrix.

2.3 Time Discretization

In order to avoid time step restrictions induced by the Courant-Friedrichs-Levy condition, a fully implicit time integration method is used. We apply the A-stable *Fractional-Step-θ* scheme [8], which is second order consistent and which uses three fractional steps ($t_{j-1} \to t_{j-1+\theta} \to t_{j-\theta} \to t_j$, $j = 1, .., N$).

With $\tilde{\mathbf{x}}^j$ denoting the approximate values of $\mathbf{x}$ at time step t_j, this scheme can be formally written as

$$F_\theta^{j,j-1}(\mathbf{x}^{j-1}, \mathbf{x}^{j-1+\theta}, \mathbf{x}^{j-\theta}, \mathbf{x}^j) = 0, \qquad \theta \in \mathbb{R} \tag{5}$$

For the choice of θ and further details see for example [14].

2.4 Solving Strategy

For each time step the system of nonlinear algebraic equations (5) has to be solved three times. The nonlinear solution is found using Newton's method. The nonlinear defect of the iterate $\tilde{\mathbf{x}}$ is defined by

$$d_\theta^j(\tilde{\mathbf{x}}) := F_\theta^{j,j-1}(\tilde{\mathbf{x}}^{j-1}, \tilde{\mathbf{x}}^{j-1+\theta}, \tilde{\mathbf{x}}^{j-\theta}, \tilde{\mathbf{x}}^j).$$

The nonlinear iteration to solve (5) is considered to have converged if

$$||d_\theta^j(\tilde{\mathbf{x}})||_2 \leq 10^{-5}||d_{\theta_1}^{j-1}(\tilde{\mathbf{x}})||_2.$$

Each nonlinear iteration requires the solution of the linear equation derived by linearizing (5). The linear system is obtained by using a Quasi-Newton linearization

$$\mathbf{u}^n \cdot \nabla \mathbf{u}^n \approx \mathbf{u}^{n-1} \cdot \nabla \mathbf{u}^n \qquad \text{and} \qquad \mathbf{u}^n \cdot \nabla T^n \approx \mathbf{u}^{n-1} \cdot \nabla T^n$$

obtaining linear convergence [5].

On an average, about 5 Newton iterations are required to solve each nonlinear problem. The linear systems are solved using the geometric multigrid method with a V(2,2)-cycle applying the ILU as a robust smoother. To realize a robust behavior of the ILU-smoother, a special ordering of the unknowns is required. It is accelerated using a Bi-CGStab Krylov-space-method. For details, see [2, 10, 9, 6].

The computations were carried out on the HP XC6000 Cluster in Karlsruhe. This parallel cluster consists of 108 nodes with two 1.5 GHz Intel Itanium2 processors with 12 GB RAM and 12 nodes with 8 1.6 GHz Intel Itanium2 processors with 64 GB RAM respectively. The time interval $T = [0..35]s$ on level l_3 was integrated in 3500 time steps with a constant time step of $\Delta t = 0.01$s. This simulation took more than 25 days pure computation time on 64 processors. As the complexity of level l_4 is about 8 times that on level l_3, only a time interval of $T = [0..3]s$ was integrated. This computation took about 24 days on 100 processors. As some parameters had to be changed on level l_4, no speedup value can be given for the evaluation of the cluster's performance.

In order to start on level l_2, l_3 and l_4 with good initial conditions, the last solution on the coarser grid was prolongated by linear interpolation onto the finer grid.

2.5 Numerical Results

As a first impression of the flow field, the axial velocity in a top section and side section is shown in Fig. 3. Before the mixing element the flow field is laminar and well ordered. Due to the narrowed cross section a stagnation area takes shape right before the mixing element. As expected, the maximum velocity is reached within the mixing element.

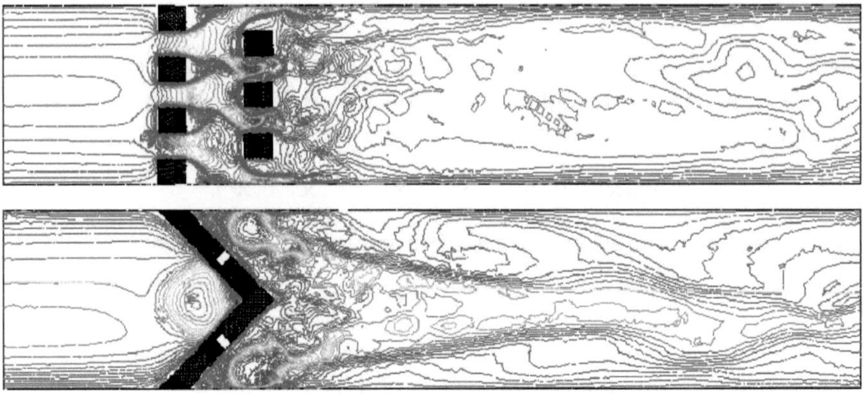

Fig. 3. Top view (*left*) and side view (*right*) of the axial velocity

Besides experimental data, the only applicable validation is grid convergence as there exists of course no analytic solution to this problem. Comparing results on different grid levels gives information about the reliability of a numerical solution. Figure 4 show the flow field (axial velocity) before the mixing element on level l_2, l_3, l_4. Despite the fact that the contour lines become smoother with increasing level, all three solution show very much the same velocity profile.

Next, the axial velocity right behind the mixing element on level l_2, l_3, l_4 is shown in Fig. 5. The flow pattern has changed completely, although coherent structures are obvious. In this case, the result for level l_3 seems not to fit really into the sequence.

The last sequence of cross sections is depicted in Fig. 6, where the axial velocity is shown in a section further downstream. Again, the flow field has changed, but the coherent structures are clearly transparent. The results on level l_3 are not satisfying again, as the structures do not really fit between the results on l_2 and l_4.

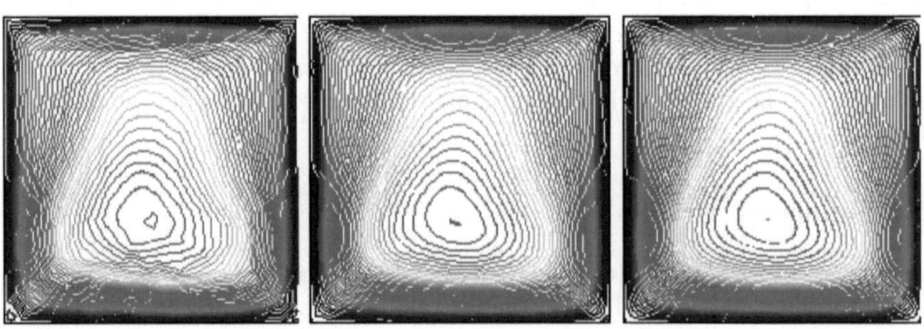

Fig. 4. Axial velocity before the mixing element: level 2 (*left*), level 3 (*middle*), level 4 (*right*)

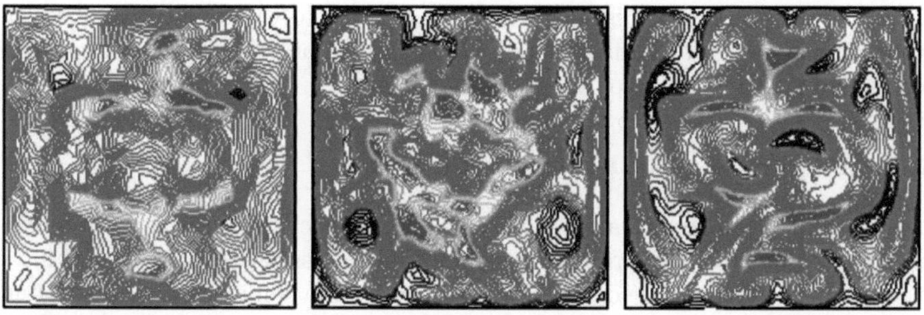

Fig. 5. Axial velocity after the mixing element: level 2 (*left*), level 3 (*middle*), level 4 (*right*)

Fig. 6. Axial velocity after the mixing element further downwards: level 2 (*left*), level 3 (*middle*), level 4 (*right*)

From a qualitative point of view it seems that the results on l_4 confirm the results on level l_2.

To get a quantitative idea, the axial velocities on the three different levels l_2, l_3, l_4 along a centered line from the inflow to the outflow boundary is taken for comparison and presented in Fig. 7. The averaged velocity $\bar{u}_a$ is scaled by

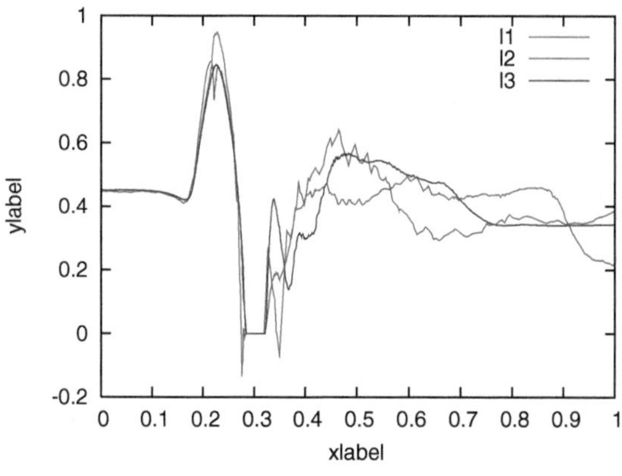

Fig. 7. Axial velocities along a centered line from the inflow to the outflow boundary

u_0, whereas the length l is scaled by l_0. Contrary to the qualitative observations made above, the difference of the velocities on l_2 and l_4 is larger than on l_3 and l_4. This behavior suggests convergence.

2.6 Comparison with Experimental Data

As the numerical results are considered with respect to consistency and convergence, the velocities are now compared in a cross-section before and after the mixing element. The detailed description of the comparison can be found in [17]. In Fig. 8 the axial velocity before the mixing element is shown. Qualitatively, both profiles are slightly rotated. Although the geometry and the inflow profile are symmetric with reference to the vertical z-axis, this phenomenon is supposed to be physical. Until now, no stringent explanation has been found. In addition, the numerical results show smoother contours then experimental data. Quantitatively, the results agree pretty good with a maximum velocity of 0.012 m/s in both cases.

Figure 9 compares the axial velocities in a cross section after the mixing element. Both data are averaged due to the turbulent fluctuations. Quali-

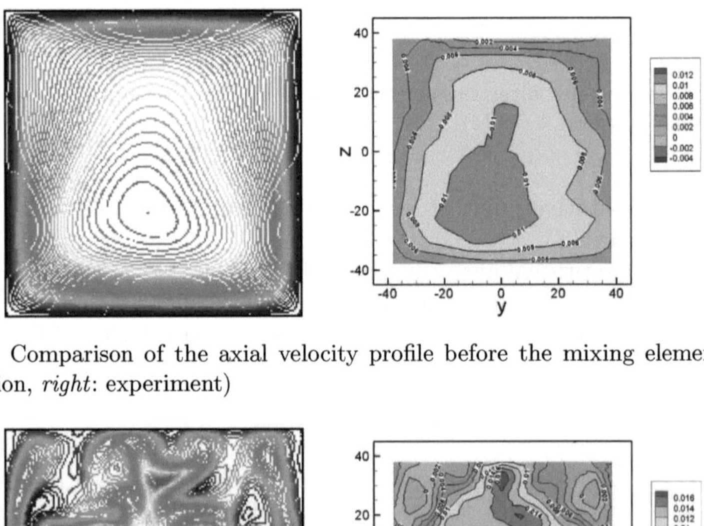

Fig. 8. Comparison of the axial velocity profile before the mixing element (*left*: simulation, *right*: experiment)

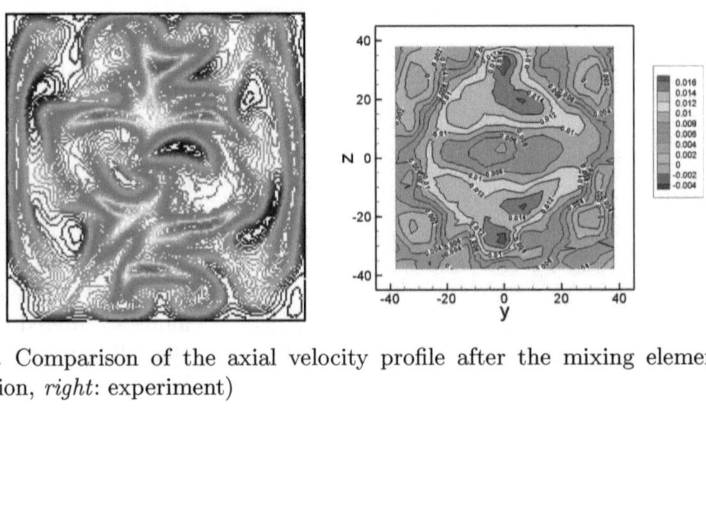

Fig. 9. Comparison of the axial velocity profile after the mixing element (*left*: simulation, *right*: experiment)

tatively, the results have some coherent structures in common. The global picture suggests agreement to a certain extent. The experimental data are symmetric according to the horizontal axis, whereas the numerical results are symmetric according to both axis, the horizontal and vertical.

Eventually, the maximum velocity and its location fit quite well. It should be mentioned that the expenses for computing the problem on l_4 pays off as the results on level l_3 do not convince.

3 Conclusion and outlook

The simulation of the fluid flow has been simulated on the levels l_2, l_3, l_4. The results on level l_2 and l_4 are plausible and suggest convergent behavior, whereas the results on l_3 does not fit in this sequence. The quantitative comparison along a line within the numerics show good agreement for all three levels. This can be explained with the fact, that the velocities in the middle of the mixer coincide in the cross section after the mixing element too. The ultimate validation with experiments show quite good agreement with respect to quality and quantity.

Finally, as uniform refinement increases the complexity of the computation tremendously, effort in adaptive methods for turbulent flow has been invested already and should be applied to the static mixer soon.

References

1. A. Hauser and O. Sterz. UG-interface for CAD geometries. Technical Report 05, Forschungsverbund WiR Baden-Württemberg, 2004.
2. R. Barrett, M. Berry, T. Chan, J. Donato, and J. Dongarra et al. Templates for the solution of linear systems: building blocks for iterative methods. *SIAM*, 1994.
3. P. Bastian, K. Birken, K. Johannsen, S. Lang, K. Eckstein, N. Neuss, H. Rentz-Reichert, and C. Wieners. UG - a flexible software toolbox for solving partial differential equations. *Computing and Visualization in Science*, 1:27–40, 1997.
4. M. Breuer and W. Rodi. Large-eddy simulation of turbulent flow through a straight square duct and a 180° bend. *Fluid Mechanics and its Applications*, 26:273–285, 1994.
5. J.E. Dennis and J.J. More. Quasi-newton methods, motivation and theory. *SIAM Review*, 19(1):46–89, 1977.
6. H. Van der Vorst. Bi-cgstab: a fast and smoothly converging variant of bicg for the solution of nonsymmetric linear systems. *SIAM J Sci Stat Comp*, 13:631–644, 1992.
7. M. Germano, M. Piomelli, U. Moin, and P. Cabot. A dynamic subgrid-scale eddy viscosity model. *Phys. Fluids A*, 3:1760–1765, 1991.
8. R. Glowinski and J. Periaux. Numerical methods for nonlinear problems in fluid dynamics. *Supercomputing. State-of-the-Art*, pages 381–479, 1987.
9. W. Hackbusch. *Multigrid Methods and Applications*. Springer, 1987.

10. W. Hackbusch. *Iterative Lösung großer schwachbesetzter Gleichungssysteme.* Teubner, 1992.
11. A.N. Kolmogorov. The local structure of turbulence in an incompressible fluid with very large reynolds numbers. *Proc. Roy. Soc. London*, 434:1890, 1991. (First published: 1941, *Dokl. Akad. Nauk SSSR*, **30, 301**).
12. S. Nägele. *Mehrgitterverfahren für die inkompressiblen Navier-Stokes Gleichungen im laminaren und turbulenten Regime unter Berücksichtigung verschiedener Stabilisierungsmethoden.* PhD thesis, IWR, Universität Heidelberg, 2004.
13. S.B. Pope. *Turbulent Flows.* Cambridge, 2000.
14. R. Rannacher. Numerische Methoden für Probleme der Strömungsmechanik, 2001. Vorlesungsskript.
15. H. Rentz-Reichert. *Robuste Mehrgitterverfahren zur Lösung der inkompressiblen Navier-Stokes Gleichung: ein Vergleich.* PhD thesis, IWR, Universität Heidelberg, 1996.
16. G.E. Schneider and M.J. Raw. Control volume finite-element method for heat transfer and fluid flux using colocated variables. *Numerical Heat Transfer*, 11:363–390, 1987.
17. G. Wittum, D. Thévenin, A. Hauser, and S. Leschka. Numerische Simulation statischer Strömungsmischer mit experimenteller Validierung, 2006. DFG Schwerpunktprogramm 1151, *Analyse, Modellbildung und Berechnung von Strömungsmischern mit und ohne chemische Reaktionen*, Abschlussbericht.

LES and DNS of Melt Flow and Heat Transfer in Czochralski Crystal Growth

A. Raufeisen[1], M. Breuer[2], V. Kumar[2], T. Botsch[1], and F. Durst[2]

[1] Process Engineering Department (VT), University of Applied Sciences Nuremberg, Wassertorstr. 10, 90489 Nuremberg, Germany
`alexander.raufeisen@fh-nuernberg.de`

[2] Institute of Fluid Mechanics (LSTM), University of Erlangen-Nuremberg, Cauerstr. 4, 91058 Erlangen, Germany
`breuer@lstm.uni-erlangen.de`

Abstract. In the present work, computations of flow and heat transfer in an idealized cylindrical Czochralski configuration are conducted using Large Eddy Simulation (LES) with the flow solver FASTEST-3D developed at LSTM Erlangen. The results match well with DNS data from the literature. However, detailed data for analysis of turbulent quantities are not available. Therefore, DNS computations are conducted using the code $\mathcal{LESOCC}$, employing explicit time marching. Preliminary simulations show the high efficiency of the solver on the NEC SX-8. Furthermore, from a study of the velocity profiles at the wall, the resolution requirements had to be corrected such that the computational grid will now consist of approximately 8×10^6 control volumes. The present run of the DNS took more than 540 hours of walltime on 8 processors. With the results, the LES computations will be thoroughly validated so that appropriate models and parameters can be chosen for efficient and accurate simulations of practically relevant cases.

1 Introduction

The Czochralski (Cz) method is the preferred process for growing large silicon single crystals for the production of electronic and photonic devices. In this process, the liquid silicon is contained in an open crucible heated from the side. The crucible is rotating, while the counterrotating crystal is slowly pulled from the melt. Due to this setup, centrifugal and Coriolis forces, buoyancy, and Marangoni convection occur in the fluid as well as thermal radiation from the surfaces and the phase change due to crystallization. The shape of the interface between melt and crystal is crucial for the quality of the resulting crystal, e.g. its purity and homogeneity. Therefore, the influences of all effects on the crystallization front need to be investigated to gain knowledge about the parameters for controlling the process.

Due to the size of the crucible, the flow and heat transfer inside the melt is three-dimensional, time-dependent and fully turbulent, which makes numerical predictions difficult. Highly accurate Direct Numerical Simulations (DNS) require high resolution and therefore use massive computational resources. This is too expensive and time-consuming for conducting parametric studies which are required by the industry. Thus, simplifications are necessary such as turbulence models. However, it was shown that the turbulent structures in this case are highly anisotropic and thus classical statistical turbulence models based on the Reynolds-Averaged Navier–Stokes (RANS) equations are not applicable. Large Eddy Simulations (LES) combine the advantages of both: The large turbulent scales are computed directly, whereas the small (subgrid) scales are modeled. Thus a relatively high accuracy is achieved with moderate computational effort, so that parametric studies become feasible. To determine the accuracy of LES computations and choose the appropriate resolution and subgrid-scale model, it is necessary to conduct reference DNS predictions for comparison.

2 Governing Equations

The flow and heat transfer in the melt are governed by the three-dimensional Navier–Stokes equations for an incompressible fluid expressing the conservation of mass, momentum and energy:

$$\int_{\Delta S} \rho\, U_i\, dS_i = 0 \tag{1}$$

$$\begin{aligned}\int_{\Delta V} \frac{\partial(\rho\, U_i)}{\partial t}\, dV + \int_{\Delta S} \rho\, U_i\, U_j\, dS_i = &\int_{\Delta S} \mu \left(\frac{\partial U_j}{\partial x_i} + \frac{\partial U_i}{\partial x_j} \right) dS_i \\ &- \int_{\Delta V} \frac{\partial P}{\partial x_i}\, dV \\ &+ \int_{\Delta V} \rho\, g_j\, \beta_T\, (T - T_0)\, dV\end{aligned} \tag{2}$$

$$\int_{\Delta V} \frac{\partial(\rho\, c_p T)}{\partial t}\, dV + \int_{\Delta S} \rho\, U_i\, c_p T\, dS_i = \int_{\Delta S} \left(\lambda \frac{\partial T}{\partial x_i} \right) dS_i \tag{3}$$

where $U_{i,j}$, P and T denote the velocities, pressure and temperature, ρ is the density, μ the dynamic viscosity, g_j the gravity component, β_T the coefficient of thermal expansion, c_p the heat capacity, and λ the heat conductivity. ΔS and ΔV are the surface area and volume of the control volume (CV) over which is integrated. Buoyancy is taken into account by the Boussinesq approximation assuming only small temperature gradients. The density changes are modeled by

$$\rho = \rho_0(1 - \beta_T(T - T_0)) \tag{4}$$

with the thermal expansion coefficient

$$\beta_T = -\frac{1}{\rho_0}\frac{\partial \rho}{\partial T} \tag{5}$$

The simulation is conducted in a rotating frame of reference. Due to Coriolis and centrifugal forces induced by rotation, additional source terms must be added to the right side of the momentum equation:

$$\int_V \rho[2\epsilon_{pqj}U_p\omega_q - \epsilon_{pqj}\omega_p(\epsilon_{rsq}\omega_r x_s)]dV \tag{6}$$

where ϵ_{pqj} denotes the Levi-Civita tensor and ω the angular velocity.

2.1 Boundary Conditions

Thermal radiation from the free surface of the melt is considered by the Stefan-Boltzmann equation:

$$\lambda\,\nabla T = \sigma\,\varepsilon\,(T^4 - T_{env}^4) \tag{7}$$

where σ is the Stefan–Boltzmann constant, ε the emissivity of the fluid, and T and T_{env} the temperatures of the surface and the surrounding environment, respectively. The corresponding properties of the silicon melt are given in Table 1.

Marangoni convection is induced by temperature gradients at the free surface of the liquid which causes changes in the surface tension and thus gives rise to fluid motion. This motion is modeled by a force balance at the free surface, which reads:

$$\mu\frac{\partial U_\xi}{\partial \eta} = \frac{d\sigma}{\mathrm{d}T}\frac{\partial T}{\partial \xi} \tag{8}$$

$$\mu\frac{\partial U_\zeta}{\partial \eta} = \frac{d\sigma}{\mathrm{d}T}\frac{\partial T}{\partial \zeta} \tag{9}$$

where η, ζ and ξ are the local normal and tangential coordinates at the free surface, and σ is the surface tension of the liquid. Here, the motion of the free surface in normal direction is neglected for simplicity. The changes in surface tension are approximated linearly:

$$\sigma = \sigma_0(1 - \gamma(T - T_0)) \tag{10}$$

with the coefficient

$$\gamma = -\frac{1}{\sigma_0}\frac{d\sigma}{dT} \tag{11}$$

Table 1. Properties of the silicon melt and the Cz configuration

Property	Symbol	Value	Unit
Density	ρ	$2,530$	kg/m^3
Dynamic viscosity	μ	8.6×10^{-4}	kg/(m s)
Kinematic viscosity	ν	3.4×10^{-7}	m^2/s
Thermal expansion coefficient	β	1.4×10^{-4}	1/K
Thermal conductivity	λ	67.0	J/(s m K)
Thermal diffusivity	α	2.65×10^{-5}	m^2/s
Heat capacity	c_p	$1,000$	J/(kg K)
Temperature coefficient of surface tension	$d\sigma/dT$	-1.0×10^{-4}	N/(m K)
Melting temperature	T_{melt}	1,685	K
Emissivity	ε	0.3	
Surrounding temperature	T_{env}	1600	K

2.2 Large Eddy Simulation

In the LES, the Navier–Stokes equations are filtered in space, i.e. the flow quantities are divided into a "grid-scale" and a "sub-grid scale" (SGS) part. Here this is done implicitly, i.e. the filter width is the cell width of the computational grid. After filtering and reformulation one obtains the Navier–Stokes equations for the large scales and additionally a subgrid scale stress tensor τ_{ij}^{SGS} and heat flux q_i^{SGS}, which have to be approximated. Using Smagorinsky's model [11], this is done by determining a turbulent eddy viscosity μ_T and a turbulent Prandtl number Pr_t. Here, the Smagorinsky constant is set to $C_s = 0.065$ and $Pr_t = 0.9$, which has proven to deliver good results in practical applications. The filter width Δ is taken as:

$$\Delta = \sqrt[3]{\Delta x \Delta y \Delta z} \tag{12}$$

In the near-wall region the SGS stress tensor must tend to zero. To achieve this, the Van-Driest damping function is used to scale the characteristic length

$$l_c = C_s \Delta \left(1 - \exp\left(\frac{-y^+}{A^+}\right)^{\gamma_1}\right)^{\gamma_2} \tag{13}$$

where y^+ is the dimensionless distance from the wall and $A^+ = 25$, $\gamma_1 = 3$, and $\gamma_2 = 0.5$ numerical parameters with optimal values.

3 Numerical Method

For the numerical simulations the general-purpose CFD package FASTEST-3D developed by LSTM Erlangen [4,5] as well as the computer code $\mathcal{LESOCC}$ (= **L**arge–**E**ddy **S**imulation **O**n **C**urvilinear **C**oordinates) [6–9] are used.

With FASTEST-3D laminar as well as turbulent steady and unsteady flows including heat and mass transfer can be simulated numerically. The three-dimensional incompressible Navier–Stokes equations are solved based on a fully conservative finite-volume discretization on non-orthogonal curvilinear grids with a colocated arrangement of the variables. In order to resolve complex geometries, block-structured grids are used, i.e. the blocks are globally unstructured, but each block consists of a curvilinear structured grid [12,13]. Second-order accurate central differences are applied for all terms. Flux blending (first-order upwind/second-order central) and deferred correction approaches for the convective fluxes are implemented. For the temporal integration of the flow field, two second-order accurate schemes are available: a fully-implicit three-point backward formulation and a Crank-Nicolson scheme. To avoid numerical oscillations a small time step $\Delta t = 0.01$ s must be chosen for the Cz predictions. The data are sampled for time averaging after every tenth time step. The averaged and fluctuating quantities are computed from these sampled data. Under-relaxation factors of 0.1, 0.9 and 0.9 were applied for velocity components, pressure and temperature, respectively. The SIMPLE algorithm is used to couple the velocity and pressure fields. The discretized conservation equations are solved in an iterative manner adopting the Strongly Implicit Procedure (SIP) of Stone [3]. In order to achieve convergence, the residual of each variable is brought down 5–6 orders of magnitude. To speed up the computation, the multigrid technique (FAS/FMG) is implemented.

$\mathcal{LESOCC}$ is based on a 3–D finite-volume method for arbitrary non-orthogonal and block-structured grids. All viscous fluxes are approximated by central differences of second-order accuracy, which fits the elliptic nature of the viscous effects. As shown in [7,9] the quality of LES predictions is strongly dependent on low-diffusive discretization schemes for the non-linear convective fluxes in the momentum equation. Although several schemes are implemented in the code, the central scheme of second-order accuracy (CDS–2) is preferred for the LES predictions in the present work.

Time advancement is performed by a predictor-corrector scheme. A low-storage multi-stage Runge–Kutta method (three sub-steps, second-order accuracy) is applied for integrating the momentum equations in the predictor step. Within the corrector step the Poisson equation for the pressure correction is solved implicitly by the incomplete LU decomposition method of Stone [3]. Explicit time marching works well for LES and DNS with small time steps which are necessary to resolve turbulence motion in time. The pressure and velocity fields on a non-staggered grid are coupled by the momentum interpolation technique of Rhie and Chow [10]. A variety of different test cases (see, e.g., [6–9]) served for the purpose of the code validation.

Both algorithms are highly vectorized and additionally parallelized by domain decomposition with explicit message-passing based on MPI (Message Passing Interface) allowing efficient computations especially on vector-parallel machines and SMP (Symmetric Multi-Processing) clusters. Due to its recursive data structure, the SIP solver for the algebraic system of equations in FASTEST-3D and $\mathcal{LESOCC}$ is not vectorizable in a straightforward manner. However, Leister and Perić [1] showed that vectorization of the SIP solver can be achieved by avoiding data dependencies through indirect addressing and sweeping through the computational domain along diagonal planes, so called hyper-planes. Thus one sweep through the entire domain consists of hyper-planes having different vector length. Due to this variable vector length and the indirect addressing used, the performance of the vectorized SIP solver is slightly lower than the other parts of the code. However, in a preceding project [18] it was shown that FASTEST-3D and $\mathcal{LESOCC}$ work extremely efficient on NEC SX-machines with sustained performances of up to 40–50% of the peak performance. Recently, a detailed study was carried out on the performance of $\mathcal{LESOCC}$ on NEC SX-6+ and SX-8 showing excellent performance values. Details of this investigation can be found in [17].

The reason why two different flow solvers are employed in the present work is that in FASTEST-3D, many features specific to crystal growth applications are implemented, such as the moving grid technique for tracking phase change interfaces. So in future, FASTEST-3D will be used to compute real Czochralski cases. In $\mathcal{LESOCC}$, however, many different LES models including sophisticated wall models are installed. So this code will be used to determine the best possible parameters for LES computations of crystal growth systems which then can be implemented in FASTEST-3D. Furthermore, $\mathcal{LESOCC}$ uses explicit time marching, which is highly advantageous for conducting DNS computations in terms of speed and demand of resource.

4 Results and Discussion

To validate the LES method for Cz simulations, computations of an idealized cylindrical case using FASTEST-3D are compared to DNS reference data [15]. The crucible and crystal diameter are 340 mm and 100 mm, respectively, and the crucible is rotating with 5 rpm, while the crystal rotation rate is – 20 rpm. A fixed temperature distribution obtained from an experiment and interpolated to the new geometry is prescribed at the crucible walls. The

Table 2. Dimensionless numbers

Number	Symbol	Formula	Value
Prandtl	Pr	$\frac{\nu}{\alpha}$	0.0128
Reynolds	Re	$\frac{R_c u_b}{\nu}$	4.7×10^4
Grashof	Gr	$\frac{\kappa g R_c^3 \Delta T}{\nu^2}$	2.21×10^9
Marangoni	Ma	$\frac{d\sigma}{dT} \frac{R_c \Delta T}{\mu\alpha}$	-2.82×10^4
Rayleigh	Ra	$Gr Pr$	2.83×10^7

crystal/melt-interface is not moved and kept at the melting temperature of silicon. Furthermore, a fixed heat flux from the free surface accounts for radiation, also calculated from experimental data. The material properties of silicon are shown in Table 1. From these quantities, the dimensionless numbers compiled in Table 2 are derived.

The computational grid for the LES consists of 627,200 control volumes, equally partitioned into 5 blocks, using the O-grid method. Near the walls the grid is refined to achieve a good resolution. A time step size of $\Delta t = 10^{-2}s$ is applied to achieve a CFL number of about 1. The simulations are run 60,000

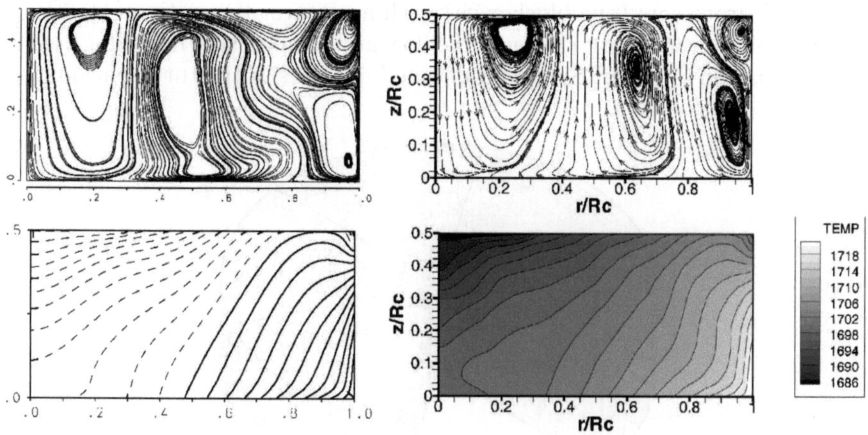

Fig. 1. Streamtraces and mean temperature by DNS [15] (*left*) and present LES computation in a vertical cut through the crucible, averaged in time and circumferential direction (*crucible axis on the left, crucible wall on the right hand side*)

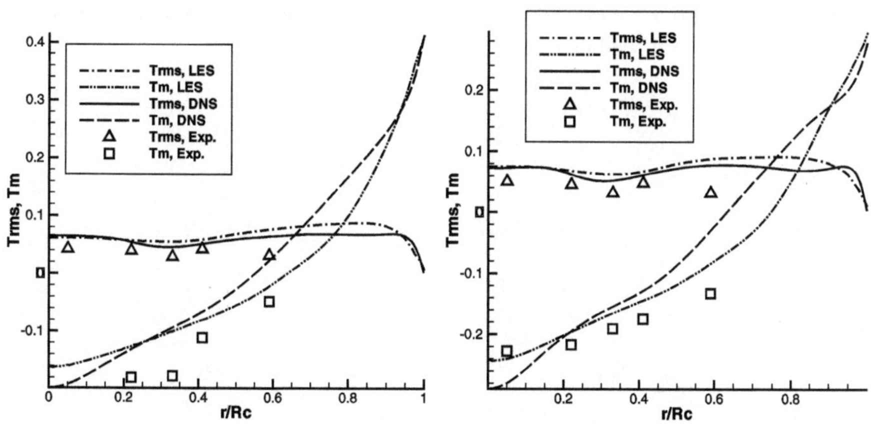

Fig. 2. Mean and rms temperature distribution along horizontal planes at $z/R_c = 0.21$ and $z/R_c = 0.32$ obtained by DNS [15], present LES prediction, and experimental measurements by Wacker [15]

steps; after allowing 6000 steps for the full development of the turbulence, time-averaging is conducted. For comparison with the DNS data set, the field is also averaged in circumferential direction.

The results are in good agreement with the reference data (Fig. 1). In the flow field, the characteristic vortices are resovled. The temperature distribution shows very little deviations, see also Fig. 2. Even simulations with a coarser resolution (192,000 CVs, $\Delta t = 5 \times 10^{-2} s$) lead to satisfactory results (not shown here). The computational effort for the LES is approximately 100 times less than for the reference DNS. On the NEC SX-8, the LES computations with FASTEST-3D achieved an average performance of 4.3 GFlop/s per CPU on 5 processors (one block each) with a vectorization ratio of 99.6%.

Besides an overall comparison of the flow and temperature fields, for a thorough validation of the LES method detailed analyses of the turbulent quanti-

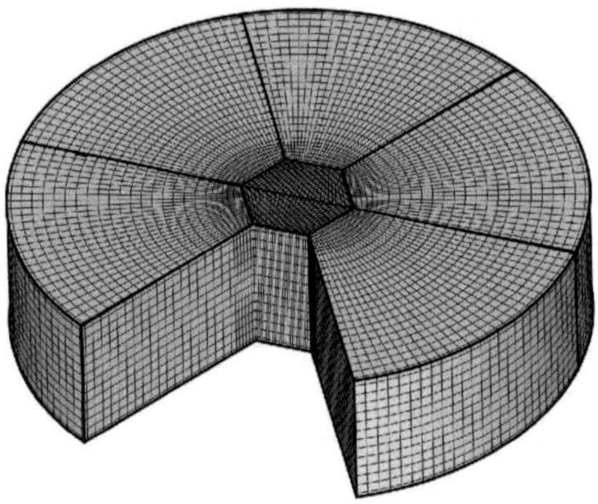

Fig. 3. Computational mesh for DNS computation, consisting of 8 blocks in O-grid configuration, containing approx. 5×10^6 control volumes. Shown here after coarsening twice for better visibility

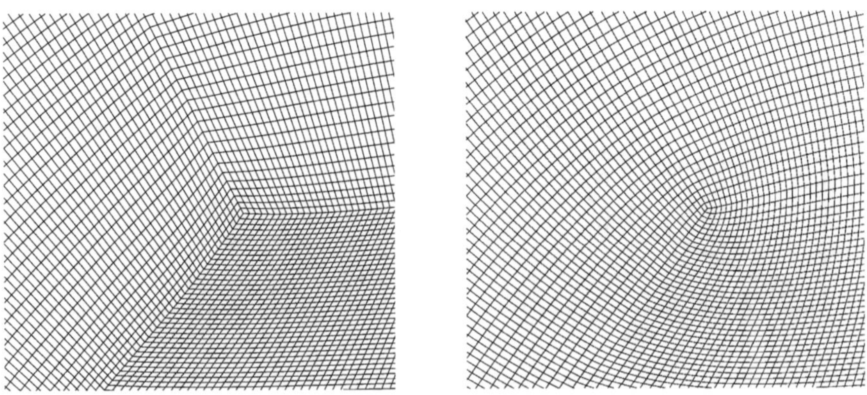

Fig. 4. Comparison of the upper left O-grid corner before (*left*) and after smoothing (*right*)

ties, especially near the wall, are necessary. Unfortunately, from the reference DNS detailed data are not available. Thus it is unavoidable to conduct new fine grid simulations. Therefore, a computational mesh consisting of 8 blocks to fit the system architecture of the NEC SX-8 was designed, see Fig. 3.

Following the criteria of Wagner [15], who adopted the resolution requirements derived by Grötzbach [16], the computational mesh was determined to consist of about 5×10^6 nodes. Test computations on this grid using $\mathcal{LESOCC}$ with a dimensionless time step size of $\Delta t = 4 \times 10^{-4}$ achieved an excellent average performance of 6.9 GFlop/s per CPU on 8 processors (1 node) with a vectorization ratio of 99.1%. However, extraction of velocity profiles at the walls revealed that the resolution is still not sufficient, contrary to the suggestions of Wagner [15], see Fig. 5. Thus the grid was refined to about 8×10^6 nodes, which almost doubles the resolution at the walls. Furthermore,

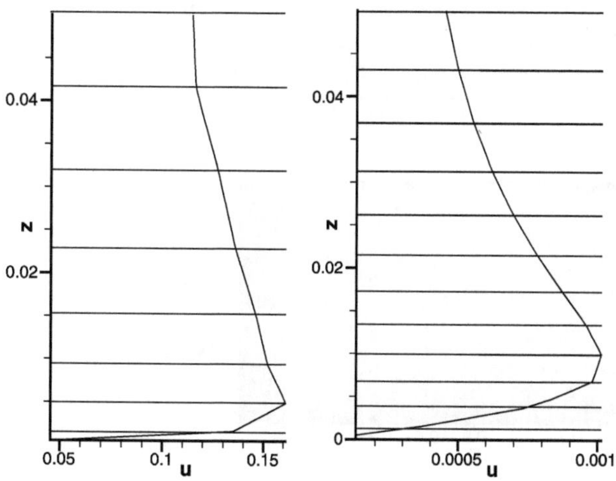

Fig. 5. Radial velocity profiles at $r/R_c = 0.05$ from the bottom of the crucible, from DNS test computation with 5×10^6 CVs (*left*) and 8×10^6 CVs (*right*), also showing the grid line distribution in z-direction

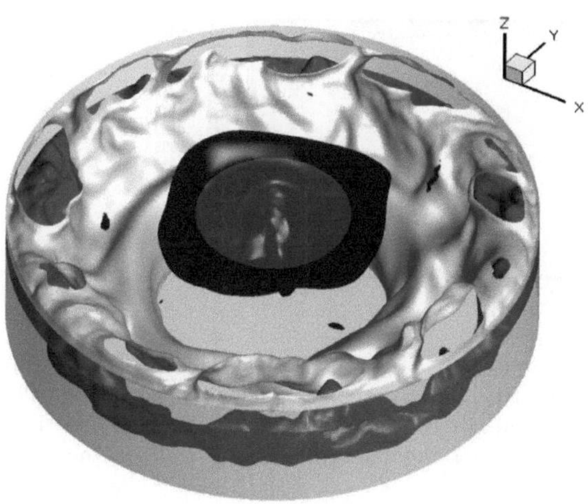

Fig. 6. Snapshot of the DNS, showing isosurfaces of instantaneous dimensionless temperature (*blue* –0.25, *white* 0, *red* 0.25). Buoyant plumes can be seen as holes at the melt surface. The crystal (*in the center at the top*) and crucible are depicted translucent

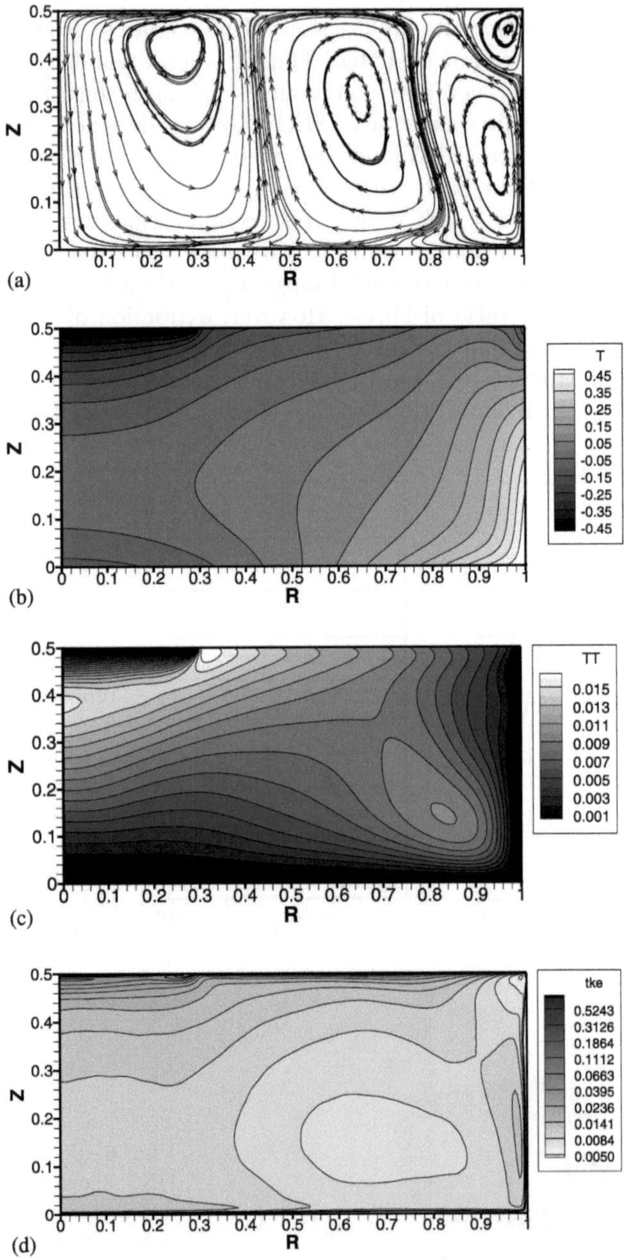

Fig. 7. (**a**) Streamlines of the mean velocities, (**b**) mean temperature field, (**c**) mean temperature fluctuations $\overline{T'T'}$ and (**d**) mean turbulent kinetic energy from the present DNS computation in a vertical cut through the crucible, averaged in time and circumferential direction (*crucible axis on the left, crucible wall on the right hand side*)

grid smoothing was applied using a Hilgenstock-Laplace and Sorensen-Laplace algorithm preserving the cell heights at the boundaries and improving orthogonality while trying to find an equal node distribution in the inner region, which is especially useful with O-grids, to achieve a better grid quality and thereby a high-quality solution and faster convergence (see Fig. 4). As can be seen from a preliminary simulation (Fig. 5, right), the velocity profile at the wall is much better resolved on this refined grid. In this computation also an outstanding average performance of 8.2 GFlop/s per CPU on 8 processors was achieved with a vectorization ratio of more than 99.6%.

A first snapshot of this new DNS can be seen in Fig. 6 showing isosurfaces of the instantaneous temperature. Furthermore, streamlines of the averaged flow field and the distributions of the mean temperature and turbulent kinetic energy are presented in Fig. 7. The flow and temperature field depict the expected form. Deviations from the DNS by Wagner [15] arise from some slightly differently chosen boundary conditions. The distribution of the turbulent kinetic energy shows its peak value at the corner of the rotating crystal, where it meets the counterrotation of the crucible and thus the bulk melt, resulting in high shear. Furthermore, the Marangoni convection is very strong due to the large temperature gradient and works against the buoyant flow causing even more shear. High values can also be seen throughout the free surface, where the same effect takes place. Moreover, the mean temperature fluctuations $\overline{T'T'}$ show a maximum at the triple point, where melt, crystal and atmosphere meet, with high values continuing under the crystal. This is caused by the temperature difference between the solid crystal and the hot melt, moved by the Marangoni convection at the free surface, and also by buoyancy under the crystal. The high level of turbulent kinetic energy at this spot adds to this effect.

5 Summary and Outlook

LES computations of flow and heat transfer in an idealized Cz case were validated against reference DNS data [15]. A good overall agreement of the velocity and temperature fields was found, however, a detailed analysis of turbulent quantities was not possible due to a lack of data. Therefore, new fine grid Direct Numerical Simulations have been conducted. For that purpose, a computational mesh was designed and test simulations were conducted to determine the necessary resolution and resources. Furthermore, in the meantime these DNS computations were performed on the NEC SX-8 at the HLRS and are presently analyzed.

It was shown that the flow solver $\mathcal{LESOCC}$ runs very efficiently on the NEC SX-8. However, due to the very small time step required for the extremely fine resolution, the DNS took more than 540 hours of walltime on 8 processors to deliver well averaged quantities over a period of nearly 3 million time steps or more than 600 dimensionless time units (equivalent to ca. 100 crucible revolutions).

References

1. Leister, H.J. and Perić, M. (1993) Vectorized Strongly Implicit Solving Procedure for a Seven-Diagonal Coefficient Matrix, *Int. Journal for Heat and Fluid Flow*, vol. **4**, pp. 159–172
2. Perić, M. (1985) A Finite-Volume Method for the Prediction of Three-Dimensional Fluid Flow in Complex Ducts, PhD thesis, Imperial College, London
3. Stone, H.L. (1968) Iterative Solution of Implicit Approximations of Multidimensional Partial Differential Equations, *SIAM Journal of Numerical Analyses*, vol. **5**, pp. 530–558
4. Durst, F., Schäfer, M. and Wechsler, K. (1996) Efficient Simulation of Incompressible Viscous Flows on Parallel Computers, In: Flow Simulation with High–Performance Computers, II, ed. E.H. Hirschel, *Notes on Numer. Fluid Mech.*, vol. **52**, pp. 87–101, Vieweg Verlag, Braunschweig
5. Durst, F. and Schäfer, M. (1996) A Parallel Block–Structured Multigrid Method for the Prediction of Incompressible Flows, *Int. Journal Num. Methods Fluids*, vol. **22**, pp. 549–565
6. Breuer, M., Rodi, W. (1996) Large–Eddy Simulation of Complex Turbulent Flows of Practical Interest, In: Flow Simulation with High–Performance Computers II, ed. E.H. Hirschel, *Notes on Numer. Fluid Mech.*, vol. **52**, pp. 258–274, Vieweg Verlag, Braunschweig
7. Breuer, M. (1998) Large–Eddy Simulation of the Sub–Critical Flow Past a Circular Cylinder: Numerical and Modeling Aspects, *Int. J. for Numer. Methods in Fluids*, vol. **28**, pp. 1281–1302
8. Breuer, M. (2000) A Challenging Test Case for Large–Eddy Simulation: High Reynolds Number Circular Cylinder Flow, *Int. J. of Heat and Fluid Flow*, vol. **21**, no. 5, pp. 648–654
9. Breuer, M. (2002) Direkte Numerische Simulation und Large–Eddy Simulation turbulenter Strömungen auf Hochleistungsrechnern, Habilitationsschrift, Universität Erlangen–Nürnberg, Berichte aus der Strömungstechnik, ISBN: 3-8265-9958-6, Shaker Verlag, Aachen
10. Rhie, C.M., Chow, W.L. (1983) A Numerical Study of the Turbulent Flow Past an Isolated Airfoil with Trailing Edge Separation, *AIAA Journal*, vol. **21**, pp. 1525–1532
11. Smagorinsky, J. (1963) General Circulation Experiments with the Primitive Equations, I, The Basic Experiment, *Mon. Weather Rev.*, vol. **91**, pp. 99–165
12. Basu, B., Enger, S., Breuer, M., and Durst, F. (2000) Three–Dimensional Simulation of Flow and Thermal Field in a Czochralski Melt Using a Block–Structured Finite–Volume Method, *Journal of Crystal Growth*, vol. **219**, pp. 123–143
13. Enger, S., Basu, B., Breuer, M., and Durst, F. (2000) Numerical Study of Three-Dimensional Mixed Convection due to Buoyancy and Centrifugal Force in an Oxide Melt for Czochralski Growth, *Journal of Crystal Growth*, vol. **219**, pp. 123–143

14. Kumar, V. (2005) Modeling and Numerical Simulations of Complex Transport Phenomena in Crystal Growth Processes, PhD thesis, Lehrstuhl für Strömungsmechanik, Universität Erlangen-Nürnberg
15. Wagner, C. (2003) Turbulente Transportvorgänge in idealisierten Czochralski-Kristallzüchtungsanordnungen, Habilitation, Lehrstuhl für Fluidmechanik, Technische Universität München
16. Grötzbach, G. (1983) Spatial Resolution Requirements for Direct Numerical Simulation of the Rayleigh-Benard Convection, *J. Comp. Phys.* vol. **49**, pp. 241–264
17. Breuer, M., Lammers, P., Zeiser, Th., Hager, G., Wellein, G.: *Towards the Simulation of Turbulent Flows Over Dimples – Code Evaluation and Optimization for NEC SX-8*, see rejected report which should be published in this book
18. Bartels, C., Breuer, M., Wechsler, K., and Durst, F. (2001) CFD-Applications on Parallel-Vector Computers: Computations of Stirred Vessel Flows, *Int. J. Computers and Fluids*, vol. **31**, pp. 69–97

Efficient Implementation of Nonlinear Deconvolution Methods for Implicit Large-Eddy Simulation

S. Hickel and N.A. Adams

Institute of Aerodynamics, Technische Universität München,
D-85747 Garching, Germany
sh@tum.de

Abstract. The adaptive local deconvolution method (ALDM) provides a systematic framework for the implicit large-eddy simulation (ILES) of turbulent flows. Exploiting numerical truncation errors, the subgrid scale model of ALDM is implicitly contained within the discretization. An explicit computation of model terms therefore becomes unnecessary. Subject of the present paper is the efficient implementation and the application to large-scale computations of this method. We propose a modification of the numerical algorithm that allows for reducing the amount of computational operations without affecting the quality of the LES results. Computational results for isotropic turbulence and plane channel flow show that the proposed simplified adaptive local deconvolution (SALD) method performs similarly to the original ALDM and at least as well as established explicit models.

1 Introduction

In Large Eddy Simulation (LES) of turbulent flows the evolution of non-universal, larger scales is computed, whereas the *effect* on the resolved scales of their nonlinear interactions with the unresolved subgrid-scales (SGS) has to be represented by an SGS model. This usually implies to assume that these effects are approximately universal.

SGS are modeled *explicitly* if the underlying conservation law is modified and subsequently discretized. In this respect, the filtering concept of Leonard [13] provides a mathematical framework for LES. Explicit subgrid scale models can be derived in this framework without reference to a computational grid and without commitment to a discretization scheme. However, numerically computed SGS stresses are strongly affected by the truncation error of the discretization method [5]. This interference can result in strange effects such as the lack of grid convergence. Employing an explicit SGS model, Schumann [18] already argued that discretization effects should be taken into account within the SGS model formulation.

As *Implicit Large Eddy Simulation* (ILES) we denote the situation when the unmodified conservation law is discretized. With ILES the numerical truncation error acts as SGS model. Since this SGS model is implicitly contained within the discretization scheme, an explicit computation of model terms becomes unnecessary. With ILES under-resolution is treated as a primarily numerical problem which can be solved by employing an appropriate discretization scheme. This approach is particularly convenient in flow regimes for which the derivation or the accurate computation of explicit SGS models is cumbersome. Many authors emphasize the potential of implicit LES for physically complex flows and for flows in complex geometries, cf. [6].

Different approaches to implicit LES can be taken. Mostly, given nonlinearly stable discretizations schemes for the convective fluxes are used as main element of implicit SGS models. A numerical analysis of Garnier *et al.* [4] dealing with several approaches to implicit LES, however, comes to the conclusion that applying off-the-shelf upwind or non-oscillatory schemes is not recommendable. In order to remedy the fact that suitable discretization schemes for ILES are found by more or less fortuitous choice, we have recently developed a systematic framework for design, analysis, and optimization of nonlinear discretization schemes for implicit LES.

The resulting so-called *Adaptive Local Deconvolution Method*, ALDM for short, represents a full merging of numerical discretization and subgrid-scale model. The efficiency of this approach to implicit LES was demonstrated in Ref. [1] for 1D conservation laws on the example of the viscous Burgers equation. The extension to three spatial dimensions and the incompressible Navier-Stokes equations is detailed in Ref. [8]. ALDM has proven itself as a reliable, accurate, and efficient method for LES of three-dimensional homogeneous isotropic turbulence [8], for plane channel flow [10], and for the complex flow in a channel with periodic constrictions [11]. It has been shown that ALDM performs at least as well as established explicit models.

In the present paper we revisit the numerical formulation of ALDM for the incompressible 3D Navier-Stokes equations. Aspects of the efficient implementation of nonlinear, solution-adaptive discretization schemes are addressed in Sect. 2. A simplification of the ALDM scheme is proposed, leading to a simplified adaptive local deconvolution (SALD) method. In Sect. 3.2 both implicit methods, ALDM and SALD, are applied to isotropic turbulence and turbulent channel flow. All simulations show good agreement with theory and experimental data and thus demonstrate the good performance of the implicit models. Comparing SALD with ALDM, the good performance of ALDM is preserved while computational costs are reduced significantly with SALD.

2 Solution-Adaptive Local Deconvolution Revisited

Systematic implicit SGS modeling requires procedures for design, analysis, and optimization of appropriate discretization schemes. We have recently developed such framework for ILES based on deconvolution methods [8].

The resulting method, ALDM, is a general nonlinear, i.e. solution adaptive, discretization scheme designed for implicit LES. The suitable framework for our method is provided by the finite-volume method implying reconstruction or deconvolution of the unfiltered solution at cell faces and the approximation of the physical flux function by a numerical flux function. With ALDM a local reconstruction of the solution is obtained from a solution-adaptive combination of deconvolution polynomials. Free parameters involved in the non-linear weight functions of the respective contributions allow for SGS modeling. Optimal model parameters were determined systematically by minimizing the difference between spectral numerical viscosity of ALDM and the eddy viscosity from the eddy-damped quasi-normal Markovian (EDQNM) theory for isotropic turbulence. With the optimized discretization parameters ALDM matches the theoretical requirements of EDQNM so that the truncation error has physical significance. The spectral eddy viscosity of the ALDM scheme exhibits a low-wavenumber plateau at the correct level and reproduces the typical cusp shape up to the cut-off wave number at the correct magnitude [8].

In the following, all relevant parts of the algorithm of ALDM as given in Ref. [8] are reviewed. We demonstrate how the original formulation can be further simplified to accelerate computation. We point out the differences between full ALDM and the *simplified* adaptive local deconvolution (SALD) method.

2.1 Proposed Simplifications of the Adaptive Local Deconvolution Method

The incompressible Navier-Stokes equations in non-dimensional form read

$$\frac{\partial \boldsymbol{u}}{\partial t} + \boldsymbol{\nabla} \cdot \boldsymbol{F}(\boldsymbol{u}) + \boldsymbol{\nabla}\, p - \nu\, \boldsymbol{\nabla} \cdot \boldsymbol{\nabla} \boldsymbol{u} = \boldsymbol{0} \, , \tag{1a}$$

$$\boldsymbol{\nabla} \cdot \boldsymbol{u} = 0 \, , \tag{1b}$$

where $\boldsymbol{u}$ is the velocity vector, p is the pressure, and ν is the molecular viscosity. For implicit modeling we only consider the nonlinear term $\boldsymbol{\nabla} \cdot \boldsymbol{F} + \boldsymbol{\nabla}\, p$ in the momentum equation (1a), whereas the linear terms, i.e. the diffusive flux, are approximated by a standard centered discretization. ALDM applies to the convective flux $\boldsymbol{F} = \boldsymbol{u}\boldsymbol{u}$. For incompressible flows a fractional-step approach is taken where the normal stresses due to the pressure p are subsequently computed by solving a Poisson equation.

Even if filtering is not performed explicitly, we can use the filtered formulation of Leonard [13] as analytical tool when designing and analyzing discrete operators. The filtering $\overline{\varphi} = \boldsymbol{G} * \varphi$ applied to Eq. (1) yields

$$\frac{\partial \overline{\boldsymbol{u}}_N}{\partial t} + \boldsymbol{G} * \boldsymbol{\nabla} \cdot \boldsymbol{F}_N(\boldsymbol{G}^{-1} * \overline{\boldsymbol{u}}_N) + \boldsymbol{\nabla}\,\overline{p} - \nu\,\boldsymbol{\nabla} \cdot \boldsymbol{\nabla}\overline{\boldsymbol{u}}_N = -\boldsymbol{G} * \boldsymbol{\nabla} \cdot \boldsymbol{\tau}_{SGS} \ , \tag{2a}$$

$$\boldsymbol{\nabla} \cdot \overline{\boldsymbol{u}}_N = 0 \ , \tag{2b}$$

where the subscript N indicates grid functions obtained by projecting continuous functions onto the numerical grid $\boldsymbol{x}_N = \{\boldsymbol{x}_{i,j,k}\}$. Note that with implicit LES the subgrid-stress tensor $\boldsymbol{\tau}_{SGS}$ is not computed explicitly but modeled by numerical truncation errors. A finite-volume discretization is based on the top-hat filter kernel $\boldsymbol{G}$ which returns the cell average of a function

$$\overline{\varphi}(\boldsymbol{x}_{i,j,k}, t) = \frac{1}{\Delta x_i \Delta y_j \Delta z_k} \iiint\limits_{I_{i,j,k}} \varphi(\boldsymbol{x}_{i,j,k} - \boldsymbol{x}, t)\ d\boldsymbol{x} \ , \tag{3}$$

where the integration domain $I_{i,j,k} = [x_{i-\frac{1}{2}}, x_{i+\frac{1}{2}}] \times [y_{j-\frac{1}{2}}, y_{j+\frac{1}{2}}] \times [z_{k-\frac{1}{2}}, z_{k+\frac{1}{2}}]$ is equivalent to a cell of the underlying Cartesian computational grid so that the filter width corresponds to the local grid size. Here and in the following half-integer indices denote cell faces and the coordinate system $\{x\,,\,y\,,\,z\}$ is synonymous with $\{1\,,\,2\,,\,3\}$.

The 3D filter kernel $\boldsymbol{G}$ can be factorized into three 1D operators

$$\boldsymbol{G}(\boldsymbol{x}) = \boldsymbol{G}_x(x) * \boldsymbol{G}_y(y) * \boldsymbol{G}_z(z) \ . \tag{4}$$

An inverse-filter operation can be defined as convolution with the inverse kernels

$$\boldsymbol{G}^{-1}(\boldsymbol{x}) = \boldsymbol{G}_x^{-1}(x) * \boldsymbol{G}_y^{-1}(y) * \boldsymbol{G}_z^{-1}(z) \ . \tag{5}$$

In the framework of a finite-volume discretization, filtering applied to the flux divergence returns the flux through the surface $S_{i,j,k}$ of cell $I_{i,j,k}$. By Gauss' theorem we obtain

$$\begin{aligned}
&\left[\boldsymbol{G} * \boldsymbol{\nabla} \cdot \boldsymbol{F}\left(\boldsymbol{G}^{-1} * \overline{\boldsymbol{u}}\right)\right]_{i,j,k} \\
&\quad= \frac{1}{\Delta x_i}\, \boldsymbol{G}_y * \boldsymbol{G}_z * \left(\overset{1}{\boldsymbol{f}}(\boldsymbol{G}^{-1} * \overline{\boldsymbol{u}})\Big|_{i+\frac{1}{2},j,k} - \overset{1}{\boldsymbol{f}}(\boldsymbol{G}^{-1} * \overline{\boldsymbol{u}})\Big|_{i-\frac{1}{2},j,k} \right) \\
&\quad+ \frac{1}{\Delta y_j}\, \boldsymbol{G}_z * \boldsymbol{G}_x * \left(\overset{2}{\boldsymbol{f}}(\boldsymbol{G}^{-1} * \overline{\boldsymbol{u}})\Big|_{i,j+\frac{1}{2},k} - \overset{2}{\boldsymbol{f}}(\boldsymbol{G}^{-1} * \overline{\boldsymbol{u}})\Big|_{i,j-\frac{1}{2},k} \right) \\
&\quad+ \frac{1}{\Delta z_k}\, \boldsymbol{G}_x * \boldsymbol{G}_y * \left(\overset{3}{\boldsymbol{f}}(\boldsymbol{G}^{-1} * \overline{\boldsymbol{u}})\Big|_{i,j,k+\frac{1}{2}} - \overset{3}{\boldsymbol{f}}(\boldsymbol{G}^{-1} * \overline{\boldsymbol{u}})\Big|_{i,j,k-\frac{1}{2}} \right) ,
\end{aligned} \tag{6}$$

where the flux vector $\overset{l}{\boldsymbol{f}} = u_l \boldsymbol{u}$ denotes the l-direction component of $\boldsymbol{F}$. As a consequence of this identity (6), finite-volume schemes require a reconstruction of data at the faces of the computational volumes.

The numerical computation of eq. (6) necessarily involves approximations. The original formulation of ALDM [8] employs a standard Gaussian quadrature rule with kernel $\boldsymbol{C}_k$ and order $\mathcal{O}\left(\Delta x^k\right)$ to approximate the filter operation $\boldsymbol{G}_l * \boldsymbol{G}_m$ over the cell faces and a solution adaptive deconvolution scheme is proposed to approximate $\boldsymbol{G}^{-1}$. In Ref. [8] a simple second-order accurate filtering scheme is recommended as approximation of $\boldsymbol{G}_l * \boldsymbol{G}_m$. The benefits of higher-order schemes were found to be negligible since second-order accurate interpolants contribute to the deconvolution operator.

Numerical experiments presented in the following section show that also the deconvolution scheme can be simplified significantly without affecting the prediction power of the implicit SGS model. Based on these results, we recommend as simplification

$$\begin{aligned}\left[\boldsymbol{G} * \boldsymbol{\nabla} \cdot \boldsymbol{F}\left(\boldsymbol{G}^{-1} * \overline{\boldsymbol{u}}\right)\right]_{i,j,k} \approx & \frac{1}{\Delta x_i}\left({}^{1}\!\boldsymbol{f}\left(\mathcal{X}_x^{+\frac{1}{2}}\overline{\boldsymbol{u}}\right) - {}^{1}\!\boldsymbol{f}\left(\mathcal{X}_x^{-\frac{1}{2}}\overline{\boldsymbol{u}}\right)\right) \\ & + \frac{1}{\Delta y_j}\left({}^{2}\!\boldsymbol{f}\left(\mathcal{X}_y^{+\frac{1}{2}}\overline{\boldsymbol{u}}\right) - {}^{2}\!\boldsymbol{f}\left(\mathcal{X}_y^{-\frac{1}{2}}\overline{\boldsymbol{u}}\right)\right) \\ & + \frac{1}{\Delta z_k}\left({}^{3}\!\boldsymbol{f}\left(\mathcal{X}_z^{+\frac{1}{2}}\overline{\boldsymbol{u}}\right) - {}^{3}\!\boldsymbol{f}\left(\mathcal{X}_z^{-\frac{1}{2}}\overline{\boldsymbol{u}}\right)\right) , \qquad (7)\end{aligned}$$

where no explicit filtering $\boldsymbol{G}_l * \boldsymbol{G}_m$ is applied. Furthermore, approximate deconvolution $\boldsymbol{G}^{-1}$ is applied only in those directions for which interpolation at cell faces is necessary. Interpolation and deconvolution is performed using the 1D reconstruction (i.e. deconvolution and interpolation) operator $\mathcal{X}_l^\lambda$ of ALDM. With this simplified method, the fully 3D scheme of ALDM is replaced by a single 1D step at the target cell face. Deconvolution is not performed in the transverse directions.

The 1D operator, denoted by $\mathcal{X}_x^\lambda$, was proposed in Ref. [1] and adapted in Ref. [8]. It is defined on a 1D grid $x_N = \{x_i\}$. Applied to the filtered grid function $\overline{\varphi}_N = \{\overline{\varphi}(x_i)\}$ it returns the approximately deconvolved grid function $\widetilde{\varphi}_N^\lambda \doteq \{\varphi(x_{i+\lambda})\}$ on the shifted grid $x_N^\lambda = \{x_{i+\lambda}\}$

$$\mathcal{X}_x^\lambda \overline{\varphi}_N = \{\varphi\left(x_{i+\lambda}\right) + \mathcal{O}\left(\Delta x_i{}^{\kappa}\right)\} = \widetilde{\varphi}_N^\lambda \ . \qquad (8)$$

The filtered data are given at the cell centers $\{x_i\}$. Reconstruction at the left cell faces $\{x_{i-\frac{1}{2}}\}$ is indicated by $\lambda = -1/2$ and at the right faces by $\lambda = +1/2$. Following Eq. (6), yet another approximation of the partially deconvolved solution would be necessary for obtaining a 3D reconstruction

$$\widetilde{\varphi}_N^{\boldsymbol{\lambda}} = \boldsymbol{\mathcal{X}}^{\boldsymbol{\lambda}}\, \overline{\varphi}_N = \mathcal{X}_z^{\lambda_3}\left(\,\mathcal{X}_y^{\lambda_2}\left(\,\mathcal{X}_x^{\lambda_1}\,\overline{\varphi}_N\,\right)\right) \qquad (9)$$

by successive 1D operations. The respective operators perform deconvolution at the cell centers and are indicated by $\lambda = 0$. These operators are replaced by identity in the simplified algorithm, reducing the number of operations by a factor of about 2.

Deconvolution and interpolation are done simultaneously using Lagrangian interpolation polynomials as proposed by Harten *et al.* [7]. Given a generic k-point stencil ranging from x_{i-r} to $x_{i-r+k-1}$ the 1D ansatz reads

$$\varphi(x_{i+\lambda}) = \sum_{l=0}^{k-1} c_{k,r,l}^{\lambda}(x_i)\overline{\varphi}_N(x_{i-r+l}) + \mathcal{O}\left(\Delta x_i{}^k\right) , \tag{10}$$

with $r \in \{0, \ldots, k\}$. The grid-dependent coefficients are

$$c_{k,r,l}^{\lambda}\left(x_i\right) = \left(x_{i-r+l+\frac{1}{2}} - x_{i-r+l-\frac{1}{2}}\right) \sum_{m=l+1}^{k} \frac{\sum\limits_{\substack{p=0\\p\neq m}}^{k} \prod\limits_{\substack{n=0\\n\neq p,m}}^{k} x_{i+\lambda} - x_{i-r+n-\frac{1}{2}}}{\prod\limits_{\substack{n=0\\n\neq m}}^{k} x_{j-r+m-\frac{1}{2}} - x_{i-r+n-\frac{1}{2}}} , \tag{11}$$

cf. [19]. They apply for grids with variable mesh width and for arbitrary target positions $x_{i+\lambda}$. In case of a staggered grid the values of x_i are different for each velocity component and the coefficients $c_{k,r,l}^{\lambda}$ have to be specified accordingly.

Selecting a particular interpolation stencil would return a linear discretization with a fixed, solution independent functional expression of the k-th order truncation error, provided the function is sufficiently smooth on the interpolation stencil. ALDM adopts the idea of the Weighted-Essentially-Non-Oscillatory (WENO) scheme of Shu [19] where interpolation polynomials of a single order $k \equiv K$ are selected and combined nonlinearly. The essential difference between ALDM and WENO is that we superpose all interpolants of order $k = 1, \ldots, K$

$$\widetilde{\varphi}_N^{\lambda}(x_{i+\lambda}) = \sum_{k=1}^{K} \sum_{r=0}^{k-1} \omega_{k,r}^{\lambda}(\overline{\varphi}_N, x_i) \sum_{l=0}^{k-1} c_{k,r,l}^{\lambda}(x_i)\, \overline{\varphi}_N(x_{i-r+l}) \tag{12}$$

to allow for lower-order contributions to the truncation error for implicit SGS modeling [1]. Usually it is $K = 3$ in our LES. The weights $\omega_{k,r}^{\lambda}(\overline{\varphi}_N, x_i)$ can be constructed as to yield an accurate approximation of order $\kappa = 2K - 1$ in smooth regions [19]. For our purpose, however, we do not need highest possible order of accuracy. Rather the superposition (12) introduces free discretization parameters which allow to control error cancelations. The sum of all weights is constrained to be unity for consistency. More restrictively we require

$$\omega_{k,r}^{\lambda}\left(\overline{\varphi}_N, x_i\right) = \frac{1}{K} \frac{\gamma_{k,r}^{\lambda}\beta_{k,r}\left(\overline{\varphi}_N, x_i\right)}{\sum\limits_{s=0}^{k-1} \gamma_{k,s}^{\lambda}\beta_{k,s}\left(\overline{\varphi}_N, x_i\right)} , \tag{13}$$

with $r = 0, \ldots, k-1$ for each $k = 1, \ldots, K$. Note that λ denotes a superscript index and not a power. The solution-adaptive behavior of ALDM is controlled by the functional

$$\beta_{k,r}\left(\overline{\varphi}_N, x_i\right) = \left(\varepsilon_\beta + \sum_{m=-r}^{k-r-2} \left(\overline{\varphi}_{i+m+1} - \overline{\varphi}_{i+m}\right)^2\right)^{-2} , \tag{14}$$

where ε_β is a small number to prevent division by zero. $\beta_{k,r}$ measures the smoothness of the grid function on the respective stencil to obtain a non-linear adaptation of the deconvolution. The advantage of definition (14) over smoothness measures proposed by Liu *et al.* [16] and by Jiang & Shu [12] is that

$$\beta_{k,r}\left(\overline{\varphi}_N, x_i\right) = \beta_{k,r-1}\left(\overline{\varphi}_N, x_{i-1}\right) . \tag{15}$$

can be exploited to improved computational efficiency. The parameters $\gamma^\lambda_{k,r}$ represent a stencil-selection preference that would become effective in the statistically homogeneous case. The requirement of an isotropic discretization for this case implies the following symmetry on the parameters

$$\gamma^{-1/2}_{k,r} = \gamma^{+1/2}_{k,k-1-r} . \tag{16}$$

The number of independent parameters is further reduced by the consistency requirement $\sum_{r=0}^{k-1} \gamma^{+1/2}_{k,r} = 1$. Using the simplified procedure with $K = 3$, three parameters $\{ \gamma^{+1/2}_{2,0} , \gamma^{+1/2}_{3,0} , \gamma^{+1/2}_{3,1} \}$ remain available for modeling. The original formulation [8] provides one additional parameter $\gamma^0_{3,1}$. Optimal parameters for the original scheme were found by means of an evolutionary optimization algorithm [1, 8, 9]. These parameters, reproduced in Table 1, result in an excellent spectral match of the numerical viscosity of ALDM with prediction of EDQNM theory [2, 14]. We have refrained from repeating the parameter optimization with the simplified scheme because the change by applying the transversal deconvolution operation was found to be negligible in terms of the effective numerical viscosity. For further computational details of evaluation and optimization of the spectral numerical viscosity of ALDM the reader should refer to Ref. [8].

It is interesting to note that the evolutionary optimization finally selected $\gamma^{+1/2}_{2,0} = 1$ and $\gamma^{+1/2}_{2,1} = 0$. Consequently, solution adaptivity is ruled out for

Table 1. Optimized values for deconvolution parameters [8]

parameter	$\gamma^{+1/2}_{2,0}$	$\gamma^{+1/2}_{2,1}$	$\gamma^{+1/2}_{3,0}$	$\gamma^{+1/2}_{3,1}$	$\gamma^{+1/2}_{3,2}$
optimal value	1.00000	0.00000	0.01902	0.08550	0.89548

all but the third-order stencils. It is therefore not necessary to compute $\beta_{1,r}$ and $\beta_{2,r}$.

Another tool which is exploited is the choice of an appropriate and consistent numerical flux function that approximates the physical flux function

$$\widetilde{\boldsymbol{F}}_N \approx \boldsymbol{F} = \boldsymbol{u}\boldsymbol{u} \qquad \text{and} \qquad {}^{l}\widetilde{\boldsymbol{f}} \approx {}^{l}\boldsymbol{f} = u_l \boldsymbol{u} \ . \tag{17}$$

A review of common numerical flux functions can be found, e.g., in LeVeque's textbook [15]. During construction of ALDM various flux functions were analyzed by MDEA [1]. Based on these findings a modification of the Lax-Friedrichs flux function was proposed [8] for the three-dimensional Navier-Stokes equations and collocated grids. A further simplification of this flux function is not possible without affecting the performance of the model. However, the computational efficiency can be improved by merging convective and diffusive flux computations.

2.2 Vectorization

Although LES does not aim at resolving all spatial scales of turbulent fluid motion, it requires large computational grids with often tens of millions of grid points. On each grid point, however, only few operations have to be performed in one time step. This implies, to our experience, a poor ratio of usable performance to peak performance on scalar machines. Applying the same operation to vast data fields is much better suited to vector machines. In the following we discuss the special requirements concerning the efficient implementation of the above detailed algorithms on vector computers, particularly with regard to the NEC SX-8 cluster at the Stuttgart High-Performance Computing Center (HLRS).

On inhomogeneous grids a large number of spatially varying coefficients is needed for deconvolution, for interpolation, for computing derivatives, etc.. These coefficients are usually computed in advance of the actual simulation. Due to memory limitations, however, one tries to avoid to store 3D coefficient arrays. Orthogonal Cartesian grids allow for a reduction of these memory requirements from $\sim (N_x \times N_y \times N_z)$ to $\sim (N_x + N_y + N_z)/3$, since a coefficient varies only in one spatial direction. The drawback associated with such an implementation is that the average computational-loop length is reduced by the same amount. With typical LES, the respective lengths are of the order of the machine vector length, resulting in a significant number of unused vector elements during computation. In order to avoid this problem, we perform an explicit vectorization, where the physical 3D arrays are projected onto a 2D computational space. In computational space, an array has at most two dimensions ξ and η, where the number of cells in the first direction N_ξ is the lowest common multiple of the machine-vector length and the respective physical vector length, e.g. N_y. To limit the required expansion, we allow 2 vector elements to be unused. The transformation does not require any additional operations at runtime because it only affects pre-computed coefficient arrays and loop-length parameters. In our implementation, this explicit vectorization is merged with a re-sorting algorithm. That is, before and after performing expensive operations, all arrays are re-sorted in such a way that the operation is then performed on the first index. The re-sorting routine consumes less then 1% of the overall computational time but accelerates in particular the deconvolution procedure significantly. As side effect, all operations need to be implemented for only one spatial direction when re-sorting is employed.

This strategy, based on transformation and re-sorting, results in efficient vectorization and memory access. The computational performance of our implementation was measured with *ftrace* on the Stuttgart NEC SX-8 cluster. Depending on grid size and shape, the measured number of floating point operations per second on a single NEC SX-8 CPU for the deconvolution operator and the numerical flux function are $12 - 13$ GFlops and $8 - 10$ GFlops, respectively. Using a Conjugate-Gradient-Method based solver for the pressure Poisson equation, the overall performance of our flow solver is ≈ 9 GFlops. Somewhat lower GFlops values are achieved with FFT-based Poisson solvers, because auxiliary parts such as communication, postprocessing, etc. become more relevant in this case.

On a single-processor NEC SX-6 more than 7.2 GFlops were measured for the deconvolution operator. The observed performance-loss relative to peak performance of NEC SX-8 is assumed to be caused by a bottleneck in the memory access.

2.3 Parallelization

Parallelization of our flow solver is generally achieved by domain decomposition. We recall, the joint stencil of deconvolution operator and numerical flux function covers 7 cells, requiring knowledge of the solution in three neighbor-cells to each side. This discretization scheme is not modified at flow boundaries or sub-domain interfaces. Rather, the computational sub-domains are covered by three layers of ghost cells that constitute the coupling of the solution with neighboring sub-domains. We follow a dual strategy for padding these ghost cells. A shared-memory *openMP* parallelization is employed for sub-domains on the same node, whereas information between different nodes is exchanged using *MPI*. We note that there is also a fully *MPI* parallelized implementation of the SALD method available by now. Compared to this implementation, the computational overhead of a shared-memory parallelization with *openMP* is negligible.

By now, LES for the majority of considered test cases are tractable on a single node using 8 SX-8 CPUs. To give an example, LES of the flow in a channel with streamwise periodic restrictions were recently performed to validate our methods for LES of flow separation in an adverse-pressure-gradient regime. A typical LES of this so-called periodic-hill flow requires about 700 CPU hours, see [11] for more details.

3 Numerical Results

3.1 Homogeneous Isotropic Turbulence

As a first test case the numerical schemes are applied to decaying grid-generated turbulence. The computations are initialized with energy spectrum and Reynolds numbers adapted to the wind-tunnel experiments of Comte-Bellot and Corrsin [3], denoted hereafter as CBC. Among other space-time correlations CBC provides streamwise energy spectra for grid-generated turbulence at three positions downstream of a mesh. Table 3 of Ref. [3] gives corresponding 3D energy spectra which were obtained under the assumption of isotropy.

In the simulation this flow is modeled as decaying turbulence in a $(2\pi)^3$-periodic computational domain, discretized on a collocated grid with 64^3 cells. Based on the Taylor hypothesis the temporal evolution in the simulation corresponds to a downstream evolution in the wind-tunnel experiment with the experimental mean-flow speed which is approximately constant. The energy distribution of the initial velocity field is matched to the first measured 3D energy spectrum of CBC. The SGS model is verified by comparing computational and experimental 3D energy spectra at later time instants which correspond to the other two measuring stations. For more details, particularly with regard to non-dimensionalization and initial-data generation, we refer to [8].

We present numerical results from implicit LES with the original ALDM scheme of Hickel *et al.* [8] and with successively simplified formulations proposed in the present paper. Figure 1a compares results obtained by 4th-order ($\boldsymbol{C}_4$) with those obtained by a second order ($\boldsymbol{C}_2$) Gaussian quadrature rule for the approximation of the filter operation $\boldsymbol{G}_l * \boldsymbol{G}_m$. It is found that the choice of the integration kernel has negligible effects on the computed energy spectra. The effect of omitting deconvolution in the transverse directions, i.e. the difference between full and simplified ALDM, is even smaller, see Fig. 1b.

3.2 Turbulent Channel Flow

As an example for anisotropic wall bounded turbulence, we simulate channel flow at $\mathrm{Re}_{bulk} = 6875$ ($\mathrm{Re}_\tau = 395$). Reference DNS data are provided by Moser *et al.* [17]. The computational domain measures $2\pi h \times 2h \times \pi h$ (streamwise $\times$ wall-normal $\times$ spanwise), where h is the channel half width. The spectral DNS of Moser *et al.* required $256 \times 193 \times 192$ grid points, whereas the computational grid of the present LES consists of $64 \times 68 \times 64$ cells. To accommodate the change of turbulence structure in the vicinity of solid walls, the grid is stretched in wall-normal direction

$$y(j) = -\frac{1}{\tanh(C_G)} \tanh\left(\frac{C_G(N_y - 2j)}{N_y}\right), \tag{18}$$

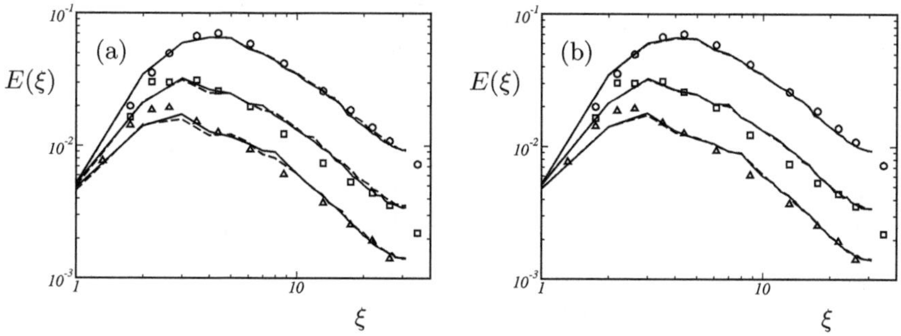

Fig. 1. Instantaneous 3D energy spectra for LES with 64^3 cells of the Comte-Bellot – Corrsin experiment. (**a**) ——— ALDM with full deconvolution and C_2, – – – – ALDM with full deconvolution and C_4. (**b**) ——— SALD, – – – – ALDM with full deconvolution and C_2. Symbols represent experimental data of Ref. [3]

where N_y is the number of cells in the wall-normal direction and $C_G = 2.0$ is the grid-stretching parameter.

Figure 2 shows profiles of mean velocity and Reynolds stresses from LES with the original and the simplified scheme. We observe a close match between both simulations and a good agreement with the reference DNS data. Inevitable differences between DNS and LES become most evident in the pressure fluctuations, see Fig. 3. Note the correct wall-asymptotic behavior of the Reynolds stresses of the LES. These results are encouraging and show that the implicit modeling approach gives good results for low-Reynolds-number wall-bounded flows although the model parameters were derived for isotropic turbulence and high Reynolds numbers.

4 Conclusion

In the present paper we have revised the numerical algorithm of the Adaptive Local Deconvolution Method (ALDM) for the incompressible 3D Navier-Stokes equations. We gave recommendations for the efficient implementation of ALDM. A simplification of the algorithm is proposed leading to the simplified adaptive local deconvolution (SALD) method. SALD allows for considerable savings of computational resources. It is about twice as fast as the original ALDM, whereas the prediction power of the implicit model is preserved. The method is computationally more efficient than a second-order central scheme with a dynamic Smagorinsky model. Note that ALDM gives at least as good predictions as the dynamic Smagorinsky model and other established explicit models [8].

For incompressible flows a fractional-step approach is pursued where a pressure correction is subsequently computed solving a Poisson equation. Flows in complex geometries with non-periodic directions require the use of

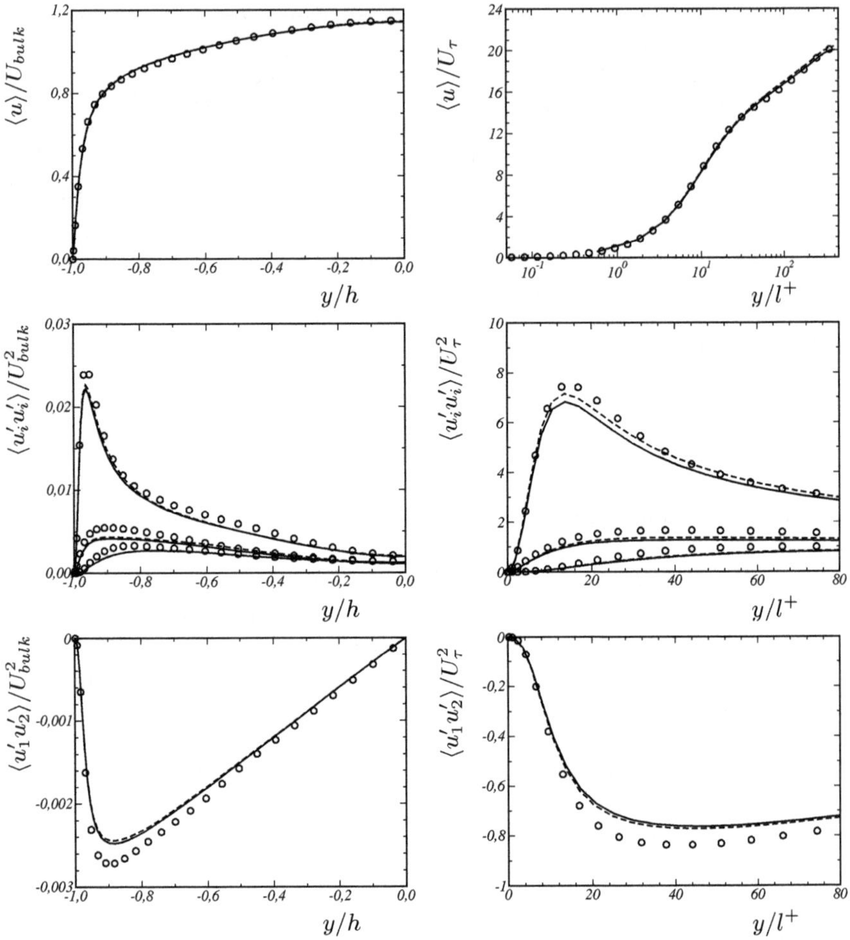

Fig. 2. Mean profiles of velocity and Reynolds stresses for LES of turbulent channel flow at $\mathrm{Re}_\tau = 395$. ——— SALD , – – – – ALDM with full deconvolution and $\boldsymbol{C}_2$, ○ DNS of Moser *et al.* [17]

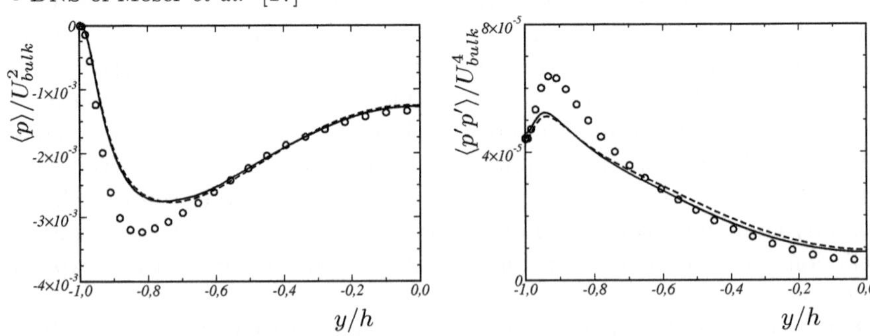

Fig. 3. Mean profiles of pressure and pressure fluctuations for LES of turbulent channel flow at $\mathrm{Re}_\tau = 395$. ——— SALD , – – – – ALDM with full deconvolution and $\boldsymbol{C}_2$, ○ DNS of Moser *et al.* [17]

iterative Poisson solvers, consuming typically about 60% and sometimes up to 95% of the computational time. It is recommended to use the SALD method whenever the CPU-time consumption of the discretization of the convective term is relevant. This holds in particular for flow configurations where efficient FFT-based solvers for the pressure-Poisson equation are available, such as homogeneous turbulence and channel flow.

References

1. N.A. Adams, S. Hickel, and S. Franz. Implicit subgrid-scale modeling by adaptive deconvolution. *J. Comp. Phys.*, 200:412–431, 2004.
2. J.-P. Chollet. Two-point closures as a subgrid-scale modeling tool for large-eddy simulations. In F. Durst and B.E. Launder, editors, *Turbulent Shear Flows IV*, pages 62–72, Heidelberg, 1984. Springer.
3. G. Comte-Bellot and S. Corrsin. Simple Eulerian time correlation of full and narrow-band velocity signals in grid-generated 'isotropic' turbulence. *J. Fluid Mech.*, 48:273–337, 1971.
4. E. Garnier, M. Mossi, P. Sagaut, P. Comte, and M. Deville. On the use of shock-capturing schemes for large-eddy simulation. *J. Comput. Phys.*, 153:273–311, 1999.
5. S. Ghosal. An analysis of numerical errors in large-eddy simulations of turbulence. *J. Comput. Phys.*, 125:187–206, 1996.
6. F.F. Grinstein and C. Fureby. From canonical to complex flows: Recent progress on monotonically integrated LES. *Comp. Sci. Eng.*, 6:36–49, 2004.
7. A. Harten, B. Engquist, S. Osher, and S. Chakravarthy. Uniformly high order accurate essentially non-oscillatory schemes, III. *J. Comp. Phys.*, 71:231–303, 1987.
8. S. Hickel, N.A. Adams, and J. A. Domaradzki. An adaptive local deconvolution method for implicit LES. *J. Comput. Phys.*, 213:413–436, 2006.
9. S. Hickel, S. Franz, N.A. Adams, and P. Koumoutsakos. Optimization of an implicit subgrid-scale model for LES. In *Proceedings of the 21st International Congress of Theoretical and Applied Mechanics*, Warsaw, Poland, 2004.
10. S. Hickel, T. Kempe, and N.A. Adams. On implicit subgrid-scale modeling in wall-bounded flows. In *Proceedings of the EUROMECH Colloquium 469*, pages 36–37, Dresden, Germany, 2005.
11. S. Hickel, T. Kempe, and N.A. Adams. Implicit large-eddy simulation applied to turbulent channel flow with periodic constrictions. *Theoret. Comput. Fluid Dynamics*, 2006. (submitted).
12. G.-S. Jiang and C.-W. Shu. Efficient implementation of weighted ENO schemes. *J. Comput. Phys.*, 126:202–228, 1996.
13. A. Leonard. Energy cascade in large eddy simulations of turbulent fluid flows. *Adv. Geophys.*, 18A:237–248, 1974.
14. M. Lesieur. *Turbulence in Fluids.* Kluwer Academic Publishers, Dordrecht, The Netherlands, 3 edition, 1997.
15. R.J. LeVeque. *Numerical methods for conservation laws.* Birkhäuser, Basel, Switzerland, 1992.

16. S.W. Liu, C. Meneveau, and J. Katz. On the properties of similarity subgrid-scale models as deduced from measurements in a turbulent jet. *J. Fluid Mech.*, 275:83–119, 1994.
17. R.D. Moser, J. Kim, and N.N. Mansour. Direct numerical simulation of turbulent channel flow up to $\mathrm{Re}_\tau = 590$. *Phys. Fluids*, 11:943–945, 1999.
18. U. Schumann. Subgrid scale model for finite-difference simulations of turbulence in plane channels and annuli. *J. Comput. Phys.*, 18:376–404, 1975.
19. C.-W. Shu. Essentially non-oscillatory and weighted essentially non-oscillatory schemes for hyperbolic conservation laws. Technical Report 97-65, ICASE, NASA Langley Research Center, Hampton, Virginia, 1997.

Large-Eddy Simulation of Tundish Flow

Nouri Alkishriwi, Matthias Meinke, and Wolfgang Schröder

Aerodynamisches Institut, RWTH Aachen University
Wüllnerstrasse 5 thr. 7 52062 Aachen, Germany
office@aia.rwth-aachen.de

Abstract. Large-eddy simulations (LES) of a continuous tundish flow are carried out to investigate the turbulent flow structure and vortex dynamics. The numerical computations are performed by solving the viscous conservation equations for compressible fluids. An implicit dual time stepping scheme combined with low Mach number preconditioning and a multigrid accelerating technique is developed for LES computations. The method is validated by comparing data of turbulent pipe flow at $Re_\tau = 1280$ and cylinder flow at $Re = 3900$ at different Mach numbers with experimental findings from the literature. The impact of jet spreading, jet impingement on the wall, and wall jets on the flow field and steel quality is investigated. The characteristics of the flow field in a one-strand tundish such as the time-dependent turbulent flow structure and vortex dynamics is analyzed and compared with experimental results.

1 Introduction

In many engineering problems compressible and nearly incompressible flow regimes occur simultaneously. For example low speed flows, which may be compressible due to surface heat transfer or volumetric heat addition. The numerical analysis of such flows requires to solve the viscous conservation equations for compressible fluids to capture the essential effects.

When a compressible flow solver is applied to a nearly incompressible flow, its performance can deteriorate in terms of both speed and accuracy [13]. It is well known that most compressible codes do not converge to an acceptable solution when the Mach number of the flow field is smaller than $\mathcal{O}(10^{-1})$. The main difficulty with such low speed flows arises from the large disparity between the wave speeds. The acoustic wave speed is $|u \pm c|$, while entropy or vorticity waves travel at $|u|$, which is quite small compared to $|u \pm c|$. In explicit time-marching codes, the acoustic waves define the maximum time step, while the convective waves determine the physical time required for the solution of the flow problem such that the overall computational time becomes large for small Mach numbers.

Different methods have been proposed to solve such mixed flow problems by modifying the existing compressible flow solvers. One of the most popular approaches is to use low Mach number preconditioning methods for compressible codes [13]. The basic idea of this approach is to modify the time marching behavior of the system of equations without altering the steady state solution. This is, however, only useful when the steady state solution is sought. A straightforward extension of the preconditioning approach to unsteady flow problems is achieved when it is combined with a dual time stepping technique.

This idea is followed in the present paper, i.e., a highly efficient large-eddy simulation method is described based on an implicit dual time stepping scheme combined with preconditioning and multigrid. This method is validated by well-known case studies [3] and finally, the preconditioned large-eddy simulation method is used to investigate the flow field in a continuous casting tundish.

In continuous casting of steel, the tundish enables to remove nonmetallic inclusions from molten steel and to regulate the flow from individual ladles to the mold. There is two types of flow conditions in a tundish: steady-state casting where the mass flow rate through the shroud m_{sh} is equal to the mass flow rate through the submerged entry nozzle into the mold m_{SEN}, and transient casting in which the mass of steel in a tundish varies in time during the filling or draining stages. The motion of the liquid steel is generated by jets into the tundish and continuously casting mold. The flow regime is mostly turbulent, but some turbulence attenuation can occur far from the inlet. The characteristics of the flow in a tundish include jet spreading, jet impingement on the wall, wall jets, and an important decrease of turbulence intensity in the core region of the tundish far from the jet [6] and [9].

In previous studies, a large amount of research has been carried out to understand the physics of the flow in a tundish mainly through numerical simulations based on the Reynolds-averaged NAVIER–STOKES (RANS) equations plus an appropriate turbulence model. To gain more knowledge about the transient turbulence process, which cannot be achieved via RANS solutions, large-eddy simulations of the tundish flow field are performed.

After a concise presentation of the governing equations the implementation of the preconditioning in the LES context using the dual time stepping technique is described. Then, the discretization and the time marching solution technique within the dual time stepping approach are discussed. After the description of the boundary conditions and the computational domain of the tundish flow, the numerical results of the validation cases and the tundish simulation are presented.

2 Governing Equations

The governing equations are the unsteady three-dimensional compressible NAVIER–STOKES equations written in generalized coordinates $\xi_i, i = 1, 2, 3$

$$\frac{\partial \mathbf{Q}}{\partial t} + \frac{\partial (\mathbf{F}_{c_i} - \mathbf{F}_{v_i})}{\partial \xi_i} = 0 \quad , \tag{1}$$

where the quantity $\mathbf{Q}$ represents the vector of the conservative variables and $\mathbf{F}_{c_i}$, $\mathbf{F}_{v_i}$ are inviscid and viscous flux vectors, respectively. As mentioned before, preconditioning is required to provide an efficient and accurate method of solution of the steady NAVIER–STOKES equations for compressible flow at low Mach numbers. Moreover, when unsteady flows are considered, a dual time stepping technique for time accurate solutions is used. In this approach, the solution at the next physical time step is determined as a steady state problem to which preconditioning, local time stepping and multigrid are applied.

Introducing a pseudo-time τ in Eq. (1), the governing equations for unsteady flow with preconditioning read

$$\mathbf{\Gamma}^{-1}\frac{\partial \mathbf{Q}}{\partial \tau} + \frac{\partial \mathbf{Q}}{\partial t} + \mathbf{R} = 0, \tag{2}$$

where $\mathbf{R}$ represents

$$\mathbf{R} = \frac{\partial (\mathbf{F}_{c_i} - \mathbf{F}_{v_i})}{\partial \xi_i} \tag{3}$$

and $\mathbf{\Gamma}$ is the preconditioning matrix, which is to be defined such that the new eigenvalues of the preconditioned system of equations are of similar magnitude. In this study, a preconditioning technique from [13] has been implemented.

It is clear that only the pseudo-time terms in Eq. (2) are altered by the preconditioning, while the physical time and space derivatives retain their original form. Convergence of the pseudo-time within each physical time step is necessary for accurate unsteady solutions. This means, the acceleration techniques such as local time stepping and multigrid can be immediately utilized to speed up the convergence within each physical time step to obtain an accurate solution for unsteady flows. The derivatives with respect to the physical time t are discretized using a three-point backward difference scheme that results in an implicit scheme, which is second-order accurate in time, which reads

$$\frac{\partial \mathbf{Q}}{\partial \tau} = \mathbf{RHS} \tag{4}$$

with the right-hand side

$$\mathbf{RHS} = -\mathbf{\Gamma}\left(\frac{3\mathbf{Q}^{n+1} - 4\mathbf{Q}^{n} + \mathbf{Q}^{n-1}}{2\Delta t} + \mathbf{R}(\mathbf{Q^{n+1}})\right) \quad .$$

Note that at $\tau \to \infty$ the first term on left-hand side of (2) vanishes such that (1) is recovered. To advance the solution of the inner pseudo-time iteration, a 5-stage Runge-Kutta method in conjunction with local time stepping and multigrid is used. For stability reasons the term $\frac{3\mathbf{Q}^{n+1}}{2\Delta t}$ is treated implicitly within the Runge-Kutta stages yielding the following formulation for the lth stage

$$\mathbf{Q^0} = \mathbf{Q^n}$$
$$\vdots$$
$$\mathbf{Q^l} = \mathbf{Q^0} + \alpha_l \Delta\tau \left[\mathbf{I} + \frac{3\Delta\tau}{2\Delta t}\alpha_l \mathbf{\Gamma} \right]^{-1} \mathbf{RHS} \tag{5}$$
$$\vdots$$
$$\mathbf{Q^{n+1}} = \mathbf{Q^5}.$$

The additional term means that in smooth flows the development in pseudo-time is proportional to the evolution in t.

3 Numerical Procedure

The governing equations are the NAVIER–STOKES equations filtered by a low-pass filter of width Δ, which corresponds to the local average in each cell volume. The monotone integrated large-eddy simulations (MILES) approach is used to implicitly model the small scale motions through the numerical scheme. Since the turbulent flow is characterized by strong interactions between various scales of motion, schemes with a large amount of artificial dissipation significantly degrade the level of energy distribution governed by the small-scale structures and therefore distort the physical representation of the dynamics of small as well as large eddies. Using a mixed central-upwind AUSM (advective upstream splitting method) scheme with low numerical dissipation could remedy this problem.

The approximation of the convective terms of the conservation equations is based on a modified second-order accurate AUSM scheme using a centered 5-point low dissipation stencil [8] to compute the pressure derivative in the convective fluxes. The pressure term contains an additional expression, which is scaled by a weighting parameter χ that represents the rate of change of the pressure ratio with respect to the local Mach number. This parameter determines the amount of numerical dissipation to be added to avoid oscillations that could lead to unstable solutions. The parameter χ was chosen in the range $0 \leq \chi \leq \frac{1}{400}$. The viscous stresses are discretized to second-order accuracy using central differences, i.e., the overall spatial approximation is second-order accurate.

The dual time stepping technique is used for the temporal integration. The solution at the next physical time step is determined as a steady state problem to which preconditioning, local time stepping and multigrid are applied. The 5-stage Runge-Kutta method is used to propagate the solution from time level n to n+1, where the Runge-Kutta coefficients $\alpha_l = (\frac{6}{24}, \frac{4}{24}, \frac{9}{24}, \frac{12}{24}, \frac{24}{24})$ are optimized for maximum stability of a centrally discretized scheme. The physical time derivative is discretized by a backward difference formula of second-order accuracy. The method is formulated for multi-block structured curvilinear grids and implemented on vector and parallel computers. A floating point performance of about 7 GFlops is achieved on each processor of the NEC SX-8 computer in parallel execution.

3.1 Validation

To validate the accuracy of the method, large-eddy simulations of turbulent pipe flow at a Reynolds number $Re_\tau = 1280$ based on the friction velocity u_τ, which corresponds to a diameter D based Reynolds number $Re_D = u_{cl}D/\nu = 25000$, are carried out. The Mach number based on the centerline velocity of the pipe is set to Ma=0.02 and the physical time step $\Delta t = 0.01$. The comparison of the pure explicit LES results from [11] and [12] and the LES

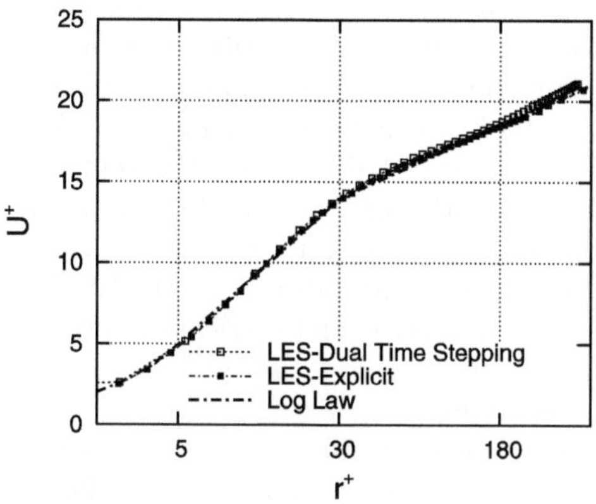

Fig. 1. Mean velocity distributions of turbulent pipe flow at $Re_\tau = 1280$

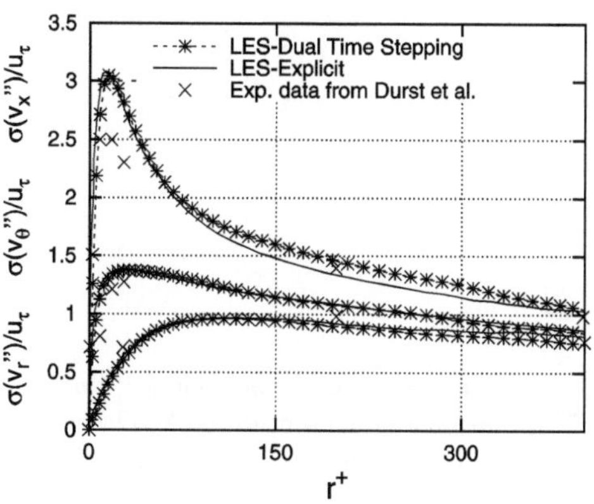

Fig. 2. Turbulence intensity distributions of turbulent pipe flow at $Re_\tau = 1280$

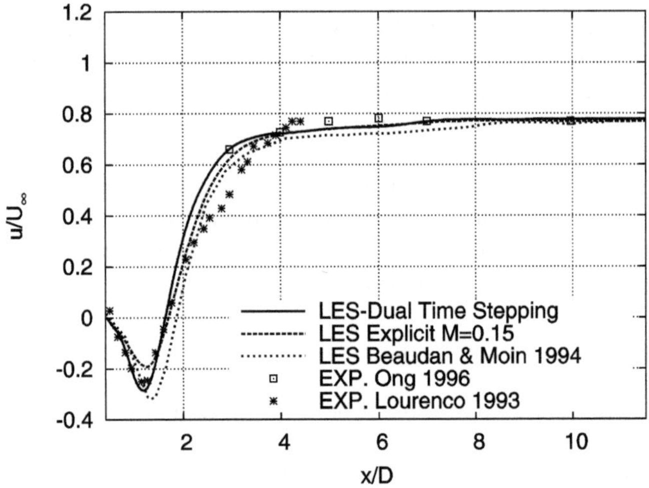

Fig. 3. Streamwise velocity as a function of X/D on the *centerline* $Y/D = 0$, $Z/D = 0$ in the cylinder wake at $Re_D = 3900$

findings of the implicit method in Figs. 1 and 2 shows a good agreement for the mean velocity profiles and the turbulence intensity distributions [5].

The flow around a cylinder at a diameter based $Re_D = 3900$ is performed at a freestream Mach number $M_\infty = 0.05$ and a physical time step $\Delta t = 0.02$. Figure 3 shows the streamwise velocity distribution on the centerline in the wake of the cylinder compared with the LES distribution of a pure explicit scheme without preconditioning at a Mach number $M_\infty = 0.1$ and with numerical and experimental data from the literature [4], [7], [10]. The profiles of the velocity fluctuations of the streamwise and vertical components at $X/D = 1.54$ as a function of Y/D presented in Figs. 4 and 5 corroborate the correspondence with the measurements from [7].

The analysis of the efficiency is performed for different parameters such as the size of the physical time step, the required residual constraint for the

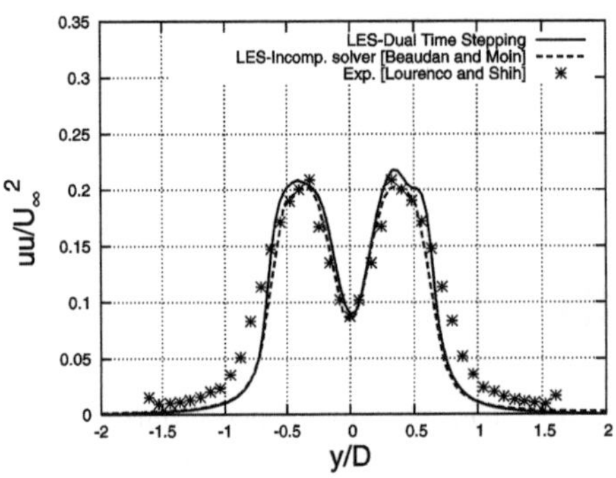

Fig. 4. Streamwise velocity fluctuations as a function of Y/D in the cylinder wake at $X/D = 1.54$

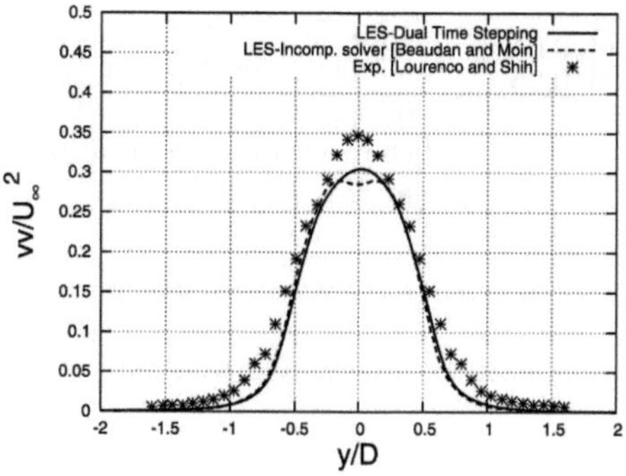

Fig. 5. Vertical velocity fluctuations as a function of Y/D in the cylinder wake at $X/D = 1.54$

inner pseudo-time iteration, the Mach number and the Reynolds number for a turbulent channel flow at $Re_\tau = 590$. Fig. 6 and 7 demonstrate the impact of the size of the physical time step and the Mach number on the efficiency of the implicit scheme. It is evident that as the size of the physical time step increases, the required number of iterations in the inner pseudo-time cycle grows. Fig. 7 demonstrates the impact of the Mach number on the efficiency of the implicit scheme. The results, which are based on the simulations of turbulent channel flow at $Re_\tau = 590$ and $Ma = 0.01$, show a speedup in the range of 9 to 60 compared to the explicit scheme for a reduction of the residual of two orders of magnitude. Note, that for a reduction of three orders of magnitude this speedup is lowered to a range of 6 to 26.

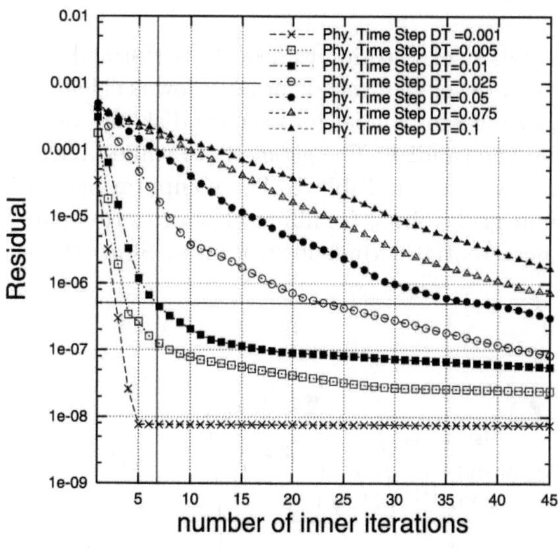

Fig. 6. Convergence of the inner iterations at $Ma = 0.01$ for channel flow computations at $Re_\tau = 590$

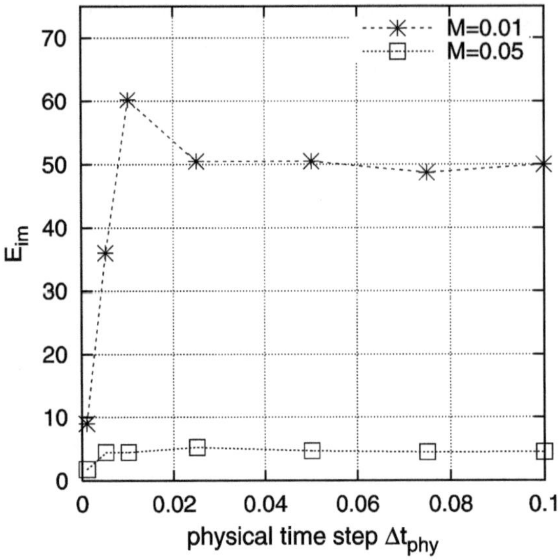

Fig. 7. Efficiency E_{im} as a function of Δt_{phy} at $Ma = 0.01$ and $Ma = 0.05$ for channel flow computations at $Re_\tau = 590$

4 Tundish Flow

4.1 Geometry and Physical Parameters

In the following the numerical setup of the tundish flow is briefly described. The geometry and the flow configuration for the simulation are shown in Fig. 1.

The numerical simulations consist of two simultaneously performed computations. On the one hand, a fully developed turbulent pipe flow is calculated to provide time-dependent inflow data for the jet into the tundish and on the other hand, the flow field within the tundish. The geometrical values and the flow parameters of the tundish are given in Table 1. The computational domain is discretized by 12 million grid points, 5 million of which are located in the jet domain to resolve the essential turbulent structures. Since the jet

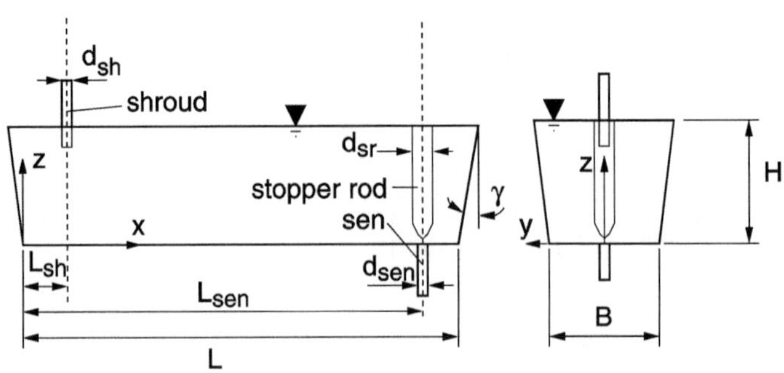

Fig. 8. Presentation of the main parameters of the tundish geometry

Table 1. Physical parameters of the tundish flow

tundish length L_1	1.847 m
tundish width B_1	0.459 m
tundish height H	0.471 m
inclination of side walls	7°
diameter of the shroud d_{sh}	0.04 m
height between bottom shroud Z_{sh}	0.352 m
diameter of the SEN d_{SEN}	0.041 m
diameter of the stopper rod d_{sr}	0.0747 m
hydraulic diameter d_{hyd}	0.6911 m
Re based on the jet diameter	25000

possesses the major impact on the flow characteristics in the tundish, it is a must to determine in great detail the interaction between the jet and the tundish flow.

The boundary conditions consist of no-slip conditions on solid walls. At the free surface, the normal velocity components and the normal derivatives of all remaining other variables are set zero. An LES of the impinging jet requires a prescription of the instantaneous flow variables at the inlet section of the jet. To determine those values a slicing technique based on a simultaneously conducted LES of a fully developed turbulent pipe flow is used (Fig. 9).

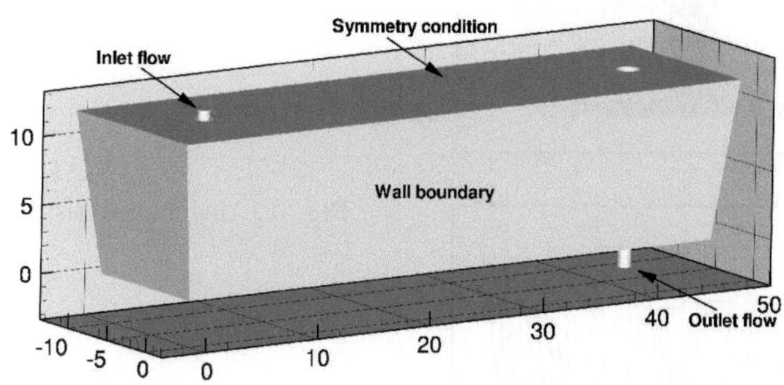

Fig. 9. Description of the integration domain and boundary conditions

4.2 Results

The typical structure of the velocity field in the tundish is shown in Figs. 10-16. These figures show the computed flow pattern in different locations within the tundish. Figs. 10 to 12 visualize the jet flow into the tundish. It is clear that the turbulent flow contains a wide range of length scales. Large eddies with a size comparable to the diameter of the pipe occur together with eddies of very small size. The figures evidence the jet spreading, jet impingement

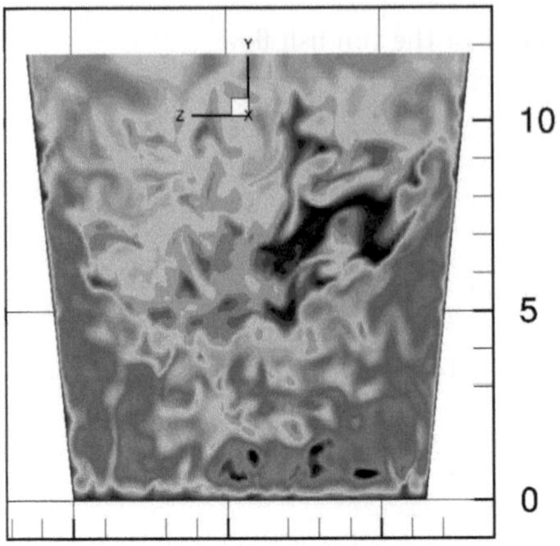

Fig. 10. Instantaneous entropy contours at $X/L_1 = 0$

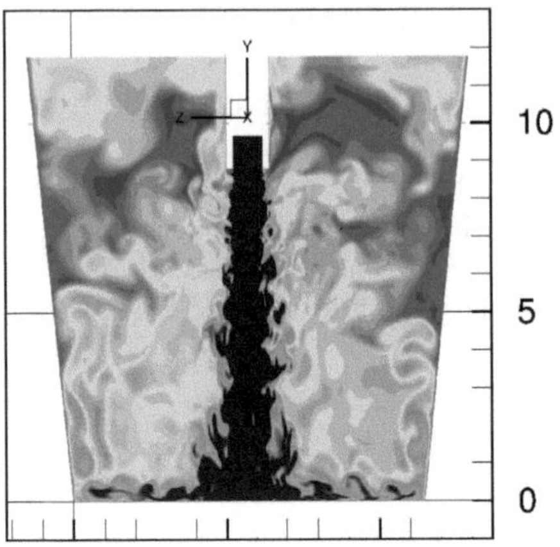

Fig. 11. Instantaneous entropy contours in the center plane of the jet

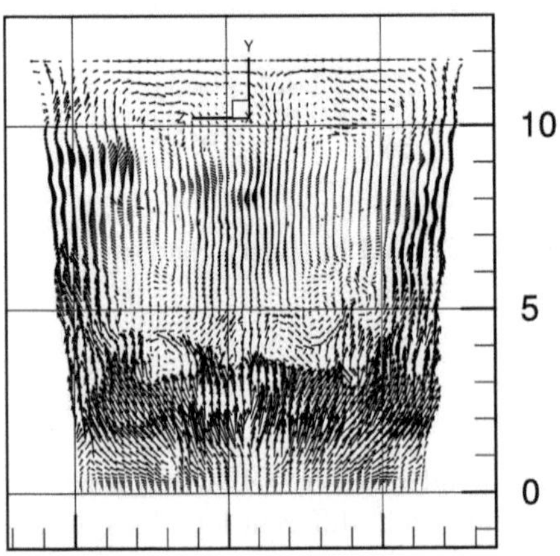

Fig. 12. Instantaneous velocity vectors at $X/L_1 = 0$

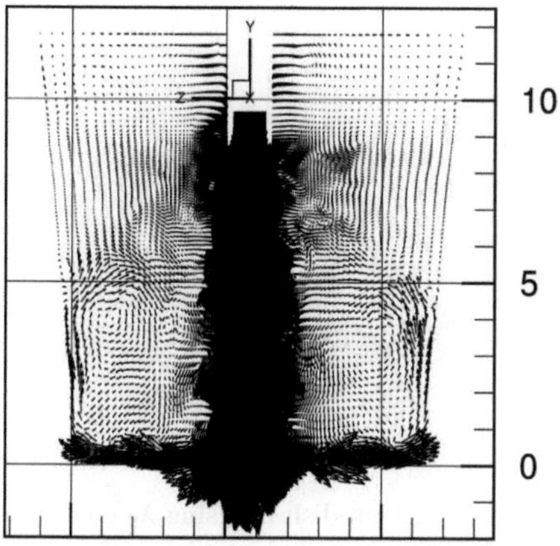

Fig. 13. Instantaneous velocity vectors in the center plane of the jet

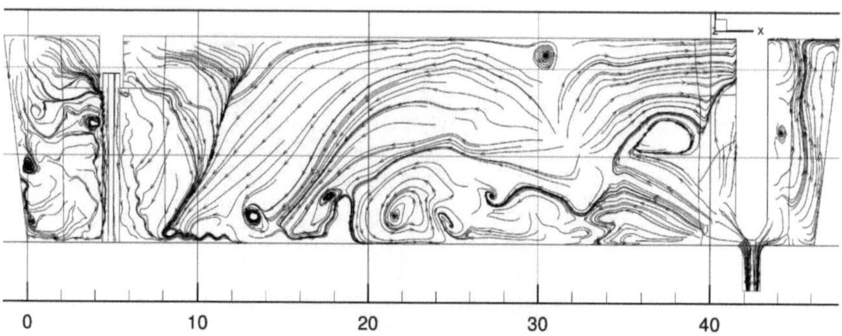

Fig. 14. Streamlines at $Z/B_1 = 0$

on the wall, and wall jets. The jet ejected from the ladle reaches the bottom of the tundish at high velocities, spreads in all directions and then mainly flows along the side walls of the tundish [1] and [2]. Such flow patterns lead to nonmetallic inclusions, which are to be avoided.

Figure 14 represents flow field structures in the center longitudinal vertical planes of the tundish. The streamlines illustrate the vortex dominated flow in the tundish. It can be seen that the flow includes strong vortices and recirculation regions mainly in the inlet region of the tundish, which is fully turbulent. In Figs. 15 and 16 the turbulent jet flow in a tundish is visualized by the λ_2 criterion.

4.3 Computational Resources

All results of the flow simulations presented were obtained on the NEC SX-8 computer of the High Performance Computing Center Stuttgart (HLRS). The vectorization rate of the flow solver is 99%, and a single processor performance of about 7 GFlops on a SX-8 processor is achieved. The memory requirement for the current simulation is around 15 GB. Approximately 250 CPU hours

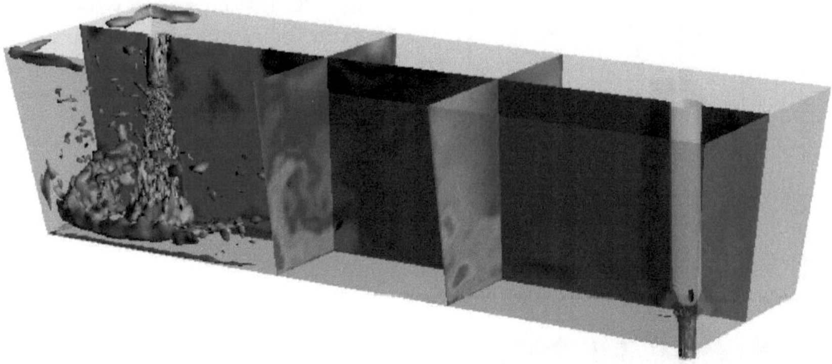

Fig. 15. Visualization of the LES of a flow field in the tundish by using λ_2 contours color coded with the local magnitude of the velocity vector

Fig. 16. λ_2 contours of the injected jet color coded with the local magnitude of the velocity vector

on 28 SX-8 CPUs for statistically converged solution for the tundish flow. Additional 150 CPU hours are required for gathering samples for statistical data. The storage requirements for the data files for the postprocessing is in the order of 500 Gigabyte. For further simulations with different tundish geometries and higher Reynolds numbers, meshes with about 25 million points will be used which are distributed on 40 processors.

5 Conclusion

An efficient large-eddy simulation method for nearly incompressible flows based on solutions of the governing equations of viscous compressible fluids has been introduced. The method uses an implicit time accurate dual time-stepping scheme in conjunction with low Mach number preconditioning and multigrid acceleration. To validate the scheme, large-eddy simulations of turbulent pipe flow at $Re_\tau = 1280$, and cylinder flow at $Re_D = 3900$ have been performed. The results show the scheme to be efficient and to improve the accuracy at low Mach number flows. Generally, the new method is 6-40 times faster than the basic explicit 5-stage Runge-Kutta scheme.

A large-eddy simulation of the flow field in a tundish is conducted to analyze the flow structure, which determines to a certain extent the steel quality. The findings evidence many intricate flow details that have not been observed before by customary RANS approaches.

References

1. N. Alkishriwi, M. Meinke, and W. Schröder. Preconditioned large eddy simulations for tundish flows. In *Proc. of turbulence and shear flow phenomena TSFP-4, Williamsburg, June 27-29*, 2005.
2. N. Alkishriwi, M. Meinke, and W. Schröder. A preconditioned LES method for nearly incompressible flows. In *Ercoftac Workshop, Direct and Large-Eddy simulation-6, Poitiers, Sept. 12-14*, 2005.
3. N. Alkishriwi, M. Meinke, and W. Schröder. A large-eddy simulation method for low mach number flows using preconditioning and multigrid. *Comp. Fluids*, 2006, in press.
4. P. Beaudan and P. Moin. Numerical experiments on the flow past a circular cylinder at sub-critical Reynolds numbers. Technical Report TF-62, Center Turb. Res., 1994.
5. F. Durst, J. Jovanović, and J. Sender. LDA measurements in the near-wall region of a turbulent pipe flow. *J. Fluid Mech.*, 295:305–335, 1995.
6. P. Gardin, M. Brunet, J. Domgin, and K. Pericleous. An experimental and numerical cfd study of turbulence in a tundish container. *Second International Conference on CFD in the Minerals and Process Industries CSIRO*, 1999, Melbourne, Australia, 6-8 December.
7. L. Lourenco and C. Shih. A particle image velocimetry study. Technical report, Data taken from Beaudan and Moin, 1993.
8. M. Meinke, W. Schr"oder, E. Krause, and T. Rister. A comparison of second- and sixth-order methods for large-eddy simulations. *Comp. Fluids*, 31:695–718, 2002.
9. H. Odenthal, R. Bölling, and H. Pfeifer. Numerical and physical simulation of tundish fluid flow phenomena. *Japan-Germany Seminar on Fundamentals of Iron and Steelmaking*, 2002, Düsseldorf.
10. L. Ong and J. Wallace. The velocity field of the turbulent very near wake of a circular cylinder. *Exp. in Fluids*, 20:441–453, 1996.
11. F. Rütten, M. Meinke, and W. Schröder. Large-eddy simulations of 90°-pipe bend flows. *Journal of Turbulence*, 2:003, 2001.
12. F. Rütten, M. Meinke, and W. Schröder. Large-eddy simulations of frequency oscillation of the dean vorticies in turbulent pipe bend flows. *phf*, 17, 2005.
13. E. Turkel. Preconditioning techniques in computational fluid dynamics. *Ann. Rev. Fluid. Mech.*, 31:385–416, 1999.

and multigrid acceleration. To validate the scheme, large eddy simulations of [illegible] pipe flow at [illegible] and cylinder flow at [illegible] have been performed. The results show the scheme to be efficient and [illegible] the [illegible] at low Mach number flows. [illegible] times [illegible] than the [illegible]

[illegible]

References

[illegible]

[illegible] and [illegible]. The [illegible] field of [illegible] [illegible], 19[illegible]

[illegible], M. [illegible] and W. [illegible]. Large eddy [illegible] of [illegible] [illegible]

[illegible], M. [illegible] and W. [illegible]. Large-eddy simulations of [illegible] [illegible] of the clean [illegible] [illegible]

[illegible] techniques in computational fluid dynamics. [illegible] [illegible]

Large Eddy Simulation of Open-Channel Flow Over Spheres

Thorsten Stoesser and Wolfgang Rodi

Institut für Hydromechanik, Universität Karlsruhe
stoesser@ifh.uka.de, rodi@ifh.uka.de

Abstract. The paper presents results of several Large Eddy Simulations (LES) of the flow in an open channel where the channel bed is roughened with one or two layers of spheres. The roughness height k, which corresponds to the sphere diameter d is 0.23 of the channel depth. The Reynolds number Re_τ, based on the average friction velocity u_τ and the channel depth h (distance from the roughness tops of the spheres to the water surface) is approximately 2820. The flow configurations were selected to correspond to recently performed laboratory experiments. Mean streamwise velocities from the LES are compared with the measured data and the distributions of the calculated turbulence intensities are evaluated by comparing them with empirical relationships for flow over rough walls suggested by Nezu [1]. The occurrence of low- and high-speed streaks is examined and their spanwise spacing is quantified. Moreover, sweeps and ejections are shown to occur as well as the amalgamation process i.e. ejection of fluid into the outer layer associated with vortex growth. It is shown that these structures occur irrespective of roughness conditions, however further studies and data analysis are needed to evaluate and quantify the effect of porosity on coherent structures.

1 Introduction

The dynamics of turbulent flow over smooth and rough walls is dominated by energetic three- dimensional organized (coherent), vortical structures. Over 4 decades of experimental work have been dedicated to the identification of the physical processes which govern these coherent structures, and much progress has been made in recent years due to the advances in measurement methods and in numerical simulation techniques associated with the growth in speed and capacity of modern supercomputers. Whereas the flow physics of coherent structures over smooth surfaces is understood fairly well (see the summary by Robinson, [2]), the flow over rough walls is still an area of active turbulence research. This is mainly due to the wide variety of possible wall roughness geometries and the fact that the details of the geometry influence the flow across the entire turbulent layer (Jimenez, [3]). The mean velocity profile over rough beds differs considerably from the profile over a smooth bed since the surface drag is significantly larger when roughness elements are present. General functional relationships between the roughness geometry and the effects of the roughness on the mean flow are still not available. There are numerous studies attacking the problem, mostly based on the concept of relating the roughness geometry to a roughness function, ΔU^+, expressing a downshift of the velocity profile in comparison to the velocity profile over a smooth bed. Further uncertainties arise in the quantification of the equivalent roughness height k_s , typically a fraction of the geometrical roughness height k, the origin of the wall normal coordinate y_0, which is usually determined empirically to optimize the fit to the logarithmic distribution, and last but not least the value of the von Karman constant κ , which may not be universal for all roughness types. The effect of roughness is of course not restricted to the mean flow properties. Flow visualizations and measurements (e.g. Grass, [4], Grass et al., [5], Defina, [6] and many others) as well as recent Direct Numerical Simulations (DNS) and Large Eddy Simulations (LES) of flow over rough-walls (Leonardi et al., [7], Stoesser and Rodi, [8]) indicate significant changes of the turbulence structure not only near the rough surface, but everywhere within the channel. The presence of organized structures near the walls, which are mainly responsible for the transport of momentum, heat and mass across the channel is, irrespective of surface condition, established from these research endeavours. The streamwise velocity field near rough walls is, similar to the velocity field over smooth walls, organized into alternating narrow streaks of high and low speed fluid that are persistent, vary only slowly, and exhibit a preferential spanwise spacing (Defina, [6]). However, Leonardi et al. [7] have shown that due to surface roughness, size and shape of the streaks change drastically as a result of enhanced momentum exchange between the near-wall region and the outer flow. Most turbulence production occurs when low-speed streaks are lifted away from the wall-layer in a violent ejection and during inrushing of high-speed fluid from the outer layer towards the wall. The complete cycle of lift-up of fluid, ejection and sweep motion makes up what is usually called the bursting phenomenon. In this paper we present the results of a LES of open-channel flow over a bed artificially roughened with one or two layers of densely packed spheres. The main purpose of this study is to provide further insight into the turbulent flow over rough boundaries and to enhance the understanding of the effect of surface roughness on the mean and instantaneous flow. Temporal and spatial averaging is used to quantify the effects on the flow velocities and the three components of turbulence intensities. Furthermore, we investigate the flow and turbulence structure near the spheres and the associated occurrence of coherent flow structures. The spanwise spacing of low speed streaks is quantified and the processes of vortex growth and amalgamation are elucidated. In here, only the statistics from the 1layer-calculations are presented, which are almost fully completed. Further work is needed to analyse the two-layer simulations. However, the purpose of this study is to provide further insight into the turbulent flow and to enhance the understanding of the effect of subsurface poreflow on the mean and instantaneous outer flow. This issue is of crucial importance for sedimentation and erosion processes in river beds, the study of which is the long-term goal of the present project.

2 Calculation Method

The LES code MGLET, originally developed at the Institute for Fluid Mechanics at the Technical University of Munich (Tremblay and Friedrich, [9]), is used to perform the Large Eddy Simulations. The code solves the filtered Navier-Stokes equations discretised with the finite-volume method on a staggered Cartesian grid. Convective and diffusive fluxes are approximated with central differences of second order accuracy and time advancement is achieved by a second order, explicit Leapfrog scheme. The Poisson equation for coupling the pressure to the velocity field is solved iteratively with the SIP method of Stone [10]. The subgrid-scale stresses appearing in the filtered Navier-Stokes equations are computed using the dynamic approach of Germano et al. [11]. The no-slip boundary condition is applied on the surface of the spheres where the immersed boundary method is employed (Verzicco et al., [12]). This method is a combination of applying body forces in order to block the cells that are fully inside the sphere and a Lagrangian interpolation scheme of third order accuracy, which is used for the cells that are intersected by the spheres' surface to maintain the no-slip condition (Tremblay and Friedrich, [9]). MGLET is fully parallelised and the geometry is split into sub-domains with an equal number of grid points. The communication between the blocks is accomplished by using MPI.

3 Flow Configuration

The setup and boundary conditions of the Large Eddy Simulation were selected in analogy to flume experiments performed by Detert [13] or Dancey et al. [14], where spheres of $d = 22mm$ diameter were placed in one or two layers on the flat bottom. A densest packing arrangement as well as a staggered arrangement was tested. The water depth h was set to $h = 94mm$ measured from the top of the spheres, which gives a roughness-height-to-depth-ratio of $d/h = 0.23$ (Fig. 1). Streamwise velocities were measured by a 1D Acoustic Doppler Current Profiler (ADCP) for Deterts experiments or with LDA in Dancey et al.'s experiments. The computational domain spanned approximately $5h$ in streamwise, $2h$ in spanwise and $1h$ in vertical directions, respectively and consisted of 16×9 spheres for the 1-layer simulations and $19 \times 9 \times 2$ for the 2-layer simulations. Several grids were tested, however in this paper we will only show the results from the very high resolution grid. As mentioned above only the single layer sphere simulations were analysed in terms of flow

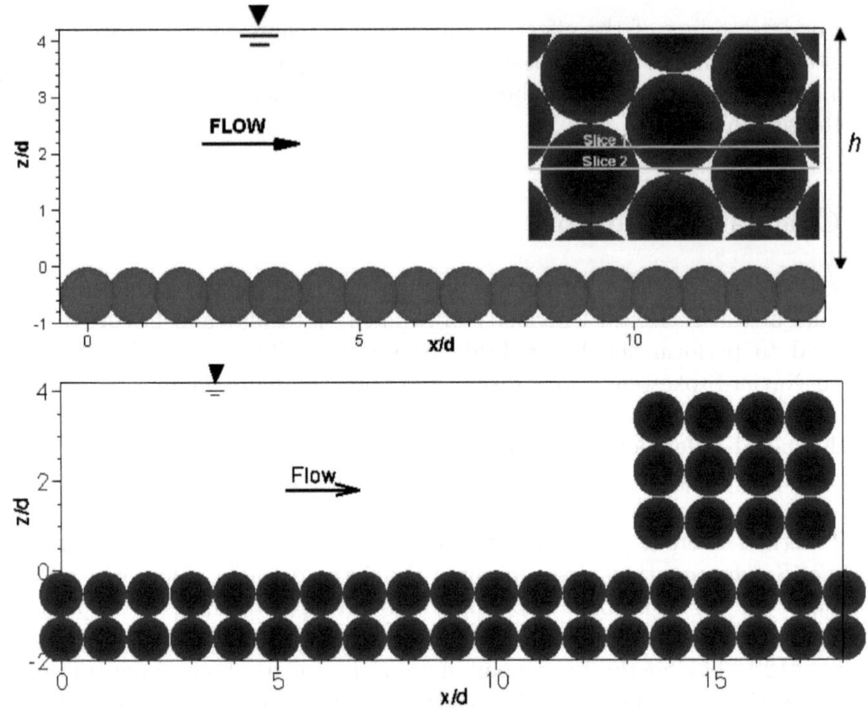

Fig. 1. Sideview of the calculation domain with a top view on the packing of the spheres for 1 layer of spheres (*top*) and 2 layers of spheres (*bottom*)

statistics and coherent structures so far This grid consists of $800 \times 320 \times 180$ points for the whole computation domain and $42 \times 36 \times 100$ points around one sphere was employed, which is approximately 46 million points in total. The grid spacings in terms of wall units were $\Delta x^+ = 5$ in streamwise direction and $\Delta y^+ = 7$ in spanwise direction. In the vertical direction the grid spacing was kept at a constant value of $\Delta z^+ = 2.5$ from the bed to the top of the spheres and was stretched above the spheres towards the surface. A part of the grid, where only every 5th grid line is plotted, and the details of the grid around one sphere are shown in Fig. 2. Periodic boundary conditions were applied in the streamwise and spanwise directions. A constant pressure gradient was maintained during the computation which yielded an average bulk velocity of $u_{bulk} = 0.8m/s$ and an average shear velocity of $u_\tau = 0.07m/s$. This gave a Reynolds number based on the channel depth h and bulk velocity u_{bulk} of $Re_h = 40,000$, a Reynolds number based on u_τ and h of $Re_\tau = 2800$ and a roughness Reynolds number based on u_τ and d of $Re_* = 650$. Assuming a similar sphere-diameter-to-roughness-height-ratio as found by Grass [4] or Defina [6], this corresponds to a dimensionless effective roughness height of $k_s^+ = 420$.

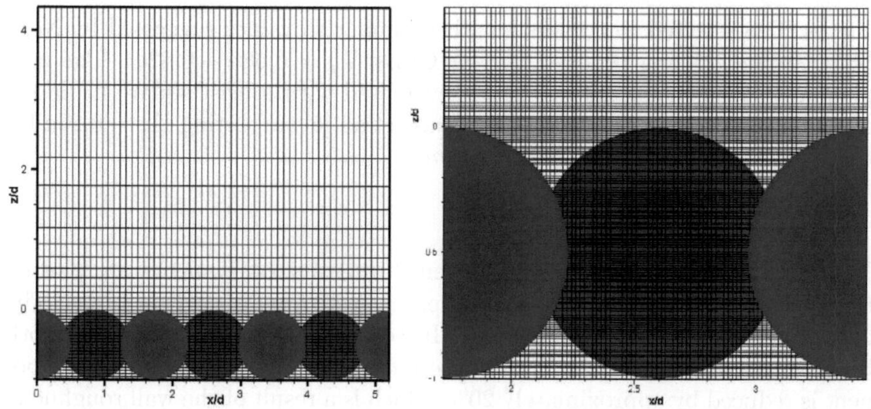

Fig. 2. Part of the computational grid (*left*) and details of the grid around one sphere

4 Preliminary Results

4.1 Mean Velocities

Figure 3 compares the time-averaged streamwise velocity profile in wall coordinates as obtained by the Large Eddy Simulations with the experiments of Detert [13]. Whereas the LES represents spatially averaged quantities, the data collected with the ADCP represent an average over the measurement volume being approximately half the size of a sphere. The overall prediction of the mean streamwise velocity is in very good agreement with the measured data. Also plotted is the log-law for rough walls i.e., where a roughness length $k_s = 0.55d$ is taken in order to get the best fit to the measured and computed

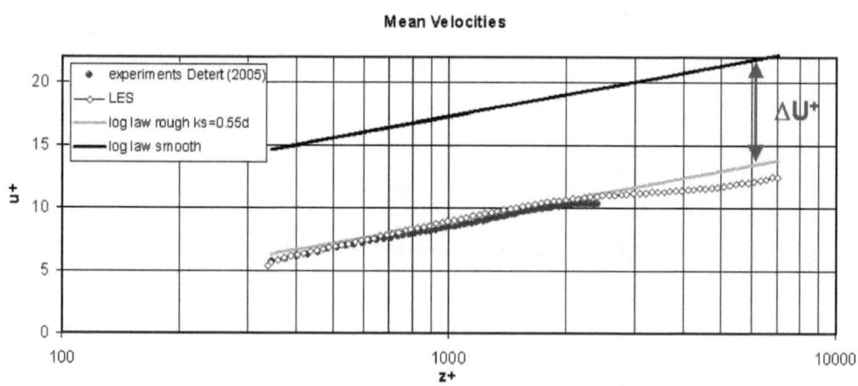

Fig. 3. Comparison of time averaged streamwiese velocities to experiments and to the log law

curves. The selected roughness length is similar to values suggested by Nezu ($k_s = 0.57d$), Defina ($k_s = 0.67d$) or Grass ($k_s = 0.68d - 0.82d$). The zero plane displacement of $\Delta z = 0.22d$ is evaluated with a least square fit and is of similar magnitude as reported by Grass et al. [5] suggesting $\Delta z = 0.22d$ for $k = 12mm$ and $\Delta z = 0.2d$ for $k = 6mm$.

4.2 Turbulence Intensities

The distributions of the three turbulence intensity components are plotted together with the empirical relationships suggested by Nezu [1] in Fig. 4. The overall agreement is fairly satisfying. In contrast to the flow over a smooth bed, the maximum value of turbulence intensities of the streamwise component is reduced by approximately 20%, which is a result of the wall roughness. Whereas most of the turbulence production over a smooth bed occurs in the buffer layer, this buffer layer disappears for the rough walls and consequently the turbulence is produced in another way. Our computed values are in excellent agreement with the measurements of Nezu [1] and Grass [4], who observed maximum turbulence intensities of around 2.1 for $u'/u*$ of 1.5 for $v'/u*$ and 1.1 for $w'/u*$ varying only slightly with the dimensionless roughness height k_s^+.

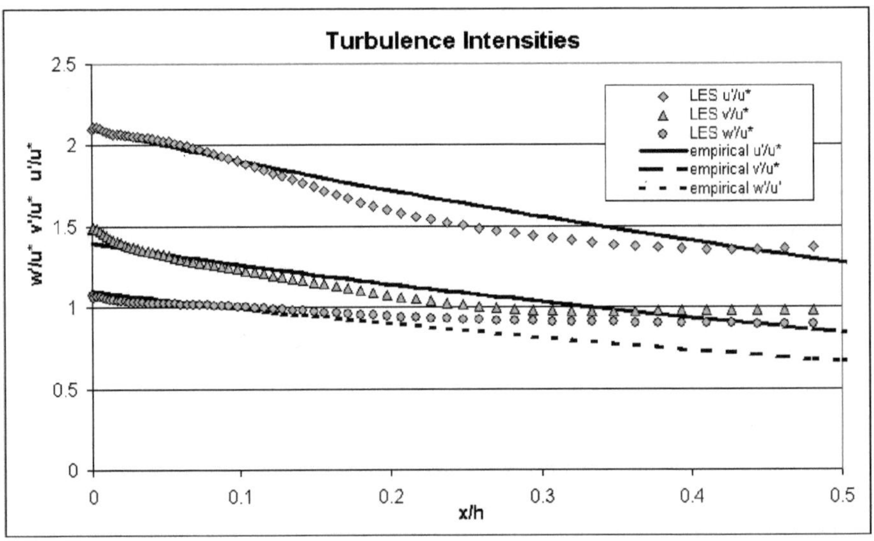

Fig. 4. Comparison of turbulence intensities to empirical relationships of Nezu (1977)

4.3 Instantaneous Flow Field

Figure 5 presents an instantaneous distribution of the streamwise velocity fluctuation together with the fluctuating velocity vector ($u' - w'$) in two se-

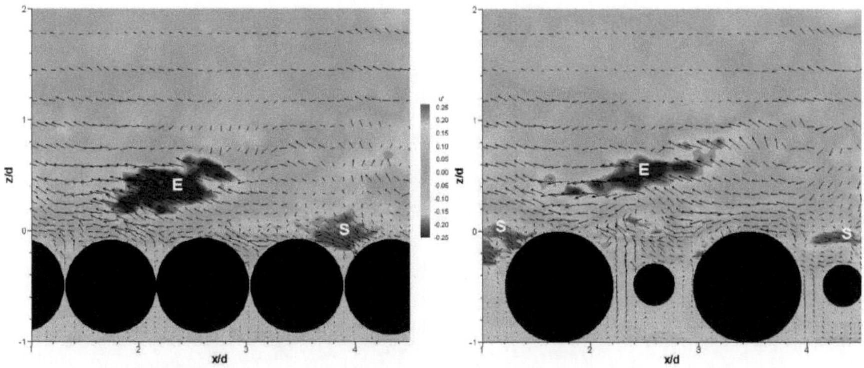

Fig. 5. Instantaneous streamwise velocity fluctuations and velocity perturbation vectors in two selected longitudinal slices as indictated in Fig. 1 – Slice 1 *left*, Slice 2 *right*.

lected x-z slices. It illustrates the presence of vortical motion, especially near the elements. Above the crest of the spheres, ejection events, where slower fluid is expelled away from the wall, and sweep events where slower fluid is pushed towards the elements, can be detected. Figure 5 further illustrates the exchange processes between the free surface flow and the interstitial region; it can be seen clearly that fluid plunges into the interstitial and moves back into the surface region at different locations.

Streaky structures are present just above the roughness elements as was shown e.g. by Defina [6] in his laboratory experiment with the help of dye. Figure 6 below shows the distribution of the instantaneous streamwise velocity fluctuation. The presence of high and low speed streaks alternating in the spanwise direction is visible. As was pointed out by several researchers in the past the streak spacing should scale with the roughness height k_s. Grass [5] used spheres with $k = 6mm$ and $k = 12mm$ and found streak spacing to roughness height ratios of $\lambda/k_s = 3.82$ and $\lambda/k_s = 3.87$. However, Defina [6] reported streak spacing to roughness height ratios just above the roughness elements of $\lambda/k_s = 4.0$ being constant irrespective of Reynolds number. He further found that the spacing increases linearly with distance from the bed to ratios up to 20. In the present LES the average streak spacing ratio is $\lambda/k_s = 3.9$ at a small distance (20 wall units) from the spheres. This corresponds very well with the findings of Defina and Grass. Figure 7 presents an isosurface of instantaneous pressure fluctuations colour coded with the instantaneous streamwise velocity. The coherence of the flow structures and the vortex growth from the bed towards the surface can be seen clearly. This mechanism is known as vortex stretching by the local velocity gradient which provides the mechanism for the energy transfer from the free stream to the near wall region (originally suggested by Theodorsen). Figure 7 furthermore indicates the amalgamation process in which wall-region vortices are ejected

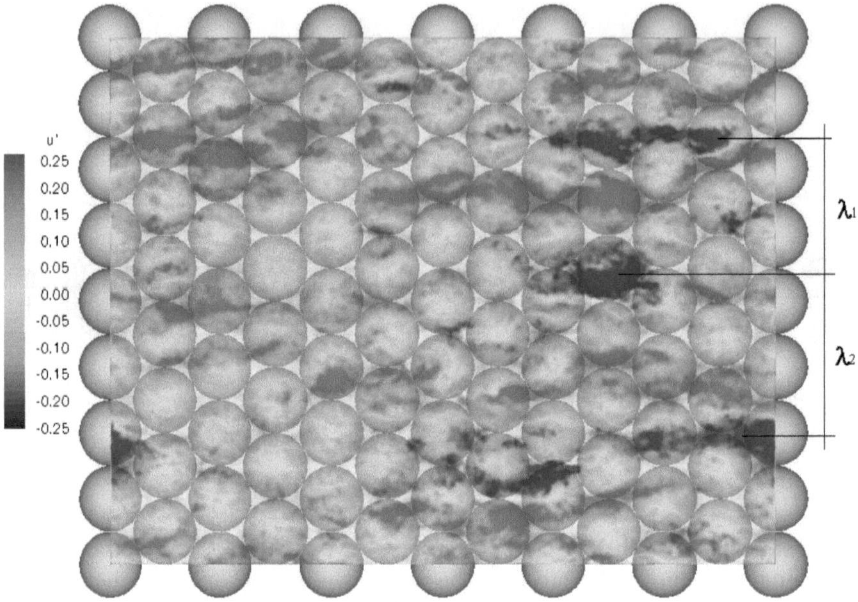

Fig. 6. Instantaneous streamwise velocity fluctuations in a wall-parallel plane just above the spheres

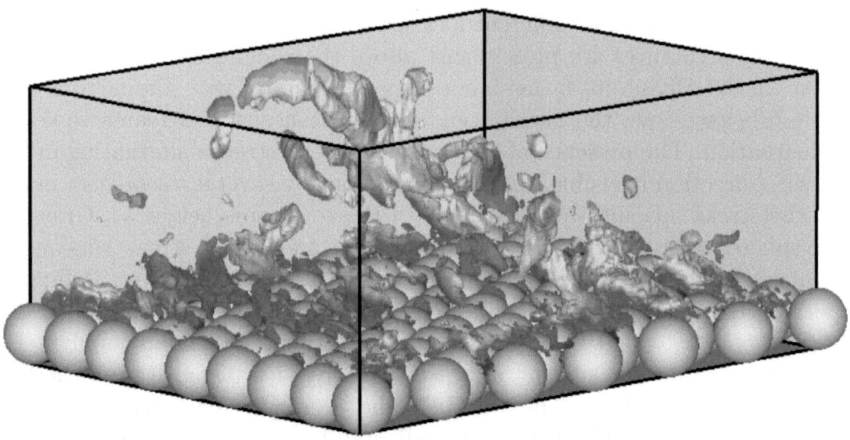

Fig. 7. Isosurfaces of instantaneous pressure fluctuations color-coded with the instantaneous streamwise velocity

and interact with the higher-speed outer region fluid during their growth and movement towards the surface. Figure 8 shows similar structures for the case of two layers of spheres; as before local velocity gradients are responsible for shearing the fluid particles transferring energy.

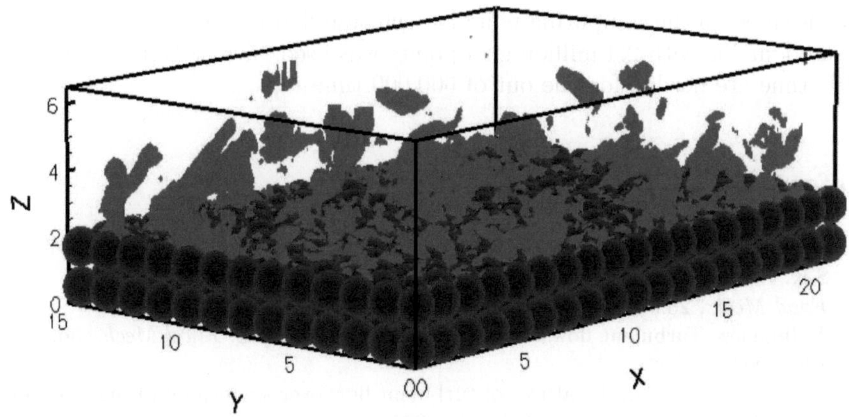

Fig. 8. Isosurfaces of instantaneous pressure fluctuations for the two-layer simulations

5 Conclusions

The paper has presented the results of a Large Eddy Simulation of open channel flow over an artificially roughened bed. The calculated mean velocities showed excellent agreement with the measured data of Detert [13] and conformity with the log law for rough walls. The predicted turbulence intensities were compared with the empirical relationships suggested by Nezu [1]. All three components found to agree well with these relationships and the data obtained by other researchers. The occurrence of low- and high-speed streaks was examined and their spanwise spacing is of similar magnitude as reported previously. Moreover, sweeps and ejections were shown to occur as well as the amalgamation process associated with vortex growth.

6 Computer Resources Used

The computations were performed on the HP-XC cluster of the SSC Karlsruhe. Various configurations were investigated, and several computations were performed. Overall the scaling of the code MGLET is very good and the overhead time used for communication between the different subdomains was in the order of less than 0.1% of the total computing time. A typical simulation of 1-layer of spheres was composed of 8 subdomains running on 8 processors using 4 "thin" nodes on the HP. Depending on the total number of gridpoints per CPU 7-14 seconds per time step, which typically consists of 80-120 inner iterations for the computation of the pressure equation. Typically around 600,000 time steps are needed for converged flow statistics. A typical

2-layer simulation used almost 3-times as many points as a 1-layer simulation because most of the computing points are clustered around the spheres. Hence 24 subdomains with 2.4 million grid points was used. Typically 11 seconds of CPU time are needed for one out of 600,000 timesteps.

References

1. I. Nezu. Turbulence intensities in open-channel flows. *Proc JSCE*, 261, 1977.
2. S.K. Robinson. Coherent motions in the turbulent boundary layer. *Ann. Rev. Fluid Mech.*, 23:601–639, 1991.
3. J. Jimenez. Turbulent flows over rough walls. *Annu. Rev. Fluid Mech.*, 36:173–196, 2004.
4. A.J. Grass. Structural features of turbulent flow over smooth and rough boundaries. *J. Fluid Mech.*, 50(2):233–255, 1971.
5. A.J. Grass, R.J. Stuart, and M. Mansour-Tehrani. Vortical structures and coherent motion in turbulent flow over smooth and rough boundaries. *Philosophical Transactions Royal Society of London A.*, 336:35–65, 1991.
6. A. Defina. Transverse spacing of low-speed streaks in a channel flow over a rough bed. In et al. Ashworth, P. J., editor, *Coherent Flow Structures in Open Channels*. Wiley, New York, 1996.
7. S. Leonardi, P. Orlandi, L. Djenidi, and A. Antonia. Structure of turbulent channel flow with square bars on one wall. In *Proceedings TSFP-3. Sendai, Japan, 24-27 June 2003*, 2003.
8. T. Stoesser and W. Rodi. Les of bar and rod roughened channel flow. In *The 6th Int. Conf. on Hydroscience and Engineering (ICHE-2004),Brisbane, Australia*, 2004.
9. F. Tremblay and R. Friedrich. An algorithm to treat flows bounded by arbitrarily shaped surfaces with cartesian meshes. In *Notes on Numerical Fluid Mechanics, Vol. 77*. Springer, 2001.
10. H.L. Stone. Iterative solution of implicit approximations of multidimensional partial differential equations for finite difference methods. *SIAM J. Numer. Anal.*, 5:530–558, 1968.
11. M. Germano, U. Piomelli, P. Moin, and W.H. Cabot. A dynamic subgrid-scale eddy viscosity model. *Physics of Fluids*, 3(7):1760–1765, 1991.
12. R. Verzicco, J. Mohd-Yusof, P. Orlandi, and D. Haworth. Large eddy simulation in complex geometric configurations using boundary body forces. *AIAA J.*, 38 (3):427–433, 2000.
13. M. Detert. Personal communication, 2005.
14. C.L. Dancey, M. Balakrishnan, P. Diplas, and A.N. Papanicolaou. The spatial inhomogeneity of turbulence above a fully rough packed bed in open channel flow. *Experiments in Fluids*, pages 402–410, 2000.

Prediction of the Resonance Characteristics of Combustion Chambers on the Basis of Large-Eddy Simulation

Franco Magagnato[1], Balázs Pritz[2], Horts Büchner[3], and Martin Gabi[4]

[1] Department of Fluid Machinery, University of Karlsruhe, 76128 Karlsruhe, Germany
magagnato@ism.uka.de
[2] Department of Fluid Machinery, University of Karlsruhe, 76128 Karlsruhe, Germany
pritz@ism.uka.de
[3] Engler-Bunte-Institute / Division for Combustion Technology, University of Karlsruhe, 76128 Karlsruhe, Germany
Horst.Buechner@vbt.uni-karlsruhe.de
[4] Department of Fluid Machinery, University of Karlsruhe, 76128 Karlsruhe, Germany
gabi@ism.uka.de

Abstract. Self-excited (thermo-acoustic) oscillations often occur in combustion systems due to the combustion instabilities. The high pressure oscillations can lead to higher emissions and structural damage of the chamber. For the disposal of the undesirable oscillations one must clearly know the mechanism of the feedback of periodic perturbations in the combustion system. In the last years intensive experimental investigations were performed at the University of Karlsruhe to develop an analytical model for the Helmholtz resonator-type combustion system. In order to understand better the flow effects in the chamber and to localize the dissipation large-eddy simulations (LES) were carried out. In this paper the results of the LES are presented, which show good agreement with the experiments. The comparison of the LES study with the experiments sheds light on the significant role of the wall roughness in the exhaust gas pipe.

1 Introduction

It is an indispensable prerequisite for the successful implementation of advanced combustion concepts to avoid periodic combustion instabilities in combustion chambers of turbines and in industrial combustors [1, 2]. For the disposal of the undesirable oscillations one must clearly know the mechanics of feedback of periodic perturbations in the combustion system. If the transfer characteristics of the subsystems (in the easiest case burner, flame, chamber) are known, one can evaluate the oscillation disposition of the combustion system during the design phase.

In the present work the resonance characteristics of a Helmholtz resonator-type combustion chamber were investigated using LES. The flow in the resonator is non-ignited (without reaction) and was excited with a sinusoidal mass flow rate. For this case the classical Helmholtz theory can be used to predict the resonance frequency and transfer characteristics. Although, a good agreement of the predicted resonance frequency and the measurements is obtained, the major drawback of this model treating the system undamped, thus the transmission magnitude becomes infinity at the resonance frequency. However, there is in real systems with small damping a finite limit of the amplitude. The mean goal of that work was the validation of an analytical model developed at the Engler-Bunte-Institute [3]. The model describes the damped resonant behavior of a Helmholtz resonator-type combustion chamber, dependent on the geometry and on the operation conditions. More about the model can be found in the reference [3].

For the validation of the model a large series of measurements were carried out with variation of the parameters of the geometry (volume of the chamber, length and diameter of the exhaust gas pipe) and of the operation conditions (fluid temperature, mean mass flow rate, amplitude and frequency of the excitation) [4, 5, 6]. An attempt was taken to determine the damping factor D by means of unsteady Reynolds averaged Navier–Stokes simulation (URANS), but the results were not satisfactory [7]. In order to understand better the flow effects in the combustor and to localize the main dissipation large eddy simulation (LES) was chosen. LES emerged as a viable and powerful tool in unsteady turbulent flow simulations at Reynolds numbers of engineering interest. Further reason of the numerical investigation will be the reduction of costs at determination the damping factor of an arbitrary combustion system.

2 Simulated Configuration

In order to compute the resonance characteristics of the system a series of LES at discrete frequencies had to be completed. These were taken for a basic configuration corresponding to the experiments, for variation of the geometry of the resonator neck and for variation of the fluid temperature (Table 1). The compressible flow in the chamber of the basic geometry was simulated for five different frequencies in the vicinity of the resonance frequency (case II). Three further LES were taken for the variation of the geometry (case III) and other three LES runs for the variation of the fluid temperature (case IV). Furthermore, one LES was performed without excitation to test the boundary conditions. In this case the computed velocity profiles were verified with measurements at the vicinity of the nozzle in eight cross sections.

In each case the mean mass flow rate was $m = 61.2$ kg/h, the diameter and length of the chamber were $d_{cc} = 0.3$ m and $l_{cc} = 0.5$ m, respectively. The rate of pulsation was 25% for the cases II-IV.

Table 1. Parameter set of the simulations

Case	T (K)	l_{egp} (m)	f_{ex} (Hz)
I	298	0.2	0
II	298	0.2	37, 39, 40, 41, 42
III	298	0.1	48, 51, 54
IV	600	0.2	48, 54, 60

2.1 Numerical Method

The simulations were carried out with the in-house developed parallel flow solver called SPARC (Structured PArallel Research Code) [8]. The code is based on 3D block structured finite volume method and parallelized with the message passing interface (MPI). In the case of the combustor the compressible Navier–Stokes equations are solved. The spatial discretization is a second-order accurate central difference formulation. The temporal integration is carried out with a second-order accurate implicit dual-time stepping scheme. For the inner iterations the 5-stage Runge–Kutta scheme was used. The time step was $\Delta t = 2 * 10^{-5}$ s. The Smagorinsky–Lilly model is chosen as subgrid-scale (SGS) model [9, 10]. By means of the full multigrid method it was possible to study the effect of the grid refinement.

2.2 Boundary Conditions

If one wants to simulate the flow only in the volume of the combustion chamber and the resonator neck (grey area in Fig. 1), there are difficulties during the definition of the boundary conditions.

At the inflow boundary the fluctuation components must be prescribed for a LES. Furthermore, the boundary should not produce unphysical reflections, if the pressure fluctuations, which move in the chamber back and forth, go through the inlet. A conventional boundary condition can reflect up to 60% of the incident waves back into the flow area. One can avoid these reflections only by application of a non-reflecting boundary condition. If the inlet would be set at the boundary of the grey area, one should solve these problems. In the experimental investigation a nozzle was used at the inflow into the chamber. The pressure drop of the nozzle ensures that the gas volume in the test rig components upstream of the combustion chamber does not affect the oscillation response of the resonator. We decided to use this nozzle in

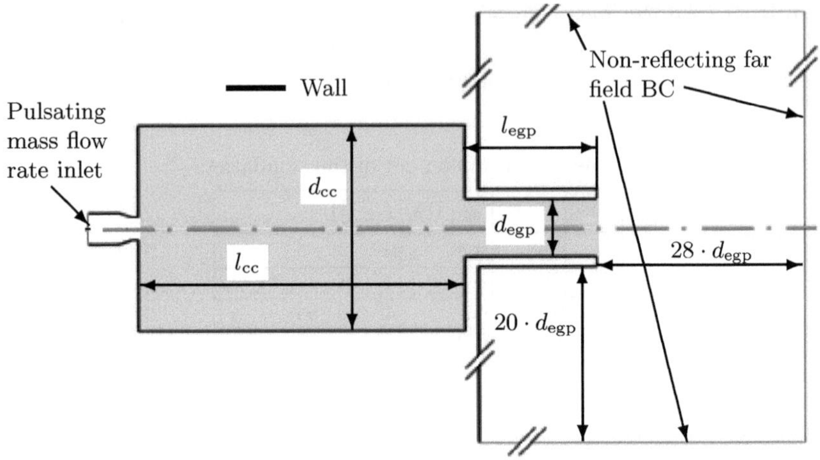

Fig. 1. Sketch of the computational domain and boundary conditions

the computation also. Although the additional volume of the nozzle increases the number of computing cells, a non-reflecting boundary condition is no more necessary. In addition, the fluctuation components at the inlet can be neglected, since the nozzle decreases strongly the turbulence level downstream.

The inflow condition prescribes the pulsating mass flow rate based on the velocity profile measured by the experiment. The pulsation was achieved by scaling the velocity profile in time.

The definition of the outflow conditions at the end of the exhaust pipe is particularly difficult. The resolved eddies produce occasionally local backflow in this cross section. In particular by excitation frequencies in the proximity of the resonant frequency there is a temporal backflow through the whole cross section, which has been observed by the experimental investigations as well. The change of the direction of flow changes the mathematical character of the set of equations. For compressible subsonic flow four boundary values must be given at the inlet, and one must be extrapolated from the flow area. At the outlet one must give one boundary value and extrapolate four others. Since these values are a function of the space and time, their determination from the measurement is impossible. Further the reflection of the waves must be avoided also at the outlet. For these reasons the outflow boundary is set not at the end of the exhaust gas pipe, but in the far field. In order to damp the waves in direction to the outlet boundary, a buffer layer with mesh stretching is used.

At the surfaces the no-slip boundary condition and an adiabatic wall are imposed. For the first grid point $y^+ < 1$ is obtained, the turbulence effect of the wall is modeled with the van Driest type damping function.

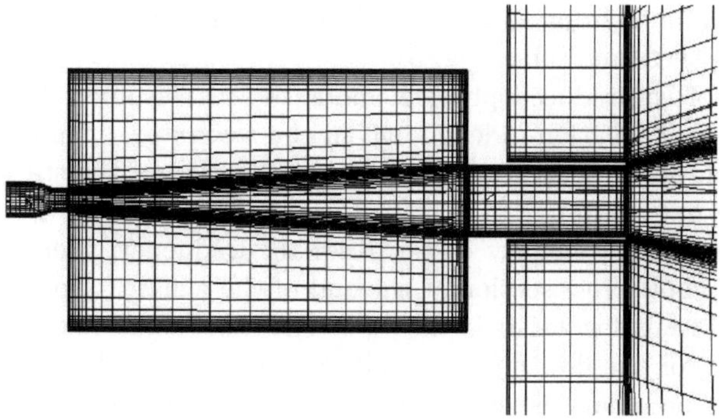

Fig. 2. Third finest mesh extracted to the symmetric plane (distortions were caused by the extracion)

The geometry of computational domain and the boundary conditions are shown in Fig. 1. The entire computational domain contains about 4.3×10^6 grid points in 111 blocks. A coarsened mesh is shown in Fig. 2.

3 Results and Discussion

3.1 Flow Features

The flow in the resonator is highly turbulent. The Reynolds number based on the mean bulk velocity is about 37000 at the nozzle and about 12000 in the exhaust pipe. The shear flow downstream of the nozzle and the edges of the resonator neck (separation bubbles) are the major vortex shedding zones (Fig. 3).

The compressibility plays a huge role at the resonance frequency in particular. The gas volume in the chamber acts as a spring and so it is enabled

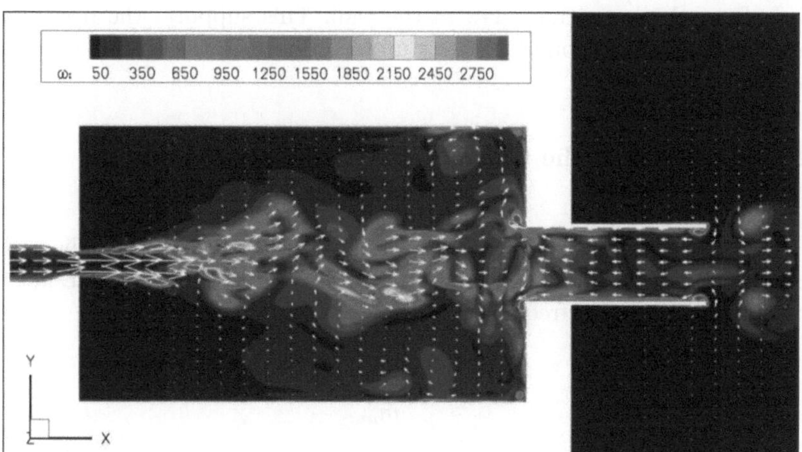

Fig. 3. Instantaneous vortex structures and velocity vectors in the symmetric plane

that gas flows in the chamber from nozzle as well as through the resonator neck in the resonator at the same time (Fig. 3).

By the formulation of the reduced physical model it was assumed, that the pressure is in phase and location-independent in the resonator. This was proven by the simulation on the basis of animated visualization of the static pressure changes during one period.

As mentioned above, in the vicinity of the resonant frequency, there is a backflow through the whole cross section at the end of the exhaust pipe. In this case a strongly oscillating flow exists in the pipe. The exact solution of the Navier–Stokes equations for the flow near an oscillating flat plate (2nd problem of Stokes) gives the boundary layer thickness for the laminar case [11]. This can serve for our turbulent flow as a rough estimation. The analytical solution predicts a boundary layer thickness of $\delta = 1.7$ mm. Despite the relatively coarse resolution (about 20 points covered in the boundary layer), the calculation with about $\delta = 2$ mm shows a good approximation. For oscillating flow in channels or in pipes the exact solution of the Navier–Stokes equations can be derived [11]. This shows that the maximum value of speed does not coincide with the axis of the pipe, but occurs near the wall. This so called annular effect was confirmed also by measurements [12, 13]. Since there is a nonzero mean flow rate in the case of the combustion chamber, this effect can be observed especially during the back flow phase. The results of the simulation agree quite well with the analytical solution.

The measurement of oscillating flows shows, that the logarithmic region of the mean velocity profile decay in about 20% of an oscillation period at an excitation of $f_{\text{ex}} = 0.1$ Hz [14]. The surveys of Tsuji and Morikawa [15] also show that when the free stream is strongly accelerated and decelerated, departures from the universal logarithmic law occur. The velocity profile in the exhaust pipe was examined at different places and for different phases. The results show that the velocity profile of the oscillatory boundary layer deviates clearly from the logarithmic law of the wall. This supports the necessity of fine resolution of the boundary layer up to the wall in the case of oscillating flow.

3.2 Comparison of the Calculations with Experiment and Analytical Model

In Fig. 4 the frequency responses of the basic configuration is indicated. It shows the variation of the amplitude ratio of the mass flow rates (Eq.(1)) in terms of the excitation frequency.

$$A = \frac{\hat{\dot{m}}_{in}}{\hat{\dot{m}}_{out}} \tag{1}$$

Experimental data and analytical model of two cases are presented here. In one case the exhaust pipe was manufactured from a turned steel tube, in the

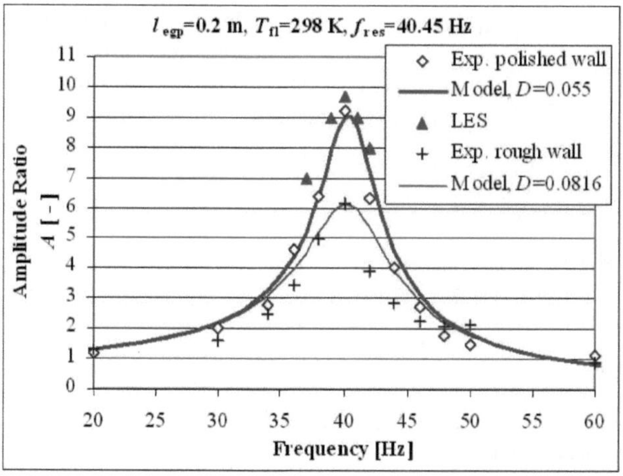

Fig. 4. Frequency response curve of the basic configuration

other case the tube was polished. The LES data compare more favorable with the experimental data of polished tube, because the wall in the simulation is aerodynamically smooth, just like the polished resonator neck. The computation shows the damping factor quite well, the deviation is about 7%. If one compares the results of measurement of the turned steel tube with the simulation, the deviation is about 40%. Earlier measurements, which were also accomplished with a pipe manufactured from PVC and with a steel pipe, showed such an influence of the wall roughness already [3].

In the basic case (case II) the mass flow rate was measured with hot wire at the inlet cross section of the chamber (excitation signal) and at the end of the exhaust gas pipe (response signal). By the experiments of the case with shorter exhaust pipe and of the case with higher fluid temperature the pressure signal in the resonator was measured as response signal. This alternating method of response signal measurement is derived in [3] and confirmed by experiments. The results of pressure measurement were converted on the basis of the equation in [3] into mass flow rate amplitudes, in order to be able to compare the results with the basic configuration.

In the case with shorter resonator neck the resonant frequency was predicted well, however, at this frequency the computed damping ratio deviates from the experiment about 15% (Fig. 5).

For the case of higher fluid temperature there is some difficulty to evaluate the system response from the computation with the mass flow rate measured in the proximity of the resonant frequency. During the backflow cold air comes in sideways from the environment (Fig. 6). Because of the cold air with higher density a larger amplitude of the mass flow rate develops at the end of the resonator neck. In the experiment the mass flow rate was measured with hot wire in the centre of the cross section at end of the exhaust pipe and a cylindrical velocity distribution was assumed. In this point the gas is always hot. This method of mass flow rate measurement agrees well with the pressure

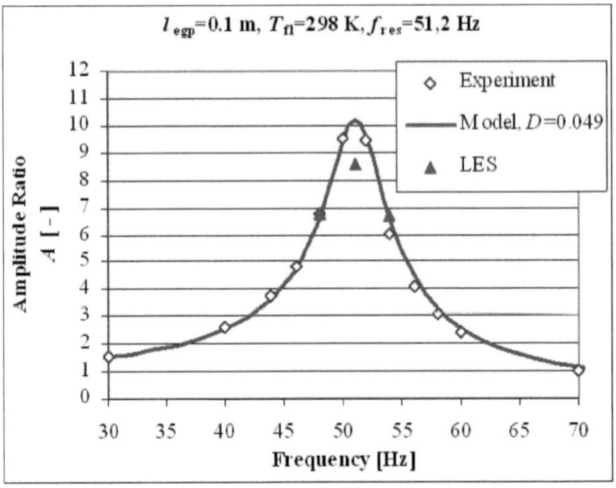

Fig. 5. Frequency response curve of the geometry variation

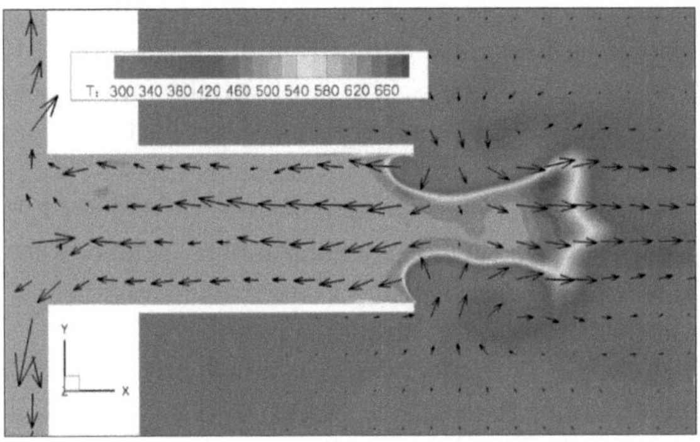

Fig. 6. Instatntaneous temperture distribution and velocity vectors in the symmetric plane in the one-third part of the backflow

measurement. In the computation the mass flow rate was established by integration over the whole cross section including the cold air backflow area also. The amplitude of the mass flow rate oscillation was unexpected large.

Therefore, the pressure signal was also stored during the simulation at the centre of the side wall of the combustion chamber, at the same place, where it was measured in the experiment. The amplitudes of the normalized pressure oscillation were converted into mass flow oscillation and compared with the mass flow oscillations determined from the simulation also. In the proximity of the resonant frequency the difference amounts to about 30%. With even larger distance from the resonant frequency this difference decreases gradually. In Fig. 7 the mass flow fluctuations determined from the pressure fluctuations are presented.

The temperature variation was accomplished in the experiment with the turned exhaust pipe, which possesses a large wall roughness. As mentioned before, the wall, however, is aerodynamically smooth in the simulation. The deviation between the measured and the simulated damping factor amounts to about 40%. This corresponds to the result of the basic configuration, where

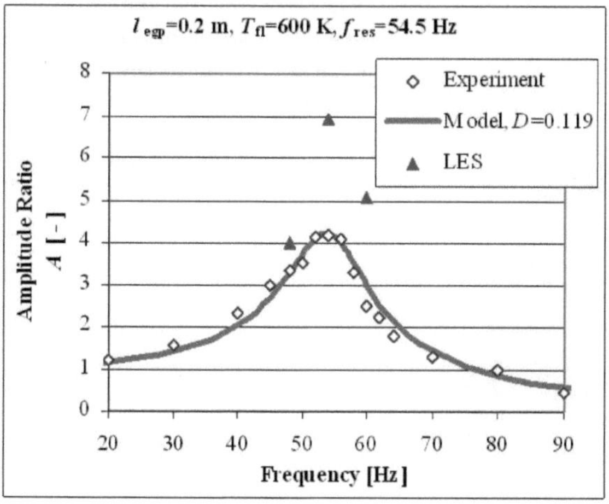

Fig. 7. Frequency response curve for variation of the fluid temperature

the values from the simulation were compared with the measurement of the turned tube.

Figures 8 and 9 demonstrate the iso-surface of the dissipation function of different flows at $\phi = 3000\ \mathrm{W/m^3}$.

Figure 8 shows the dissipation of the mean flow without pulsation and Fig. 9 the dissipation of the pulsating flow (excitation is near to the resonant frequency), which was averaged over one period. Here one can see that the place of the oscillation damping dissipation is mainly the oscillatory boundary layer in the resonator neck. Beyond that the separation bubbles at the ends of the exhaust pipe play an important role. A further part of the dissipation takes the shear layer of the jet, which develops at the inflow into the chamber. This shear layer is responsible for the major part of the dissipation for the flow without pulsation.

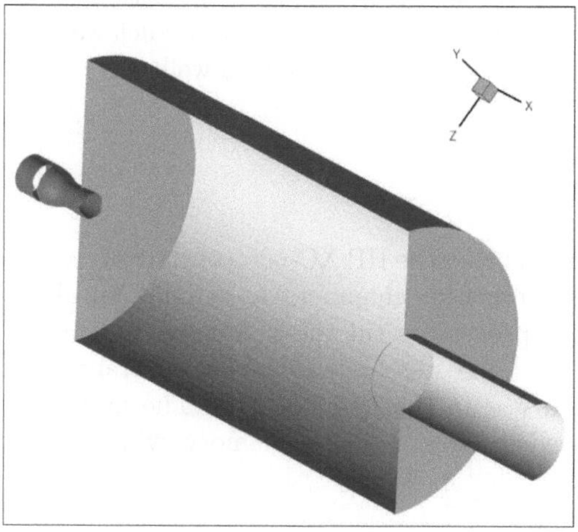

Fig. 8. Dissipation of the flow without excitation

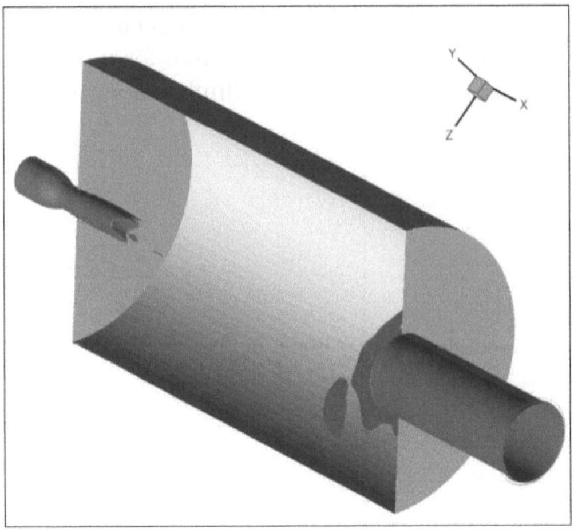

Fig. 9. Over one period averaged dissipation of the flow with excitation near to the resonant frequency

4 Conclusions

In this paper the resonance characteristics of a Helmholtz resonator-type combustion chamber were examined using large-eddy simulation. The results are in good agreement with the experimental data and with an analytical model. The simulations confirm the basic assumptions made by the reduced physical model, which was developed to describe the resonant behavior of a damped Helmholtz resonator-type combustion chamber. It was assumed, that the pressure is location-independent in the chamber and the main place of the dissipation is the oscillatory boundary layer in the exhaust gas pipe. The effects of the oscillating flow are good predicted by the simulations compared with literature. The comparison of the LES results with the measurements sheds light on the role of the wall roughness in the resonator neck, which affects significantly the damping factor D of the resonator system.

The next goal is to investigate the coupled system of the combustion chamber and a premixed swirl burner. It is believed that LES can predict well the resonance characteristics of the coupled resonator system as well.

5 Computational Efficiency

The calculations have been performed using the HP XC-6000 of the SSC Karlsruhe. A typical calculation of the combustion chamber was running for about 300 hours (elapsed time) using 32 processors till the maximum magnitude of the pulsation in the exhaust gas pipe was approximately constant. The parallel efficiency with this system was in the order of 95%. The numerical simulations on the finest mesh needed about 6 GB of memory, while during an averaging process about 9 GB.

Nomenclature

Roman Symbols	
A	amplitude ratio
d	diameter (m)
D	damping factor
f	frequency (Hz)
l	length (m)
$\dot{m}$	mass flow rate (kg/h)
T	temperature (K)
y^+	dimensionless distance from the wall
Δt	time step (s)
Greek Symbols	
δ	boundary layer thickness
ω	vorticity
ϕ	dissipation function (W/m^3)
Subscript	
cc	combustion chamber
egp	exhaust gas pipe
ex	excitation
in	inflow
out	outflow
res	resonance

References

[1] Külsheimer, C., Büchner, H., Leuckel, W., Bockhorn, H., Hoffmann, S.: Untersuchung der Entstehungsmechanismen für das Auftreten periodischer Druck-/Flammenschwingungen in hochturbulenten Verbrennungssystemen. VDI-Berichte 1492, pp.463 (1999)

[2] Büchner, H., Bockhorn, H., Hoffmann, S.: Aerodynamic Suppression of Combustion-Driven Pressure Oscillations in Technical Premixed Combustors. Proceedings of Symposium on Energy Engineering in the 21st Century (SEE 2000), **4**, pp.1573 (2000)

[3] Büchner, H.: Strömungs- und Verbrennungsinstabilitäten in technischen Verbrennungssystemen. Professorial dissertation, University of Karlsruhe (2001)

[4] Lohrmann, M., Büchner, H.: Scaling of Stability Limits of Lean-Premixed Gas Turbine Combustors. Proceedings of, ASME Turbo Expo, Wien, Austria (2004)

[5] Arnold, G., Büchner, H.: Modelling of the Transfer Function of a Helmholtz-Resonator-Type combustion chamber. Proceedings of the European Combustion Meeting (2003)

[6] Lohrmann, M., Arnold, G., Büchner, H.: Modelling of the Resonance Characteristics of a Helmholtz-Resonator-Type Combustion Chamber with Energy Dissipation. Proceedings of the International Gas Research Conference (IGRC) (2001)

[7] Rommel, D.: Numerische Simulation des instationären, turbulenten und isothermen Strömungsfeldes in einer Modellbrennkammer. Master thesis, Engler-Bunte-Institute, University of Karlsruhe (1995)
[8] Magagnato, F.: KAPPA – Karlsruhe Parallel Program for Aerodynamics. TASK Quarterly **2** (2), pp.215–270 (1998)
[9] Smagorinsky, J.: General Circulation Experiments with the Primitive Equations. Monthly Weather Review, **91**, pp.99–164 (1963)
[10] Lilly, D.K.: The Representation of Small-Scale Turbulence in Numerical Simulation Experiments. Proc. IBM Scientific Computing Symposium on Environmental Sciences, pp.195–210 (1967)
[11] Schlichting, H., Gersten, K.: Grenzschicht-Theorie. Springer, 9th ed. (1997)
[12] Sexl, T.: Über den von E.G. Richardson entdeckten "Annulareffekt". Zeitschrift für Physik, **61**, pp.349 (1930)
[13] Richardson, E.G., Tyler, E.: The Transverse Velocity Gradient Near the Mouth of Pipes in which an Alternating or Continuous Flow of Air is Established. The Proceedings of the Physical Society, **42**, pp.1 (1929)
[14] Jensen, B.L., Sumer, B.M., Fredsoe, J: Turbulent Oscillatory Boundary Layers at High Reynolds Numbers. J. Fluid Mech., **206**, pp.265–297 (1989)
[15] Tsuji, Y., Morikawa, Y.: Turbulent Boundary Layer with Pressure Gradient Alternating in Sign. Aero. Quart. 27, pp.15–28 (1976)

Investigations of Flow and Species Transport in Packed Beds by Lattice Boltzmann Simulations

Thomas Zeiser

Regionales Rechenzentrum Erlangen, Universität Erlangen-Nürnberg
Martensstraße 1, 91058 Erlangen, Germany
thomas.zeiser@rrze.uni-erlangen.de

Abstract. This report summarizes selected results of investigations of the flow and species transport in packed beds. First of all, the difficulty of segmenting image data with respect to the correct choice of the threshold value and thus the resulting porosity is discussed. Then, the accuracy of lattice Boltzmann flow simulations is compared with CFX-5 simulations. The 3-D flow data is furthermore used to show how the pressure drop is made up by shear forces and dissipation owing to elongation and deformation. For the species transport, a random walk particle tracking algorithm is used to complement the lattice Boltzmann method thus allowing a wide range of Peclet numbers. In the last part, preliminary performance results of a new *1-D list based* lattice Boltzmann implementation ("sparse lattice") are summarized which soon will replace the currently used *full array based* code. It is shown that outstanding performance on vector as well as cache based parallel computers can be achieved with this 1-D list based "sparse lattice" code, too. Despite sophisticated optimizations for cache based microprocessors, the sustained application performance of a single NEC SX-8 CPU is about 10–20 times higher than that of any commodity CPU. For parallel calculations, this gap even grows further.

1 Introduction

Fundamental knowledge of the transport phenomena taking place on very different length scales ranging from molecular dimensions to gigascopic extents is of major importance for divers areas of science and engineering. Within this research project, the focus is set on detailed studies of momentum transport (and to some extent also the transport of chemical species) in fixed bed reactors using modern numerical and experimental techniques.

Fixed bed reactors are used for numerous large-scale chemical processes in particular heterogeneously catalyzed reactions in continuously operated multi-tubular packed bed reactors. Single tubes typically have a diameter in the range of some millimeters and several centimeters, and a length of up to a few meters. The packing within the tubes may consist of any shape, however, in practice usually either monolithic structures (e.g. [6, 1]) or, more often, random packings of spherical or cylindrical particles are used. For numerical investigations of the transport processes, the shape of the particles and the structure of the resulting *porous medium* does not make much difference as long as the pore space (and/or the solid matrix) can be described or obtained on the pore scale level to allow for direct numerical simulations.

Random packings of spheres in narrow tubes and random packings of spheres in periodic domains are used in the present investigations owing to their technical relevance, but mainly because Monte Carlo methods can easily be used to synthetically generate the geometrical structure [21, 13, 20]. But the same approaches can be applied to any other structure as well [15, 23].

Tremendous progress during the last decades in the area of numerical methods, the available computational power, but also new numerical approaches nowadays allow – at least to some extent [8] – detailed numerical studies of coupled transport phenomena in arbitrarily complicated geometries using only spatially resolved three-dimensional data of the geometrical structure as well as basic material properties like the kinematic viscosity or molecular diffusion coefficients. Thus, methods from computational fluid dynamics can provide reliable time dependent data for any spatial location. This information more or less is independent from experimental investigations and therefore can be used for mutual validation. Moreover, the *numerical experiments* may provide much more insight in local phenomena than traditional experiments offer. However, handling the large amount of data generated by the numerical simulations has already become another challenge.

In the framework of the present research project, detailed numerical flow simulations in packings of spheres are carried out using the lattice Boltzmann method (LBM) [22, 25, 7, 17]. The flow data is subsequently used as input to a random walk method for investigations of the mass transport [2]. On the experimental side, magnetic resonance imaging (MRI) is applied to obtain images of the structure as well as the flow field [15].

The basics of the applied methods have been described in detail elsewhere (see the references given above). Therefore, the following sections will not unnecessarily repeat the theory but give references to appropriate work and focus on selected results and possible pitfalls encountered when applying these techniques. In the last part, preliminary performance results of a new *1-D list based "sparse lattice"* lattice Boltzmann implementation are summarized which soon will replace the currently still used *full array based $\mathcal{BEST}$* code of LSTM-Erlangen. The aim of both codes is to achieve outstanding performance on vector as well as cache based parallel computers [24, 19].

2 Segmentation of MRI Data as Input for CFD Simulations

Since recent years, MRI (magnetic resonance imaging) has been discovered in engineering as an interesting method for investigating a variety of materials and processes [14]. In the preset research project, MRI first of all is used as an alternative to computed X-ray tomography (CT) to digitize 3-D objects. With MRI and CT similar spatial resolutions of up to only a few micrometers can be obtained. The big advantage of MRI, however, is that it not only can be used for measuring the geometry but also for obtaining information on the transport processes taking place [15].

The result of all imaging techniques are gray scale or intensity bitmaps (Fig. 1(a)). To further utilize the data, a binary segmentation (reconstruction) is required. A major difficulty for the present application is the correct choice of the threshold value which is directly related to the resulting global porosity ϵ (Fig. 1(c)). As all relevant dimensionless quantities like Ergun's Reynolds number

$$Re_E = \frac{u_0\, d}{\nu} \frac{1}{(1-\epsilon)} \tag{1}$$

or the friction factor and friction coefficient

$$f_E = \frac{\Delta p}{L} \frac{d}{\rho u_0^2} \frac{\epsilon^3}{(1-\epsilon)}, \tag{2}$$

$$\Lambda_E = f_E\, Re_E = \frac{\Delta p}{L} \frac{d^2}{\mu u_0} \frac{\epsilon^3}{(1-\epsilon)^2} \tag{3}$$

heavily depend on the porosity, a small error in defining the porosity can have a big impact on the fluid mechanical results in non-dimensional form. Therefore, it is usually necessary to determine the global porosity independently as visual inspection does not help much (Fig. 1(d)–1(f)). However, also measuring just the global porosity may be tricky [26].

3 Analysis of the Flow Field and Pressure Drop

It is well known that lattice Boltzmann methods reliably predict the fluid flow and the pressure drop in arbitrary geometries. Figure 2 shows the dimensionless pressure drop Λ_E as a function of the Reynolds number Re_E for the flow through an ordered packing of spheres[1]. The results from own experiments [16], correlations (e.g. [6]) as well as numerical simulations using our lattice Boltzmann solver and CFX-5 with two different types of grids[2]

1 square channel with 2x2 spheres per cross-section and sufficient length

2 unstructured body adapted grid and the same Cartesian voxel mesh as used for the lattice Boltzmann simulations

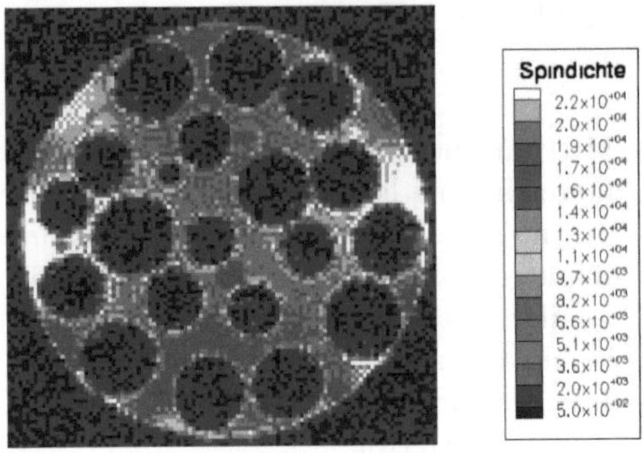

(a) Measured spin density

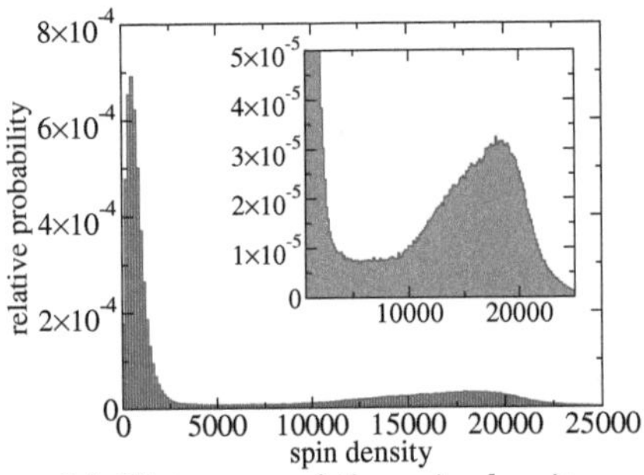

(b) Histogram of the spin density

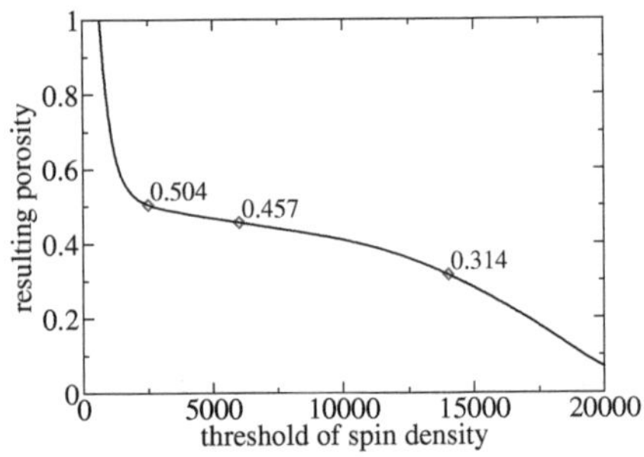

(c) Porosity as function of the threshold

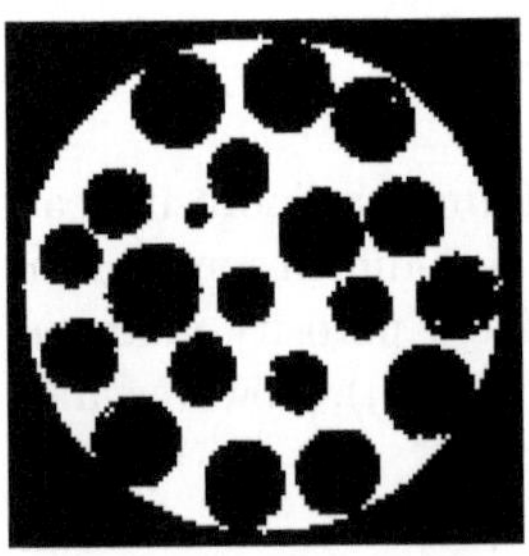

(d) threshold = 2500

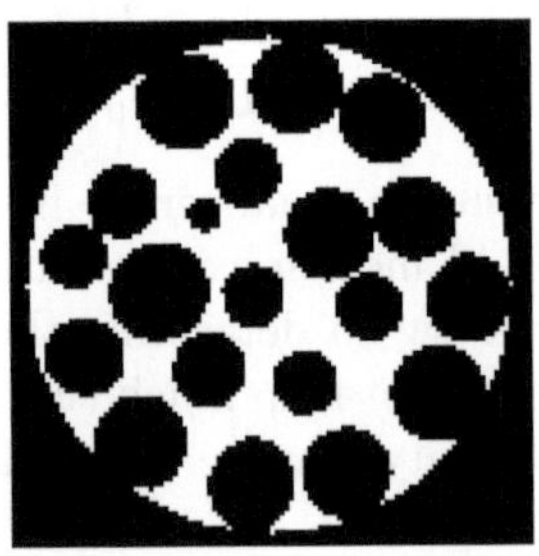

(e) threshold = 6000

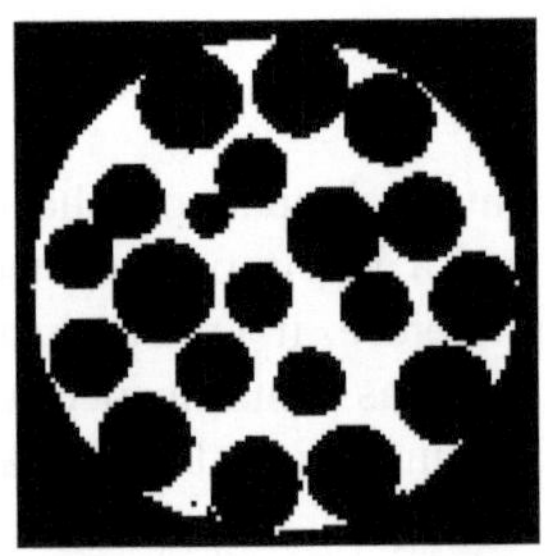

(f) threshold = 14000

Fig. 1. Dependence of the global porosity on the threshold used for segmentation

are shown. It is remarkable that the lattice Boltzmann simulations predict the correct pressure drop even on a rather coarse Cartesian voxel mesh. With CFX-5 much finer meshes (not shown here) are required to achieve the same level of accuracy [3]. Using the results of different resolutions and Richardson's extrapolation method [11] all numerical results match the experimental data.

In Fig. 2 as well as for the flow through a random packing of spheres in a circular tube with a tube-to-sphere diameter ratio of 5 as shown in Fig. 3,

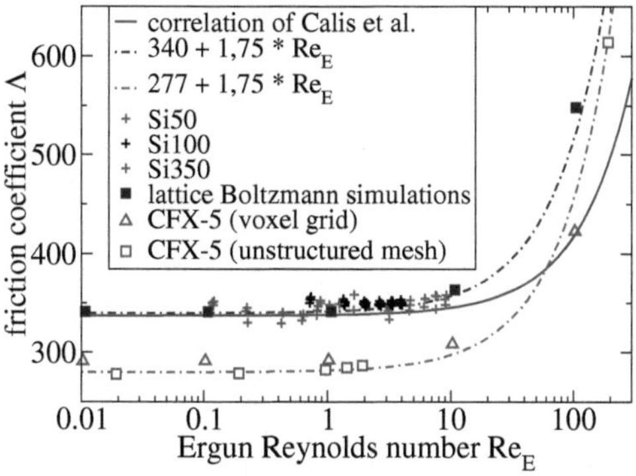

Fig. 2. Comparison of the simulated pressure drop with experimental data and correlations. Flow through an ordered packing a a channel with 2×2 spheres per cross-section

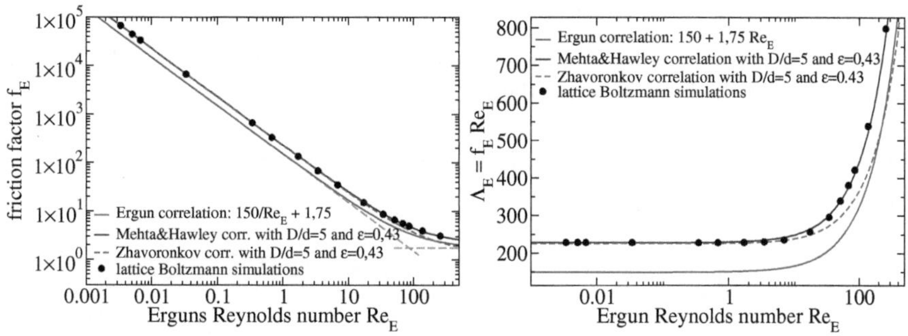

Fig. 3. Dimensionless pressure drop as function of the Reynolds number for the flow through the MRI data set `ch_020815-256`, i.e. a random packing with a tube-to-sphere diameter ratio of 5

the obtained dimensionless pressure drop Λ_E is much higher than predicted by Ergun's equation for the flow through infinitely large random packings. The increased resistance is caused by the additional friction and deformation of fluid elements at the walls which is not completely counterbalanced by the increased porosity near the walls of the tube [26, 27].

The spatially resolved velocity data can also be used to verify that the dimensionless pressure drop correlates to the integral viscous dissipation [5, 26] and thus fulfills the mechanical energy balance (Fig. 4).

The pressure drop is made up of two contributions: shear and elongation/deformation (Fig. 5). The capillary theory neglects the contribution by elongation [4, 10] and instead introduces a fudge factor, the tortuosity. However, the contribution of shear forces make up only about 60%, independently of the Reynolds number or the distance from the wall.

The tortuosity[3] takes rather low values of about 1.2 (Fig. 6). With increasing Reynolds number, the distribution spreads as recirculation zones are formed locally.

[3] ratio between the actual flow path and the length of the packing

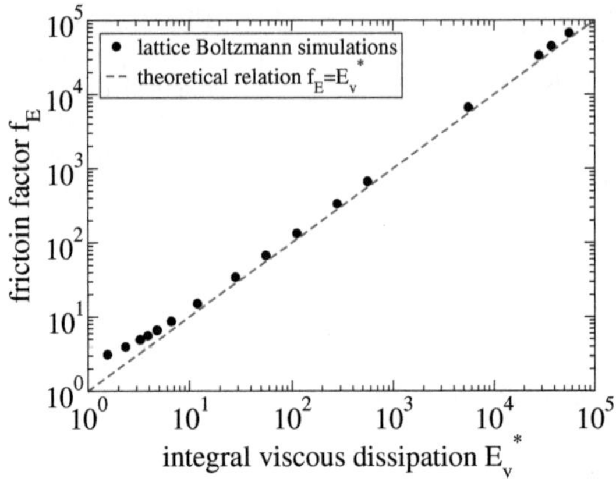

Fig. 4. Correlation of the dimensionless pressure drop with the non-dimensional integral viscous dissipation (same data set as in Fig. 3)

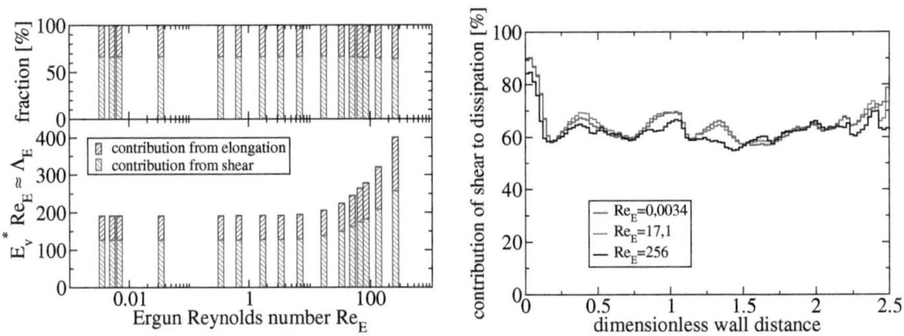

Fig. 5. Ingredients of the pressure drop: dissipation by shear and elongation/deformation (same data set as in Fig. 3)

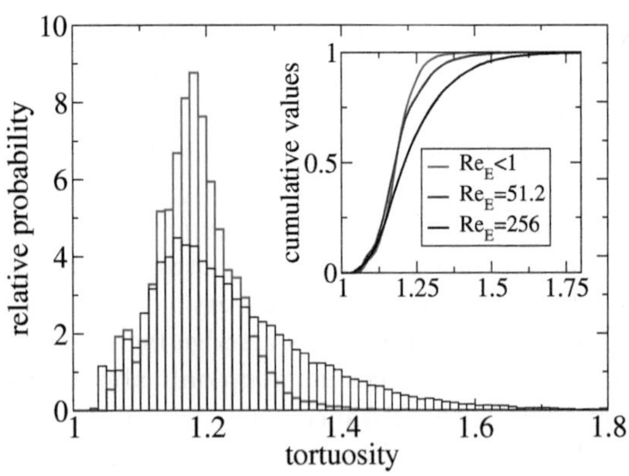

Fig. 6. Distribution of the tortuosity values for different Reynolds numbers (same data set as in Fig. 3)

4 Analysis of the Mass Transport

From the flow field, some information about the mass transport can be derived directly. Figure 7 shows, as an example, the residence time distribution as a function of the Reynolds number when neglecting molecular diffusion. To include molecular diffusion over a wide range of Peclet numbers, a random walk particle tracking method was applied in the present research project (Fig. 8, [2, 12]). As this part of the code is not (yet) vectorized, the simulation procedure was split into two parts: the flow field was determined using the lattice Boltzmann code on the NEC SX while the random walk particle tracking was carried out afterwards on a local PC cluster. For the present resolutions, it was still feasible to transfer the flow results from HLRS to RRZE. With increasing domain size (either owing to better resolution or larger regions) also this part will be carried out at HLRS.

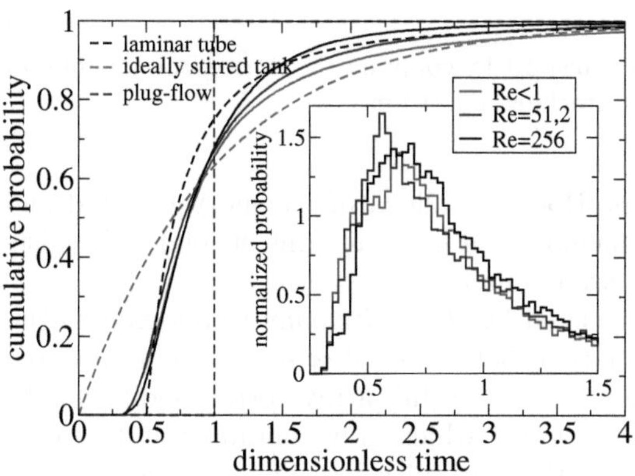

Fig. 7. Probability distribution of the residence time for different Reynolds numbers in the limiting case of $Pe \to \infty$ and $Bo \to \infty$ (same data set as in Fig. 3). For reference, the behavior of plug flow, an ideally stirred tank and a laminar tube are plotted as well

5 New 1-D List-Based Lattice Boltzmann Solver

Most lattice Boltzmann implementations known today, directly take the geometrical bounding box as basis for the implementation, i.e. the bounding box is mapped just as it is to a 2-D or 3-D array of cells containing the values of the lattice Boltzmann distributions functions and an additional mask is used to block out occupied cells during the calculation steps only. From the computational point of view, this *full array* representation (together with ghost or halo cells at the domain boundaries) is advantageous as it results in very simple data structures[4] and the neighbors of all cells can easily be obtained

[4] The impact of the order if the indexes, i.e. *collision optimized* (array of structures) or *propagation optimized* (structure of arrays) has been discussed in detail elsewhere [24, 9]. In most cases the propagation optimized layout is advantageous.

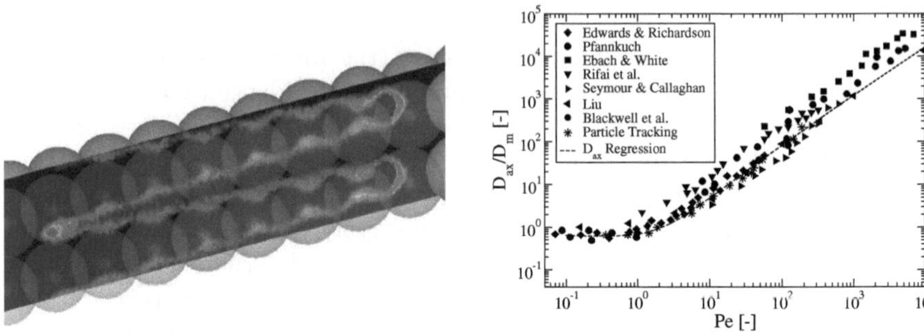

(a) Visualized tracer experiments in the simple cubic sphere array at Pe = 750

(b) Simulated axial dispersion coefficients in a fixed-bed ($Re = 4.13$, $L/d_p = 27$, $D/d_p = 9.3$, $\epsilon = 0.40$) compared to experimental data

Fig. 8. Investigations of the mass transport by coupling the lattice Boltzmann flow solver with a random walk particle tracking algorithm

by elementary index arithmetics. However, allocating memory for fluid and solid cells results in a waste of memory if a large amount of solid cells exists where no distribution functions are required.

In the framework of an international *Lattice Boltzmann Development Consortium*[5] a new general purpose lattice Boltzmann flow solver is currently being developed which is no longer based on a full array corresponding to the geometrical bounding box, but a 1-D list which only contains the fluid cells (and information about the connectivity of all cells), i.e. a "sparse lattice". Neighboring cells are no longer accessible via simple index arithmetics. Instead, a lookup in the connectivity data and indirect addressing is required. Details of the implementation will be described elsewhere.

For simple geometries like an empty channel, this new 1-D list representation requires about 25% more memory than the full array variant as no solid cells can be eliminated but additional information about the no longer obvious offsets of the geometrical neighboring cells within the list has to be stored.

On the NEC SX-6/SX-8 there is not much difference in the application performance[6] of the full array implementation and the 1-D list based "sparse lattice" version when applied to simple geometries despite the fact that indirect addressing is required with the 1-D algorithm. This is understandable

[5] including NEC CCRLE, HLRS, RRZE, the computer science department of the University of Amsterdam (NL), the Medical Physics and Clinical Engineering Department of the University of Sheffield (UK), the Lehrstuhl für Computeranwendungen im Bauingenieurwesen der TU-Braunschweig

[6] measured as the number in millions of fluid lattice cell updates per second, MLUPS

as the performance on the NEC SX series was never limited by the available memory bandwidth but the arithmetic units [24, 28].

On cache based architectures, an additional advantage of the 1-D list can be exploited. In the preprocessing step, when the 1-D list and the connectivity of the entries is built, the cells can be processed in arbitrary order, i.e. complicated cache blocking schemes no longer have to be integrated into the computational kernel but only into the initial setup routine. If, for example, a normal 3-D blocking is applied within the computational kernel, this usually inevitably results in nested loops of rather short length, thus, in a high loop overhead. With this 1-D list on the other hand, the computational kernel still only sees the long one dimensional list. The different access patterns are completely hidden by the algorithm used to map geometrical cells to the node list and the additional connectivity list. As the connectivity data is always read linearly, this does not result in major cache pollution or consumption of memory bandwidth. However, the ability of cache based microprocessors to cope with the indirect memory access heavily depends on the prefetching technique used. For the Intel IA64 compiler it is rather difficult to predict the memory access. Together with the in-order execution model of the Intel Itanium2, this results – at least for simple geometries like an empty channel – in a significant reduction of the performance of the 1-D list representation compared to a full array implementation. The Intel Xeon on the other hand with its hardware prefetcher which loads all data up to the page boundary to the cache if two successive cache hits occurs, delivers a similar performance for the two approaches. The AMD Opteron lies somewhere in between.

For complicated geometries with larger amounts of solid cells, the saving in CPU time and memory is considerable on all platforms. For reference, a few examples measured on an NEC SX-6i are shown Table 1.

The MPI parallelisation of the new 1-D list based code is not yet finished. But it is expected that the trends will be similar to $\mathcal{BEST}$ (Fig. 9) . Probably, the dependency on a fast network interconnect will increase as $\mathcal{BEST}$ uses a simple 1-D or 3-D domain decomposition with flat surfaces whereas the new 1-D code will use METIS as partitioner.

Table 1. Comparison of $\mathcal{BEST}$ and the new 1-D list-based code on a NEC SX6-i

testcase / code	runtime	MLUPS
empty channel / BEST	32s	32
empty channel / 1-D DC	36s	27
porous foam 50% / BEST	40s	11
porous foam 50% / 1-D DC	18s	25
packed bed 60% / BEST	187s	13
packed bed 60% / 1-D DC	98s	26

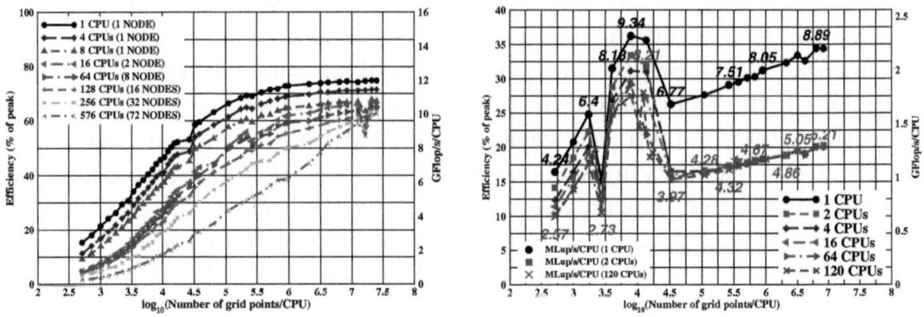

(a) Up to 72 nodes (576 CPUs) of a NEC SX-8

(b) Up to 120 CPUs of an SGI Altix3700Bx2

Fig. 9. Efficiency, GFlop/s and MLUPS of the full array lattice Boltzmann solver $\mathcal{BEST}$ depending on the domain size of an empty channel and the number of processors [18]

6 Summary and Conclusions

Numerical simulations nowadays can complement experimental investigations and deliver a large amount of detailed information with high accuracy. The time-dependent local data may be used to better understand the physical processes taking place on the pore-scale level as demonstrated for the different contributions to the total dissipation, i.e. the pressure drop, in a random packing of spheres.

Cache based microprocessor architectures tremendously suffer from the growing gap between the speed of the processor cores and the available memory bandwidth. Therefore, vector systems not only have the ability to parry the attack of commodity-off-the-shelf cluster – in many cases, they are the only chance to significantly reduce the turn-around time when the total domain size is fixed.

Acknowledgements

This work is financially supported by the German Research Foundation (DFG) and the Competence Network for Technical, Scientific High Performance Computing in Bavaria (KONWIHR). The NMR measurements have been carried out by Claudia Heinen and Joachim Tillich from the *Institut für Mechanische Verfahrenstechnik und Mechanik* at the University of Karlsruhe. Helpful discussions with and support through my colleagues at LSTM, RRZE and the *Lehrstuhl für chemische Reaktionstechnik* are gratefully acknowledged. The new 1-D list based code is the result of joined work of the partners within the *Lattice Boltzmann Development Consortium*.

References

1. ABB Lummus Global Inc. Fixed catalytic bed reactor. International patent WO 99/48604, 1999.
2. J. Bauer. Ortsaufgelöste Simulation von Transportprozessen in porösen Medien zur Untersuchung von Dispersion und Verweilzeitverhalten. Master's thesis, Lehrstuhl für Chemische Reaktionstechnik, Universität Erlangen-Nürnberg, 2004.
3. V. Bhandari. Detailed investigations of transport properties in complex reactor components. Masters thesis, Lehrstuhl für Strömungsmechanik, Universität Erlangen-Nürnberg, 2002.
4. R.B. Bird, W.E. Stewart, and E. N. Lightfoot. *Transport Phenomena*. John Wiley & Sons, New York, 2 edition, 2002.
5. G. Brenner. *Numerische Simulation komplexer Transportvorgänge in der Verfahrenstechnik*. Habilitationsschrift, Technische Fakultät, Universität Erlangen-Nürnberg, 2002.
6. H.P.A. Calis, J. Nijenhuis, B.C. Paikert, F.M. Dautzenberg, and C.M. van den Bleek. CFD modelling and experimental validation of pressure drop and flow profile in a novel structured catalytic reactor packing. *Chem. Eng. Sci.*, 56(7):1713–1720, 2001.
7. S. Chen and G. D. Doolen. Lattice Boltzmann method for fluid flows. *Annu. Rev. Fluid Mech.*, 30:329–364, 1998.
8. A.G. Dixon and M. Nijemeisland. CFD as a design tool for fixed-bed reactors. *Ind. Eng. Chem. Res*, 40:5246–5254, 2001.
9. S. Donath. On optimized implementations of the lattice Boltzmann method on contemporary high performance architectures. Bachelor thesis, Lehrstuhl für Informatik 10 (Systemsimulation) / Regionales RechenZentrum Erlangen, Universität Erlangen-Nürnberg, 2004.
10. F. Durst, R. Haas, and W. Interthal. The nature of flows through porous media. *J. Non-Newtonian Fluid Mech.*, 22:169–189, 1987.
11. J. Ferziger and M. Perić. *Computational Methods for Fluid Dynamics*. Springer, Berlin, Heidelberg, New York, 1996.
12. H. Freund, J. Bauer, T. Zeiser, and G. Emig. Detailed simulation of transport processes in fixed-beds. *Ind. Eng. Chem. Res*, 44(16):6423–6434, 2005.
13. H. Freund, T. Zeiser, F. Huber, E. Klemm, G. Brenner, F. Durst, and G. Emig. Numerical simulations of single phase reacting flows in randomly packed fixed-bed reactors and experimental validation. *Chem. Eng. Sci.*, 58(3-6):903–910, 2003.
14. L.F. Gladden. Magnetic resonance: Ongoing and future role in chemical engineering research. *AIChE Journal*, 49(1):2–9, 2003.
15. C. Heinen. *MRI Untersuchungen zur Strömung newtonscher und nicht-newtonscher Fluide in porösen Strukturen*. PhD thesis, Universtität Karlsruhe (TH), 2004.
16. F. Huber. Simulation und experimentelle Untersuchungen lokaler Transportprozesse in Festbettreaktoren. Masters thesis, Lehrstuhl für Technische Chemie I / Lehrstuhl für Strömungsmechanik, Universität Erlangen-Nürnberg, 2002.
17. C. Körner, T. Pohl, U. Rüde, N. Thürey, and T. Zeiser. Parallel lattice Boltzmann methods for CFD applications. In A.M. Bruaset and A. Tveito, editors,

Numerical Solution of Partial Differential Equations on Parallel Computers, volume 51 of *LNCSE*, pages 439–465. Springer, 2005.
18. P. Lammers, G. Wellein, T. Zeiser, and M. Breuer. Have the vectors the continuing ability to parry the attack of the killer micros? In M. Resch, T. Bönisch, K. Benkert, T. Furui, Y. Seo, and W. Bez, editors, *High Performance Computing on Vector Systems. Proceedings of the High Performance Computing Center Stuttgart, March 2005*, pages 25–39. Springer, 2006.
19. T. Pohl, F. Deserno, N. Thürey, U. Rüde, P. Lammers, G. Wellein, and T. Zeiser. Performance evaluation of parallel large-scale lattice Boltzmann applications on three supercomputing architectures. In *Proceedings of Supercomputing Conference SC04, Pittsburg*, 2004.
20. W. Soppe. Computer simulation of random packings of hard spheres. *Powder Technol.*, 62:189–196, 1990.
21. M. Steven. Detaillierte Simulation und Analyse der Struktur von Katalysatorschüttungen und der lokalen Transportprozesse. Diplomarbeit, Lehrstuhl für Technische Chemie I / Lehrstuhl für Strömungsmechanik, Universität Erlangen-Nürnberg, 2001.
22. S. Succi. *The Lattice Boltzmann Equation – For Fluid Dynamics and Beyond.* Clarendon Press, 2001.
23. V. Vassilev. Analyse experimentell (mittels MRI/NMR) oder numerisch (durch LBM) ermittelter Geschwindigkeitsfelder poröser Strukturen. Bachelor thesis, Lehrstuhl für Strömungsmechanik, Universität Erlangen-Nürnberg, 2003.
24. G. Wellein, T. Zeiser, S. Donath, and G. Hager. On the single processor performance of simple lattice Boltzmann kernels. *Computers & Fluids*, 35:910–919, 2006.
25. D.A. Wolf-Gladrow. *Lattice-Gas Cellular Automata and Lattice Boltzmann Models*, volume 1725 of *Lecture Notes in Mathematics.* Springer, Berlin, 2000.
26. T. Zeiser. *Simulation von durchströmten Schüttungen auf Hochleistungsrechnern.* PhD thesis, Technische Fakultät, Universität Erlangen-Nürnberg, laufende Arbeit.
27. T. Zeiser, M. Steven, H. Freund, P. Lammers, G. Brenner, F. Durst, and J. Bernsdorf. Analysis of the flow field and pressure drop in fixed bed reactors with the help of lattice Boltzmann simulations. *Phil. Trans. R. Soc. Lond. A*, 360(1792):507–520, 2002.
28. T. Zeiser, G. Wellein, G. Hager, S. Donath, F. Deserno, P. Lammers, and M. Wierse. Optimized lattice Boltzmann kernels as testbeds for processor performance. Technical report, Regionales Rechenzentrum Erlangen, May 2004.

Rheological Properties of Binary and Ternary Amphiphilic Fluid Mixtures

Jens Harting[1] and Giovanni Giupponi[2]

[1] Institut für Computerphysik, Pfaffenwaldring 27, 70569 Stuttgart, Germany
[2] Centre for Computational Science, Chemistry Department, University College London, 20 Gordon Street, London WC1H 0AJ, UK

Abstract. Within this project, we perform lattice Boltzmann simulations of spinodal decomposition and structuring effects in binary immiscible and ternary amphiphilic fluid mixtures under shear. We use a highly scalable parallel Fortran 90 code for the implementation of the lattice Boltzmann method. We demonstrate that the domain growth mechanisms in ternary amphiphilic fluid mixtures strongly depend on the amphiphile concentration. For systems under constant and oscillatory shear we analyze domain growth rates in directions parallel and perpendicular to the applied shear and find that these systems undergo structural transitions with tubular and lamellar structures appearing.

1 Introduction

In recent years there has emerged a class of fluid dynamical problems, called "complex fluids", which involve both hydrodynamic flow effects and complex interactions between fluid particles. Computationally, such problems are too large and expensive to tackle with atomistic methods such as molecular dynamics, yet they require too much molecular detail for continuum Navier–Stokes approaches.

Algorithms which work at an intermediate or "mesoscale" level of description in order to solve these problems have been developed within the last two decades. These include Dissipative Particle Dynamics [5, 14, 7], Lattice Gas Cellular Automata [19], Stochastic Rotation Dynamics [17, 11, 20], and the Lattice Boltzmann method [23, 1, 16]. In particular, the Lattice Boltzmann method has been found highly useful for the simulation of complex fluid flows in a wide variety of systems. This algorithm is extremely well suited to be implemented on parallel computers, which permits very large systems to be simulated, reaching hitherto inaccessible physical regimes.

"Simple fluids" usually can be described to a good degree of approximation by macroscopic quantities only, such as the density $\rho(\mathbf{x})$, velocity $\mathbf{v}(\mathbf{x})$, and temperature $T(\mathbf{x})$. Such fluids are governed by the Navier–Stokes Eq. [6], which are nonlinear and difficult to solve in the most general case, with the result that numerical solution of the equations has become a common tool for understanding the behaviour of "simple" fluids, such as water or air. "Complex fluids" on the other hand are fluids where the macroscopic flow is affected by microscopic properties. A good example of such a fluid is blood: as it flows through vessels, it is subjected to shear forces, which cause red blood cells to align with the flow so that they can slide over one another more easily. This effect causes a change in viscosity which in turn affects the flow profile. Hence, the macroscopic blood flow is affected by the microscopic alignment of its constituent cells. Other examples of complex fluids include fluids such as paint, milk, cell organelles, as well as polymers and liquid crystals. In all of these cases, the density and velocity fields are insufficient to describe the fluid behaviour, and in order to understand this behaviour, it is necessary to treat effects which occur over a very wide range of length and time scales. This length and time scale gap makes complex fluids even more difficult to model.

In a mixture containing various fluid components, an amphiphile is a molecule which is composed of two parts, each part being attracted towards a different fluid component. For example, soap molecules are amphiphiles, containing a head group which is attracted towards water, and a tail which is attracted towards oil. If many amphiphile molecules are collected together in solution, they can exhibit highly varied and complicated behaviour, often assembling to form amphiphile mesophases, which are complex fluids of significant theoretical and industrial importance. Some of these phases have long-range order, yet remain able to flow, and are called liquid crystal mesophases.

Within this project we have performed large scale lattice-Boltzmann simulations in order to study the bahaviour of binary immiscible or ternary amphiphilic fluid mixtures under shear. Since in our case the amphiphilic molecules can be seen as surfactant, we will also use the latter term in this paper.

1.1 The Model

A standard LB system involving multiple species is usually represented by a set of equations [13]

$$n_i^\alpha(\mathbf{x}+\mathbf{c}_i,t+1) - n_i^\alpha(\mathbf{x},t) = -\frac{1}{\tau_\alpha}(n_i^\alpha(\mathbf{x},t) - n_i^{\alpha eq}(\mathbf{x},t)),\ i = 0,1,\ldots,b\ , \tag{1}$$

where $n_i^\alpha(\mathbf{x},t)$ is the single-particle distribution function, indicating the density of species α (for example, oil, water or amphiphile), having velocity $\mathbf{c}_i$, at site $\mathbf{x}$ on a D-dimensional lattice of coordination number b, at time-step t. The collision operator Ω_i^α represents the change in the single-particle distribution function due to the collisions. We choose a single relaxation time τ_α, 'BGK' form [2] for the collision operator. In the limit of low Mach numbers, the LB equations correspond to a solution of the Navier-Stokes equation for isothermal, quasi-incompressible fluid flow whose implementation can efficiently exploit parallel computers, as the dynamics at a point requires only information about quantities at nearest neighbour lattice sites. The local equilibrium distribution $n_i^{\alpha eq}$ plays a fundamental role in the dynamics of the system as shown by Eq. (1). In this study, we use a purely kinetic approach, for which the local equilibrium distribution $n_i^{\alpha eq}(\mathbf{x},t)$ is derived by imposing certain restrictions on the microscopic processes, such as explicit mass and total momentum conservation [4]

$$n_i^{\alpha eq} = \zeta_i n^\alpha \left[1 + \frac{\mathbf{c}_i \cdot \mathbf{u}}{c_s^2} + \frac{(\mathbf{c}_i \cdot \mathbf{u})^2}{2c_s^4} - \frac{u^2}{2c_s^2} + \frac{(\mathbf{c}_i \cdot \mathbf{u})^3}{6c_s^6} - \frac{u^2(\mathbf{c}_i \cdot \mathbf{u})}{2c_s^4}\right] , \quad (2)$$

where $\mathbf{u} = \mathbf{u}(\mathbf{x},t)$ is the macroscopic bulk velocity of the fluid, defined as $n^\alpha(\mathbf{x},t)\mathbf{u}^\alpha \equiv \sum_i n_i^\alpha(\mathbf{x},t)\mathbf{c}_i$, ζ_i are the coefficients resulting from the velocity space discretization and c_s is the speed of sound, both of which are determined by the choice of the lattice, which is D3Q19 in our implementation. Immiscibility of species α is introduced in the model following Shan and Chen [21, 22]. Only nearest neighbour interactions among the immiscible species are considered. These interactions are modelled as a self-consistently generated mean field body force

$$\mathbf{F}^\alpha(\mathbf{x},t) \equiv -\psi^\alpha(\mathbf{x},t)\sum_{\bar{\alpha}} g_{\alpha\bar{\alpha}} \sum_{\mathbf{x}'} \psi^{\bar{\alpha}}(\mathbf{x}',t)(\mathbf{x}' - \mathbf{x}) , \quad (3)$$

where $\psi^\alpha(\mathbf{x},t)$ is the so-called effective mass, which can have a general form for modelling various types of fluids (we use $\psi^\alpha = (1 - e^{-n^\alpha})$[21]), and $g_{\bar{\alpha}\alpha}$ is a force coupling constant whose magnitude controls the strength of the interaction between components α, $\bar{\alpha}$ and is set positive to mimic repulsion. The dynamical effect of the force is realized in the BGK collision operator by adding to the velocity $\mathbf{u}$ in the equilibrium distribution of Eq. (2) an increment

$$\delta\mathbf{u}^\alpha = \frac{\tau^\alpha \mathbf{F}^\alpha}{n^\alpha} . \quad (4)$$

As described above, an amphiphile usually possesses two different fragments, each having an affinity for one of the two immiscible components. The addition of an amphiphile is implemented as in [3]. An average dipole vector $\mathbf{d}(\mathbf{x},t)$ is introduced at each site $\mathbf{x}$ to represent the orientation of any amphiphile present there. The direction of this dipole vector is allowed to vary continuously and no information is specified for each velocity $\mathbf{c}_i$, for reasons of computational efficiency and simplicity. Full details of the model can be found in [3] and [18].

In order to inspect the rheological behaviour of multi-phase fluids, we have implemented Lees-Edwards boundary conditions, which reduce finite size effects if compared to moving solid walls [15]. This computationally convenient method imposes new positions and velocities on particles leaving the simulation box in the direction perpendicular to the imposed shear strain while leaving the other coordinates unchanged. Choosing z as the direction of shear and x as the direction of the velocity gradient, we have

$$z' \equiv \begin{cases} (z+\Delta_z) \bmod N_z & ,x > N_x \\ z \bmod N_z & ,0 \le x \le N_x \\ (z-\Delta_z) \bmod N_z & ,x < 0 \end{cases} \quad u'_z \equiv \begin{cases} u_z + U & ,x > N_x \\ u_z & ,0 \le x \le N_x \\ u_z - U & ,x < 0 \end{cases} , \quad (5)$$

where $\Delta_z \equiv U\Delta t$, U is the shear velocity, u_z is the $z-$component of $\mathbf{u}$ and $N_{x(z)}$ is the system length in the $x(z)$ direction. We also use an interpolation scheme suggested by Wagner and Pagonabarraga [24] as Δ_z is not generally a multiple of the lattice site. Consistent with the hypothesis of the LB model, we set the maximum shear velocity to $U = 0.1$ lattice units. This results in a maximum shear rate $\dot{\gamma}_{xz} = \frac{2\times 0.1}{64} = 3.2 \times 10^{-3}$ For oscillatory shear, we set

$$U(t) = U\cos(\omega t) , \quad (6)$$

where $\omega/2\pi$ is the frequency of oscillation.

2 The Simulation Code

We use LB3D [9], a highly scalable parallel LB code, to implement the model. LB3D is written in Fortran 90 and designed to run on distributed-memory parallel computers, using MPI for communication. In each simulation, the fluid is discretized onto a cubic lattice, each lattice point containing information about the fluid in the corresponding region of space. Each lattice site requires about a kilobyte of memory per lattice site so that, for example, a simulation on a 128^3 lattice would require around 2.2GB memory. The code runs at over $3 \cdot 10^4$ lattice site updates per second per CPU on a recent machine, and has been observed to have roughly linear scaling up to order $3 \cdot 10^3$ compute nodes. Larger simulations have not been possible so far due to the lack of access to a machine with a higer processor count. The largest simulation we performed used a 1024^3 lattice to describe a mesophase consisting of two immiscible and one amphiphilic phase. The output from a simulation usually takes the form of a single floating-point number for each lattice site, representing, for example, the density of a particular fluid component at that site. Therefore, a density field snapshot from a 128^3 system would produce output files of around 8MB. Writing data to disk is one of the bottlenecks in large scale simulations. If one simulates a 1024^3 system, each data file is 4GB in size. LB3D is able to benefit from the parallel filesystems available on many large machines today, by using the MPI-IO based parallel HDF5 data format [12]. Our code is very robust regarding different platforms or cluster interconnects: even with moderate inter-node bandwidths it achieves almost linear scaling for large processor counts with the only limitation being the available memory per node. The platforms our code has been successfully used on include various supercomputers like the NEC SX8, IBM pSeries, SGI Altix and Origin, Cray T3E, Compaq Alpha clusters, as well as low cost 32- and 64-bit Linux clusters.

3 Complex Fluids under Shear

In many industrial applications, complex fluids are subject to shear forces. For example, axial bearings are often filled with fluid to reduce friction and transport heat away from the most vulnerable parts of the device. It is very important to understand how these fluids behave under high shear forces, in order to be able to build reliable machines and choose the proper fluid for different applications. In our simulations we use Lees–Edwards boundary conditions, which were originally developed for molecular dynamics simulations in 1972 [15] and have been used in lattice Boltzmann simulations by different authors before [25, 24, 10]. We apply our model to study the behaviour of binary immiscible and ternary amphiphilic fluids under constant and oscillatory shear. In the non-sheared studies of spinodal decomposition it has been shown that lattice sizes need to be large in order to overcome finite size effects: 128^3 was the minimum acceptable number of lattice sites [8]. For high shear rates, systems also have to be very long because, if the system is too small, the domains interconnect across the $\mathbf{z} = \mathbf{0}$ and $\mathbf{z} = \mathbf{nz}$ boundaries to form interconnected lamellae in the direction of shear. Such artefacts need to be eliminated from our simulations. Figure 1 shows an example from a simulation with lattice size 128x128x512. The volume rendered blue and red areas depict the different fluid species and the arrows denote the direction of shear.

The focus of our current project is on the behaviour of ternary amphiphilic fluids. We are interested in the effect an amphiphilic phase has on the demixing of two immiscible fluids. In all simulations we keep the total density of the system at $\rho^{tot} = \rho^A + \rho^B + \rho^s = 1.6$ and $\rho^A = \rho^B$.

We study a cubic 256^3 system with a total density of 1.6 and surfactant densities $\rho^s = 0.00, 0.10, 0.20, 0.30$. As can be seen in Fig. 2, after 10000 timesteps the phases have separated to a large extend if no surfactant is present. If one adds surfactant, the domains grow more slowly and the growth process might even come to arrest for high amphiphile concentrations.

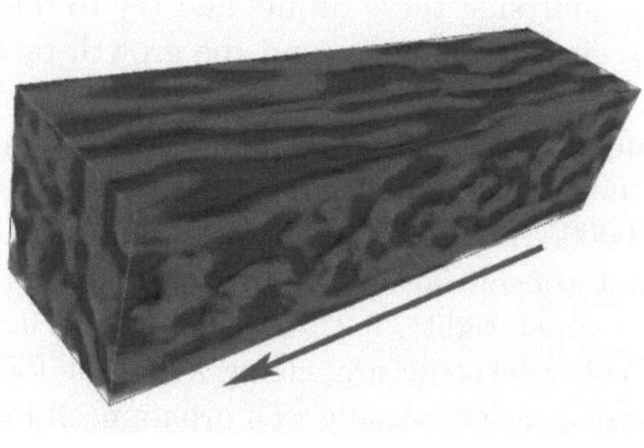

Fig. 1. Spinodal decomposition under shear. *Differently coloured regions* denote the majority of the corresponding fluid. The *arrows* depict the movement of the sheared boundaries

Fig. 2. Volume rendered fluid densities of 256^3 systems at $t = 10000$ for surfactant densities $\rho^s = 0.00, 0.10, 0.20, 0.30$ (*from left to right*)

In order to quantitatively compare between simulations with different surfactant densities, we define the time dependent lateral domain size $L(t)$ along direction $i = x, y, z$ as

$$L_i(t) \equiv \frac{2\pi}{\sqrt{\langle k_i^2(t) \rangle}}, \tag{7}$$

where

$$\langle k_i^2(t) \rangle \equiv \frac{\sum_{\mathbf{k}} k_i^2 S(\mathbf{k}, t)}{\sum_{\mathbf{k}} S(\mathbf{k}, t)} \tag{8}$$

is the second order moment of the three-dimensional structure function

$$S(\mathbf{k}, t) \equiv \frac{1}{V} \left| \phi'_{\mathbf{k}}(t) \right|^2 \tag{9}$$

with respect to the cartesian component i. $\langle \rangle$ denotes the average in Fourier space, weighted by $S(\mathbf{k})$ and V is the number of nodes of the lattice, $\phi'_{\mathbf{k}}(t)$ the Fourier transform of the fluctuations of the order parameter $\phi' \equiv \phi - \langle \phi \rangle$, and k_i is the ith component of the wave vector.

In Fig. 3 the time dependent lateral domain size $L(t)$ is shown for a number of surfactant densities $\rho^s = 0.00, 0.10, 0.15, 0.20, 0.25, 0.30, 0.35, 0.40$, and 0.45. The plots in Fig. 3 confirm what we already expected from the three-dimensional images presented in Fig. 2: For $\rho^s = 0.00$ domain growth dows not come to an end until the domains span the full system. Only by adding surfactant, we can slow down the growth process and for high surfactant densities $\rho^s > 0.25$, the domain growth stops after a few thousand simulation timesteps. By adding even more surfactant, the final average domain size becomes very small. We are currently analysing these results and try to find scaling laws for the dependence of the final domain size and the growth rate on the surfactant density.

Another interesting topic is the influence of shear on the phase separation. Figure 4 shows four examples of systems which are sheared with a constant shear rate $\dot{\gamma} = 1.56 \cdot 10^{-3}$ at $t = 10000$. Again, one observes a decreasing domain size with increasing surfactant concentration. We vary the concentration as $\rho^s = 0.0$ (upper left), 0.1 (upper right), 0.2 (lower left), and 0.3 (lower right). In addition one can also observe structural changes like the formation of lamellae in the system. Lamellae are especially well pronounced for

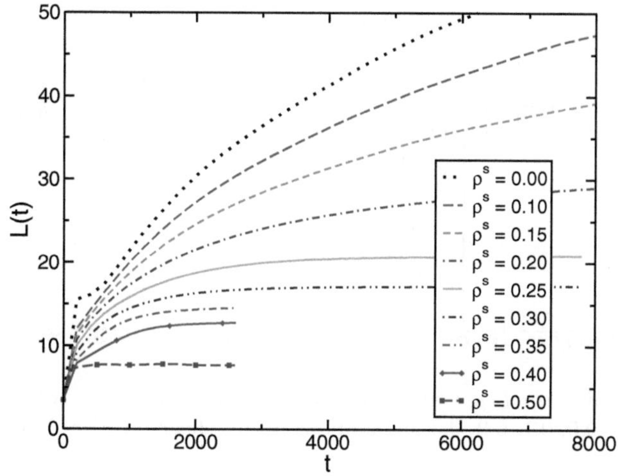

Fig. 3. Average domain size $L(t)$ for various surfactant densities $\rho^s = 0.00$, 0.10, 0.15, 0.20, 0.25, 0.30, 0.35, 0.40, 0.45

Fig. 4. Volume rendered fluid densities for surfactant densities $\rho^s = 0.0$ (*upper left*), 0.1 (*upper right*), 0.2 (*lower left*), 0.3 (*lower right*), a constant shear rate $\dot{\gamma} = 1.56 \cdot 10^{-3}$ and $t = 10000$

low surfactant concentrations, while for larger concentrations the domains try to form more tube-like structures. The time dependent lateral domain size shows an even richer behaviour as can be seen in Fig. 5. Here, we show the domain size in all three directions, where x denotes the direction perpendicular to the shear plane (Fig. 5a), y the direction parallel to the shear plane, but perpendicular to the direction of shear (Fig. 5b) and z is the direction of shear.

For high surfactant concentrations ($\rho^s = 0.4$ all three directions behave very similar, i.e. the growth comes to an end after less than 2000 timesteps and the final domain size is between 10 and 15 in lattice units. For $\rho^s = 0.3$, domains grow faster in x-direction, while in the other directions the growth process comes to an end after $t = 2000$ at $L(t) = 15$. For lower surfactant

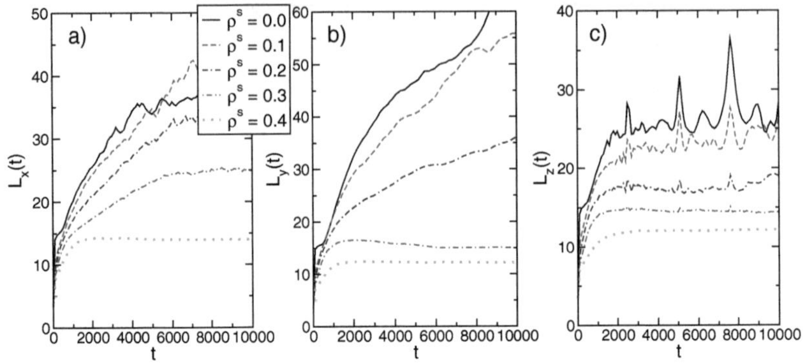

Fig. 5. Domain size $L(\rho^s)$ in x- (**a**), y- (**b**), and z-direction (**c**) for surfactant densities $\rho^s = 0.0, 0.1, 0.2, 0.3, 0.4$ and a constant shear rate $\dot{\gamma} = 1.56 \cdot 10^{-3}$

concentrations, the lateral domain size behaves similar in x- and y-directions, except for late times where the growth is limited by the shear. In z-direction, small peaks start to occur for $\rho^s < 0.4$ which increase with decreasing ρ^s. These peaks are due to lamellae forming parallel to the shear and which start to get tilted by the movement of the walls. After a given time, these lamellae are not parallel to the shear anymore and get broken up. Then, the process of lamaellae formation and breaking up starts again.

In Fig. 6 we show an example of the time dependent lateral domain size for a system undergoing oscillatory shear. The shear rate is $\dot{\gamma} = 1.56 \cdot 10^{-3}$, and the frequency of the oscillation is given by $\omega = 0.01$. Here, the upper and lower planes are moved periodically as given by Eq. 6. Here, one can observe a number of new phenomena: For low surfactant concentrations, domain growth is limited in x-direction, but in y direction the domains continue growing until the end of the simulation. This means that we can find tubular structures

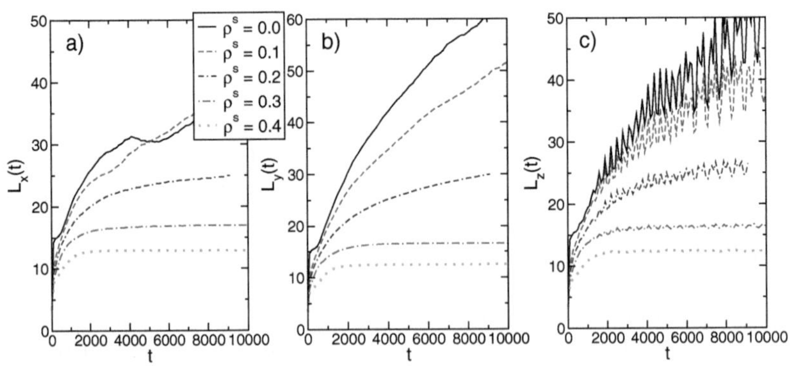

Fig. 6. Domain size $L(\rho^s)$ in x- (**a**), y- (**b**), and z-direction (**c**) for surfactant densities $\rho^s = 0.0, 0.1, 0.2, 0.3, 0.4$ and oscillatory shear with $\dot{\gamma} = 1.56 \cdot 10^{-3}$, $\omega = 0.01$

surpassing our system. Also, in z-direction, strong oscillations start to appear due to the continuous growing and destruction of elongated domains.

4 Conclusion

In this report we have presented our first results from ternary amphiphilic lattice-Boltzmann simulations performed on the NEX SX-8 at the HLRS. We have shown that our simulation code performs well on the new machine and that we are able to investigate spinodal decomposition with and without shear. In addition, we have studied the influence of the surfactant concentration on the time dependent lateral domain size.

Acknowledgements

We are grateful for the support of the HPC-Europa programme, funded under the European Commission's Research Infrastructures activity, contract number RII3-CT-2003-506079 and the Höchstleistungsrechenzentrum Stuttgart for providing access to their NEC SX8. We would especially like to thank H. Berger, R. Keller, and P. Lammers for their technical support and J. Chin, and P.V. Coveney for fruitful discussions.

References

1. R. Benzi, S. Succi, and M. Vergassola. The lattice Boltzmann equation: theory and applications. *Phys. Rep.*, 222(3):145–197, 1992.
2. P.L. Bhatnagar, E.P. Gross, and M. Krook. Model for collision processes in gases. I. Small amplitude processes in charged and neutral one-component systems. *Phys. Rev.*, 94(3):511–525, 1954.
3. H. Chen, B.M. Boghosian, P.V. Coveney, and M. Nekovee. A ternary lattice Boltzmann model for amphiphilic fluids. *Proc. R. Soc. Lond. A*, 456:2043–2047, 2000.
4. S. Chen, H. Chen, D. Martínez, and W. Matthaeus. Lattice Boltzmann model for simulation of magnetohydrodynamics. *Phys. Rev. Lett.*, 67(27):3776–3779, 1991.
5. P. Español and P. Warren. Statistical mechanics of dissipative particle dynamics. *Europhys. Lett.*, 30(4):191–196, 1995.
6. T.E. Faber. *Fluid Dynamics for Physicists.* Cambridge University Press, 1995.
7. E.G. Flekkøy, P.V. Coveney, and G.D. Fabritiis. Foundations of dissipative particle dynamics. *Phys. Rev. E*, 62(2):2140–2157, 2000.
8. N. González-Segredo, M. Nekovee, and P.V. Coveney. Three-dimensional lattice-Boltzmann simulations of critical spinodal decomposition in binary immiscible fluids. *Phys. Rev. E*, 67(046304), 2003.
9. J. Harting, M. Harvey, J. Chin, M. Venturoli, and P.V. Coveney. Large-scale lattice Boltzmann simulations of complex fluids: advances through the advent of computational grids. *Phil. Trans. R. Soc. Lond. A*, 363:1895–1915, 2005.

10. J. Harting, M. Venturoli, and P. V. Coveney. Large-scale grid-enabled lattice-Boltzmann simulations of complex fluid flow in porous media and under shear. *Phil. Trans. R. Soc. Lond. A*, 362:1703–1722, 2004.
11. Y. Hashimoto, Y. Chen, and H. Ohashi. Immiscible real-coded lattice gas. *Comp. Phys. Comm.*, 129(1–3):56–62, 2000.
12. 2003. HDF5 – a general purpose library and file format for storing scientific data, http://hdf.ncsa.uiuc.edu/HDF5.
13. P.J. Higuera, S. Succi, and R. Benzi. Lattice gas dynamics with enhanced collisions. *Europhys. Lett.*, 9(4):345–349, 1989.
14. S. Jury, P. Bladon, M. Cates, S. Krishna, M. Hagen, N. Ruddock, and P. Warren. Simulation of amphiphilic mesophases using dissipative particle dynamics. *Phys. Chem. Chem. Phys.*, 1:2051–2056, 1999.
15. A. Lees and S. Edwards. The computer study of transport processes under extreme conditions. *J. Phys. C.*, 5(15):1921–1928, 1972.
16. P.J. Love, M. Nekovee, P.V. Coveney, J. Chin, N. González-Segredo, and J.M.R. Martin. Simulations of amphiphilic fluids using mesoscale lattice-Boltzmann and lattice-gas methods. *Comp. Phys. Comm.*, 153:340–358, 2003.
17. A. Malevanets and R. Kapral. Continuous-velocity lattice-gas model for fluid flow. *Europhys. Lett.*, 44(5):552–558, 1998.
18. M. Nekovee, P.V. Coveney, H. Chen, and B. M. Boghosian. Lattice-Boltzmann model for interacting amphiphilic fluids. *Phys. Rev. E*, 62:8282, 2000.
19. J.-P. Rivet and J. P. Boon. *Lattice Gas Hydrodynamics.* Cambridge University Press, 2001.
20. T. Sakai, Y. Chen, and H. Ohashi. Formation of micelle in the real-coded lattice gas. *Comp. Phys. Comm.*, 129(1–3):75–81, 2000.
21. X. Shan and H. Chen. Lattice Boltzmann model for simulating flows with multiple phases and components. *Phys. Rev. E*, 47(3):1815–1819, 1993.
22. X. Shan and H. Chen. Simulation of nonideal gases and liquid-gas phase transitions by the lattice Boltzmann equation. *Phys. Rev. E*, 49(4):2941–2948, 1994.
23. S. Succi. *The Lattice Boltzmann Equation for Fluid Dynamics and Beyond.* Oxford University Press, 2001.
24. A. Wagner and I. Pagonabarraga. Lees-edwards boundary conditions for lattice Boltzmann. *J. Stat. Phys.*, 107:521, 2002. cond-mat/0103218].
25. A.J. Wagner and J. M. Yeomans. Phase separation under shear in two-dimensional binary fluids. *Phys. Rev. E*, 59(4):4366–4373, 1999.

The Effects of Vortex Generator Arrays on Heat Transfer and Flow Field

C.F. Dietz, M. Henze, S.O. Neumann, J. von Wolfersdorf, and B. Weigand

Institute of Aerospace Thermodynamics, University of Stuttgart
christopher.dietz@itlr.uni-stuttgart.de

Abstract. The effect of arrays of single-body, delta shaped vortex generators on heat transfer and flow field has been investigated numerically using RANS (Reynolds averaged Navier–Stokes) methods. The Reynolds number based on the hydraulic diameter of the channel in which the vortex generators are positioned is fixed at 300,000. For the closure of the equation system of the flow field a full differential Reynolds stress model has been used to capture the anisotropic effects of the induced vortex structures. To gain realistic results for the heat transfer the common approach for the closure of the Reynolds-averaged energy equation using a turbulent Prandtl number has been abandoned for explicit algebraic models which deliver more realistic results for complex flows. Simultaneously to the calculations measurements have been performed on some of the geometries to validate the numerical results.

1 Introduction

The focal point of this work is set to the augmentation of convective heat transfer, which is very important for the optimization of compact heat exchangers or in gas turbine cooling systems. In order to enhance the convective heat transfer turbulence promoters or vortex generators (VGs) can be used to manipulate the flow field, resulting in a local reduction of the thermal boundary layer thickness, and to benefit of their effect on thermal performance.

Longitudinal vortex structures have been the interest of research work for some time and quite some experimental results have been published while relatively few numerical studies for high Reynolds numbers are available so far. Wedge and delta shaped vortex generators in a square channel have been analysed regarding internal cooling purposes in gas turbine blades by Han et al. [6] among others. The applied Reynolds numbers ranged up to 80,000, based on the hydraulic diameter of the channel. It was found that the delta shaped configurations performed better than the wedge shaped arrangements concerning heat transfer as well as total pressure loss. Another noteworthy result was that the backward flow direction – which is the topic of this research – with the delta shaped vortex generator facing upstream (as shown

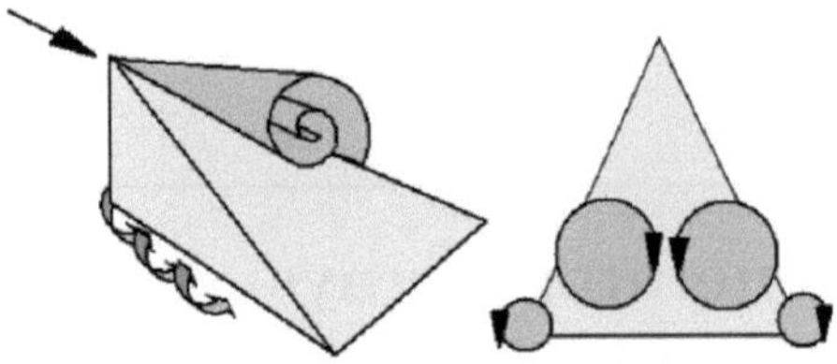

Fig. 1. Schematic view of a single delta shaped vortex generator

in Fig. 1) provided a 30% higher heat transfer than the forward flow configuration, while at the same time producing only an 11% higher total pressure loss. Liou et al. [16] reported similar findings for a single delta shaped vortex generator in a square channel.

An overview on the generation of longitudinal vortices and related heat transfer enhancement can be found in Jacobi & Shah [7] and Fiebig [4]. In these reviews, wing type vortex generators with applications in heat exchangers at lower Reynolds numbers were described and the heat transfer characteristics were compared with the flow structures. Details about the vortex structures behind vortex generators and their divergence along the flow path can be found in the reports of Pauley & Eaton [18] as well as Wendt et al. [20]. These also account for the interaction of the vortex structures behind arrays of vortex generators.

Other interesting areas of application for vortex generator structures are for example boundary layer separation control as reported by Schubauer & Spangenberg [19], lift enhancement on aircraft wings or noise reduction purposes. Those two topics are covered by the report of Lin [15].

This work deals with the impact of a multielement array of delta shaped full-body vortex generators on heat transfer and flow field in a 2:1 (bottom width to channel height) rectangular channel for a Reynolds number based on the hydraulic diameter of the channel of 300,000. A single vortex generator of this kind induces two main vortices moving downstream with diverging paths. The vorticity decays with the downstream motion and depends on vortex generator dimensions as well as on the main flow velocity. In vortex generator arrays these vortices interact with each other, either increasing or decreasing in intensity or changing the distance to the bottom wall (see Fig. 2). Both effects influence the heat transfer augmentation.

In order to find configurations which are suitable for application more knowledge regarding the heat transfer mechanism, especially the dependance on flow and turbulence field, has to be gathered. Numerical investigations have been performed on a set of configurations of two and three vortex generators, which will show the implications of varying distance between the VGs and different positioning (side-by-side or in line). To reflect the highly complex structure of the flow as accurately as possible a complete, differential Reynolds stress turbulence model was used to close the set of Reynolds averaged equations. For the energy equation an explicit, algebraic approach for the turbulent heat flux has been chosen, as those models are able to reflect

Fig. 2. Streamlines behind a pair of vortex generators colored by turbulent kinetic energy

the physical occurences more closely than the widely used approach using a turbulent Prandtl number.

In parallel to the numerical investigations heat transfer measurements have been performed using a transient method for some of the described configurations. Therefore, special attention is paid to a comparison between experimental and numerical results.

2 Theory

2.1 Flow Field Computation

The transport equations for the flow field computations used by *FLUENT* (see Fluent Inc. [5]) are quoted in the following for the sake of completeness. If the continuity equation and the Navier–Stokes equation are Reynolds-averaged – that is to say they are split into a mean and a fluctuating part – the set of equations

$$\frac{\partial}{\partial x_i}\left(\rho U_i\right) = 0 \tag{1}$$

$$\frac{\partial}{\partial x_j}\left(\rho U_i U_j\right) = -\frac{\partial p}{\partial x_i} + \frac{\partial}{\partial x_j}\left[\mu\left(\frac{\partial U_i}{\partial x_j} + \frac{\partial U_j}{\partial x_i} - \frac{2}{3}\delta_{ij}\frac{\partial U_l}{\partial x_l}\right)\right] + \frac{\partial}{\partial x_j}\left(-\rho\overline{u_i' u_j'}\right) \tag{2}$$

(with $i, j = 1, 2, 3$) is obtained for steady, incompressible flow without body forces (given for example by Kays et al. [9]). As this process introduces some new unknowns, more precisely the Reynolds stresses $-\overline{u_i' u_j'}$, turbulence models are needed to close the system of equations.

While there are many different approaches for the closure of the RANS[1] equations, for example the popular k-ϵ family of turbulence models (which is based on the original proposal of Launder & Spalding [14]), throughout this work a Reynolds Stress closure (e.g. Launder et al. [12], Launder [10, 11]) is used. This specific type of turbulence model has the advantage that anisotropies in the flow field can be captured. But at the same time it

[1] RANS: Reynolds-averaged Navier–Stokes

introduces six transport equations for the Reynolds stresses themselves plus one for the turbulent length scale (usually the dissipation rate ϵ of the turbulent kinetic energy) in the case of a three dimensional setup (vs. two for the k-ϵ models), which results in a total of 11 transport equations which have to be solved simultaneously. This heavily increases the computational demand.

For all computations the solver *FLUENT* in version 6.2 has been used, the discretization scheme was always of second order accuracy, and the option "enhanced wall treatment" was activated. This option deactivates wall functions in areas where the grid resultion is sufficiently fine to resolve the viscous sublayer. In the near wall area the Reynolds stresses are not used anymore to close the RANS equation set (however, the transport equations are solved throughout the complete domain). Instead the one-equation turbulence model of Wolfshtein [21] is applied. Details of the models can be seen in Fluent, Inc. [5], where additionally the low-Reynolds number extension to the Reynolds stress model – which is based on a proposal of Launder & Shima [13] to modify the coefficients in the pressure-strain modelling – is presented in detail.

2.2 Solution of the Energy Equation

The equation which has to be solved in the first place for heat transfer determination is the Reynolds-averaged energy equation. *FLUENT* uses the formulation

$$\frac{\partial}{\partial x_i}\left[(\rho E + p)\, U_i\right] = \frac{\partial}{\partial x_i}\left(k_f \frac{\partial T}{\partial x_i} - \rho c_p \overline{u_i' T'}\right) + S_h \tag{3}$$

(with $i = 1, 2, 3$) which incorporates the total specific energy $E = h - {}^{p}/_{\rho} + {}^{U^2}/_{2}$ as a variable. The part on the left hand side of Eq. (3) is heat transport through convection while the terms on the right hand side represent conductive heat transfer plus the turbulent heat flux and an arbitrary source term which is usually zero but can be used to implement external heat sources. The product $-\overline{u_i' T'}$ is the turbulent part of convective heat transfer and – in analogy to the Reynolds-averaged Navier–Stokes equations – arises from the process of Reynolds-averaging. The exact value is unknown and has to be modelled in order to solve the energy equation.

Boussinesq Analogy

The most common approach for determining the turbulent heat flux in today's computational fluid dynamics applications is to apply an analogy to the eddy viscosity concept for the Reynolds stresses. Here, the turbulent heat flux $-\overline{u_i' T'}$ is modelled with the assumption (Ferziger & Peric [3])

$$-\overline{u_i' T'} = \frac{{}^{\mu_t}/_{\rho}}{Pr_t} \frac{\partial T}{\partial x_i}, \tag{4}$$

(with $i = 1,2,3$) where μ_t is the turbulent viscosity, as delivered by the turbulence model. Pr_t is the turbulent Prandtl number that, for computations with turbulence models of higher order, is customarily taken as constant for a particular value of the Prandtl number Pr and is set to

$$Pr_t = 0.85 \tag{5}$$

for wall-bounded flows.

While the implemented Reynolds stress turbulence model for the flow field is able to describe the anisotropic effects present in the flow correctly, this clearly is not the case for the turbulent energy transport. The reason is that the turbulent viscosity, computed from the turbulent kinetic energy k and its dissipation rate ϵ, is only a scalar value and does not contain any dependencies on direction. Furthermore the turbulent heat flux in one direction depends only on the corresponding gradient while disregarding the other gradients.

Model of Younis et al.

The model of Younis et al. [22] enhances the isotropic approach with influences from Reynolds stresses as well as velocity gradients. The model is given as

$$\begin{aligned} -\overline{u_i'T'} &= C_1 \frac{k^2}{\epsilon} \frac{\partial T}{\partial x_i} + C_2 \frac{k}{\epsilon} \overline{u_i' u_j'} \frac{\partial T}{\partial x_j} + C_3 \frac{k^3}{\epsilon^2} \frac{\partial U_i}{\partial x_j} \frac{\partial T}{\partial x_j} \\ &+ C_4 \frac{k^2}{\epsilon^2} \left(\overline{u_i' u_k'} \frac{\partial U_j}{\partial x_k} + \overline{u_j' u_k'} \frac{\partial U_i}{\partial x_k} \right) \frac{\partial T}{\partial x_j} \end{aligned} \tag{6}$$

with $i, j, k = 1,2,3$ and the coefficients

$$C_1 = -0.0455 \qquad C_2 = 0.373$$
$$C_3 = -0.00373 \qquad C_4 = -0.0235.$$

The model in the presented form is not suited for usage near solid walls. The application of wall functions would be necessary, which is not desirable for the applications considered here. A proposal to solve this introduces a variable coefficient C_1 which is derived from Lumley's flatness factor and a turbulent Reynolds number as a criterion for the wall distance. This modified model performs well for complex flows, as is shown in Dietz et al. [1].

3 Numerical Setup

3.1 Vortex Generator Geometry

Figure 3 shows the base geometry of one of the vortex generators which have been the center of interest in this work. The vortex generator is delta shaped, and was used as a backward-facing wedge. The dimensions for all test cases are

$$H = 0.026\,\mathrm{m} \qquad W = 0.065\,\mathrm{m} \qquad L = 0.065\,\mathrm{m}.$$

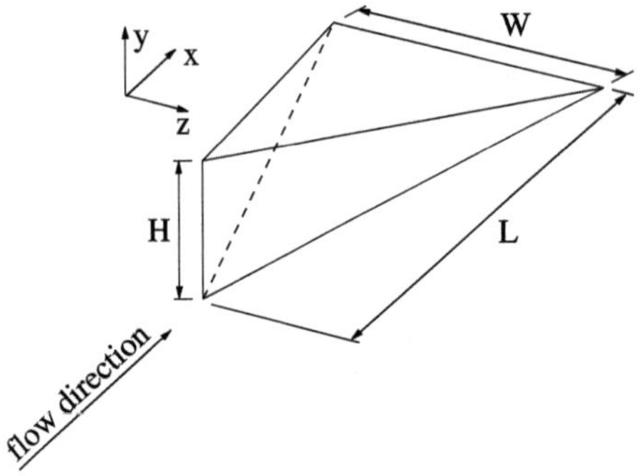

Fig. 3. Geometry of a single vortex generator

The channel in which the vortex generator were placed has a width of 0.4 m and is 0.2 m high, so the flow behind the vortex generator is not influenced by the side walls.

3.2 Description of the Test Cases

Different setups of two and three vortex generators with lateral as well as longitudinal displacement have been tested for a Reynolds number of 300,000. Calculations have been performed for $\Delta W/W = 0.5$, $\Delta W/W = 1.0$, and $\Delta W/W = 1.5$, as well as for $\Delta L/L = 1.0$, $\Delta L/L = 2.0$, and $\Delta L/L = 3.0$. As a special case a three VG setup with $\Delta L/L = 1.0$ also has been included in this report. Figure 4 shows the configurations for the two groups which were investigated.

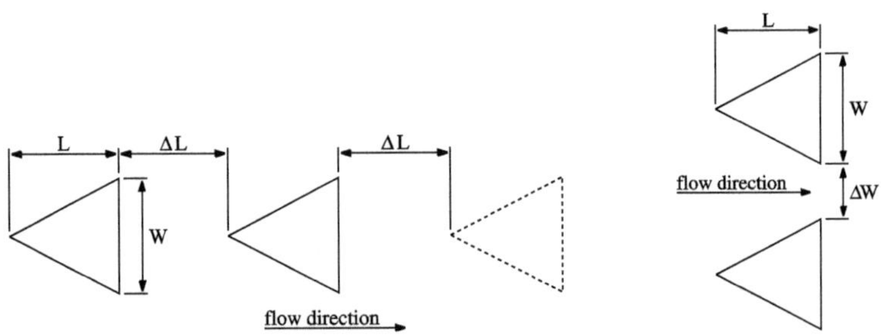

Fig. 4. Test case definitions, longitudinal (*left*) and lateral (*right*) displacement

3.3 Coordinate System

The part farthermost downstream of the channel represents the main test area. The main coordinate system is based on the floor panel at the position of the beginning of this test area (see Fig. 5). Looking in the direction of the main flow, which represents the x-coordinate (that starts at the beginning of

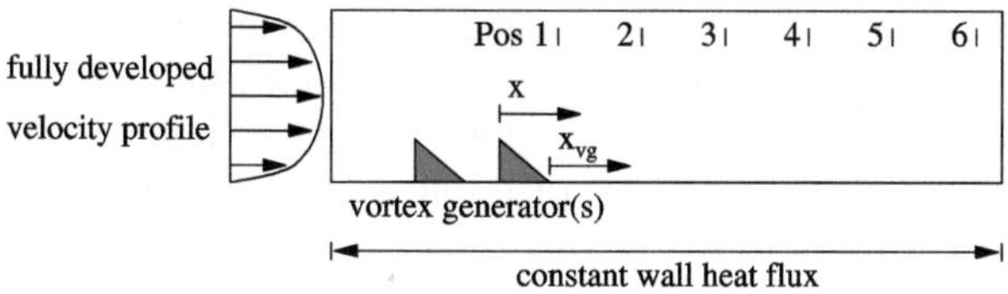

Fig. 5. Coordinate system definition

the vortex generator), the y-coordinate points up and by pointing to the right the z-coordinate completes the system. An additional coordinate x_{vg} is used which has its origin at the trailing edge of the vortex generator. Dealing with the length and mounting position of the devices, the point of origin of x_{vg} is 6.5 mm further downstream of the one for x. If multiple vortex generators have been used in a row, they have been placed in front of the vortex generator at which the coordinate system has its origin as can be seen in Fig. 5. Different positions of constant x have been defined for evaluation of the flow field. The first is positioned at $x_{vg} = 10$ mm, the following have a distance of 100 mm to their predecessor.

3.4 Measurement Technique

A transient measurement technique using Thermochromic Liquid Crystals (TLC) is used. The dimensions for the cross section of the test channel and the VG correspond to the numerical calculations. A bypass, connecting two separate circuits (cold and hot circuit) of the wind tunnel, leads hot air into the test section, resulting in a prompt change of the fluid temperature. For optical access the channel walls are made of Perspex with a low heat conductivity and a thickness of 20 mm, to fulfil the assumption of a semi-infinite plate, which is necessary for the evaluation procedure. For a semi-infinite plate with a convective boundary condition and a step change of the fluid temperature, the solution of the one-dimensional, transient heat conduction equation leads to a relation of the wall temperature T_W and the heat transfer coefficient h

$$\frac{T_W - T_0}{T_B - T_0} = 1 - \exp\left(\frac{h^2 t}{(k\rho c_p)_s}\right) \operatorname{erfc}\left(\frac{h\sqrt{t}}{\sqrt{(k\rho c_p)_s}}\right) \tag{7}$$

T_0 is the initial temperature of the test facility (channel wall and fluid temperature) and T_B is the fluid temperature after the step change. Since the temperature change is not an ideal step, the real temperature evolution is adapted via the Duhamel Principle, see Metzger & Larson [17]. The progression of the wall temperature indicated by the TLC is captured via CCD-cameras. The time history for one indicated temperature (here the maximum for the green value is used) allows the calculation of the heat transfer coefficient in Eq. (8).

Measurement uncertainties are dependent on the range of the heat transfer coefficient. High values for h imply a short time t for the change of colours during the transient measurement, which results in a high measurement error in time. The uncertainty for long times until the change of colours takes place is dominated by the measurement error of the temperature and the material properties of the channel. The present investigation leads to a maximum standard deviation of 11.4%, the values for the averaged heat transfer coefficient are given with a standard deviation of 5.5%.

3.5 Grid Generation & Numerical Accuracy

The computational grids have been generated using the grid generator *Centaur*, as described by Kallinderis et al. [8]. The grids are hybrid, with prismatic elements near the walls and unstructured tetrahedral elements in the far field. This ensures a fine resolution of the boundary layer, while still taking advantage of the characteristics of unstructured tetrahedral grids. The grids feature a dimensionless wall distance of $y^+ \leq 1$ for all applied Reynolds numbers, so no wall-functions had to be used. This requirement as well as the complexity of the differential Reynolds stress turbulence model made it necessary to use very finely resolved grids with a total number of cells ranging from 4.5 million cells for the single vortex generator to 5.8 million cells for the setup consisting of three VGs. The computational cost for each test case amount to approximately 360 CPU hours on a Cray Opteron Cluster using 2GHz CPUs and use up to 7GB of memory.

The discretization scheme was always of second order accuracy. To ensure grid independence various computations have been performed with different parameters until a grid independent solution was found. As inlet boundary condition a fully developed channel flow has been used where all relevant variables (the velocity field, as well as the turbulent quantities) have been taken into account. With this procedure better convergence rates could be achieved. While in the experiments the channel flow was not fully developed at the position where the VG was placed, the VG was nonetheless fully embedded in the boundary layer at this point, so the differences in the boundary conditions are negligible. As thermal boundary condition a constant heat flux has been applied at the lower wall as can be seen in Fig. 5.

4 Results

4.1 Effect of a Single Vortex Generator

Figure 6 shows the flow field at positions 1 and 3 for the reference case ($Re =$ 300,000) using one single vortex generator. The two main vortex streaks and their dissipation along the flow path can be clearly seen. The shown normal velocities are normalized by the bulk velocity for the corresponding case.

In Fig. 7 the heat transfer enhancement for the baseline case is presented. All values are given in normalized form to provide a comparison of the heat transfer enhancement between different Reynolds numbers. As a reference the

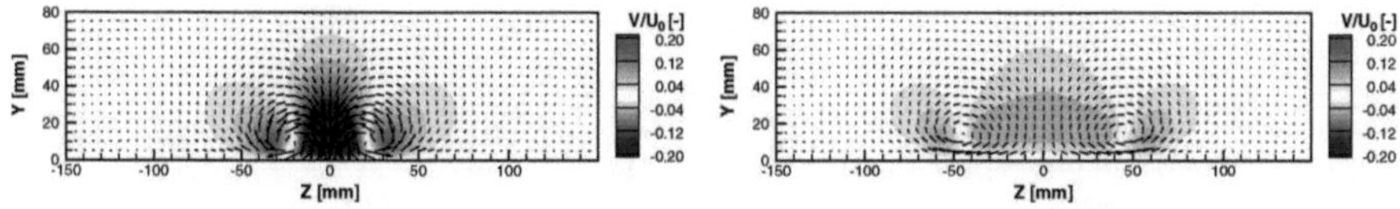

Fig. 6. Velocity vectors and normal velocity at positions 1 (*left*) and 3 (*right*) for a single VG

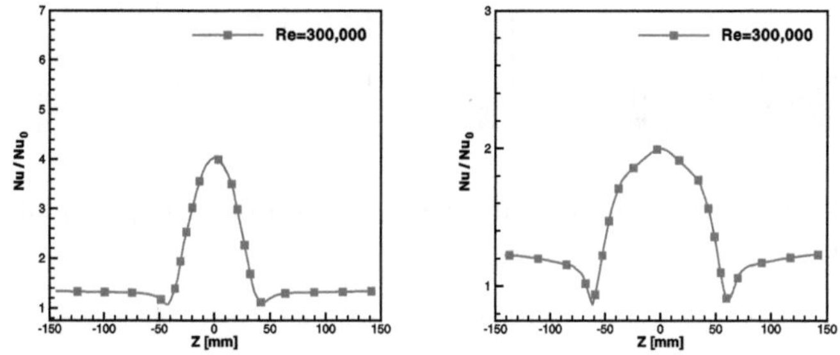

Fig. 7. Normalized Nusselt numbers at positions 1 (*left*) and 3 (*right*) for a single VG

definition of Dittus & Boelter [2] for heat transfer in a fully developed smooth channel flow

$$Nu_0 = 0.023Re^{0.8}Pr^{0.4} \tag{8}$$

has been used.

4.2 Effect of Lateral Displacement

Figures 8 to 10 show the velocity field for different distances between vortex generators being placed side by side for a Reynolds number of 300,000. The distances are ${}^{\Delta W}/_{W} = 0.5$, ${}^{\Delta W}/_{W} = 1.0$, and ${}^{\Delta W}/_{W} = 1.5$ for the Figs. 8, 9, and 10, respectively. It can be observed that for the small distance between the vortex generators the vortex centers move away from the bottom wall by a small distance, while an area with a very pronounced upwash zone is formed by the two colliding vortices. In the same context the strength of the inner downwash areas increases. For the larger distances this effect is not visible in the same degree, although for a distance of ${}^{\Delta W}/_{W} = 1.5$ the offshoots of the vortices still collide and increase in strength compared to the outer vortex pairs.

For the heat transfer this means that the peak Nusselt number should increase with decreasing distance of the vortex generators to each other, which is due to the stronger downwash area which is formed. On the other hand the heat transfer has to be lower directly between the VGs as the upwash region is constricting the heat exchange here. In Fig. 11 it can indeed be seen that the maximum Nusselt number is slightly higher directly behind the VG for the

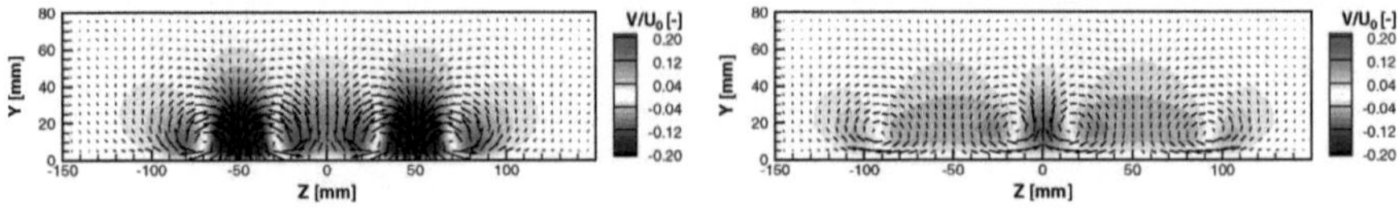

Fig. 8. Velocity vectors and normal velocity at positions 1 (*left*) and 3 (*right*) for $\Delta W/W = 0.5$

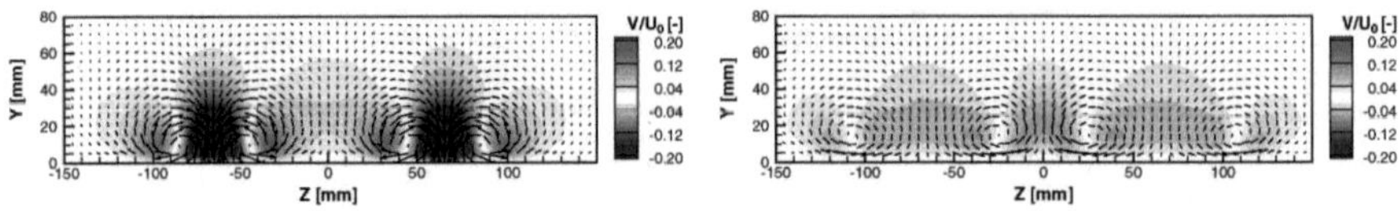

Fig. 9. Velocity vectors and normal velocity at positions 1 (*left*) and 3 (*right*) for $\Delta W/W = 1.0$

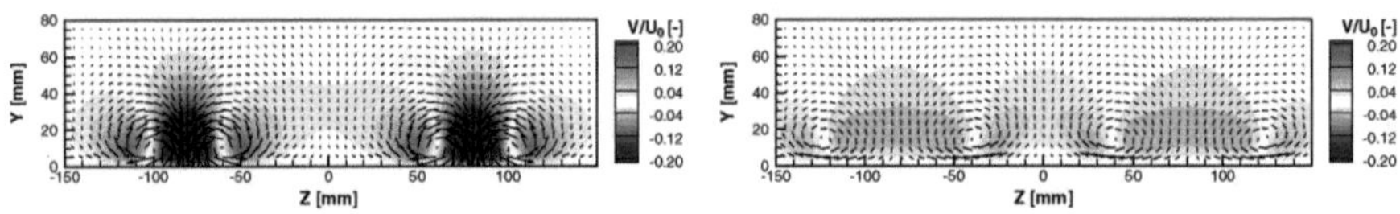

Fig. 10. Velocity vectors and normal velocity at positions 1 (*left*) and 3 (*right*) for $\Delta W/W = 1.5$

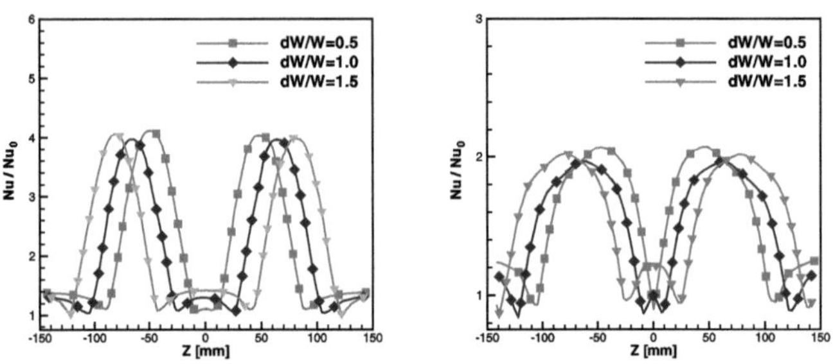

Fig. 11. Normalized Nusselt numbers at positions 1 (*left*) and 3 (*right*) for $\Delta W/W =$ 0.5 to 1.5

lower vortex generator distance, while the heat transfer distributions becomes unsymmetric towards the centerline with increasing distance from the VG.

The decrease of heat transfer between the VGs is not as obvious, as the low heat enhancement rate for the $\Delta W/W = 0.5$ case accords to the low heat transfer area which can also be found in the $\Delta W/W = 1.5$ case.

4.3 Effect of Longitudinal Displacement

With increasing logitudinal distance from the preceeding vortex generator the vortex streaks have more space to dissipate from each other until the position of the next vortex generator is reached. This can be observed in Figs. 12

to 14, where the velocity vectors are shown for different distances between the vortex generators ($\Delta L/L = 1.0$ for Fig. 12, $\Delta L/L = 2.0$ for Fig. 13, and $\Delta L/L = 3.0$ for Fig. 14) at a Reynolds number of $Re = 300{,}000$. It becomes clear that the downwash region gets condensed directly behind the vortex generator with increasing distance between the VGs, the region in which the fluid moves tangential to the wall grows, and the upwash area splits up. The last mentioned process starts for the $\Delta L/L = 2.0$ case, where the vortices produced by the different vortex generators can be clearly distinguished. As the fluid moves further downstream to position three, the intensity of the down and upwash regions seem to decrease faster with increasing distance between the vortex generators.

Figure 15 depicts a more special case. Here, three vortex generators in a row, each with a distance of $\Delta L/L = 1.0$ from the preceeding one, have been examined. It can be seen that the third VG accelerates the dissipation

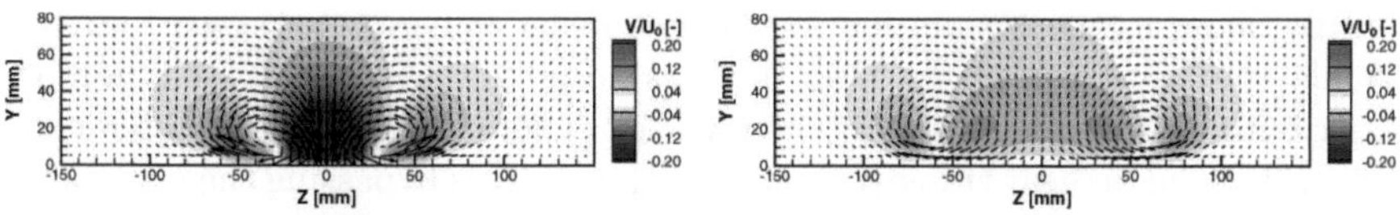

Fig. 12. Velocity vectors and normal velocity at positions 1 (*left*) and 3 (*right*) for $\Delta L/L = 1.0$

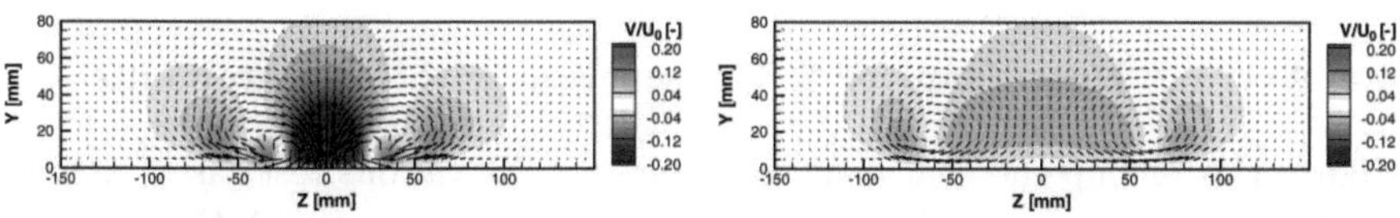

Fig. 13. Velocity vectors and normal velocity at positions 1 (*left*) and 3 (*right*) for $\Delta L/L = 2.0$

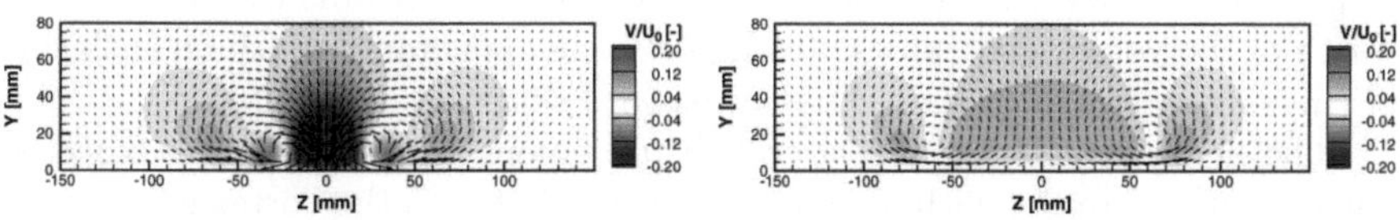

Fig. 14. Velocity vectors and normal velocity at positions 1 (*left*) and 3 (*right*) for $\Delta L/L = 3.0$

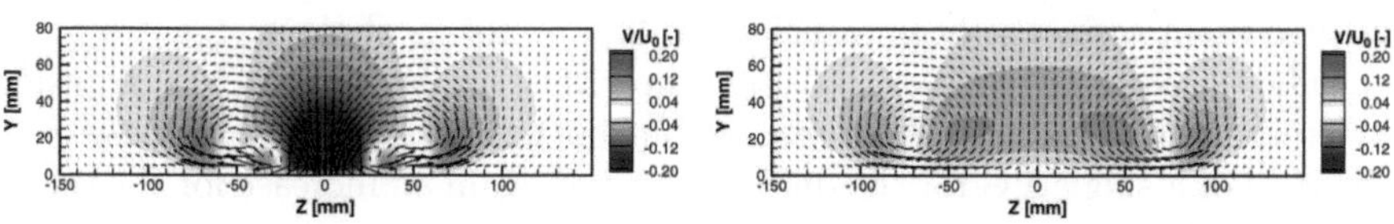

Fig. 15. Velocity vectors and normal velocity at positions 1 (*left*) and 3 (*right*) for three VGs

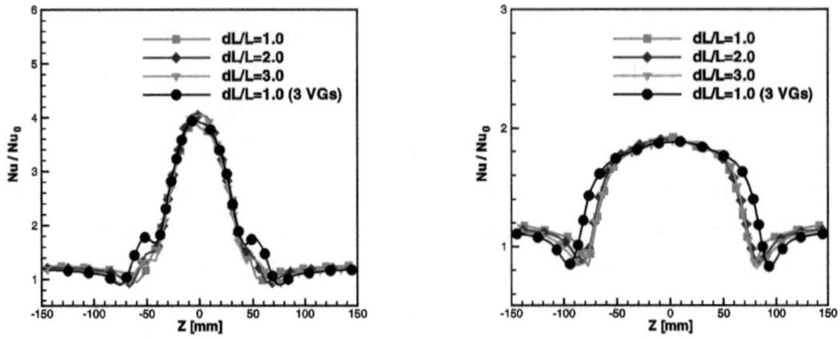

Fig. 16. Normalized Nusselt numbers at positions 1 (*left*) and 3 (*right*) for $\Delta L/L =$ 1.0 to 3.0 and three VGs

mechanism of the vortex streaks, which thus cover a larger area at position three in comparison with the dual VG alignments.

Surprisingly the very differential behaviour of the four configurations does hardly show in the results for the heat transfer enhancement (see Fig. 16). While the cases using two VGs show essentially the same Nusselt number distributions, the results for the three vortex generators show some differences. First, the peak in heat transfer caused by the vortices bypassing the last vortex generators is clearly visible. Second, the increasing area covered by the vortices results in a larger area of high heat transfer in lateral direction, altough not by as much as the results for the flow field lead to believe.

4.4 Average Heat Transfer & Comparison with Measurements

Figure 17 sums up the heat transfer results for the various test cases. A comparison with experimental results is also given. As a representative area for the averaged Nusselt number an area with the length of $4L$ and a width of $2W$ was chosen, as the vortex streaks do not leave this area. For the lateral displacement cases the definition of the test area width was extended to the width of the vortex generators plus half a vortex generator width on each side of the vortex generator pair. These areas are sketched in Fig. 18.

The heat transfer for the lateral displacement cases shows the expected pattern and decreases with increasing distance between the vortex generators. As could be seen earlier in Fig. 11 the heat transfer on the centerline decreases but as the areas of high heat transfer directly behind the vortex generators move close to each other, the averaged heat transfer increases. For the lateral displacement cases the heat transfer results are not as obvious.

The $\Delta L/L = 1.0$ to $\Delta L/L = 3.0$ cases show very similar levels of heat transfer in experimental and numerical results, but both show a distinct increase over the level of a single VG for the dual setups and again an increase for the setup of three VGs. If comparing numerical and experimental results it can be stated that the total level of heat transfer is overpredicted by 25%. Nonetheless the trend is predicted correctly. In fact, if the uncertainty in the experimental

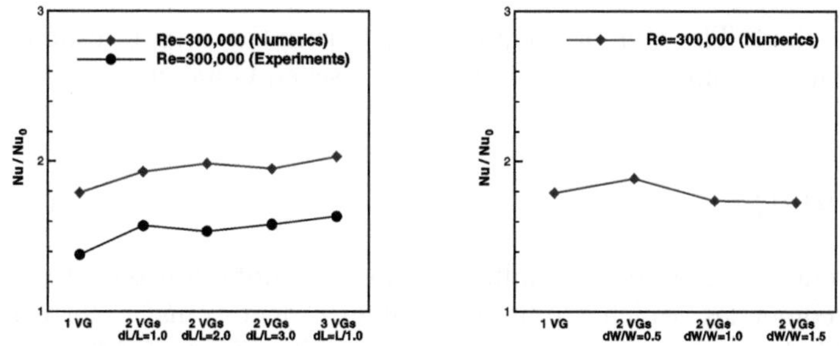

Fig. 17. Normalized average Nusselt numbers for all test cases, longitudinal (*left*) and lateral (*right*) displacement

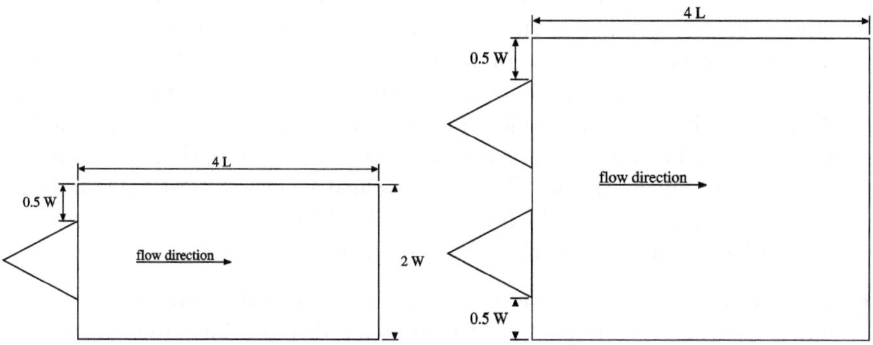

Fig. 18. Definition of the areas used to evaluate the average Nusselt numbers

results (about 5.5 %) is considered the numerical and experimental data sets seem to be connected through a constant factor. On analyzing the results in detail, it is observed that the heat transfer rate obtained through experiments drops at a lower level in between the areas which are occupied by the vortices. This behaviour is not shown in the numercal results as could be seen the results presented earlier, which is likely to be evoked by the near-wall treatment of the used turbulence model, which does not yet resolve the drop in turbulence in the area between the vortices.

5 Computational Resources

All results presented in this paper have been computed on the CRAY Opteron Cluster at the HLRS. The used grids contain between 4.8 and 5.6 million cells to allow a fine resolution of the boundary layer at the channel walls and to achieve the wall distance of the first grid point which is necessary in conjunction with low Reynolds capable turbulence models.

The time required for each computation is usually 90 hours real-time on 4 CPUs depending on the time which is needed until a stable, stationary solution is reached. The memory requirement during the computation on hybrid meshes with cell numbers this high is about seven Gigabytes.

6 Summary

The present paper presents results for different configurations of vortex generators geometries, which are used to enhance heat transfer, for example in turbine blades or in a combustion chamber. The effect of different configurations with lateral and longitudinal displacement have been tested. For the computations a complete differential Reynolds stress turbulence model with enabled near-wall treatment has been used to resolve the complex turbulence structure which is present in the areas occupied by the vortices.

To improve the prediction of heat transfer the commercial flow solver *FLUENT* has been enhanced by a user defined routine which implements different explicit algebraic models for the turbulent heat flux. Nonetheless some of the effects which are present in the flow field prediction do seemingly not influence the heat transfer. The lack of sensitiveness is most likely a result of the near-wall treatment which is applied by *FLUENT*.

The following key observations have been made:

- The interaction of the vortices generated by the delta shaped vortex generators has a clear impact on the velocity distribution after the vortex generators. This effect is dependant on the distance between the vortex generators (lateral as well as longitudinal), as well as on the flow velocity.
- The influence on the heat transfer yields similar results, and is in general aligned with the results for the flow field.
- Comparison with experimental results are possible in comparing different cases relatively to each other. Results for absolute values could not be reproduced.

Further numerical studies will comprise of more complex setups of several vortex generators as well as investigations of periodic setups to get a clear overview of the usability of these structures in practical applications. As the near-wall treatment leads to non-optimal results for the heat transfer, work will be done in this area to improve the prediction quality for the heat transfer while maintaining the good results for the flow field.

Acknowledgements

The authors greatly appreciate the High Performance Computing Center Stuttgart (HLRS) for support and supply of computational time on the high performance computers. Further, the financial support of this project by the DFG (Deutsche Forschungsgemeinschaft) is greatly acknowledged.

References

1. C.F. Dietz, S.O. Neumann, and B. Weigand. A comparative study of the performance of explicit algebraic models for the turbulent heat flux. *Submitted to Numerical Heat Transfer, Part A*, 2006.
2. F.W. Dittus and L.M.K. Boelter. Heat transfer in automobile radiators of the tubular type. *Int. Comm. Heat Transfer*, 12:3–22, 1985.
3. J.H. Ferziger and M. Perić. *Computational Methods for Fluid Dynamics.* Springer, 2002.
4. M. Fiebig. Vortices, generators and heat transfer. *Trans IChemE*, 76:108–123, 1998.
5. Fluent, Inc. *Fluent 6.2 – User's guide*, 2005.
6. J.-C. Han, J.J. Huang, and C.P. Lee. Augmented heat transfer in square channels with wedge-shaped and delta-shaped turbulence promoters. *Enhanced Heat Transfer*, 1(1):37–52, 1993.
7. A.M. Jacobi and R.K. Shah. Heat transfer surface enhancement through the use of logitudinal vortices: A review on recent progress. *Experimental Thermal and Fluid Science*, 11:295–309, 1995.
8. Y. Kallinderis, A. Khawaja, and H. McMorris. Hybrid prismatic/tetrahedral grid generation for viscous flows around complex geometries. *AIAA Journal*, 34(2):291–298, 1996.
9. W.M. Kays, M.E. Crawford, and B. Weigand. *Convective Heat and Mass Transfer.* McGraw-Hill Series in Mechanical Engineering, 2004.
10. B.E. Launder. Second-moment closure and its use in modelling turbulent industrial flows. *International Journal for Numerical Methods in Fluids*, 9:963–985, 1989.
11. B.E. Launder. Second-moment closure: Present... and future? *International Journal of Heat & Fluid Flow*, 10(4):282–300, 1989.
12. B.E. Launder, G.J. Reece, and W. Rodi. Progress in the development of a reynolds-stress turbulence closure. *Journal of Fluid Mechanics*, 68(3):537–566, 1975.
13. B.E. Launder and N. Shima. Second-moment closure for the near-wall sublayer: Development and application. *AIAA Journal*, 27(10):1319–1325, 1989.
14. B.E. Launder and D.B. Spalding. The numerical computation of turbulent flows. *Computer methods in applied mechanics and engineering*, 3:269–289, 1974.
15. J.C. Lin. Review of research on low-profile vortex generators to control boundary-layer separation. *Progress in Aerospace Sciences*, 38:389–420, 2002.
16. T.-M. Liou, C.-C. Chen, and T.-W. Tsai. Heat transfer and fluid flow in a square duct with 12 different shaped vortex generators. *Journal of Heat Transfer*, 122:327–335, 2000.
17. D. Metzger and D. Larson. Use of melting point surface coatings for local convection heat transfer measurements in rectangular channel flow with 90-deg turns. *Journal of Heat Transfer*, 108:48–54, 1986.
18. W.R. Pauley and J.K. Eaton. Experimental study of the development of logitudinal vortex pairs embedded in a turbulent boundary layer. *AIAA Journal*, 26(7):816–823, 1988.
19. G. Schubauer and W. Spangenberg. Forced mixing in boundary layers. *Journal of Fluid Mechanics*, 8:10–32, 1960.

20. B.J. Wendt, I. Greber, and W.R. Hingst. Structure and development of streamwise vortex arrays embedded in a turbulent boundary layer. *AIAA Journal*, 31(2):319–325, 1993.
21. M. Wolfshtein. The velocity and temperature distribution of one-dimensional flow with turbulence augmentation and pressure gradient. *International Journal of Heat & Mass Transfer*, 12:301–318, 1969.
22. B.A. Younis, C.G. Speziale, and T.T. Clark. A rational model for the turbulent scalar fluxes. *Proceedings of the Royal Society*, 461:575–594, 2005.

Investigation of the Influence of the Inlet Geometry on the Flow in a Swirl Burner

Manuel García-Villalba[1], Jochen Fröhlich[2], and Wolfgang Rodi[1]

[1] Institut für Hydromechanik, Universität Karlsruhe
villalba@ifh.uka.de, rodi@ifh.uka.de
[2] Institut für Techische Chemie und Polymerchemie, Universität Karlsruhe
froehlich@ict.uni-karlsruhe.de

Abstract. A series of Large Eddy Simulations (LES) of non-reacting flow in a swirl burner has been performed. The configuration consists of two unconfined co-annular jets at a Reynolds number of 81500. The flow is characterized by a Swirl number of 0.93. Two cases are studied differing with respect to the axial location of the inner pilot jet. It was observed in a companion experiment (Bender and Büchner, 2005) [1] that when the inner jet is retracted the flow oscillations are considerably amplified. The present simulations allow to understand this phenomenon: the recirculation zone and the jet interact in such a way that large scale coherent structures are generated. The resulting spectra correspond well to the experimental data.

1 Introduction

In recent years, there has been increased demand for gas turbines that operate in a lean premixed mode of combustion with the objetive of meeting stringent emission goals. Highly turbulent swirl-stabilized flames are often used in this context. However, swirling flows are prone to flow instabilities which can trigger combustion oscillations and cause damage to the device. Lean premixed burners in modern gas turbines often make use of a richer pilot flame which is typically introduced near the symmetry axis.

In order to prevent the appearance of undesired flow instabilities, it is necessary to understand the underlying physical phenomena. Several mechanisms have been identified in the literature as potential triggers of combustion instabilities. There is, however, no consensus about the real importance of each of them. Lieuwen *et al.* [2] suggested that heat-release oscillations excited by fluctuations in the composition of the reactive mixture entering the combustion zone are the dominant mechanism responsible for the instabilities observed in the combustor. Other authors [3, 4] favour the in-phase formation and combustion of large-scale coherent vortical structures. In premixed combustors, these large-scale structures play an important role in combustion and heat-release processes by controlling the mixing between the fresh mixture and hot combustion products [5].

The formation of large-scale coherent structures is a fundamental fluid-dynamical problem, which must be understood also in the absence of combustion. Large Eddy Simulation (LES) is a particularly suitable approach for studying this problem. It allows the treatment of high-Reynolds-number flows and at the same time the explicit computation of these structures. If properly conducted, LES should have only limited sensitivity to modelling assumptions. In the context of swirling flows, LES was first applied by Pierce and Moin [6]. Wang *et al.* [7] performed LES of swirling flow in a dump combustor and studied the influence of the level of swirl on the mean flow and on turbulent fluctuations. LES has also been used in combination with other techniques like acoustic analysis; for example, Roux *et al.* [8] studied the interaction between coherent structures and acoustic modes and found important differences between iso-thermal and reactive cases.

In the present paper LES is used to study the iso-thermal flow in a swirl burner for two different configurations. The analysis of the results focuses on the strength and sensitivity of flow instabilities generating large-scale coherent structures. In [9, 10] the present authors performed LES of an unconfined annular swirling jet without pilot jet and with a pilot jet issuing into the flow at the same axial position as the main jet. Instabilities leading to large-scale coherent structures were detected and identified to be responsible for the oscillations observed in the corresponding experiment.

2 The Swirl Burner

The present study has been undertaken within project A6 of the Collaborative Research Centre SFB606 [3] established at the University of Karlsruhe. In the companion project C1 of SFB606 a new co-annular swirl burner was developed [1]. The advantage of this burner is its versatility: Swirl generation in the main jet and the pilot jet can be realized in different manners. Figure 1 presents the configuration with tangential swirl generation for the main jet. Furthermore, the axial position of the pilot jet can be changed (Fig. 1).

A large number of experiments were performed with this burner in several configurations including isothermal and reactive cases [1, 11]. For the present numerical investigation the most suitable ones have been selected. In particular, measurements taken with LDA are available for the configuration with tangential swirl generation and co-rotating swirl. Three different axial positions of the inner jet were experimentally investigated.

In all cases the co-annular jets issue into an ambient of the same fluid which is at rest in the experiment. The outer radius of the main jet, $R = 55$ mm, is used as the reference length here. The reference velocity is the bulk

[3] http://www.sfb606.uni-karlsruhe.de

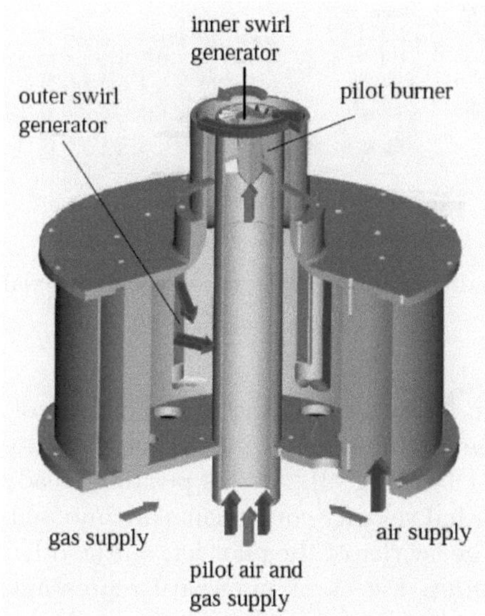

Fig. 1. Sketch of the burner (taken from [1])

velocity of the main jet $U_b = 22.1$ m/s and the reference time is $t_b = R/U_b$. The inner radius of the main jet is $0.63R$. For the pilot jet the inner radius is $0.27R$ and the outer radius is $0.51R$. The mass flux of the pilot jet is 10 % of the total mass flow. The Reynolds number based on the bulk velocity of the main jet U_b and R is $Re = 81000$. The swirl number is defined as

$$S = \frac{\int_0^R \rho \langle u_x \rangle \langle u_\theta \rangle r^2 \, dr}{R \int_0^R \rho \langle u_x \rangle^2 r \, dr}, \tag{1}$$

where $\langle u_x \rangle$ and $\langle u_\theta \rangle$ are the mean axial and azimuthal velocities respectively. Its value at the burner exit is $S = 0.93$.

3 Computational Details

3.1 Computational Grid and Boundary Conditions

The computational domain was defined using the experience of previous studies [10]. It extends $32R$ downstream of the burner exit located at $x/R = 0$ and $12R$ in radial direction. It also covers part of the inlet ducts as illustrated by the zooms in Fig. 2. The block-structured mesh consists of about 8.5 million cells with 160 cells in azimuthal direction. The grid is stretched in both the axial and radial direction to concentrate points close to the nozzle and the inlet duct walls. The stretching factor is everywhere less than 5 %.

The specification of the inflow conditions for both jets requires a strong idealization. For the main jet, the way the swirl is introduced is not so critical

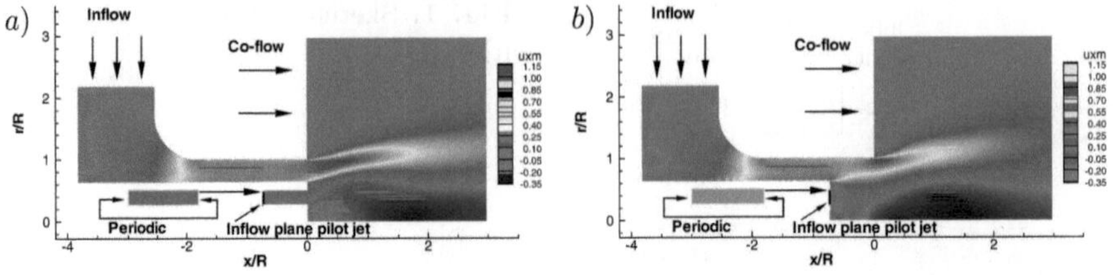

Fig. 2. Numerical setup and boundary conditions. Gray scale represents mean axial velocity **a)** $x_{pilot} = 0$. **b)** $x_{pilot} = -0.73R$

because the swirler is located upstream, far away from the region of interest. Therefore, the flow is prescribed at the circumferential inflow boundary located at the beginning of the inlet duct (see Fig. 2). At this position steady top-hat profiles for the radial and azimuthal velocity components are imposed. This procedure was validated in [10]. The swirler of the pilot jet, on the other hand, is located directly at the jet outlet, Fig. 1. A numerical representation of this swirler would be very demanding because of the large number of blades and was therefore not considered in the present investigation. Instead, the inflow conditions for the pilot jet were obtained by performing simultaneously a separate, streamwise periodic LES of developed swirling flow in an annular pipe (see Fig. 2) using body forces to generate co-rotating swirl with $S_{pilot} = 2$ as described in [12], where S_{pilot} is the swirl number of the pilot jet only. Recall that S_{pilot} has little impact on the swirl number of the entire flow due to the small mass flux of this stream. No-slip conditions were applied at solid walls. The fluid entrained by the jet is fed in by a mild co-flow of 5 % of U_b. By using different values of the co-flow velocity it was shown in [13] that the flow development is not sensitive to this conditions. Free-slip conditions were applied at the open lateral boundary. A convective outflow condition was used at the exit boundary.

3.2 The Computer Simulation Tool and Computational Effort

The simulations were performed with the in-house code LESOCC2 (Large Eddy Simulation On Curvilinear Coordinates). The code has been developed at the Institute for Hydromechanics. It is the successor of the code LESOCC developed by Breuer and Rodi [14] and is described in its most recent status in [15]. The code solves the Navier–Stokes equations on body-fitted, curvilinear grids using a cell-centered Finite Volume method with collocated storage for the cartesian velocity components. Second order central differences are employed for the convection as well as for the diffusive terms. The time integration is performed with a predictor-corrector scheme, where the explicit predictor step for the momentum equations is a low-storage 3-step Runge–Kutta method. The corrector step covers the implicit solution of the Poisson equation for the pressure correction (SIMPLE). The scheme is of second order accuracy in time because the Poisson equation for the pressure correction is not solved during the sub-steps of the Runge–Kutta algorithm in order to save CPU-time. The Rhie and Chow momentum interpolation [16] is applied to avoid pressure-velocity decoupling. The Poisson equation for the pressure increment is solved iteratively by means of the 'strongly implicit procedure' [17]. Parallelization is implemented via domain decomposition, and explicit message passing is used with two halo cells along the inter-domain boundaries for intermediate storage. One of the most important features of LESOCC2 is the possibility to use block-structured grids with an unstructured arrangement of the blocks, which was not possible with the previous version of the code. This feature allows the simulation of very complex geometries, like the co-annular burner in the present study.

The computations were carried out using 32 processors on the HP XC6000. As mentioned above, the total number of grid nodes for a typical run was 8.5×10^6, with 3.4×10^7 unknowns. The problem was split into 263 sub-domains which were distributed into the 32 processors used. Such a large number of sub-domains was needed to be able to balance the load of the processors. The geometry of the domain is such that blocks of very different sizes would have been obtained if a more reduced number of blocks were selected. The inter-processor communication was performed with standard MPI, and each one of the 32 processors has approximately 0.26×10^6 nodes, which easily fit into the memory. Of the various possible options to perform the domain decomposition, the one adopted does not necessarily minimise the number of inter-processor communication, but it was found to produce the minimum reduction of the convergence rate. On average, each simulation required a total number of 300.000 time steps. For each run it was possible to get 32 processors for nearly 24 hours a day for an elapsed time of approximately 1495 hours ($\sim$ 62 days). The total CPU time for each run on 32 processors was therefore of the order of 47.000 hours.

4 Results

In the experiments [1] three different axial positions of the pilot jet were studied. For the isothermal flow without external forcing, it was observed that axial retraction of the central jet into the duct leads to an increased amplitude of flow oscillations reflected by audible noise. In order to investigate this phenomenon, two of the cases were selected. In the first one both jets exit at the same position. In the second case, the pilot jet is retracted by 40 mm, i.e. $-0.73R$. This retraction of the pilot jet generates a double expansion for the main jet visualized in Fig. 2*b*. LDA measurements are available for both cases. In particular, radial profiles of mean and RMS velocities have been measured at four axial stations in the near field of the burner. Only for the case with retraction, power spectra of velocity fluctuations were recorded.

Table 1. Overview of the simulations performed

Sim	Grid	x_{pilot}	S in pilot jet
1	2.5 Mio	0	1 at $x/R = 0$
2	8.5 Mio	0	2 at $x/R = -0.73R$
3	8.5 Mio	$-0.73R$	2 at $x/R = -0.73R$

The simulations are summarized in Table 1. Sim 1 was calculated previously [9] and it is only included here for comparison. Some discrepancies between the results of Sim 1 and the experiment were observed close to the burner exit at $x/R = 0.1$. This was mainly due to the model used for the pilot jet. An attempt to improve the agreement was made by slightly changing the boundary conditions. Figure 2*a* displays a zoom of the inflow region for Sim 2. In Sim 2 the inflow plane for the pilot jet was shifted upstream to $x/R = -0.73$, i.e. the annular duct is included in the main LES. To compensate for the additional relaxation of swirl between $x/R = -0.73$ and $x/R = 0$ the swirl number has been increased to $S = 2$ in Sim 2, while in Sim 1 it was $S = 1$. The same boundary conditions were employed in Sim 2 and Sim 3. Figure 2*b* displays a zoom of the inflow region for $x_{pilot} = -0.73R$. In the case $x_{pilot} = 0$ the inflow region differs because the wall separating main and pilot jet and the cylindrical centre body reach until $x = 0$, with the inflow plane for the pilot jet still located at the same position $x/R = -0.73$.

The same grid was employed in Sims 2 and 3. In Sim 2, the wall separating the annular ducts for main and pilot jet, and the cylindrical centre body were introduced by blocking corresponding cells. The grid employed for Sim 1 is coarser by a factor of about 3.

Figure 3 shows the improvement of the results at $x/R = 0.1$ in Sim 2 compared to Sim 1. As a result of the change in boundary conditions and the grid refinement both mean velocity components and the corresponding turbulent fluctuations are substantially improved in the region of the pilot jet.

A comparison of Sims 2 and 3 with the corresponding experiments was reported in [18] and is not repeated here. It is important to mention that the agreement was in general good for the mean flow with only a small overprediction of the recirculation zone for Sim 3. For the turbulent fluctuations on the other hand the agreement was excellent.

Figure 4 shows the power spectrum of axial velocity fluctuations at $x/R = 0.1$, $r/R = 0.73$. In the case with retraction of the pilot jet a pronounced peak appears. The frequency of the principal peak is $f_{peak} = 0.25U_b/R$, which in the dimensional units of the experiment corresponds to a value of $f_{peak} = 102Hz$, which is well audible. The amplitude of the peak is very large, covering almost two decades in logarithmic scale. The total fluctuating kinetic energy is substantially larger than for $x_{pilot} = 0$, reflected by the larger integral under this curve. In the case $x_{pilot} = 0$, no pronounced peak is observed which

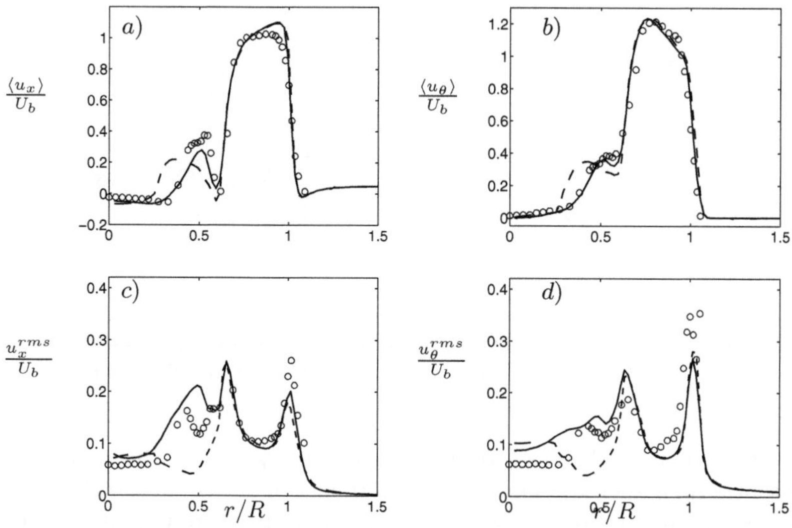

Fig. 3. Comparison of velocity profiles from Sim 1 and Sim 2 at $x/R = 0.1$. *Dashed line*, Sim 1. *Solid line*, Sim 2. *Symbols*, experiment. **a**) Mean axial velocity. **b**) Mean tangential velocity. **c**) RMS axial velocity. **d**) RMS tangential velocity

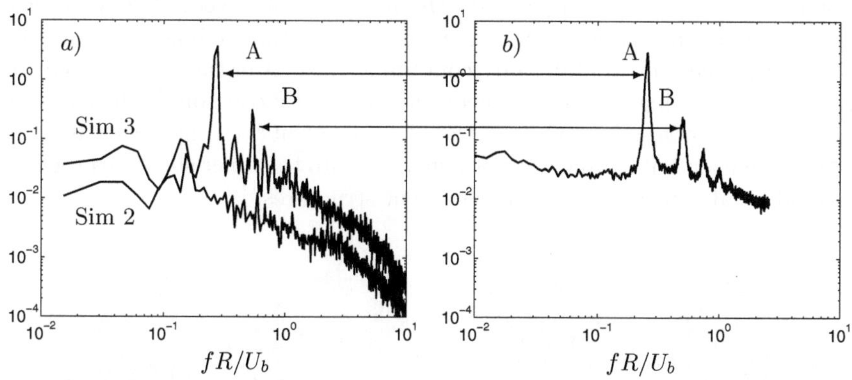

Fig. 4. **a**) Power spectrum of axial velocity fluctuations at $x/R = 0.1$, $r/R = 0.73$ from computations 1 and 2. **b**) the same data for $x_{pilot} = -0.73R$ from the experiment

confirms the preliminary experimental tests in which no flow instability was detected.

Figure 5 shows iso-surfaces of pressure fluctuations ($p'' = p - \langle p \rangle = -0.3$) for both cases visualizing the coherent structures of the flow. Pronounced large-scale coherent structures are observed in the case of the retracted pilot jet, Fig. 5*b*. As in the case without inner jet reported in [10], two types of structures can be observed in these pictures. Most of the time only one inner structure is visible and animations show that its rotation around the symmetry

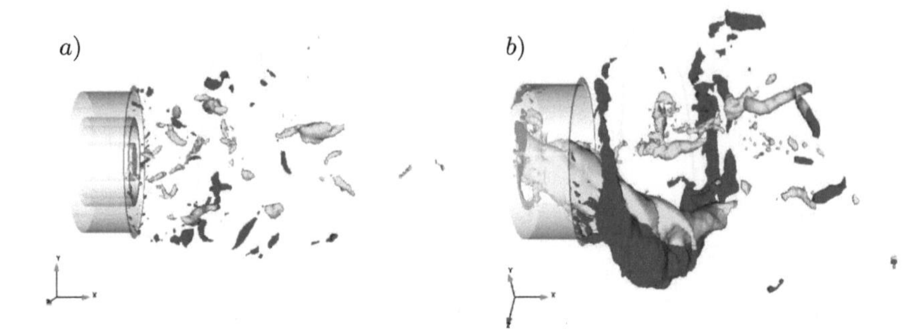

Fig. 5. Coherent structures visualized using an iso-surface of pressure fluctuations. **a**) Sim 2 with $x_{pilot} = 0$. **b**) Sim 3 with $x_{pilot} = -0.73R$. Figure taken from [18]

axis is very regular. In the case without retraction, $x_{pilot} = 0$, the structures are substantially smaller and more irregular. In fact, if one compares the same level of pressure fluctuations $p'' = -0.3$, hardly any structure is visible in the flow, Fig. 5*a*. Increasing the pressure level to $p'' = -0.15$, small structures are visible, which exhibit small coherence (not shown here). Hence, with $x_{pilot} = 0$ in Sim 2, the pilot jet destroys the large-scale structures [9]. When the pilot jet is retracted to $x_{pilot} = -0.73R$ this is not observed. The cylindrical tube enclosing the main jet prevents the recirculation bubble from moving upstream to the central bluff body containing the exit of the pilot jet. This is also illustrated using the instantaneous axial velocity component displayed in Fig. 6 for both cases. The pilot jet therefore only "hits" the upstream front of the recirculation bubble but cannot penetrate into the inner shear layer where it would be able to impact on the coherent structures.

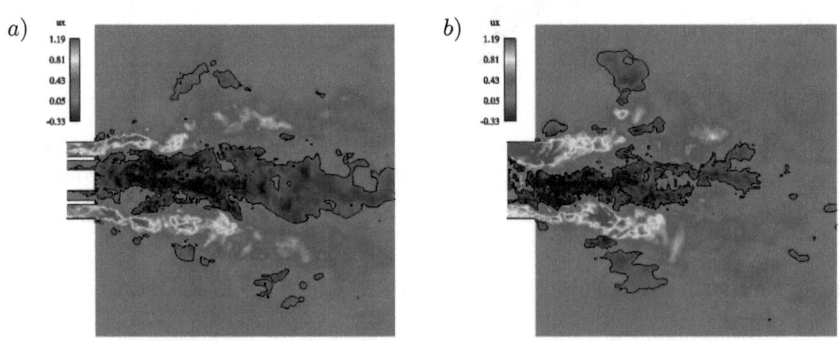

Fig. 6. Instantaneous axial velocity. **a**) Sim 2 with $x_{pilot} = 0$. **b**) Sim 3 with $x_{pilot} = -0.73R$

5 Conclusions

The possibility to perform a series of LES on the HP XC6000 supercomputer in Karlsruhe, allowed a comprehensive investigation of the effects of the inlet geometry on the turbulent flow in a swirl burner.

From the simulations, the following conclusions can be drawn:

- Good agreement is obtained between experiments and simulation for mean flow, turbulent fluctuations and power spectral density of velocity fluctuations.
- The flow presents very different coherent structures depending on the geometry of the inlet duct.
- If the inner jet is not retracted only short-living structures are formed.
- The retraction of the inner jet leads to the formation of very strong coherent structures which produce pronounced peaks in the spectra of velocity fluctuations.

Acknowledgements

This work was funded by the German Research Foundation (DFG) through the Collaborative Research Centre SFB-606 at the University of Karlsruhe. The authors are grateful to Dr.H. Büchner and Mr.O. Petsch for providing the experimental data and to the steering committee of the supercomputing facilities in Karlsruhe for granting computing time on the HP XC6000.

References

1. C. Bender and H. Büchner. Noise emissions from a premixed swirl combustor. In *Proc. 12th Int. Cong. Sound and Vibration, Lisbon, Portugal*, 2005.
2. T. Lieuwen, H. Torres, C. Johnson, and B.T. Zinn. A mechanism of combustion instability in lean premixed gas turbine combustors. *J. Eng. Gas Turbines and Power*, 123:182–189, January 2001.
3. C.O. Paschereit, E. Gutmark, and W. Weisenstein. Excitation of thermoacustic instabilities by interaction of acoustics and unstable swirling flow. *AIAA J.*, 38:1025–1034, 2000.
4. C. Külsheimer and H. Büchner. Combustion dynamics of turbulent swirling flames. *Comb. Flame*, 131:70–84, 2002.
5. C.M. Coats. Coherent structures in combustion. *Prog. Energy and Comb. Sci.*, 22:427–509, 1996.
6. C.D. Pierce and P. Moin. Large eddy simulation of a confined coaxial jet with swirl and heat release. AIAA paper no. 98-2892, 1998.
7. P. Wang, X.S. Bai, M. Wessman, and J. Klingmann. Large eddy simulation and experimental studies of a confined turbulent swirling flow. *Phys. Fluids*, 16(9):3306–3324, 2004.
8. S. Roux, G. Lartigue, T. Poinsot, U. Meier, and C. Bérat. Studies of mean and unsteady flow in a swirled combustor using experiments, acoustic analysis and large eddy simulations. *Comb. Flame*, 141:40–54, 2005.

9. M. García-Villalba and J. Fröhlich. On the sensitivity of a free annular swirling jet to the level of swirl and a pilot jet. In W. Rodi and M. Mulas, editors, *Engineering Turbulence Modelling and Experiments 6*, pages 845–854. Elsevier, 2005.
10. M. García-Villalba, J. Fröhlich, and W. Rodi. Identification and analysis of coherent structures in the near field of a turbulent unconfined annular swirling jet using large eddy simulation. *Phys. Fluids*, 18:055103, 2006.
11. P. Habisreuther, C. Bender, O. Petsch, H. Büchner, and H. Bockhorn. Prediction of pressure oscillations in a premixed swirl combustor flow and comparison to measurements. In W. Rodi and M. Mulas, editors, *Engineering Turbulence Modelling and Experiments 6*. Elsevier, 2005.
12. C.D. Pierce and P. Moin. Method for generating equilibrium swirling inflow conditions. *AIAA J.*, 36(7):1325–1327, 1998.
13. M. García-Villalba. *Large Eddy Simulation of turbulent swirling jets.* PhD thesis, University of Karlsruhe, 2006.
14. M. Breuer and W. Rodi. Large eddy simulation of complex turbulent flows of practical interest. In E.H. Hirschel, editor, *Flow simulation with high performance computers II*, volume 52 of *Notes on Numerical Fluid Mechanics*, pages 258–274. Vieweg, Braunschweig, 1996.
15. C. Hinterberger. *Dreidimensionale und tiefengemittelte Large-Eddy-Simulation von Flachwasserströmungen.* PhD thesis, University of Karlsruhe, 2004.
16. C.M. Rhie and W.L. Chow. Numerical study of the turbulent flow past an airfoil with trailing edge separation. *AIAA J.*, 21(11):1061–1068, 1983.
17. H.L. Stone. Iterative solution of implicit approximations of multidimensional partial differential equations for finite difference methods. *SIAM J. Numer. Anal.*, 5:530–558, 1968.
18. M. García-Villalba, J. Fröhlich, and W. Rodi. Numerical simulation of isothermal flow in a swirl burner. ASME paper GT2006-90764, 2006.

Numerical Investigation and Simulation of Transition Effects in Hypersonic Intake Flows

Martin Krause[1], Birgit Reinartz[2], and Josef Ballmann[3]

[1] Lehr- und Forschungsgebiet für Mechanik (Mechanics Laboratory), RWTH Aachen University, Templergraben 64, 52062 Aachen, Germany
martin@lufmech.rwth-aachen.de
[2] birgit@lufmech.rwth-aachen.de
[3] ballmann@lufmech.rwth-aachen.de

Abstract. A numerical and experimental analysis of a Scramjet intake flow has been initiated at RWTH Aachen University as part of the Research Training Group GRK 1095: "Aerothermodynamic Design of a Scramjet Engine for a Future Space Transportation System". This report presents an overview of the ongoing work on numerical simulations using Reynolds averaged Navier–Stokes solvers. Two different geometry concepts in 2D and 3D are investigated using several turbulence models to point out the influence of the geometry on the flow behaviour. One with a double ramp/convex curve configuration, the other with a double ramp/convex corner configuration. The data obtained will be compared with results from experiments which will be started in autumn 2006. It has to be said that not all results presented here were achieved using the NEC computing cluster. For comparison several calculations were conducted on the IBM Jump system of the Jülich Research Centre and at the SUN cluster of RWTH Aachen University. At the end of this report the computational performance will be compared.

1 Introduction

The intake of an air breathing hypersonic propulsion system mostly consists of exterior compression ramps followed by a so-called isolator/diffusor assembly. Important features of the flow field can be studied assuming two-dimensional flow. Oblique shock waves without a final normal shock are performing the compression of the incoming flow. Concerning the flight conditions two main difficulties of a hypersonic intake are evident. The first one is the interaction of strong shock waves with thick hypersonic boundary layers, which cause large separation zones that are responsible for a loss of mass flow and some unsteadinesses of the flow, like e.g. shock movement. Thereby the engine performance decreases. The second difficulty is that the high total enthalpy of the flow causes severe aerodynamic heating, further enhanced by turbulent heat flux. Up to now the following cases were studied: i) the influence of ge-

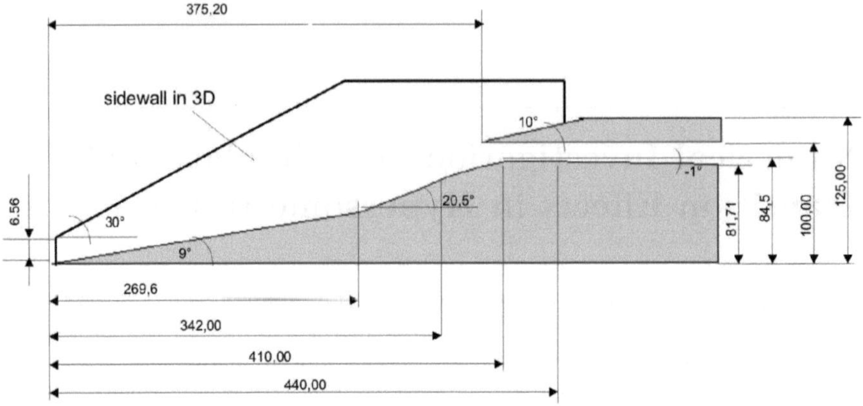

Fig. 1. Hypersonic inlet model 1, non-heated model with curved wall part ahead of the isolator entrance

ometry changes on the flow, especially on the separation regions and ii) the impact of the used turbulence modeling within the numerical method on the flow solution iii) the difference between 2D approach and 3D simulations.

Figure 1 shows a configuration with sharp leading edges. But in practice, sharp leading edges will not withstand the high thermal loading in hypersonic flow, i.e. rounded leading edges are more realistic. This geometric change has a remarkable influence on the flow in several aspects, for example the shapes and positions of front and cowl shock. These will be detached and exhibit strong curvature around the blunt edges generating entropy layers. Along the ramp the boundary layer grows inside the entropy layer causing an increase in aerodynamic heating [1]. Another effect of detached bow shocks is that farther downstream unfortunate changes of the position of the leading edge shock can diminish the captured mass flow and worsen the flow conditions in the isolator inlet.

Quantitatively, the results of 2D computations will not be sufficient for the final design of the compression ramps because these will have sidewalls to prevent losses of mass by sideflow. Therefore, 3D computations on fine grids will be more adequate to achieve a trustworthy quantitative prediction of the intake flow. With sidewalls an increase of mass flow shall be achieved, but the additional boundary layers and corner flow phenomena create additional difficulties due to shock-shock and shock-boundary-layer interactions which cause strong longitudinal vortices [2].

The numerical work reported here is complementary work to experiments performed within the frame of the GRK 1095 at the German Aerospace Center DLR Cologne and in the TH2 hypersonic shock tunnel of RWTH. Presently two models of a compression ramp/isolator configuration are under investigation. The first one is a not-heated model, within which suction can be installed inside and behind the flow expansion region. Furthermore, it is possible to change the isolator geometry and the position of the cowl. The experiments with this configuration are done at DLR. In the second model, temperature controlled ramp heating is installed to simulate the wall temperature change during hypersonic flight and to investigate its influence on the performance of the air intake and compression. These experiments are conducted in the Shock Wave Laboratory of the RWTH Aachen University.

2 Physical Model

2.1 Conservation Equations

The governing equations for high-speed turbulent flow are the unsteady Reynolds-averaged Navier–Stokes equations for compressible fluid flow in integral form

$$\frac{\partial}{\partial t}\int_V \mathbf{U}\ \mathrm{d}V + \oint_{\partial V} \left(\mathbf{F}^{\mathrm{c}} - \mathbf{F}^{\mathrm{d}}\right)\mathbf{n}\ \mathrm{d}S = \mathbf{0} \tag{1}$$

where

$$\mathbf{U} = [\ \bar{\rho}\ ,\ \bar{\rho}\tilde{\mathbf{v}}\ ,\ \bar{\rho}\tilde{e}_{tot}\ ]^{\mathrm{T}} \tag{2}$$

is the array of the mean values of the conserved quantities: density of mass, momentum density, and total energy density. The tilde and the bar over the variables denote the mean value of Reynolds-averaged and Favre-averaged variables, respectively. The quantity V denotes an arbitrary control volume with the closed surface ∂V and the outer normal $\mathbf{n}$. The fluxes are splitted into the inviscid part

$$\mathbf{F}^{\mathrm{c}} = \begin{pmatrix} \bar{\rho}\tilde{\mathbf{v}} \\ \bar{\rho}\tilde{\mathbf{v}}\circ\tilde{\mathbf{v}} + \bar{p}\,\mathbf{1} \\ \tilde{\mathbf{v}}(\bar{\rho}\tilde{e}_{tot} + \bar{p}) \end{pmatrix} \tag{3}$$

and the diffusive part

$$\mathbf{F}^{\mathrm{d}} = \begin{pmatrix} \mathbf{0} \\ \bar{\boldsymbol{\sigma}} - \overline{\rho\mathbf{v}''\circ\mathbf{v}''} \\ \tilde{\mathbf{v}}\bar{\boldsymbol{\sigma}} + \overline{\mathbf{v}''\boldsymbol{\sigma}} - \bar{\mathbf{q}} - c_p\,\overline{\rho\mathbf{v}''T''} - \frac{1}{2}\,\overline{\rho\mathbf{v}''\,\mathbf{v}''\circ\mathbf{v}''} - \tilde{\mathbf{v}}\,\overline{\rho\mathbf{v}''\circ\mathbf{v}''} \end{pmatrix}, \tag{4}$$

where $\mathbf{1}$ is the unit tensor and $\circ$ denotes the dyadic product[4]. The air is considered to be a perfect gas with constant ratio of specific heats, $\gamma = 1.4$, and a specific gas constant of $R = 287\,\mathrm{J/(kgK)}$. Correspondingly the expression for the specific total energy reads:

$$\tilde{e}_{tot} = c_v\tilde{T} + \frac{1}{2}\tilde{\mathbf{v}}\tilde{\mathbf{v}} + k\ . \tag{5}$$

The last term represents the turbulent kinetic energy

[4] Scalar Products of dyadics formed by two vectors $\mathbf{a}$ and $\mathbf{b}$ with a vector $\mathbf{c}$ are defined as usual, i.e., $\mathbf{a}\circ\mathbf{b}\,\mathbf{c} = \mathbf{a}(\mathbf{bc})$, $\mathbf{c}\,\mathbf{a}\circ\mathbf{b} = (\mathbf{ca})\mathbf{b}$.

$$k := \frac{1}{2} \frac{\overline{\rho \mathbf{v}'' \mathbf{v}''}}{\bar{\rho}} \quad . \tag{6}$$

For isotropic Newtonian fluids, the mean molecular shear stress tensor is a linear, homogeneous and isotropic function of the strain rate

$$\bar{\boldsymbol{\sigma}} = 2\bar{\mu}\,\bar{\mathbf{S}} - \frac{2}{3}\bar{\mu}\,\mathrm{tr}(\bar{\mathbf{S}})\ \mathbf{1}, \tag{7}$$

if bulkviscosity is neglected. The mean strain rate tensor is

$$\bar{\mathbf{S}} := \frac{1}{2}\left[\mathrm{grad}(\bar{\mathbf{v}}) + (\mathrm{grad}(\bar{\mathbf{v}}))^{\mathrm{T}}\right], \tag{8}$$

and the molecular viscosity $\bar{\mu} = \bar{\mu}(\bar{T})$ obeys Sutherland's law. Similarly, the molecular heat flux is considered a linear, homogeneous, isotropic function of the temperature gradient

$$\bar{\mathbf{q}} = -\frac{c_{\mathrm{p}}\bar{\mu}}{Pr}\,\mathrm{grad}(\bar{T}), \tag{9}$$

with the Prandtl number $Pr = 0.72$.

2.2 Turbulence Closure

To close the above system of partial differential equations, the Boussinesq hypothesis is used where the remaining correlations are modeled as functions of the gradients of the mean conservative quantities and turbulent transport coefficients. The Reynolds stress tensor thus becomes

$$-\overline{\rho \mathbf{v}'' \circ \mathbf{v}''} = 2\mu_{\mathrm{t}}\,(\bar{\mathbf{S}} - \frac{1}{3}\,\mathrm{tr}(\bar{\mathbf{S}})) - \frac{2}{3}\bar{\rho}k\,\mathbf{1} \quad , \tag{10}$$

with the eddy viscosity μ_{t}, and the turbulent heat flux is

$$c_p\,\overline{\rho \mathbf{v}'' T''} = -\frac{c_p \mu_{\mathrm{t}}}{Pr_{\mathrm{t}}}\,\mathrm{grad}(\bar{T}), \tag{11}$$

with the turbulent Prandtl number $Pr_{\mathrm{t}} = 0.89$. Finally, for hypersonic flows the molecular diffusion and the turbulent transport are modeled as functions of the gradient of the turbulent kinetic energy

$$\overline{\mathbf{v}''\boldsymbol{\sigma}} - \frac{1}{2}\overline{\rho \mathbf{v}''\,\mathbf{v}'' \circ \mathbf{v}''} = (\mu + \frac{\mu_{\mathrm{t}}}{Pr_k})\,\mathrm{grad}(k), \tag{12}$$

with the model constant $Pr_k = 2$.

The turbulent kinetic energy and the eddy viscosity are then obtained from the turbulence model. In case of laminar flow, both variables are set to zero to regain the original transport equations. As mentioned, several turbulence models were used for the numerical simulations within this paper. For description of the models used, we refer to [3], [4] and [5].

3 Numerical Method

The computations have been performed using the DLR FLOWer code, Versions 116.+, using a finite-volume formulation on structured multi-block grids [6]. The code had been originally developed for the simulation of flows around airfoils in subsonic and transonic regimes. It has been extended for the simulation of hypersonic, turbulent flow problems by implementing upwind discretizations and advanced compressible turbulence models [7].

3.1 Spatial Discretization

A finite-volume discretization is applied to (1) which results in a consistent approximation to the conservation laws. The computational domain is divided into non-overlapping hexahedra in general curvilinear coordinates ξ, η, ζ, and the integral formulation (1) is then applied to each cell (i, j, k) separately. Standard central discretization schemes are used for the convective and diffusive terms in the presented supersonic computations. Then semidiscretization of Eq. (1) results in a set of equations for the time rates of change of the volume-averaged conserved quantities $\mathbf{W}_{i,j,k}$ which are in balance with the sum of the corresponding area-averaged fluxes, $\mathbf{R}^{\mathrm{c}}_{i,j,k}$ and $\mathbf{R}^{\mathrm{d}}_{i,j,k}$, across the cell faces and the artificial dissipation $\mathbf{D}_{i,j,k}$:

$$\frac{\mathrm{d}\mathbf{W}_{i,j,k}}{\mathrm{d}t} = -\frac{1}{V_{i,j,k}}\left(\mathbf{R}^{\mathrm{c}}_{i,j,k} - \mathbf{R}^{\mathrm{d}}_{i,j,k} + \mathbf{D}_{i,j,k}\right) = \mathbf{Res}_{i,j,k} \ . \tag{13}$$

However, to account for the directed propagation of information in the inviscid part of the equations, the AUSM (Advection Upstream Splitting Method) scheme will be used for the approximation of the convective flux functions in hypersonic flows [8]. Higher-order accuracy and consistency with the central differences used for the diffusive terms is achieved by MUSCL (Monotonic Upstream Scheme for Conservation Laws) Extrapolation, and TVD (Total Variation Diminishing) property of the scheme is ensured by a modified van Albada limiter function.

3.2 Time-Stepping Scheme

The system of ordinary differential Eq. (13) is solved by an explicit five-stage Runge–Kutta time-stepping scheme of fourth order in combination with different convergence acceleration techniques like multigrid and local time stepping for asymptotically steady-state solutions [9]. Additionally, for inviscid stationary flow, the total enthalpy is a constant throughout the flow field and its numerical deviation is applied as forcing function to accelerate convergence. For turbulent flow, the time integration of the turbulence equations is decoupled from the mean equations and the turbulence equations are solved using a Diagonal Dominant Alternating Direction Implicit (DDADI) scheme. The implicit scheme increases the numerical stability of turbulent flow simulations which is especially important since the low Reynolds number damping terms as well as the high grid cell aspect ratios near the wall make the system of turbulent conservation equations stiff. Due to the CFL-condition for explicit schemes, the CFL number of the Runge–Kutta scheme has an upper limit of 4. Implicit residual smoothing allows to increase the explicit stability limit by a factor of 2 to 3 [9]. For the implicitly solved turbulence equations, the CFL number can be ten times higher.

3.3 Boundary Conditions

At the inflow, outflow and other farfield boundaries, a locally one-dimensional inviscid flow normal to the boundary is assumed. The governing equations are linearized based on characteristic theory and the incoming and outgoing number of characteristics are determined. For incoming characteristics, the state variables are corrected by freestream values using the linearized compatibility equations. Else the variables are extrapolated from the interior [9]. For turbulent flow, the turbulent freestream values are determined by specifying the freestream turbulence intensity Tu_∞: $k_\infty = 0.667 \cdot Tu_\infty v_\infty$ and $\omega_\infty = k_\infty/(0.001 \cdot \mu)$.

For steady inviscid flow, it is sufficient to set the normal velocity component to zero at slip surfaces. In the viscous case, the no-slip condition is enforced at solid walls by setting all velocity components to zero. Additionally, the turbulent kinetic energy and the normal pressure gradient are set to zero. The specific dissipation rate is set proportional to the wall shear stress and the surface roughness. The energy boundary condition is directly applied through the diffusive wall flux: either by driving to zero the contribution of the diffusive flux for adiabatic walls or by prescribing the wall temperature or wall heat flux when calculating the energy contribution of wall faces. At the symmetry plane of the half configuration, the conservation variables are mirrored onto the ghost cells to ensure symmetry.

4 Farfield Flow Conditions

The farfield flow conditions are the same for all computations shown in this report. The conditions belong to an altitude of 25000m and are as follows: $M = 7.0$, $Re = 5.689 \cdot 10^6$, $T_\infty = 222K$ and $p_\infty = 2511Pa$. A sketch of the solution domain is shown in Fig. 1. All gradients in the third space direction are assumed to be zero. In the corresponding boundary conditions on the sides, symmetry conditions are used for fictitious points.

5 Results

5.1 Grid Generation for Geometries with Sharp Leading Edges and Rounded Leading Edges

The grid generation for both considered intake geometries were performed using the MegaCads program, which has been developed and used at DLR and was further developed in the MEGAFLOW project. For validation and grid convergence study several grids were generated for sharp leading edges and rounded leading edges. All of them were split up into several blocks to allow the use of MPI and OpenMP. The results presented here have been generated on a grid with 170000 points in 2D for sharp leading edges. For rounded leading edges grids with 77000 points in 2D and 6 million points in 3D were generated. All grids were densified at solid walls and at the inflow boundary as well as at the isolator inlet. The minimum resolution is 10^{-6} in x and y direction. It has to be mentioned that the grid for the intake geometry 2 with the convex corner configuration has the same number of points and the same block distribution as mentioned before.

5.2 Sharp Leading Edges

Several results of 2D computations are available for geometries with sharp leading edges (Fig. 1) which were achieved with different turbulence models. Simulations have been performed using the original Wilcox k–ω-Turbulence Model, the Spalart–Allmaras and a Reynolds stress model, called SSG–ω-model (according to Speziale, Sarkar, Gatski). It should be mentioned that also inviscid flow computations were carried out. The inviscid flow showed the shock system that is generated within the intake as it was expected during the first design by using a 2D characteristic method for steady supersonic flow. In the following, we report on 2D turbulent flow computations. At first, the differences between an adiabatic wall and an isothermal wall ($T_{wall} = 300K$) have been investigated. The flow conditions were described in Chap. 4. Figure 2 shows contour lines of Mach number in the Scramjet intake (geometry 1) for an adiabatic wall. Figure 3 shows results for the same geometry with an isothermal wall at 300 K wall temperature. It can be seen that the wall temperature has a remarkable effect on the flow field, especially on the separation regions between the double ramp and the inlet of the isolator. There the cowl shock interacts with the boundary layer on the lower isolator wall. This interaction creates a separation bubble, which fills about 35% of the overall intake height for an isothermal wall. In the case of an adiabatic wall (Fig. 2) the separation fills about 55% of the intake. A higher wall temperature produces thicker boundary layers, which increases the shock angle and causes larger separation. Furthermore, the mass flow entering the isolator is reduced. Concerning these results it is obvious that the wall temperature strongly influences the flow field.

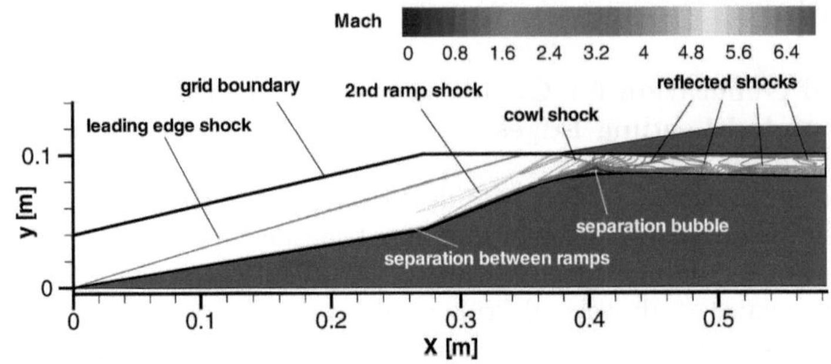

Fig. 2. Mach contours for scramjet intake (geometry 1) adiabatic wall, SA turbulence model (M_∞= 7.0, $Re_l = 5.689 \cdot 10^6$ [1/m], T_∞=222 K)

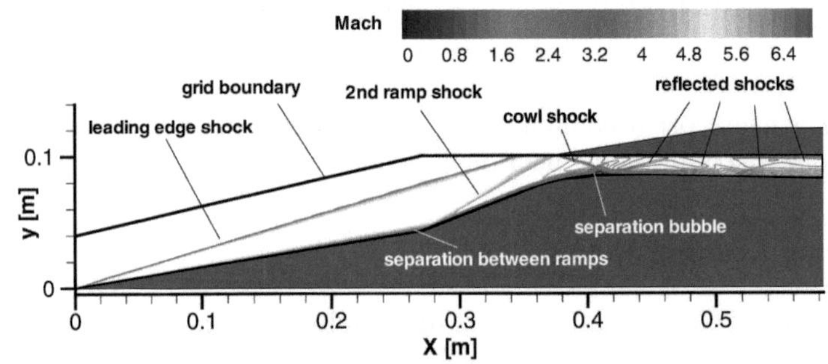

Fig. 3. Mach contours for scramjet intake (geometry 1) isothermal wall $T_{wall} = 300K$, SA turbulence model (M_∞= 7.0, $Re_l = 5.689 \cdot 10^6$ [1/m], T_∞=222 K)

The importance of considering the influence of the wall temperature on the flow field can also be seen in Fig. 4 and Fig. 5. It can be asserted that by reaching a certain wall temperature the separation bubble in the isolator inlet becomes so big, that it may block the flow. A strong shock in front of the isolator will be the consequence and no supersonic combustion can be realised anymore.

Contemplating Fig. 4 and Fig. 2, which present results computed with different turbulence models, differences can be seen, although the boundary and flow conditions are identical. That means these are produced simply and solely by the turbulence models.

The comparison between the simulations for geometry 1 and 2 showed, that the expansion of the flow over a convex corner is inferior then over a convex curved wall. The emerging Prandtl–Meyer-expansion above the corner has much more influence on the cowl shock, which is bended upstream so that the shock angle increases. This leads to higher static pressure and smaller velocity behind it. Therefore, the separation bubble grows and the flow is much more inhomogeneous. For comparison have a look at Fig. 5 and 6.

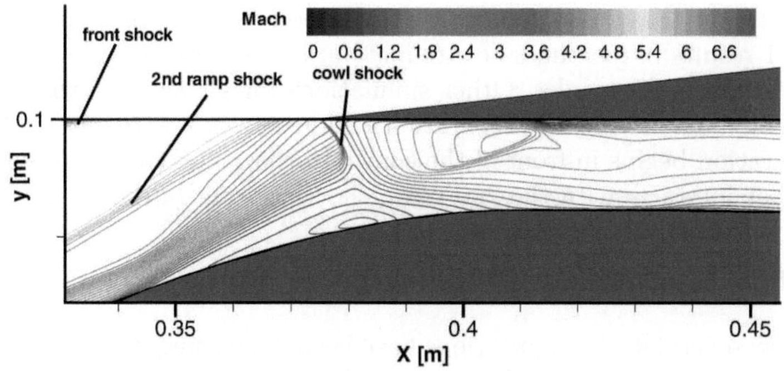

Fig. 4. Mach contours for scramjet intake (geometry 1) adiabatic wall, k–ω turbulence model ($M_\infty = 7.0$, $Re_l = 5.689 \cdot 10^6$ [1/m], T_∞=222 K)

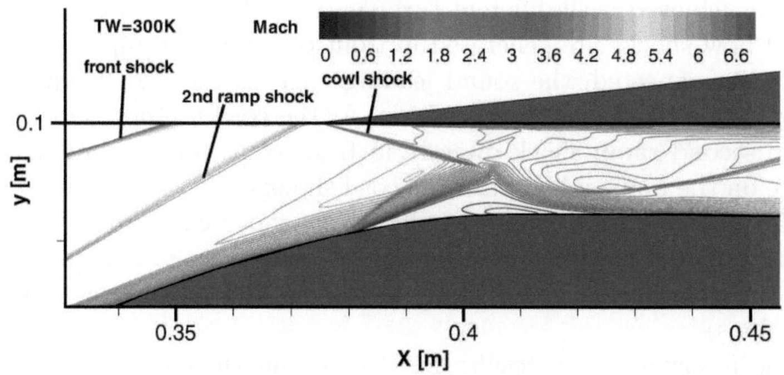

Fig. 5. Mach contours for scramjet intake (geometry 1) isothermal wall $T_{wall} = 300K$, k–ω turbulence model ($M_\infty = 7.0$, $Re_l = 5.689 \cdot 10^6$ [1/m], $T_\infty = 222$ K)

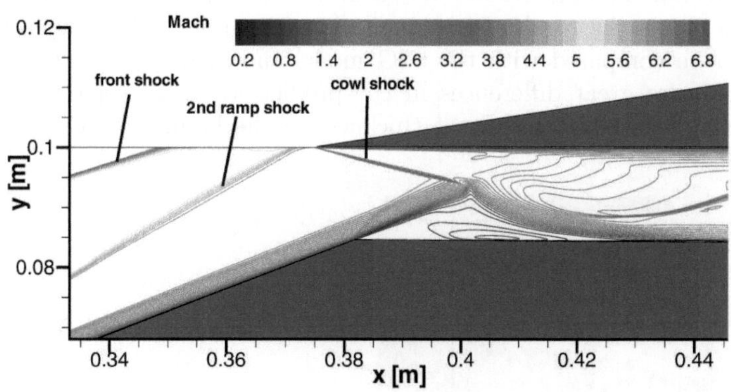

Fig. 6. Mach contours for scramjet intake (geometry 2) isothermal wall $T_{wall} = 300K$, k–ω turbulence model ($M_\infty = 7.0$, $Re_l = 5.689 \cdot 10^6$ [1/m], $T_\infty = 222$ K)

Relying on these first results it can be pointed out, that the use of geometry 1 results in less mass flow reduction, because of the minor seperation bubble in the isolator inlet. Other simulations for geometry 2 with an adiabatic wall showed, that the separation region moves upstream. That means the separation begins in front of the convex corner. This induces a separation shock that interacts with the 2nd ramp shock in front of the cowl.

5.3 Influence of Rounded Leading Edges on the Intake Flow Field

Up to now several flow computations have been done using different turbulence models (SA, LEA, LLR, SST, Standard Wilcox k-ω model and a Reynolds stress model (SSG-ω)). Within this chapter the differences to the geometry with sharp leading edges will be pointed out, followed by a comparison of the results achieved with different turbulence models. It can be asserted that detached bow shocks are generated in front of the first ramp and in front of the cowl. As expected, the round leading edge at the first ramp forces the shock to take a more upstream position. At the cowl the bow shock interacts with the second ramp shock directly in front of it. This interaction has no influence on the body side part of the cowl shock and thus not on the isolator flow. The entropy layers caused by the strong shock curvature at the leading edges of ramp and cowl will be a matter of further study. Figure 7 shows the inlet of the isolator and Fig. 8 shows a close up view of the cowl (both Mach distributions).

In the following some results for 2 equation turbulence models will be shown. Four models were tested (SST, LEA, LLR and original k–ω from Wilcox). It was asserted, that even using the same grid, boundary- and flow conditions not the same results were achieved for different turbulence models. Figure 9 shows the results for a simulation with the LLR turbulence model that can be compared with the SSG model shown in Fig. 10. One can see that there are great differences in the prediction of the separation bubble size in the isolator inlet and the thickness of the boundary layers. Followed

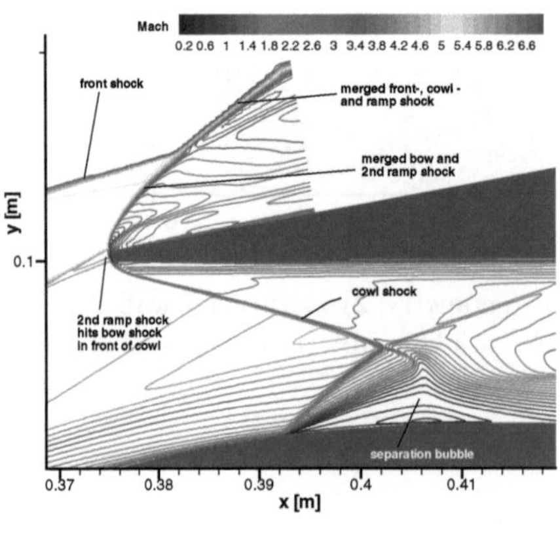

Fig. 7. Mach contours for isolator intake (FLOWer) isothermal wall, $T_{Wall} = 300K$, SA turbulence model, rounded leading edges (M_∞= 7.0, $Re_l = 5.689 \cdot 10^6$ [1/m], T_∞=222 K)

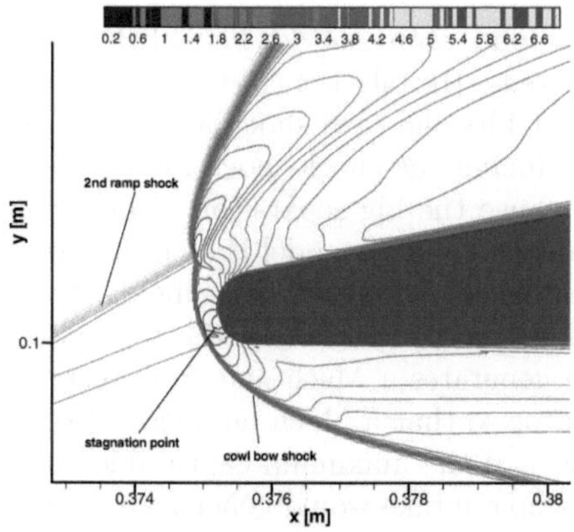

Fig. 8. Mach contours for cowl (FLOWer) isothermal wall, $T_{Wall} = 300K$, SA turbulence model, (M_∞= 7.0, Re_l = 5.689 · 10^6 [1/m], T_∞=222 K)

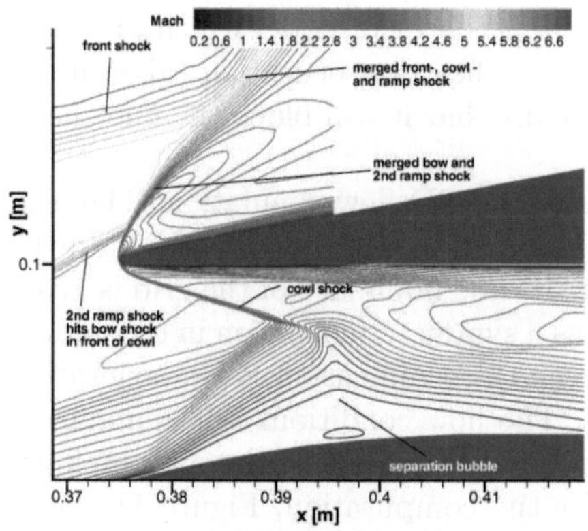

Fig. 9. Mach contours isolator intake (FLOWer) isothermal wall, $T_{Wall} = 300K$, LLR turbulence model, (M_∞= 7.0, Re_l = 5.689 · 10^6 [1/m], T_∞=222 K)

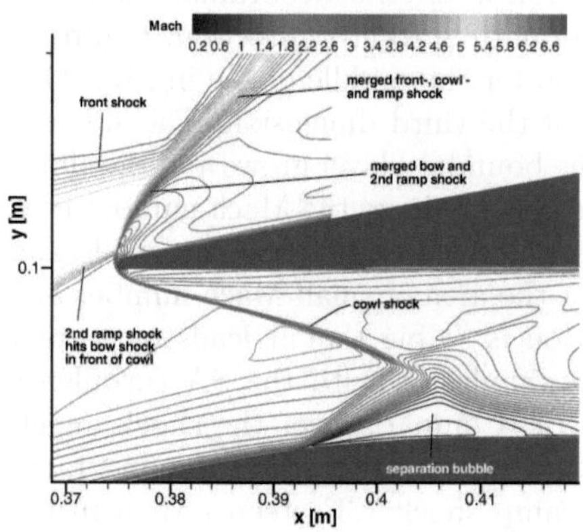

Fig. 10. Mach contours for isolator intake (FLOWer) isothermal wall, $T_{Wall} = 300K$, SSG turbulence model, rounded leading edges (M_∞= 7.0, Re_l = 5.689 · 10^6 [1/m], T_∞=222 K)

by different positions of the first and second ramp shock. The LLR model produces thicker boundary layers that increase the offset of the first bow shock from the round leading edge. Therefore it hits the cowl shock as well as the second ramp shock, further upstream reducing the shock-shock interactions. Nevertheless the thick boundary layers cause the big separation resulting in a blockage of the isolator inlet. The result of the SST model is quantitatively comparable with the ones of the SA model and the SSG-ω-Reynolds stress model. Considering the results of the LEA and k–ω-models, the separation within the isolator intake blocks it and generates a Mach one cross section. Afterwards the flow expands in the isolator so that a Mach number of two is achieved at its exit. In that case only 60% of the maximum capturable mass flow exits the isolator. That means the whole intake would generate a spill of 40%. The spill for the other turbulence models is lower than 20% (SA: 14.2%; SST: 19.2%; SSG–ω: 15.4%). An exception is the LLR model which produces a spillage of 25%. But in Fig. 9 one can see that the separation takes about 80% of the overall intake height. That means that it will block the inlet here, too.

In the following we will present results of a 3D flow simulation. The grid is based on the one used for 2D. It was made up by multiple repetition in the third dimension to create a 3D intake. In 3D one boundary of the grid is given by the sidewall the other one is assumed as a symmetry condition in the middle of the intake. The number of points in the third dimension is 80, generating a grid with a total of 6.1 million points. The flow conditions still remain the same. Considering the 2D results and the smaller amount of simulation effort, the SA turbulence model was chosen for the computation. Figure 11 shows slices of Mach contours for the 3D simulation where the x-coordinate had been held constant which means, the slices are normal to flow direction. Figure 12 shows the Mach contours in flow direction for the middle of the intake. One can see, that there is a great influence of the third dimension. The sidewall creates a boundary layer and a shock. The boundary layer grows and the shock surface interacts with these of the ramps. This reduces the Mach number near the wall dramatically as expected in the corner between the side walls and the ramps. Upstream of the isolator inlet the area of small Mach number and the size of the subsonic corner flow region is so big that it leads to a great separation bubble. Contrary to the 2D simulation with the SA turbulence model, this separation blocks the isolator. Comparison of the shock angels between 2D and 3D show a difference, that means the angle of the first ramp shock in 3D is 0.2° and of the second ramp shock 1.3° steeper than in the 2D result. This leads to a decrease of the Mach number at the isolator inlet, leading to a decreased velocity after the expansion and, therefore, to a greater separation. It can be asserted that the sidewall shock bends the ramp shocks upstream. Future studies will show if this is the case when using the other turbulence models.

Finally, it can be asserted that geometry changes have to be introduced. The separation bubble in the isolator inlet is too big and must be reduced.

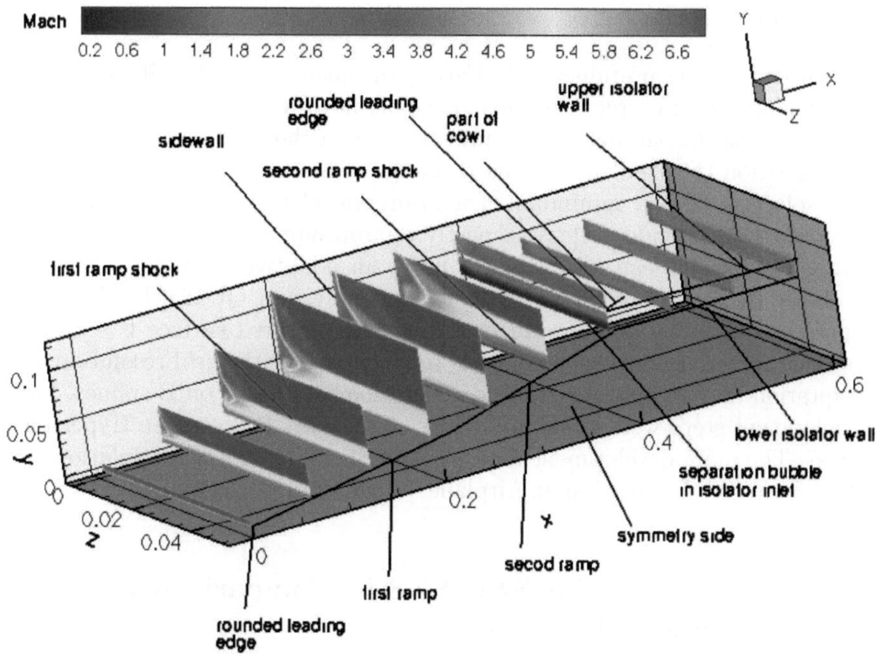

Fig. 11. Mach contours isolator intake (FLOWer) isothermal wall, $T_{Wall} = 300K$, SA turbulence model, (M_∞= 7.0, $Re_l = 5.689 \cdot 10^6$ [1/m], T_∞=222 K)

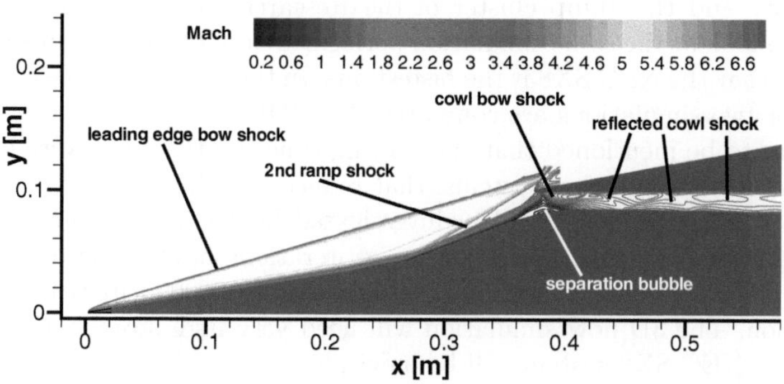

Fig. 12. Mach contours for isolator intake (FLOWer) isothermal wall, $T_{Wall} = 300K$, SA turbulence model, rounded leading edges (M_∞= 7.0, $Re_l = 5.689 \cdot 10^6$ [1/m], T_∞=222 K)

Furthermore, the point where the 2nd ramp shock hits the bow shock has to be moved downstream, that means, this convergence point must stand above the cowl, so that the influence on the isolator flow is as small as possible. To achieve that it might be necessary to extend the isolator height, move the cowl lip downstream or change the ramp angles.

5.4 Future Work

The study on the usefulness of existing turbulence models will be continued implementing corrections for hypersonic flow. Main focus will be on 3D computations. In parallel it is planned to extend the flow solver for simulating transition to turbulence. A disadvantage of the presently used structured flow solver is that a minimum of a priori knowledge about the flow field is required to provide at least a partly appropriate grid for a good numerical solution. This can be overcome using an adaptive solver. The method of choice in this project will be thc program system QUADFLOW, which represents an integrated system of a fully unstructured surface based finite volumen solver, a multi-scale analysis tool controlling the grid resolution and adaptation and an appropriate grid generation tool based on B-splines. First development steps were made in GRK 5: "Transport Processes in Hypersonic flows". The main development took place in SFB 401: "Flow Modulation and Fluid-Structure Interaction at Airplane Wings" [10].

6 Performance of the NEC SX8 in Comparison with Parallel Computer Systems

As already mentioned not all simulations were performed on the NEC cluster. Additionally parallel computer systems like the SUN cluster of RWTH Aachen University and the Jump cluster of the Research Centre Jülich were used. Having a look at the same simulation performed on all three systems it can be asserted that the NEC SX8 is the fastest one. In the following table computing times for two simulations are compared (2D/3D).

It has to be mentioned that the waiting times on the NEC were longer as on the other computation systems, that reduces the factor a little.

The hypersonic flow problem investigated here requires a great amount of computing resources, because a lot of iterations have to be done in each run to get a stationary result e.g. approximately 200000 iterations for one simulation. The 3D flow simulation will need very high storage as well. The use of the NEC SX8 system will be inevitable.

The QUADFLOW solver which is presently running on PC- and Workstation clusters will be transposed to NEC SX 8 as well. The flow computations

Table 1. Performance of the different computing systems

	SUN	Jump	NEC SX8
Iterations(2D)/(3D)	20000/10000	20000/10000	20000/10000
number of processors(2D/3D)	10/20	10/20	1/24
computing time in minutes(2D/3D)	1920/13000	840/6400	300/260
factor (NEC = 1)(2D/3D)	6.4/42	2.8/21	1/1

in 2D and, thereafter, in 3D will then performed with both solvers FLOWer and QUADFLOW.

7 Summary

Results of 2D and first 3D flow computations were presented. Two different intake geometries have been studied numerically, one with a convex curve and the other one with a convex corner configuration to deflect the flow into the isolator. The difference between sharp leading edges and rounded leading edges was also pointed out. It was found, that there are great differences in the results created by different turbulence models. These differences are much bigger than those produced by the change of the geometry from sharp to rounded leading edges. It was also asserted that the SA, SST and SSG–ω model qualitatively yielded similar results, where the LEA and original k–ω turbulence model from Wilcox showed much greater separation and thicker boundary layers. The LLR model produced a result somewhere in between. Therefore these results have to be compared with experimental ones in future when available. So far, one result for a 3D computation was presented. It was found out, that the side wall shock and the corner flow interactions bend the ramp shocks upstream resulting in reduced velocity, followed by a greater separation bubble that blocks the isolator inlet.

Further numerical investigation is of great importance, because one aim of the research is to garantee the functioning of the hypersonic Scramjet intake. The future experiments within the frame of GRK1095 will allow to check and validate the computational predictions, so that the confidence on these numerical predictions of the functioning Scramjet can be improved.

References

1. Anderson, J., *Hypersonic and High Temperature Gas Dynamics*, MacGraw–Hill, 1989.
2. Coratekin, T.A., van Keuk, J., and Ballmann, J., "Preliminary Investigations in 2D and 3D Ramjet Inlet Design," AIAA Paper 99–2667, June 1999.
3. Wilcox, D.C., "Turbulence Energy Equation Models," *Turbulence Modeling for CFD*, Vol. 1, DCW Industries, Inc., La Canada, CA, 2nd ed., 1994, pp. 73–170.
4. Spalart, P.R. and Allmaras, S.R., "A One-Equation Turbulence Model for Aerodynamic Flows," AIAA Paper 92–0439, January 1992.
5. Menter, F.R., "Two-Equation Eddy-Viscosity Turbulence Models for Engineering Applications," *AIAA Journal*, Vol. 32, No. 8, Aug. 1994, pp. 1598–1605.
6. Kroll, N., Rossow, C.-C., Becker, K., and Thiele, F., "The MEGAFLOW Project," *Aerospace Science and Technology*, Vol. 4, No. 4, 2000, pp. 223–237.
7. Reinartz, B.U., van Keuk, J., Coratekin, T., and Ballmann, J., "Computation of Wall Heat Fluxes in Hypersonic Inlet Flows," AIAA Paper 02-0506, January 2002.

8. Kroll, N. and Radespiel, R., "An Improved Flux Vector Split Discretization Scheme for Viscous Flows," DLR-Forschungsbericht 93–53, 1993.
9. Radespiel, R., Rossow, C., and Swanson, R., "Efficient Cell-Vertex Multigrid Scheme for the Three-Dimensional Navier–Stokes Equations," *AIAA Journal*, Vol. 28, No. 8, 1990, pp. 1464–1472.
10. Bramkamp, F.D., Lamby, P., and Müller, S., "An adaptive multiscale finite volume solver for unsteady and steady flow computations," *Journal of Computational Physics*, Vol. 197, 2004, pp. 460–490.

Aeroelastic Simulations of Isolated Rotors Using Weak Fluid-Structure Coupling

M. Dietz, M. Kessler, E. Krämer

Institut für Aerodynamik und Gasdynamik (IAG), Universität Stuttgart,
Pfaffenwaldring 21, 70550 Stuttgart, Germany

Abstract. In this paper we present a weak fluid-structure coupling method for the aeroelastic simulation of isolated helicopter main rotors. The CFD Code FLOWer (by DLR) is coupled to the flight mechanics code HOST (by Eurocopter). HOST is used to compute the blade dynamics and the rotor trim, whereas the aerodynamic loads are determined by FLOWer. The method has been applied to two different rotors: the advanced EC145 rotor in fast forward flight and the well known Bo105 rotor in slow descent flight.

1 Introduction

Within the past years helicopter main rotor aerodynamics and aeroelasticity has been an extensive field of research at the Institut für Aerodynamik und Gasdynamik (IAG). Although even the plain aerodynamic simulation of a helicopter main rotor in forward flight remains a challenging task for CFD, it has become evident that aeroelastic effects have to be taken into account in order to even get qualitatively correct results [7, 8]. Different concepts can be applied in order to couple aerodynamics and structural dynamics. In the past years the institute has focussed on the time-accurate coupling which is also denoted as strong coupling [1, 10].

In the case of strong coupling the data exchange between fluid and structure is performed on a time-accurate basis, i.e. the transfer of fluid loads from CFD to the structure code and the transfer of the blade deformation from the structure code to CFD is carried out every time step. A special treatment becomes necessary in order to maintain the timewise order of the coupled aeroelastic scheme: So called staggered coupling schemes have been developed which avoid the reduction of the timewise order below the order of the participating codes [9]. The advantage of the time-accurate coupling method is, that the real physics of the coupling process is reproduced. Therefore the method can be applied to any flight condition including non-periodic flight cases like manoeuvering flight. On the other hand this is also a disadvantage for flight conditions where periodicity of the solution can be implied. The method reproduces the complete transient phenomenon which is mostly not of interest. This transient process can take up to eight to ten rotor revolutions until a periodic state has been obtained.

A method allowing for a faster convergence in case of a periodic flight condition is the so called weak or loose coupling. In the past months the FLOWer code and the HOST code have been extended to this coupling method and the technique has been applied to the computations presented in the present paper. In contrast to the strong coupling technique the data exchange strategy of the weak coupling method is not performed on a time step basis but uses a Fourier description for both the blade deformation and the aerodynamic loading. This method enforces the periodicity of the solution, and thus it allows for a faster convergence of the CFD computation and a faster rotor trim. The weak coupling method and results obtained from this technique will be presented in the following sections.

2 Mathematical Formulation and Numerical Scheme

2.1 Governing Flow and Structure Models

Aerodynamics. FLOWer solves the three-dimensional, unsteady Reynolds-averaged Navier–Stokes equations (RANS) in integral form in the subsonic, transonic and supersonic flow regime. The equations are derived for a moving frame of reference and are given by

$$\frac{\partial \mathbf{Q}}{\partial \tau} + \frac{\partial \mathbf{E}}{\partial \xi} + \frac{\partial \mathbf{F}}{\partial \eta} + \frac{\partial \mathbf{G}}{\partial \zeta} - \frac{1}{Re_0}\left(\frac{\partial \mathbf{E}_v}{\partial \xi} + \frac{\partial \mathbf{F}_v}{\partial \eta} + \frac{\partial \mathbf{G}_v}{\partial \zeta}\right) = \mathbf{R}\,. \tag{1}$$

$\mathbf{Q}$ represents the solution vector containing the conservative variables. Centrifugal and coriolis accelerations are included in the source term $\mathbf{R}$ and Re_0 is the reference Reynolds number evolving from non–dimensionalization.

The parabolic–hyperbolic system of partial differential equations can be closed by assuming a thermally and calorically perfect gas as well as Newtonian fluid properties. Turbulence can be modelled either by algebraic or by transport equation models.

The numerical procedure is based on structured meshes. The spatial discretization uses a central cell-vertex, cell-centered or an AUSM (Advection Upstream Splitting Method) finite volume formulation. Dissipative terms are explicitly added in order to damp high frequency oscillations and to allow sufficiently sharp resolutions of shock waves in the flow field. On smooth meshes, the scheme is formally of second order in space. The time integration is carried out by an explicit Runge–Kutta scheme featuring convergence acceleration by local time stepping and implicit residual smoothing. The solution procedure is embedded into a sophisticated multigrid algorithm, which allows standard single grid computations as well as successive grid refinement. Unsteady calculations are carried out using the the implicit Dual Time Stepping Scheme which reduces the solution of a physical time step to a steady–state solution in pseudo time. This approach is very effective as all convergence acceleration methods mentioned above can be used.

The code is written in a flexible block structured form enabling treatment of complex aerodynamic configurations with any mesh topology. It is fully portable for either scalar or vector architectures on sequential and parallel computers.

FLOWer is capable of calculating flows on moving grids (arbitrary translatory and rotatory motion). For this purpose the RANS equations are transformed into a body fixed rotating and translating frame of reference. Furthermore the Arbitrary-Lagrangian-Eulerian (ALE) method allows the usage of flexible meshes which is essential in the context of fluid structure coupling. Arbitrary relative motion of grid blocks is made possible by the Chimera technique of overlapping grids.

Structure Dynamics. The Eurocopter flight mechanics tool HOST represents a computational environment for simulation and stability analysis of the complete helicopter system. It enables the study of single helicopter components like isolated rotors as well as complete configurations with related substructures. As a general purpose flight mechanics tool, HOST is capable of trimming the rotor based on a lifting line method with 2D airfoil tables.

The elastic blade model in HOST considers the blade as a quasi one–dimensional Euler–Bernoulli beam. It allows for deflections in flap and lag direction and elastic torsion along the blade axis. In addition to the assumption of a linear material law, tension elongation and shear deformation are neglected. However, possible offsets between the local cross-sectional center of gravity, tension center and shear center are accounted for, thus coupling bending and torsional degrees of freedom.

The blade model is based on a geometrically non-linear formulation, connecting rigid segments through virtual joints. At each joint, elastic rotations are permitted around the lag, flap and torsion axes. Since the use of these rotations as degrees of freedom would yield a rather large system of equations, the number of rotations is reduced by a modal Rayleigh–Ritz approach. A limited set of mode-like deformation shapes together with their weighting factors are used to yield a deformation description. Therefore any degree of freedom can be expressed as

$$h(r,\psi) = \sum_{i=1}^{n} q_i(\psi) \cdot \bar{h}_i(r) \tag{2}$$

where n is the number of modes, q_i the generalized coordinate of mode i (a function of the azimuth angle ψ), and $\bar{h}_i$ is the modal shape (a function of the radial position r).

2.2 Weak Coupling Strategy

The idea of the weak coupling scheme is as follows: HOST uses CFD loads to correct its internal 2D aerodynamics and re-trims the rotor. The blade dynamic response is introduced into the CFD calculation in order to obtain updated aerodynamic loads. This cycle is repeated until the CFD loads match the blade dynamic response evoked by them. A criterion for this converged state is given by the change of the collective and cyclic pitch angles with respect to the preceding cycle. Convergence has been reached after the changes in these free control angles have fallen below an imposed limit. The specific steps of the coupling procedure are thus given as follows:

1. HOST determines an initial trim of the rotor based on its internal 2D aerodynamics derived from airfoil tables. The complete blade dynamic response for a given azimuth angle is fully described by the modal base and the related generalized coordinates.
2. The blade dynamic response is taken into account in the succeeding CFD calculation by the reconstruction of the azimuth angle dependent blade deformation from the modal base and the respective grid deformation of the blade grid.
3. The CFD calculation determines the 3D blade loads in the rotating rotor hub system ($F_x[N/m]$, $F_y[N/m]$, $F_z[N/m]$, $M_x[Nm/m]$, $M_y[Nm/m]$, $M_z[Nm/m]$) for every azimuth angle and radial section of the blade.
4. For the next trim HOST uses a load given by
$$\bar{F}^n_{HOST} = \bar{F}^n_{2D} + \bar{F}^{n-1}_{3D} - \bar{F}^{n-1}_{2D} \tag{3}$$
$\bar{F}^n_{2D}$ represents the free parameter for the actual HOST trim. A new dynamic blade response is obtained which is expressed by an update of the generalized coordinates.
5. Steps (2) to (4) are repeated until convergence has been reached, i.e. when the difference
$$\Delta\bar{F}^n = \bar{F}^n_{2D} - \bar{F}^{n-1}_{2D} \longrightarrow 0 \tag{4}$$
tends to zero and the trim-loads depend only on the 3D CFD aerodynamics.

It is mandatory that the updated CFD loads for each successive trim are periodic with respect to the azimuth angle. After the CFD calculation has been restarted from the previous run, a certain number of time steps (i.e. a certain azimuth angle range) is neccessary until the perturbation introduced by the updated set of generalized coordinates has been damped down and a periodic answer is obtained again. In fact, it is not necessary to continue the calculation until one fully periodic 1/rev response of an individual blade is obtained, as this can be composed from the last quarter revolution of all four rotor blades. It is therefore sufficient to run the CFD calculation until a periodic 4/rev behaviour of the complete rotor can be observed. Clearly, this state is reached more quickly, the smaller the initial disturbance. For this reason the azimuth angle range covered by the CFD calculation can be reduced with an increasing number of re-trims. The changes in the free controls and the blade dynamic response become smaller from one re-trim to the next.

For all computations presented in the present paper four re-trims were sufficient to obtain the desired accuracy for the converged (trimmed) solution.

2.3 Grid Deformation Tool

For aeroelastic computations, a robust algebraic grid deformation tool utilizing Hermite polynomials is applied prior to each time step in order to update the structured aerodynamic mesh according to the surface deformation provided by the structure solver. The deformed surface is determined by bending and twisting the blade quarter–chord line. In order to minimize the amount of grid deformation and thus maintain a high level of grid quality, the entire virgin grid is rotated into the root–tip secant of the deformed quarter–chord line prior to updating the 3D grid according to the current deformed blade surface.

The grid deformation tool has been originally designed for monoblock blade grids. Recently this drawback has been eliminated as the deformation tool has been extended towards the treatment of multiblock grids. The different blocks of the blade grid may now be distributed onto different processes, thus allowing for an effective parallelization of the flow computation.

3 Results

3.1 EC145 Rotor in Fast Forward Flight

The performance of a helicopter rotor with respect to power consumption, noise and vibration can be improved by introducing higher harmonic control. One possibility to achieve this is to use a rotor which is equipped with active trailing-edge flaps. In order to correctly capture the effect of the flap control it is absolutely mandatory to perform an aeroelastic analysis. As a rotor blade is a torsionally flexible structure, the flap acts as a servo flap. This means a positive flap angle (downward deflection) will cause a negative (nose down) twist of the blade due to the negative pitching moment in the flap region. This leads to a reduction of the overall blade lift. Therefore qualitatively wrong results will be obtained if aeroelastic coupling effects are neglected.

The intention of this investigation was the performance evaluation of the active rotor compared to the passive rotor. In order to perform such a comparison both rotors need to be trimmed towards an identical flight condition. The weak coupling strategy between HOST and FLOWer has emerged to be well suited for this purpose [11].

In the case of the active rotor the servo flap has to be modelled both in the HOST code and in the FLOWer code. In HOST the flap is taken into account by using modified 2D airfoil polars in the flap range. On the CFD side the flap is modelled by a local deformation of the blade surface in the flap region. At the inner and outer flap segment boundary the deflection angle is reduced to zero within a certain smoothing area, neglecting a possible side gap. The deformed blade surface is then generated in two steps: In the first step the blade surface is locally deformed according to the flap deflection. In the second step the blade deformation due to the coupling process is applied to this pre-deformed surface. For the present investigation we restricted the flap effect to the aerodynamic model, i.e. the structure model has been left unchanged. If the flap is supposed to be taken into account on the structure side as well, the structure description (in terms of bending and torsional stiffness etc.) has to be adapted to the current flap deflection.

For the present investigation a forward flight case with an advance ratio of $\mu = 0.3$ was selected. For both the passive and the active rotor the rotor shaft angle was held fixed at $\alpha_q = -4.9°$ and the calculations were trimmed for thrust, lateral and longitudinal mast moment by adaption of the free controls θ_0, θ_c, θ_s. The active blade features three adjoining flap segments with a chordwise extent of 15% chord and the radial positions $r/R = 0.69 - 0.75$, $r/R = 0.75 - 0.8$ and $r/R = 0.8 - 0.85$. For the present calculations a common control law was used for the innermost and the central flap segment, whereas the outermost flap segment remains fixed at zero deflection. The 2/rev flap control law is given by

$$A(t) = A_0 \cdot \cos(2 \cdot \Omega t - 2 \cdot \phi). \tag{5}$$

The flap amplitude was prescribed to $A_0 = 6°$. With an increment in azimuth of $\Delta\phi = 30°$ for the rotor, a $\Delta\phi' = 60°$ resolution of the phase shift in the control law has been investigated.

Figures 1 to 3 show the trim convergence in the control angles θ_0, θ_c and θ_s for all flap phase angles and the passive rotor. For all computations all free controls converged to the required accuracy within four re-trim cycles. Even though a systematic deviation of the 0th HOST trim to the final trimmed state can be observed for all three controls, HOST is able to predict the influence of the variation of the flap phase on the rotor trim.

Figure 4 shows the unsteady aerodynamic rotor loads for the complete weak coupling process exemplarily for the passive rotor. Each re-trim is marked off with respect to the preceding trim by the line type change from solid to dash. It can be clearly seen that the disturbance introduced by the update of the blade dynamic response decreases from one re-trim to the next as the procedure converges towards the trimmed state. After four re-trims (seven rotor revolutions) the calculation has reached the trimmed state with the required accuracy. Similar figures arise for the computations of the active rotor.

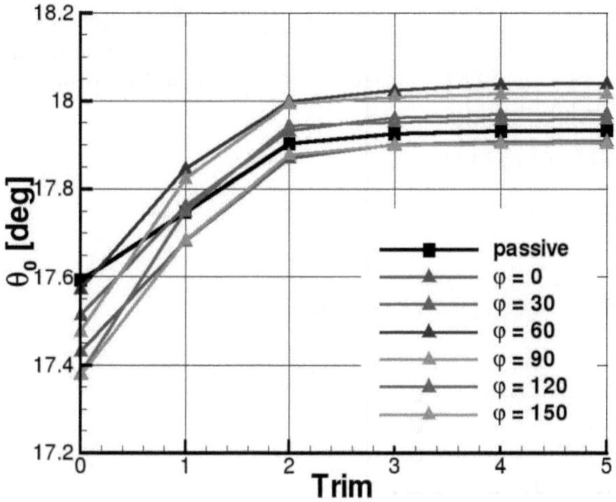

Fig. 1. Convergence of collective pitch

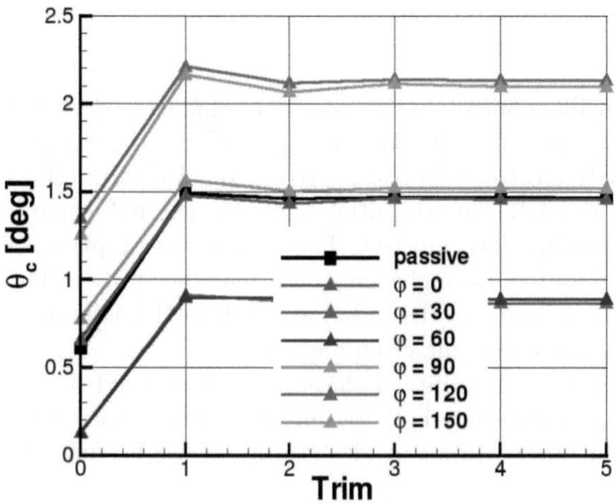

Fig. 2. Convergence of longitudinal cyclic pitch

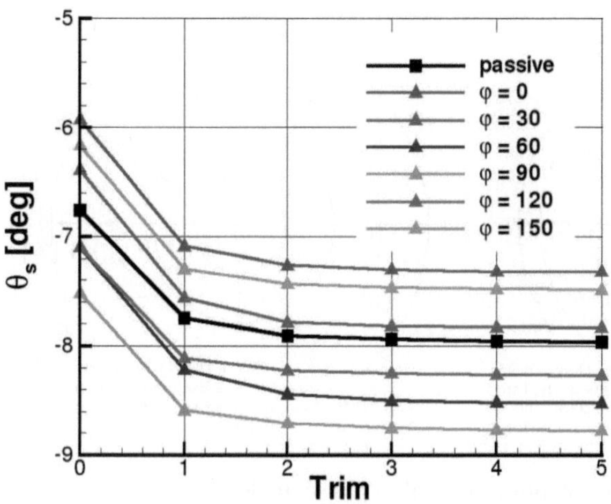

Fig. 3. Convergence of lateral cyclic pitch

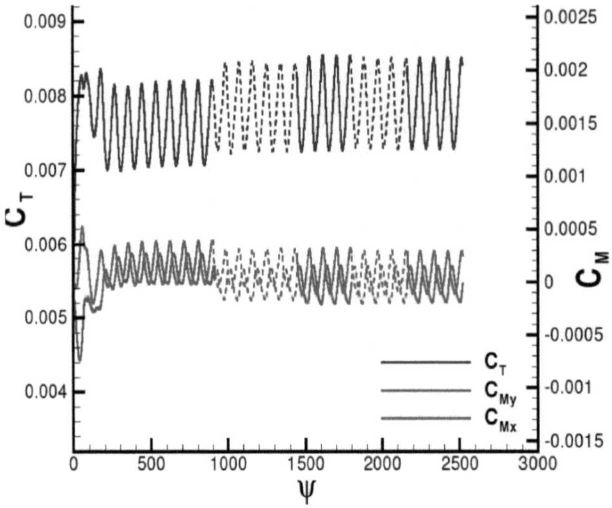

Fig. 4. Unsteady rotor coefficients during the coupling process

The influence of the 2/rev flap control on the rotor performance is depicted in Fig. 5. The polar diagram on the left side of Fig. 5 shows the relative power consumption of the active rotor with respect to the passive rotor. Both the performance predicted by HOST at its initial trim and the peformance after the weak coupling procedure are plotted. The power consumption of the passive rotor at the 0th trim has been chosen as the reference power. The relative power comsumption is plotted on the radial axis, whilst the azimuthal increment of the flap phase is given on the circumferential axis.

Using this plot style we obtain an elliptical shape for the required power, due to the fact that the flap control law is 2/rev and the plot is point-wise symmetrical. Looking at the phase variation of the initial HOST trim, the

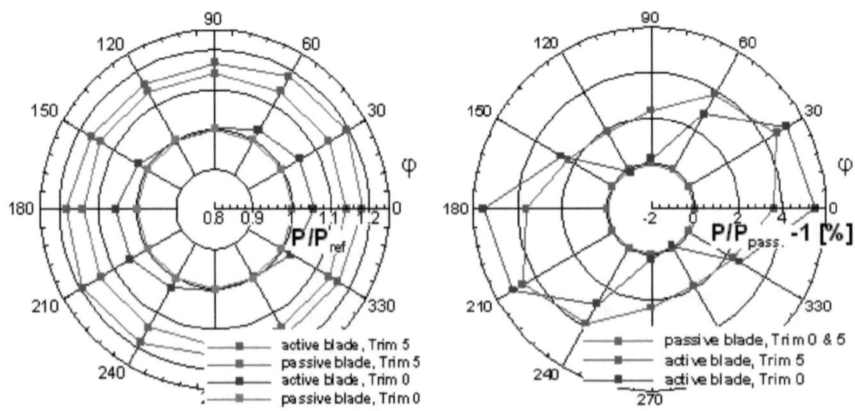

Fig. 5. Rotor power over flap phase angle

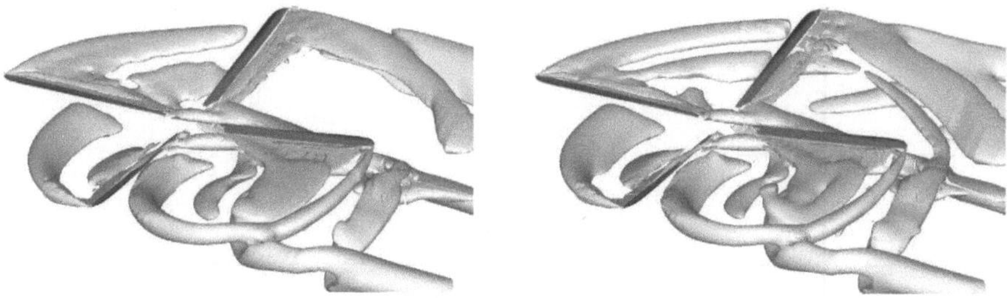

Fig. 6. Vortex system of passive rotor (*left*) and active rotor (*right*)

active rotor requires approximately the same power as the passive one between $\phi = 90°$ and $\phi = 120°$. Outside of this region a higher power consumption can be observed for the active rotor. At $\phi = 110°$ one might even expect a decreased power requirement. This is shown in detail by the trim 0 curve on the right side of Fig. 5.

The trimmed solution (trim 5), i.e. with 3D loads, generally predicts a higher power requirement compared to that for the initial trim (trim 0). A 14% increase is observed for the passive rotor. The active rotor performance diagram still maintains its elliptical shape, although the power requirement now entirely exceeds that of the passive rotor. Further, the phase angle for minimum power has slightly increased to about $\phi = 120°$. The power of the active rotor relative to the passive one, both for the initial and the final trim, is given on the right side of Fig. 5. As previously mentioned, the initial HOST trim predicts roughly 0% power increase for the active rotor compared to the passive one, at the optimum phase angle. The trimmed solution (with 3D loads) predicts an increase of 2% for the optimum phase angle.

The qualitative flow field of the passive rotor is shown on the left side of Fig. 6. The vortex system generated by the rotor is visualized by the well known λ_2-criterion of Jeong and Hussain [15]. An iso-surface of $\lambda_2 = -0.0001$ has been chosen for visualization. Both the blade tip vortex system and the inboard wake are visible. The qualitative flow field of the active rotor (phase angle $\phi = 120°$) is shown on the right side of Fig. 6. Significant differences can be observed compared to the passive rotor. As expected additional vortices are shed from the blade at the inner and outer flap boundaries. The strenght of these vortices must not be underestimated. The interaction of the flap vortices with the blade tip vortex system and the inboard wake leads to a highly complex flow field which is difficult to predict.

3.2 Bo105 Rotor in Slow Descent Flight

The slow descent flight condition of a helicopter rotor is of particular interest as for this flight condition Blade Vortex Interaction (BVI) may occur. The interaction of the blade tip vortices with the rotor blades is the main noise

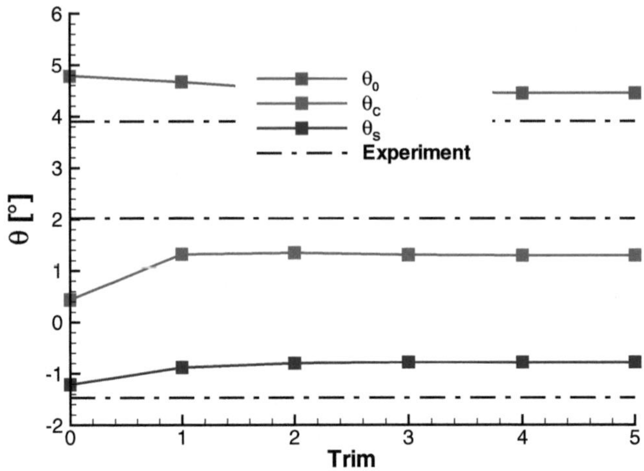

Fig. 7. Convergence of control angles

source of a helicopter in descent flight. For the Bo105 rotor in descent flight extensive measurements have been performed during the HART-II test campaign. The measurements have been performed on a 40% Mach-scaled model of the Bo105 rotor in the open-jet test section of the German-Dutch wind tunnel (DNW). The measurements include aerodynamic forces, blade dynamics, vortex locations and acoustics. The experimental results have been provided to us by the HART consortium [13].

Again, the computation has been trimmed towards the rotor hub loads, now given from the experimental trim condition, by adaption of the free controls θ_0, θ_c, θ_s. The rotor shaft angle has been held fixed at the experimental value of $\alpha_q = 4.5°$, taking 1° wind tunnel interference effect into account. Figure 7 shows the trim convergence in the control angles θ_0, θ_c and θ_s. Altough the Bo105 rotor blades are torsionally very soft the computation converges quite quick towards the required accuracy and the resulting control angles are close to the experimental values.

Figures 8 and 9 show a comparison of the experimental blade dynamics and the computed blade dynamics. The elastic blade tip torsion is given in Fig. 8. The dashed line denotes the tip torsion predicted by the 0th HOST trim, whereas the solid line shows the final trimmed result using CFD loads. It can be seen that both computational results predict a too low mean value compared to the experiment. The trimmed solution (with CFD loads) provides a better agreement in the azimuthal torsion variation than the pure HOST result (0th trim). Except from the wrong mean value the agreement of the trimmed solution to the experiment can be denoted as good. The wrong mean value might be due to the trailing edge tab of the NACA23012 airfoil of the Bo105 rotor. The influence of the tab on the airfoil's pitching moment is

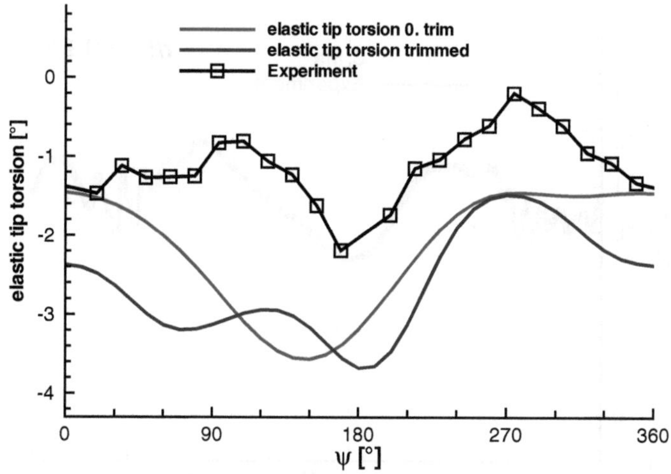

Fig. 8. Blade tip torsion

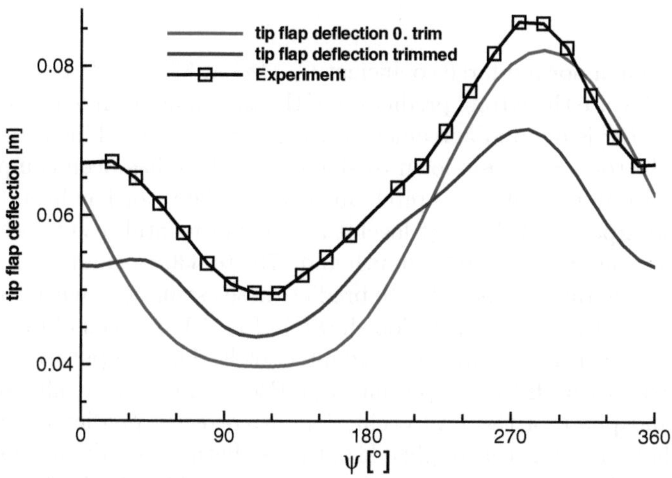

Fig. 9. Blade tip flap deflection

significant and small deviations from its actual geometry (e.g. the tab angle) lead to considerable changes in the pitching moment. Together with the low torsional stiffness of the blade this leads to large effects in the tip torsion prediction. Figure 8 shows the the blade tip flap deflection (including precone). Again, the mean value is not correctly predicted, but with respect to the azimuthal variation the trimmed solution matches the experimental result very well.

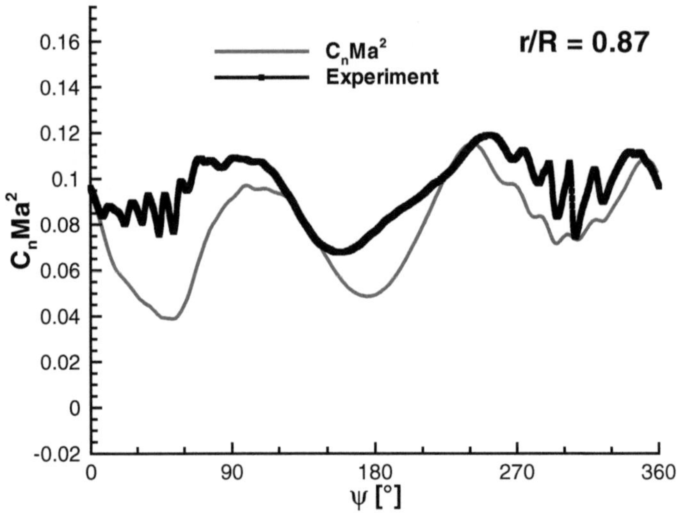

Fig. 10. Sectional normal force coefficient at 87% blade radius

The sectional normal force coefficient C_nMa^2 at 87% blade radius is given in Fig. 10. Despite the wrong prediction of the torsional mean value, the mean value of C_nMa^2 is in good agreement with the experiment. This is due to the fact, that the too large nose-down torsion of the blade has been compensated by an increase in the collective pitch angle θ_0, as the rotor has been trimmed towards the experimental thrust level. In the experimental results oscillations due to BVI around 30° to 90° azimuth and 270° to 330° azimuth can be seen. The CFD computation is not able to predict these oscillations with the correct amplitude. The C_nMa^2-distribution shown in Fig. 10 has been obtained using a grid setup featuring special vortex-adapted grids. The computation with the standard grid setup did not reproduce any BVI oscillations at all. Now both direction and phase of the BVI oscillations is correctly predicted for the retreating blade side, but the amplitude of the oscillations is still underpredicted by about a factor of five. For the advancing blade side, the BVI oscillations are still completely missed. Our current work focuses on the improvement of the reproduction of these BVI oscillations [12].

4 Computational Performance

All computations of the EC145 rotor in fast forward flight have been performed as single-node calculations on the NEC SX-8. At this point of time coupled computations were limited to single block blade grids due to the implementation of the grid deformation algorithm. A computation on more than one node was not reasonable, as this would have resulted in a poor load bal-

Table 1. Computational performance

Rotor	EC145	Bo105
Platform	SX8	SX8
Number of cells	3.552.768	19.687.424
Number of nodes	1	3
Number of CPUs	8	24
GFLOPS	21	67.6
Vector operation ratio	97.0%	98.7
Wall clock time per rotor revolution	7h	6.5h
Required memory	approx. 2.0 kB/cell	approx. 2.0 kB/cell

ancing. The computational performance of the EC145 computations is given in Table 1. Due to the extension of our grid deformation algorithm multi-block blade grids could be used for the Bo105 computation, thus allowing for an effective parallelization of the computation on more than one node. Table 1 contains information on the performance of a computation with a refined background grid that has been performed on three nodes of the NEC SX-8.

5 Conclusions and Outlook

We presented a weak coupling method for the aeroelastic simulation of isolated helicopter main rotors. The method has been applied to both a fast forward flight configuration and a slow descent flight case. It allows for a fast trim of the rotor towards given hub loads and is well suited for performance evaluations on active rotors. Our current activities deal with the improvement of the reproduction of BVI oscillations in coupled CFD/CSD computations. In this context we plan to further increase the number of grid cells and to further parallelize the computation on more CPUs. Furthermore we plan to extend our activities from stand–alone main rotor simulations towards complete helicopter configurations.

Acknowledgements

This work has been funded by BMWa (Bundesministerium für Wirtschaft und Arbeit). The authers would like to thank the system administrators of HLRS for their technical support. Furthermore we would like to thank the HART consortium (DLR, ONERA, NASA Langley, DNW, US Army AFDD) to provide us with the HART-II experimental results.

References

1. ALTMIKUS, A.R.M.: **Nichtlineare Simulation der Strömungs-Struktur-Wechselwirkung am Hubschrauber.** Dissertation, Universität Stuttgart, ISBN 3-18-346607-4, 2004.
2. POMIN, H.: **Simulation der Aerodynamik elastischer Rotorblätter mit einem Navier–Stokes-Verfahren.** Dissertation, Universität Stuttgart, ISBN 3-8322-2276-6, 2003.
3. FISCHER, A.K.: **Untersuchungen zur Aeromechanik eines flugfähigen Hubschraubermodells.** Dissertation, Universität Stuttgart, ISBN 3-89963-126-9, 2005.
4. BUCHTALA, B.: **Gekoppelte Berechnung der Dynamik und Aerodynamik von Drehflüglern.** Dissertation, Universität Stuttgart, ISBN 3-8265-9732-X, 2002.
5. WEHR, D.: **Untersuchungen zum Wirbeltransport bei der Simulation der instationären Umströmung von Mehrblattrotoren mittels der Euler-Gleichungen.** Dissertation, Universität Stuttgart, ISBN 3-89722-285-X, 1999.
6. STANGL, R.: **Ein Euler-Verfahren zur Berechnung der Strömung um einen Hubschrauber im Vorwärtsflug.** Dissertation, Universität Stuttgart, ISBN 3-89675-141-7, 1996.
7. WAGNER, S.: **On the Numerical Prediction of Rotor Wakes Using Linear and Non-Linear Methods.** AIAA-Paper 2000-0111, January 2000.
8. WAGNER, S.: **Flow Phenomena on Rotary Wing Systems and their Modeling.** ZAMM 79 (1999) 12, pp. 795–820, 1999.
9. ALTMIKUS, A.R.M.: **On the timewise accuracy of staggered aeroelastic simulations of rotary wings.** AHS Aerodynamics, Acoustics, and Test and Evaluations Technical Specialist Meeting, San Francisco, CA, 2002.
10. ALTMIKUS, A.R.M., WAGNER, S., BEAUMIER, P., SERVERA, G.: **A comparison: Weak versus strong modular coupling for trimmed aeroelastic rotor simulations.** American Helicopter Society 58th Annual Forum, San Francisco, CA, 2002.
11. DIETZ, M., KRÄMER, E., WAGNER, S., ALTMIKUS, A.R.M.: **Weak coupling for active advanced rotors.** Proceedings of the 31st European Rotorcraft Forum, Florence, Italy, 2005.
12. DIETZ, M., KESSLER, M., KRÄMER, E.: **Advanced Rotary Wing Aeromechanics.** High Performance Computing in Science and Engineering, pp. 197–208, Springer Verlag, 2005.
13. LIM, J.W. ET AL: **HART-II: Prediction of Blade-Vortex Interaction Loading.** Proceedings of the 29th European Rotorcraft Forum, Friedrichshafen, Germany, 2003.
14. SERVERA, G., BEAUMIER, P., COSTES, M.: **A weak coupling method between the dynamics code HOST and the 3D unsteady Euler code WAVES.** Proceedings of the 26th European Rotorcraft Forum, The Hague, The Netherlands, 2000.
15. JEONG, J., HUSSAIN, F.: **On the Indentification of a Vortex.** Journal of Fluid Mechanics, Vol. 285, pp. 69–94, 1995.

Computational Study of the Aeroelastic Equilibrium Configuration of a Swept Wind Tunnel Wing Model in Subsonic Flow

L. Reimer[1], C. Braun[2], and J. Ballmann[2]

[1] Lehrstuhl für Computergestützte Analyse Technischer Systeme (CATS), RWTH Aachen, Steinbachstrasse 53b, D-52074 Aachen, Germany,
`http://www.cats.rwth-aachen.de`
`reimer@cats.rwth-aachen.de`

[2] Lehr- und Forschungsgebiet für Mechanik (LFM), RWTH Aachen, Templergraben 64, D-52062 Aachen, Germany,
`http://www.lufmech.rwth-aachen.de`,
`carsten@lufmech.rwth-aachen.de`

Abstract. In the Collaborative Research Center SFB 401 at RWTH Aachen University, the numerical aeroelastic method SOFIA for direct numerical aeroelastic simulation is being progressively developed. Numerical results obtained by applying SOFIA were compared with measured data of static and dynamic aeroelastic wind tunnel tests for an elastic swept wing in subsonic flow.

1 Introduction

In the Collaborative Research Center SFB 401 "Flow Modulation and Fluid-Structure Interaction at Airplane Wings" at Aachen (SOlid Fluid InterAction) for direct numerical aeroelastic simulation is being progressively developed. SOFIA is based on a coupled field formulation, in which distinct numerical solvers for fluid and structure are coupled. Since the aerodynamic performance, maneuverability and flight shape of aircrafts operating in the transonic regime depend highly on the deformation of their wings under aerodynamic loads, computational aeroelastic methods must be applied in order to predict the aerodynamic state of the coupled problem accurately.

In the present method a coupling module performs the exchange of information at the boundaries of the different fields, i. e. the exchange of energy via the aerodynamic surface and information about the change of the surface shape. Since the boundaries of the computational grid for the fluid solver must coincide with the aerodynamic surface of the structure, a flow grid deformation tool has to be applied in every step of the aeroelastic computation. In the present version of SOFIA, the flow field is modeled by the Reynolds-averaged

Navier–Stokes equations, which are solved using the finite volume flow solver FLOWer, developed under the direction of the German Aerospace Center DLR. The wing structure is modeled by a multi-axial Timoshenko-like beam and the governing equations are discretised by a finite element approach.

Newly developed codes for computational aeroelastic analysis need extensive sets of wind tunnel data for elastic 3D wing models for validation. Due to the limited number of elastic wing models for public aero-structural research (at least in Europe), one objective of the Collaborative Research Center SFB 401 is to create data sets for benchmarking. In this context a flexible swept wing model has been designed and manufactured at the Institute for Lightweight Structures (ILB) at Aachen University [2] and was tested in the German–Dutch Wind Tunnels (DNW-LST) in Emmeloord (NL). The performed wind tunnel tests included static aeroelastic experiments (aeroelastic equilibrium configurations) as well as dynamic response tests in the subsonic flow regime. In all tests the root forces and moments, the spanwise deformation (torsional twist and bending) and the pressure at one spanwise location have been recorded and are available for comparison with numerical results.

The aeroelastic method SOFIA has been tested for its static aeroelastic prediction capability by computing the aeroelastic equilibrium configuration along with the surrounding flow field of the swept wind tunnel wing model for several experimental test cases. Since the evaluation of results from simulating the performed unsteady aeroelastic experiments using SOFIA is currently in progress, the present paper will concentrate on the comparison of results obtained by static aeroelastic simulations to the corresponding experiments. In detail the computed bending displacement, the torsional twist of the wing, and the root forces and moments will be compared to the measured data for two different flow velocities within the subsonic flow regime and for the whole range of experimentally rigging angles of incidence. It will be shown that the computed bending deformation and the root forces will be in excellent agreement with the measured data, the torsional twist deformation and the root moments have been slightly underpredicted by the computations.

2 Physical Models and Numerical Methods

The aeroelastic method SOFIA solves the coupled problem consisting of the flow field, the displacement field of the structure and the deformation of the flow grid. Due to its coupled multi-field formulation each of the identifiable fields is represented within SOFIA by an independent program component, which is specialized regarding the particular demands of the respective field. The essential communication between flow solver, structural solver and flow grid deformation method is managed via an aeroelastic coupling module. Each program component including its underlying physical model and numerical method is described below.

2.1 Fluid Dynamics

The governing equations for the flow field involving viscosity and heat conduction are the three-dimensional Reynolds-averaged Navier–Stokes (RANS) equations which are derived from the Navier–Stokes equations by applying an averaging process. Although this paper concentrates on static aeroelastic problems the governing equations for the flow field will be derived for the general unsteady case. The integral form of the RANS equations, which form the basis for the finite-volume technique, read

$$\frac{\partial}{\partial t}\int\limits_{\Omega(t)} \mathbf{W}\,\mathrm{d}\Omega + \oint\limits_{\partial\Omega(t)} \mathbf{F}(\mathbf{W})\cdot\mathbf{n}\,\mathrm{d}S = \mathbf{0}\,, \tag{1}$$

where $\mathbf{W} = [\rho, \rho u, \rho v, \rho w, \rho e]^T$ is the algebraic vector of conserved quantities: mass density, Cartesian components of momentum and specific total energy. Ω represents the control volume and $\partial\Omega$ denotes its closed surface. The flux function $\mathbf{F} = \mathbf{F}^c - \mathbf{F}^d$ can be expressed by a summation of a convective part

$$\mathbf{F}^c(\mathbf{W}) = \begin{bmatrix} \rho(\mathbf{v}-\mathbf{v}_b) \\ \rho u(\mathbf{v}-\mathbf{v}_b) + p\mathbf{e}_x \\ \rho v(\mathbf{v}-\mathbf{v}_b) + p\mathbf{e}_y \\ \rho w(\mathbf{v}-\mathbf{v}_b) + p\mathbf{e}_z \\ \rho e(\mathbf{v}-\mathbf{v}_b) + p\mathbf{v} \end{bmatrix} \tag{2}$$

including pressure and a diffusive part caused by viscous stresses and heat conduction

$$\mathbf{F}^d(\mathbf{W}) = \begin{bmatrix} 0 \\ \tau_{xx}\mathbf{e}_x + \tau_{xy}\mathbf{e}_y + \tau_{xz}\mathbf{e}_z \\ \tau_{yx}\mathbf{e}_x + \tau_{yy}\mathbf{e}_y + \tau_{yz}\mathbf{e}_z \\ \tau_{zx}\mathbf{e}_x + \tau_{zy}\mathbf{e}_y + \tau_{zz}\mathbf{e}_z \\ \psi_x\mathbf{e}_x + \psi_y\mathbf{e}_y + \psi_z\mathbf{e}_z \end{bmatrix}\,, \tag{3}$$

using the abbreviations

$$\begin{aligned} \psi_x &= u\tau_{xx} + v\tau_{xy} + w\tau_{xz} - q_x\,, \\ \psi_y &= u\tau_{yx} + v\tau_{yy} + w\tau_{yz} - q_y\,, \\ \psi_z &= u\tau_{zx} + v\tau_{zy} + w\tau_{zz} - q_z\,. \end{aligned} \tag{4}$$

$\mathbf{v}_b$ represents the velocity of the boundary of the control volumes Ω. The consideration of $\mathbf{v}_b$ is necessary for unsteady aeroelastic applications, where the computational mesh has to be updated according to the deformation of the aerodynamic surface, which leads to time dependent control volumes. The pressure p is calculated applying the equations of state for a perfect gas. The temperature dependance of viscosity is approximated in case of laminar flow by Sutherland's formula as a function of the static temperature only. The vector of heat conduction is modeled by Fourier's law with the heat conductivity based on a constant Prandtl number.

In the present method, the flow solver FLOWer, developed under the direction of DLR [3], is employed to solve the governing flow equations on block-structured grids. Because a finite volume technique with control volumes dependent on time is used, FLOWer is applicable to aeroelastic problems where a temporal deformation of the computational mesh is required. For the results presented in this report, central differences were used for the spatial discretization. For steady state calculations, time integration is performed by a 5-step Runge–Kutta method. To achieve a good convergence rate, acceleration techniques such as local time stepping, implicit residual smoothing, and multigrid algorithms are used. Several turbulent models are available in FLOWer. For the computation of the results shown in the present work only the one-equation model according to Spalart–Allmaras has been applied. FLOWer is optimized especially for vector computers and highly parallelized using the MPI standard.

2.2 Structural Dynamics

In SOFIA the elastic wing is modeled by a Timoshenko-like beam structure with six degrees of freedom for a material cross-section and non-coinciding centerlines of mass, bending and torsion. The governing equations for the motion of the cross-section can be found by applying Hamilton's principle. This leads to the variational form of the action integral

$$\delta\mathcal{I}(\mathbf{u}_s,\boldsymbol{\varphi}) = \delta\int_{t_a}^{t_e}\int_{0}^{l} \ell_b \,\mathrm{d}\xi\,\mathrm{d}t + \int_{t_a}^{t_e} \delta W^a \,\mathrm{d}t = 0. \tag{5}$$

with the Lagrangian density per unit length

$$\begin{aligned}\ell_b = &\left(\frac{1}{2}\varrho A^{\varrho}\dot{\mathbf{u}}_s\cdot\dot{\mathbf{u}}_s + \frac{1}{2}\dot{\boldsymbol{\varphi}}\,\boldsymbol{\Theta}_s\dot{\boldsymbol{\varphi}}\right)\\ &-\left(\frac{1}{2}EA(\frac{\partial u_{1,b}}{\partial\xi})^2 + \frac{1}{2}GA\boldsymbol{\gamma}\mathbf{K}\boldsymbol{\gamma} + \frac{1}{2}\frac{\partial\boldsymbol{\varphi}}{\partial\xi}\mathbf{C}^*\frac{\partial\boldsymbol{\varphi}}{\partial\xi}\right).\end{aligned} \tag{6}$$

Notations are $\mathbf{u}_s$ for displacement vector, $\boldsymbol{\varphi}$ for bending and torsion vector, $\boldsymbol{\gamma}$ for shear vector, A for cross section, $\boldsymbol{\Theta}$ for mass inertia tensor of a material cross-section, $\mathbf{K}$ for equivalent shear distribution tensor, $\mathbf{C}^*$ for stiffness tensor containing the resistance against bending, torsion, and warping. Indices s and b represent the centers of gravity and bending, respectively. Applying Timoshenko's theory the shear angles are represented as

$$\begin{aligned}\gamma_2 &= \frac{\partial u_{2,s}}{\partial\xi} - \varphi_3 - \frac{\partial\left(\zeta_{sd}\varphi_1\right)}{\partial\xi},\\ \gamma_3 &= \frac{\partial u_{3,s}}{\partial\xi} + \varphi_2 + \frac{\partial\left(\eta_{sd}\varphi_1\right)}{\partial\xi},\end{aligned} \tag{7}$$

where ζ_{sd} and η_{sd} are the Cartesian co-ordinate differences between the center of gravity and the shear center of the cross-section.

Based on the resulting variational formulation, a system of coupled ordinary differential equations (ODEs) of second order in time

$$\mathbf{M}\,\ddot{\mathbf{q}}(t) + \hat{\mathbf{D}}\,\dot{\mathbf{q}}(t) + \mathbf{K}\,\mathbf{q}(t) = \mathbf{R}^a(t) \tag{8}$$

is derived for the time dependent generalized displacements $\mathbf{q}(t)$ of the material cross-section by applying the finite element method in the sense of Ritz/ Kantorowitsch [4, 5]. The generalised external forces are represented by $\mathbf{R}^a(t)$. In order to accelerate the computation of steady problems as the asymptotic solution of the unsteady problem, the artificial damping term $\hat{\mathbf{D}}\,\dot{\mathbf{q}}(t)$ was added. The damping matrix $\hat{\mathbf{D}}$ is built in the sense of modal damping. Discretization is done by iso-parametric, two-noded elements. A reduced integration scheme avoids shear locking. The solution of the generalised eigenvalue problem

$$\mathbf{K}\,\mathbf{z} = \omega^2\,\mathbf{M}\,\mathbf{z}, \tag{9}$$

leads to n eigenvalues ω_i^2 and the corresponding eigenvectors $\mathbf{z}_i$ of the undamped system. n is equivalent to the number of degrees of freedom of the supported structure. The set of ODEs is diagonalized by applying subsequently the transformation formula

$$\mathbf{q} = \mathbf{Z}\,\mathbf{x} \tag{10}$$

to (8), where $\mathbf{Z} = [\mathbf{z}_1, \cdots, \mathbf{z}_n]$ is the matrix consisting of the eigenvectors and $\mathbf{x} = [x_1(t), \cdots, x_n(t)]^T$ denotes the vector of the modal displacements. This results in n uncoupled equations of type

$$\ddot{x}_i(t) + 2\hat{\xi}_i\omega_i\dot{x}_i(t) + \omega_i^2 x_i(t) = r_i^a(t), i = 1, \cdots, n \tag{11}$$

with the modal damping coefficients $\hat{\xi}_i$. The time integration is done by evaluating Duhamel's integral

$$\begin{aligned} x_i(t) = {} & \frac{1}{\hat{\omega}_i}\int_0^t r_i^a(\tau)\,\mathrm{e}^{-\hat{\xi}_i\omega_i(t-\tau)}\;\mathrm{d}\tau \\ & + \mathrm{e}^{-\hat{\xi}_i\omega_i t}\left(\alpha\sin(\hat{\omega}_i t) + \beta\cos(\hat{\omega}_i t)\right). \end{aligned} \tag{12}$$

while using the definition of the angular frequency of the damped system $\hat{\omega} = \omega\sqrt{1-\hat{\xi}^2}$. The coefficients α and β are determined by the initial conditions. The solution of (12) is based on the assumption that the generalised external forces vary linearly during every time-step.

2.3 Flow Grid Deformation Method

In every time step of an aeroelastic computation the computational mesh for the flow solver has to be updated according to the change of shape of the wing surface. Therefore an algorithm has been developed, in which the block boundaries and a certain number of grid lines, which depend on the grid topology, are modeled as a fictitious framework of elastic beams [6, 7]. These beams are considered rigidly fixed together in points of intersection and to the aerodynamic surface as well, such that angles are preserved where beams (= grid lines) intersect or emerge from a solid surface. The deformation of the framework, which is due to displacements of the surface grid nodes, is calculated by an FE solver. The new positions of grid points in the interior of the domain which are not included in the beam-framework are determined by interpolation.

2.4 Aeroelastic Coupling Module

When using an aeroelastic method, which is based on a coupled multi-field formulation, the natural interface between the elastic structure and the surrounding fluid is the aerodynamic surface. Along this surface information corresponding to forces and displacements have to be exchanged between the concerned solvers. Within the last year, a largely independent coupling module has been developed and implemented, which performs the data transfer between fluid solver and structural solver. It satifies following essential requirements:

- Coupling of different two- and three-dimensional structural dynamics and fluid dynamics solvers, structured or unstructured, e. g. FLOWer, TAU [8], QUADFLOW [9].
- Control of solution steps for different loose and tight staggered coupling algorithms.
- Conservative load and displacement transfer for different aerodynamic configurations.
- Ability to handle steady and time accurate unsteady aeroelastic simulations.

The implemented aeroelastic coupling module is available as a set of library functions or as an independent program version and can be called from the flow solver as a single routine.

3 Computational Results

Within the framework of the collaborative research center SFB 401 the aeroelastic method SOFIA has been applied to compute the aeroelastic equilibrium configuration of a swept wing model. The aeroelastic equilibrium configuration is achieved when the aerodynamic loads and the structural reaction forces are in a state of equilibrium. The studied wind tunnel model was manufactured with a backward sweep angle of 34°, a half span of $1.5m$ and a chord length of $0.333m$. The chosen profile of the wing corresponds to the reference

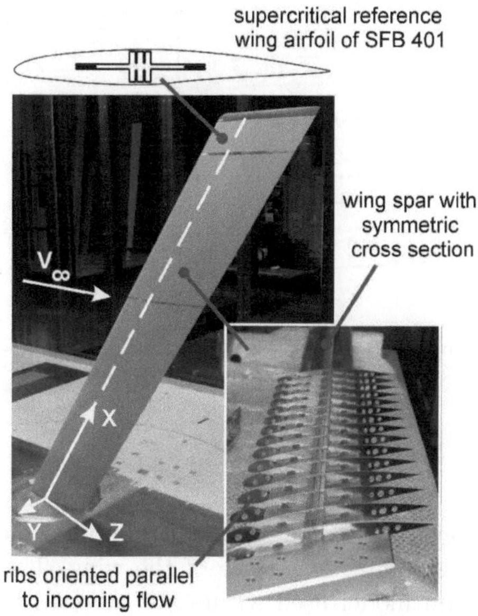

Fig. 1. Swept wind tunnel wing model mounted in German–Dutch Wind Tunnels (DNW-LST) (©ILB 2001)

airfoil (BAC 3-11/RES/30/21) in cruise configuration defined by the SFB 401. The structural design of the wing had to consider the demands of a desirable flexible structure with large deflections within the limits of the DNW-LST wind tunnel, relatively low eigenfrequencies, very low structural damping and a wide stability range which encloses the planned wind tunnel test conditions. Therefore the load supporting structure is composed of a cross-shaped wing spar as shown in Fig. 1 being appropriate to fulfill the requirements of aeroelastic experiments of a particularly low torsional stiffness in combination with a sufficiently high bending stiffness. The transfer of aerodynamic loads to the wing spar is realized by ribs installed on the spar with positive locking and foam segments filling the space between the ribs. The orientation of the ribs is parallel to the incoming flow. The spar and the ribs are made of an aluminium alloy which garanties low structural damping and large deflections without occurence of plasticity. The structural cross-sectional properties along the wing were determined by laboratory test series at ILB. Due to the symmetric cross-section of the spar and the homogeneity of its mass distribution the centerlines of gravity, bending and torsion coincide with the centerline of symmetry. The structural dataset obtained from laboratory tests was used to identify the reduced structural model (Timoshenko-like beam model) corresponding to the representation of the structure within the aeroelastic method SOFIA. Detailed descriptions regarding the properties of the material, the cross-sectional stiffnesses along the wing and the experimental measurement instrumentation and techniques can be found in [2].

First a modal analysis was performed in order to examine the quality of the identified beam model by means of eigenshapes and eigenvalues in

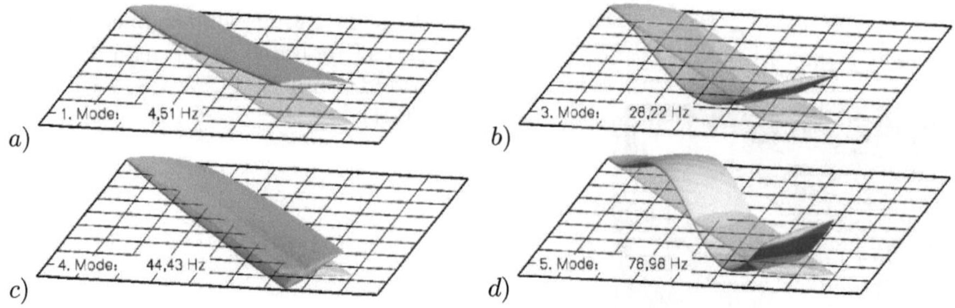

Fig. 2. Natural vibration modes of the simulated wing as a projection of the natural modes of the Timoshenko-like beam on the wing, **a**) 1st mode, **b**) 3rd mode, **c**) 4th mode and **d**) 5th mode

comparison to the wind tunnel model. Avoiding to present the eigenshapes related to purely horizontal wing motions, Fig. 2 shows the natural vibration modes of the beam projected on the aerodynamic surface of the wing. The eigenshapes and eigenvalues of the beam model used within the aeroelastic simulation applying SOFIA are in excellent agreement with the results from experimental modal analysis for the equipped wind tunnel wing model. The presented eigenfrequencies of the beam differ from the eigenfrequencies of the experimental assembly less than 0.5%.

The following section is attended to a description of results achieved by applying SOFIA to the coupled problem of the flexible swept wing in subsonic flow. Reference values for comparison with the experimental results are spanwise bending displacements and twist deformations due to torsion, reaction forces and moments within the clamping plane of the wing and pressure distributions on the surface of the deformed wing. The following comparison is focused on results related to flow velocities of $V_\infty = 65m/s$ and $75m/s$. The Reynolds numbers in the computations were set according to the experimental conditions to $Re_\infty = 1.55 \cdot 10^6$ and $Re_\infty = 1.75 \cdot 10^6$, respectively. Starting from the rigging root angle of incidence leading to a vanishing lift of the aeroelastic equilibrium configuration, the rigging angle was decreased and increased in the experiments in steps of two degrees, one step below the angle with vanishing lift and five steps above. Consequently, the experimental procedure was imitated in the computations resulting in a range of rigging angles from about $\alpha_R = -3.5°$ to $+8.5°$.

Figure 3 a) and b) show computational results for the deformation at the wing tip, i. e. the displacement u_Y^{tip} and the twist due to torsion φ_X^{tip} of a cross-section perpendicular to the beam axis, against the angle of incidence α_R for a flow velocity of $V_\infty = 65m/s$. The symbols are related to the experimental results, whereas the lines belong to results from solving the Euler equations on the one hand and the Navier–Stokes equations on the other hand. A solution to the Navier–Stokes equations was found by using the one-equation turbulence model of Spalart–Allmaras. Thereby the location of transition was assumed to be at the leading edge in all considered test cases. Similarly, Fig. 3 c) and d)

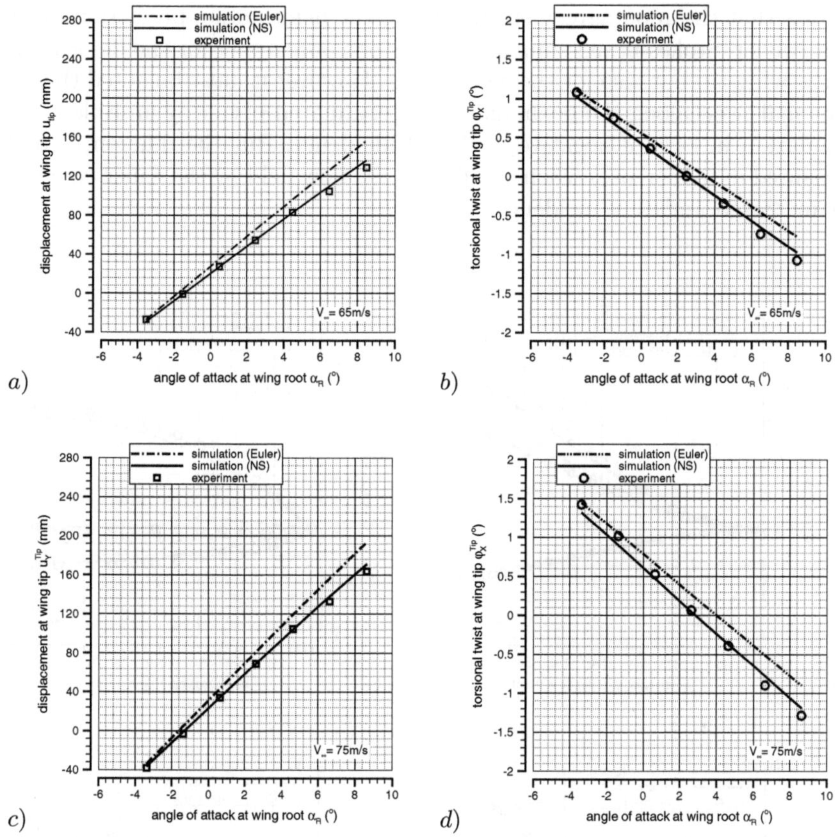

Fig. 3. Comparison of measured and computed displacement (subfigures **a**) and **c**)) and torsional twist (**b**) and **d**)) at the wing tip as a function of the rigging angle of incidence at wing root α_R for two different flow velocities $V_\infty = 65m/s$ and $75m/s$

present the resulting deformations at the wing tip for the flow velocity of $V_\infty = 75m/s$. The plotted results have in common that the solutions based on the Navier–Stokes equations lead to a precise prediction of the tip displacement for all angles of incidence unlike the computations without considering the influence of boundary layers. The agreement for the twist deformation due to torsion is insignificantly incorrectly predicted. The main difference is that the slope of change of the torsional twist against the angle of incidence is not reproduced exactly. But the differences remain below 0.15°.

In Fig. 4 a) to d) the displacements and torsional twists are plotted against the beam axis co-ordinate X for three exemplary angles of incidence. Again, the computational results from solving the Euler and Navier–Stokes equations for the flow field are compared to the measured data. The comparison reveals that the good agreement regarding the deformation prediction in case of a Navier–Stokes-based simulation does not only apply for the deformation of the wing tip but also for the whole wing. However it must be noted that the torsional twist along the beam axis is approximated better by an Euler-based simulation in case of negative angles of incidence. But the difference to the measured data increases when simulating the flow field by the Euler equations for increased angles of incidence.

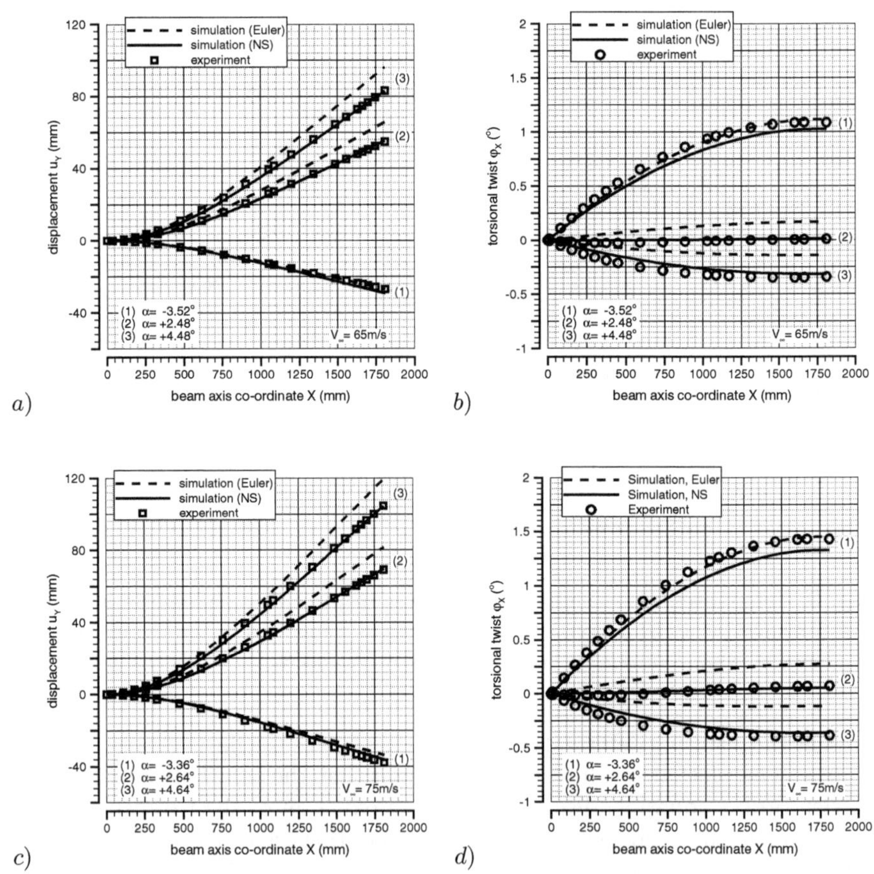

Fig. 4. Comparison of measured and computed displacement and torsional torque as a function of the beam axis co-ordinate for three different rigging angles of incidence and two different flow velocities $V_\infty = 65m/s$ and $75m/s$

The reaction forces of the wing structure in the clamping plane were recorded by a six-component balance during the experiments. Figure 5 a) to d) show the comparison of measured vertical force F_Y^{Root} and torsional moment M_X^{Root} to the corresponding quantities in the simulation over the whole range of angles of incidence. One can observe again that the overall agreement is well, but the differences between experimental and computional results slightly increase with an increasing angle of incidence. Anyhow those differences do not exceed $11N$ for the vertical forces and $4Nm$ for the torsional moments within the considered range of root angles of incidence.

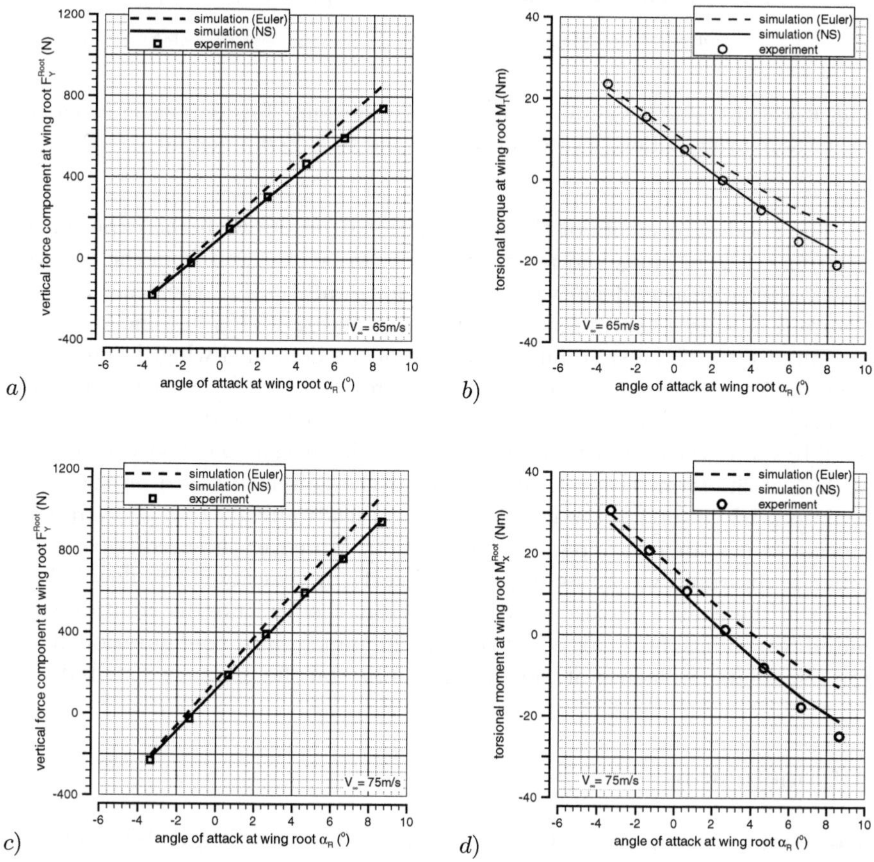

Fig. 5. Comparison of measured and computed vertical force component (**a**) and **c**)) and torsional torque (**b**) and **d**)) at the wing root as a function of the rigging angle of incidence at wing root α_R for two different flow velocities $V_\infty = 65$ and $75m/s$

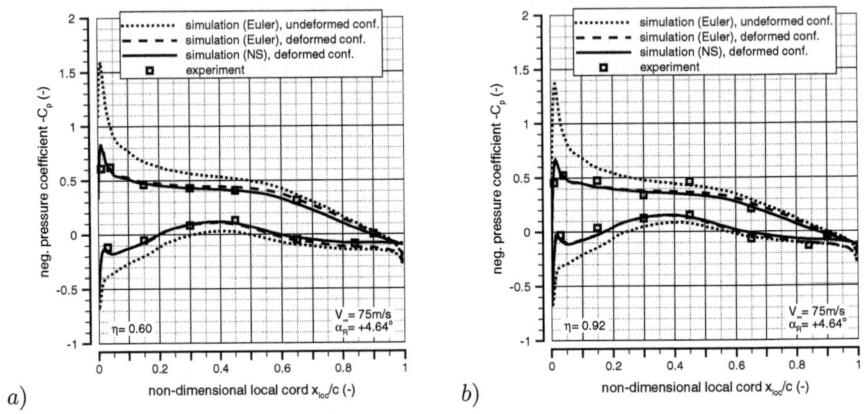

Fig. 6. Comparison of measured and computed pressure distributions in two different measurement sections located at **a**) 60% and **b**) 92% of the span

The good agreements of the computational results with the experiments regarding deformations and reaction forces can be observed from the pressure distributions measured at two sections along the span of the wing as well. Figure 6 a) shows the pressure distribution acting within the measurement section at 60% of the span on the surface of the deformed wing. Exemplarily a comparison to the experimental data is provided for the test case at a flow velocity of $V_\infty = 75m/s$ and a root angle of incidence of $\alpha_R = +4.64°$. Figure 6 b) shows the corrensponding results for the measurement section located at 92% of the span. Using a description of the flow field based on the Euler equations while solving the coupled problem leads already to a good approximation concerning the pressure distribution. Moreover the prediction of the pressure distribution can be improved slightly by describing the flow field in terms of the Navier–Stokes equations. Adding the pressure distribution computed by solving the Euler equations for the undeformed configuration shows the remarkable influence of the wing flexibility on the pressure distribution.

4 Conclusions

Within the framework of the Collaborative Research Center SFB 401 the aeroelastic method SOFIA has been applied to compute the wing deformation in combination with the surrounding flow field for a swept wind tunnel wing model in subsonic flow. SOFIA uses a coupled field formulation, the flow is modeled optionally by either the Euler or the Navier–Stokes equations, and the wing structure is described by a generalised quasi one-dimensional theory based on Timoshenko's beam theory. The computational results concerning deformations and integral reaction forces and moments obtained by applying SOFIA were compared extensively to experimental results obtained from static aeroelastic wind tunnel tests in DNW-LST.

An excellent agreement between experiment and simulation has been found in terms of the spanwise displacement of the wing which dominates the local angle of attack by the kinematic coupling of bending and twist of aerodynamic sections in main flow direction. The differences in bending displacement do not exceed $8.0mm$ even for the high angles of incidence. The agreement for the twist deformation due to torsion is insignificantly incorrectly predicted. The main difference is that the slope of change of the torsional twist against the angle of incidence is not reproduced exactly. But the differences remain below $0.15°$ for all studied test cases. Comparison with computed pressure distributions for the undeformed wing show that the neglection of the wing deformation leads to completely wrong results. It must be noted here that the good agreement between experiment and simulation could only be achieved on conditions, which include the viscosity of the fluid in the aeroelastic computation.

5 Code Performance

During aeroelastic computations using a beam model to represent the supporting wing structure, approximately less than 5% of the cpu time is consumed for computing the structural and mesh deformations and more than 95% for the computation of the flow field. Since the FLOWer code has been highly optimised for vector architectures, a code performance of about 1300 MFLOPS and a vector operation ratio not less than 95% is achieved on a single NEC SX-8 processor. Steady aeroelastic analysis of a wing configuration using Navier–Stokes equations to model the fluid flow, as presented in this paper, takes less than 1.5 hours and about 1.2 GB memory on 4 NEC SX-8 processors for a computational mesh of about one million grid points. An extensive utilisation of the FLOWer code with its hybrid parallelisation capabilities using MPI and OpenMP standard will be start on NEC SX-8 soon. Within the HIRENASD (HIgh REynolds Number Aero-Structural Dynamics) project [10], a subproject within the Collaborative Research Center SFB 401, experiments will be performed with an elastic wing model under cryogenic conditions at transonic Mach numbers and Reynolds numbers up to 70 millions in the European Transonic Windtunnel (ETW) in 2006. While preparing those tests it is essential to perform a whole series of dynamic aeroelastic response simulations under consideration of viscous effects using the aeroelastic method SOFIA presented in this paper. The wind tunnel model corresponds to the SFB 401 clean wing reference configuration with a planform typical for a wing of high speed transport aircrafts. In order to obtain acceptable turnaround times for computational intensive unsteady aeroelastic simulations, as they are planned in the HIRENASD project, hardware architectures like the

NEC-facilities at the High-Performance Computing-Center Stuttgart (HLRS) with powerful processors are of a great importance for this project.

Acknowledgements

This work has been partly supported by the Deutsche Forschungsgemeinschaft (DFG) in the Collaborative Research Center SFB 401 "Flow Modulation and Fluid-Structure Interaction at Airplane Wings" at Aachen University. Computations were performed using the NEC-facilities of HLRS. We would like to express our gratitude to the members of the German Aerospace Research Center (DLR/Braunschweig) and all partners within the project MEGAFLOW developing the FLOWer code, as well as our partners within the SFB from the Institute of Lightweight Structures at Aachen University, who provided the opportunity for testing SOFIA against aeroelastic experimental data.

References

1. Ballmann, J.: *Flow Modulation and Fluid-Structure Interaction at Airplane Wings – Research Results of the Collaborative Research Center SFB 401 at RWTH Aachen.* Ballmann, J. (Ed.): Notes on Numerical Fluid Mechanics and Multidisciplanary Design (NNFM), Springer Verlag, Vol. 84, 2003
2. Kämpchen, M., Dafnis, A., Reimerdes, H.-G., *Aero-Structural Response of a Flexible Swept Wind Tunnel Wing Model in Subsonic Flow.*, Proceedings of the International Forum on Aeroelasticity and Structural Dynamics (IFASD) in Amsterdam, 2003
3. Kroll, N., Rossow, C.-C., Becker, K. Thiele, F.: *The MEGAFLOW project.*, Aerosp. Sci. Technol. 4, pp. 223–237, 2000
4. Nellessen, D.: *Schallnahe Strömungen um elastische Tragflügel.*, VDI Fortschrittsberichte Reihe 7: Strömungstechnik, Nr. 302, 1996
5. Britten, G.: *Numerische Aerostrukturdynamik von Tragflügeln großer Spannweite.*, Doctoral Thesis, RWTH Aachen, Shaker Verlag, Aachen, 2002
6. Boucke, A.: *Kopplungswerkzeuge für aeroelastische Simulationen.*, Doctoral Thesis, RWTH Aachen, 2003
7. Hesse, M.: *Entwicklung eines automatischen Gitterdeformationsalgorithmus zur Strömungsberechnung um komplexe Konfigurationen auf Hexaeder-Netzen.*, Doctoral Thesis, RWTH Aachen, 2006
8. Heinrich, R., Dwight, R., Widhalm, M., Raichle, A.: *Algorithmic Developments in TAU.* In: Kroll, N., Faßbender, J. K. (Eds.): *MEGAFLOW – Numerical Flow Simulation for Aircraft Design*, Notes on Numerical Fluid Mechanics and Multidisciplinary Design (NNFM), Vol. 89, Springer Verlag, pp. 93–108, 2005
9. Bramkamp, F.D., Lamby, Ph., Müller, S.: *An adaptive multiscale finite volume solver for unsteady and steady flow computations.* Journal of Comp. Physics, Vol. 197, pp. 460–490, 2004.
10. Ballmann. J.: *The HIRENASD Elastic Wing Model and Aeroelastic Test Program in the European Transonic Windtunnel (ETW).* DGLR conference proceedings, paper no. DGLR-2005–264, 2005

Structural Mechanics

P. Wriggers

Institut für Baumechanik und Numerische Mechanik, University of Hannover, Appelstr. 9a, 30167 Hannover
wriggers@ibnm.uni-hannover.de

1 Preface

Besides classical applications in structural mechanics, which can be solved on standard computers, there exist many problem classes which can only be numerically simulated using high performance computing environments. These involve usually highly nonlinear system behaviour which can be due to inelastic materials or due to finite deformations. When these problems are coupled with dynamics then large finite element discretizations with several hundred thousand of unknowns have to be solved within several thousand time steps. The three contributions in this section fall in this category of applications.

The first project by M. Zimmermann, M. Klemenz abd V. Schulze is concerned with shot peening processes. Here hardening effects of surfaces due to shot peening are investigated. By using a finite element software the authors obtain results for different materials which can be used to understand the influence of geometry and constitutive behaviour on the development of residual stresses near the surface of workpieces.

The second project by S. Mattern, G. Blankenhorn and K. Schweizerhof is targeted to blast processes. The controlled destruction of buildings by blasting can be viewed as an economic technique to demolish the structures. Here a strategy for such processes is been developed based on numerical simulation of the structural response to the blast loading process. These computations of a complete failure of a structure require enormous computational resources and can only be performed using parallel computers.

The third project by S. Mattern and K. Schweizerhof is related to the prediction of the propagation of high frequency oscillations by numerical simulations. The investigations are essential for the design of safety components in automotive industry, such as the construction of sensors for airbag activation. The numerical simulations techniques are developed to reduce and to support experimental investigations. Due to the involved high frequencies very fine meshes need to be used and hence the project requires high performance computing on parallel computers.

All projects show that the use of numerical simulation techniques can enhance the engineering analysis and the design of new constructions. On the other side such simulations reduce costs for experiments and also support experimental investigations.

Numerical Prediction of the Residual Stress State after Shot Peening

Manuel Klemenz, Marc Zimmermann, Volker Schulze, and Detlef Löhe

Institut für Werkstoffkunde I, Universität Karlsruhe (TH),
Kaiserstraße 12, 76131 Karlsruhe
marc.zimmermann@iwk1.uni-karlsruhe.de

Abstract. Shot peening is a mechanical surface treatment with the purpose to modify the surface state of a material in order to improve fatigue strength of a component subjected to cyclic loading. The material state after a shot peening treatment is governed by various shot peening parameters. In order to comprehend the complex interaction between process parameters and material state time and money consuming experiments are usually accomplished. Promising alternatives are numerical simulation methods such as FEM[1] combined with similarity mechanics having the possibility to predict the shot peening results for arbitrary combinations of shot peening parameters. In this work the residual stress development during shot peening for various shot peening parameters was simulated successfully with a FEM model. Additionally the method of similarity mechanics was used in combination with the FEM simulation results in order to predict residual stress states for a wide parameter field with almost no additional computing effort.

Key words: Shot Peening, FEM Simulation, Similarity Mechanics, Material Modeling, Residual Stresses

1 Shot Peening

Shot peening is a mechanical surface treatment with the purpose to modify the surface state of a material in order to improve the fatigue strength or the fatigue life, respectively, of a component when subjected to cyclic loading. During a shot peening treatment peening media with a specific shape and a sufficiently high hardness are accelerated in peening devices of various kinds and interact with the surface of the treated workpiece. The main focus of shot peening is to induce work hardening and compressive residual stresses in the regions close to the surface in order to achieve a suppression or deceleration of crack initiation and/or crack propagation during cyclic loading. During the impact of the peening medium most of the kinetic energy of the

[1] Finite Element Method

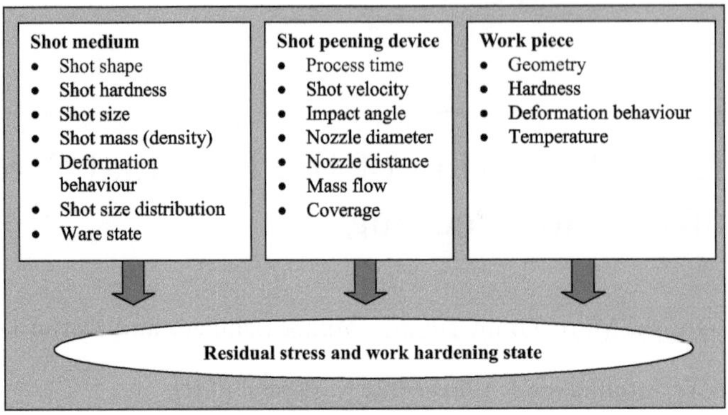

Fig. 1. Influencing parameters on the results of the shot peening process (according to (6))

shot transforms into elastic and plastic deformation work, preferably in the workpiece, but possibly also in the shot, yielding to the generation of a dimpled work piece surface. Due to the inhomogeneous plastic deformations of the workpiece surface residual compressive stresses in the regions close to the surface and tensile residual stresses in the deeper work piece areas remain after a shot peening treatment. A representative residual stress profile resulting from a shot peening treatment is shown in Fig. 4.

Shot peening is successfully applied in surface areas that are critical concerning crack initiation due to high tensile loads during service. Components which are typically shot peened in technical mass production are springs, conrods, gears, stepped or grooved shafts and axles, turbine vanes, blade bases, and heat-affected zones of welded joints.

Several influencing shot peening parameters determine the residual and work hardening state after shot peening. These parameters are shown in Fig. 1, classified into the three categories shot medium, shot peening device, and work piece. In order to achieve a desired peening result by means of a compressive residual stress field reaching to a certain work piece depth for a given work piece material and geometry several of these process parameters have to be adjusted adequately. Nowadays this adjustment is done on the basis of more or less profound empirical knowledge in combination with time and money consuming experiments. Herefrom arises the motivation to predict the peening result by means of the residual stress state using tools like FEM simulation and similarity mechanics.

2 Process Simulation

The accuracy of the prediction of the residual stress state after shot peening highly depends on the FEM model and the applied boundary conditions, respectively, as well as on the material model describing the complex material behavior of the workpiece during shot peening. In the following it will be shown how the shot peening process and the material behavior were modeled and how close the FEM simulation results meet reality.

2.1 Shot Peening Model

The model used for the shot-peening simulations with ABQUS/Explicit consists of a 3-dimensional rectangular body with defined dimensions and is based on the work presented in (8). Its mesh is set up by 360000 8-node linear brick elements with reduced integration and hourglass control. The plate is surrounded by so-called half-infinite elements that provide "quiet" boundaries by minimizing the reflection of dilatational and shear waves back into the region of interest. The boundary conditions on the target's base fix the model in z-direction. Half spherical rigid surfaces with parameterized diameter, velocity and direction are used to model the shot. Each rigid surface is connected to a point mass and a rotary inertia element providing the properties of a full sphere. During the contact of the shots with the surface of the plate isotropic Coulomb friction is assumed with a constant experimentally measured friction coefficient $\mu = 0.4$. Figure 2 shows the mesh and one half-sphere after the impact of 19 shots.

To achieve a realistic modeling of the shot peening process an arrangement of the spheres was chosen that provides a dimple pattern of full coverage on the surface with the impact sequence and arrangement shown in Fig. 3. The grey marked inner area, which is the circumscribed circle of the inner 7 impacts, was used for the calculation of the residual stress profiles. For each element layer the residual stress values in x and y direction of all elements within the marked area were averaged.

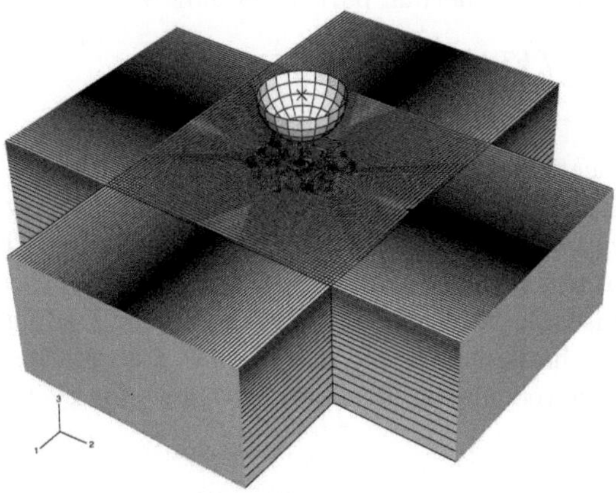

Fig. 2. Model geometry

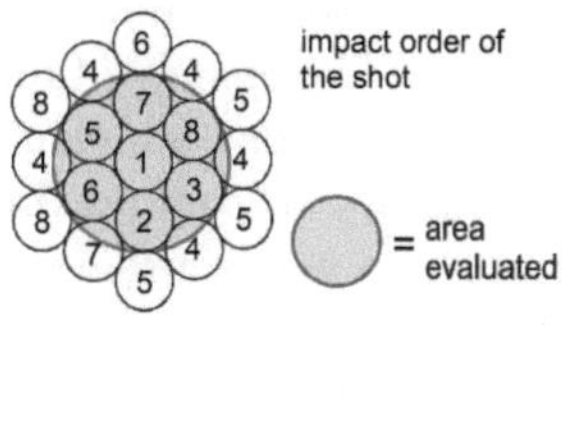

Fig. 3. Shot arrangement

2.2 Material Model

ABAQUS/Explicit allows the use of user-defined material laws in case of the standard material models provided by ABAQUS are insufficient. Therefore efforts were drawn to the development of an adequate mathematical representation to describe the elastic-plastic behavior of the investigated material, the steel AISI 4140[2] in a state quenched and tempered at 450°C (5). Mainly three characteristics of this material were taken into account: the influence of temperature and strain-rate on the flow stress based on thermally activated dislocation slip (7), thc work hardening by initial plastic deformation and the intensive Bauschinger-effect during reversed loading. This leads to an elastoviscoplastic, combined isotropic and kinematic material model in which the yield criterion is reached, when the von Mises norm of the effective stress equals the amount of the isotropic parts of the flow stress that build up the radius of the flow surface:

$$\sqrt{\frac{3}{2}\left(S-\xi\right)\left(S-\xi\right)} = k_0 + K + \sigma^* \tag{1}$$

Here the effective stress tensor is the deviatoric stress S reduced by the back-stress ξ. The Bauschinger-effect can only be described with a translation of the flow surface during plastic deformation. The back-stress ξ defines the center of the flow surface and follows the evolution term

$$\dot{\xi_i} = c_i \dot{\epsilon_p} - b_i \xi_i \dot{p} \qquad \text{, with} \qquad \dot{\xi} = \dot{\xi_1} + \dot{\xi_2} \qquad \text{and} \qquad \dot{p} = \sqrt{\frac{2}{3}}\, |\dot{\epsilon_p}| \tag{2}$$

according to (4). The back-stress can be understood as a resistance force against the outside loading due to dislocation pile-ups. This resistance force acts as a driving force for much earlier dislocation slip when the loading is reversed (2). The right side of Eq. 1 consists of a constant athermal part k_0, an evolutional athermal part K and the thermal part of the flow stress

$$\sigma^* = \sigma_0^* \left[1 - \left(\frac{kT \ln\left(\dot{\epsilon_0}/\dot{\epsilon}\right)}{\Delta G_0}\right)^n\right]^m \tag{3}$$

due to thermally activated dislocation slip (7). Increasing values of temperature T in Kelvin and decreasing strain rates $\dot{\epsilon}$ will reduce the thermal flow stress σ^*. The change of temperature due to plastic deformation is considered adiabatically for each element of the geometric model.

Contrary to common elastoviscoplastic material models for cyclic loading with cyclic work-hardening the athermal isotropic part K has a decreasing evolution with increasing accumulated plastic strain which allows for cyclic work-softening effects.

$$\dot{K} = -\beta\left(Q + K\right)\dot{p} \qquad \text{, with} \qquad K(t=0) = 0 \tag{4}$$

[2] German grade: 42CrMo4

Only such a type of model allows to describe both, the initial plastic deformation behavior and the Bauschinger-effect during unloading and following loadings in other directions. Q is the asymptotic value for the magnitude of the saturated K value. In order to take into account a constant size of the flow surface for cyclic loading at constant strain amplitudes and high accumulated plastic strains while showing a decrease of K in case of increasing strain amplitudes, Q is not a constant but also an evolutional term with strain memory:

$$\dot{Q} = 2\mu\eta\,(Q_{max} - Q)\,\dot{q} \qquad \text{, with} \qquad Q(t=0) = Q_0 \tag{5}$$

where $\dot{q}$ is an additional internal variable related to $\dot{p}$ that is zero as long as the plastic strain does not exceed the highest strain $\epsilon_{p,max}$ applied yet (4). The material parameters k_0, c_1, b_1, c_2, b_2, σ_0^* , $\dot{\epsilon_0}$, n, m, ΔG_0, β, μ, η, Q_0 and Q_{max} were determined by tensile tests at different temperatures and strain-rates and by push-pull tests at different strain amplitudes.

2.3 Results

In Fig. 4 simulated residual stress profiles are compared with measurements by X-ray diffraction (5). The simulations were performed using the above presented combined isotropic kinematic elastoviscoplastic material model as well as an isotropic material model presented in (8). Following characteristic values are pointed out exemplarily on the curve of the isotropic material model: the residual stress at the surface σ_0^{RS}, the maximum residual stress below the surface σ_{max}^{RS}, the distance from the surface to the maximum residual stress z_{max} and the distance from the surface to the change of sign of the residual stress values z_0.

The experimentally determined profile with compressive residual stresses at the surface and a compressive stress maximum close to the surface is typ-

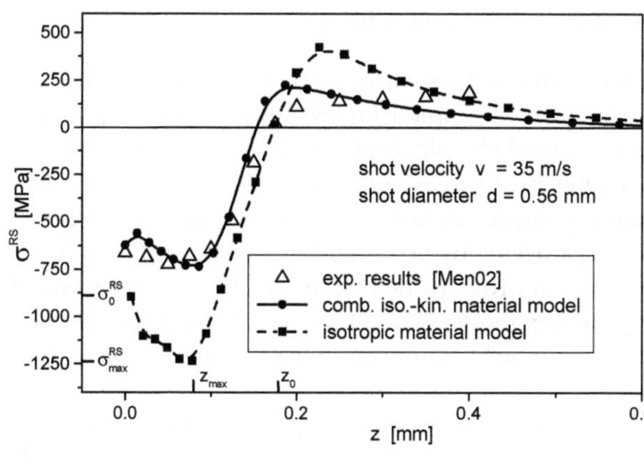

Fig. 4. Simulated and experimentally determined residual stress profiles

ical for this treatment. Both simulated curves show qualitatively the same course. While the simulation with the isotropic material model highly overestimates the residual stresses at the surface and at its compressive maximum, the simulation with the combined isotropic-kinematic material model predicts almost exactly the residual stresses at the surface and at its maximum. The depth value of the maximum compressive stress is slightly overestimated but does not lead to a big residual stress aberration because the residual stress gradients in the specimen are quite flat up to the depth of 0.1 mm. The depth value of the changeover from compressive to tensile residual stresses is slightly underestimated by the simulation with the refined material model. In case of using the simulation as a dimensioning tool this aberration would have an conservative effect. The local minimum in the compressive residual stresses close to the surface in the simulated curve with the presented material model may be due to the fact that the smallest values of the isotropic hardening K are combined with the highest applied plastic strain amplitudes which occur in an area slightly below the surface.

It can be seen that the implementation of a combined isotropic-kinematic material model that takes into account the cyclic deformation behavior of the material was a major improvement in the reliability of the shot peening simulation.

3 Similarity Mechanics

For the investigation of a wide field of different process parameters the use of FEM has already the advantage of being independent of costly experiments. Having its origins in the research field of scaling problems similarity mechanics can also be an elegant method in order to reduce the time for the prediction of desired results for an arbitrary combination of the process parameters.

3.1 Fundamentals and Methodology

According to the Buckingham-theorem (1) a dimensionless output value a depends only on $p = n - q$ dimensionless input values $D_1, \ldots, D_p$ with n the number of parameters of the physical problem and q the rank of the dimension matrix. So a scaling problem is completely described, when the dependency of all dimensionless input values is known for an interesting output value: $a = a(D_1, \ldots, D_p)$. Even by a reduction of the input values from n to p the necessary effort in order to determine $a(D_1, \ldots, D_p)$ would by far be too extensive for a complex model. Therefore, as a further simplification, a product ansatz (3) is chosen, which requires only the dependencies of one input value.

$$a(D_1, \ldots, D_p) = K \cdot a(D_1) \cdot a(D_p) \qquad \text{with} \qquad K = \frac{1}{a_0^{p-1}} \tag{6}$$

a_0 represents the output value at standard parameters of the process.

3.2 Applied Similarity Mechanics

The total number of the parameters in the shot peening simulation is 28, with 7 parameters for the process and 22 for the material description. Length, mass, time and temperature are the 4 dimensions of this problem, q is therefore 4. All $28 - 4 = 24$ dimensionless input values will not be presented here but only those that are of further interest and that are necessary to describe the process.

Table 1. Selection of dimensionless input values D in order to describe the process

	Shot diameter	Shot velocity	Impact angle	Pre–stress
D_i	$Si = \frac{\dot{\epsilon}_0 d}{\sqrt{E/\rho_P}}$	$Ca = \frac{\nu}{\sqrt{E/\rho_P}}$	α	$PS = \frac{\sigma_{pre}}{E}$

The dimensionlessness of the size number Si, the Cauchy number Ca and the pre-stress number PS is attained by a suitable choice of the material parameters. The impact angle α being already dimensionless is the angle between the normal of the plate surface and the velocity vector of the shots. The interesting dimensionless output values that characterize the surface close region are shown in Table 2. $\sigma_{0||}^{RS}/E$, $\sigma_{0\perp}^{RS}/E$, $\sigma_{max||}^{RS}/E$ and $\sigma_{max\perp}^{RS}/E$ represent the dimensionless surface residual stresses and the maximum residual stresses respectively in parallel and orthogonal direction to the pre-stress and/or the projection of the velocity vector on the surface. In case the distinction is irrelevant it is only written σ_0^{RS}/E or σ_{max}^{RS}/E. Because of no significant difference between the depth values in parallel and orthogonal direction the dimensionless depth values simplify to z_{max}/d and z_0/d. The angle between the pre-stress vector and the vector of the shot velocity is kept constant in this investigation.

Table 2. Dimensionless output values

	surf. RS parallel	surf. RS orthogonal	max. RS parallel	max. RS orthogonal	Depth of max. RS	Depth of RS = 0	Roughness
a_i	$\sigma_{0\|\|}^{RS}/E$	$\sigma_{0\perp}^{RS}/E$	$\sigma_{max\|\|}^{RS}/E$	$\sigma_{max\perp}^{RS}/E$	z_{max}/d	z_0/d	R_t/d

The dependencies of the dimensionless input values on the chosen dimensionless output values are presented as results of simulation calculations by variation of only one parameter and by keeping all other parameters constant at standard values. The symbols marked with a circle represent the results at standard parameters which are also listed in Table 3.

It can be seen that for each variation of one dimensionless input parameter the results can be described by linear or non-linear formulae in which only one of the dimensionless output values serves as variable. These formulae fit

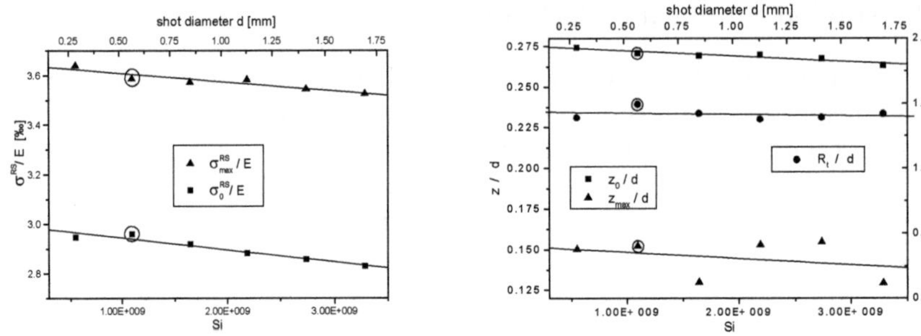

Fig. 5. Dimensionless output values vs. Si for constant $Ca = Ca_0$, $\alpha = \alpha_0$, and $PS = PS_0$

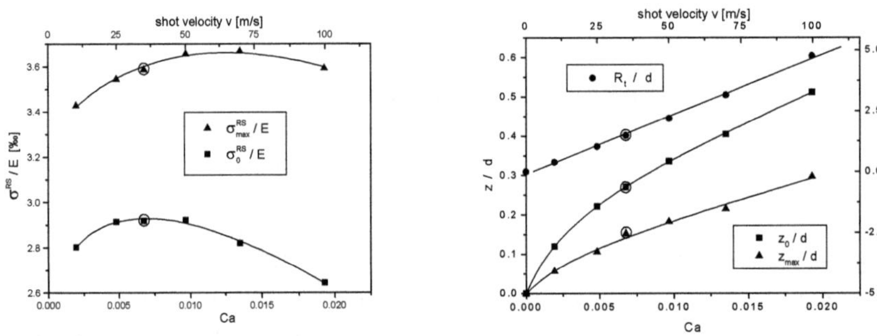

Fig. 6. Dimensionless output values vs. Ca for constant $Si = Si_0$, $\alpha = \alpha_0$, and $PS = PS_0$

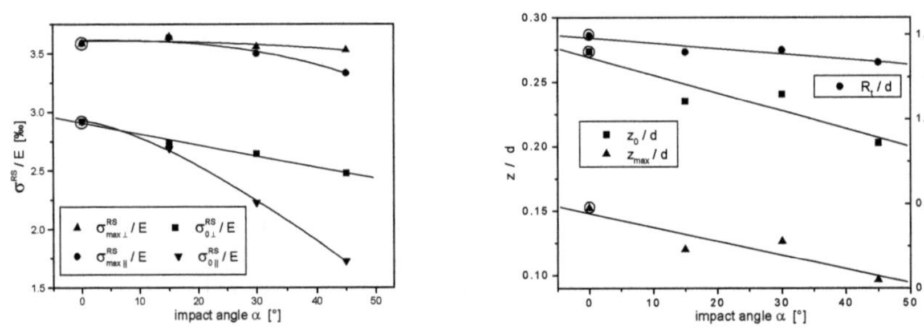

Fig. 7. Dimensionless output values vs. α for constant $Si = Si_0$, $Ca = Ca_0$, and $PS = PS_0$

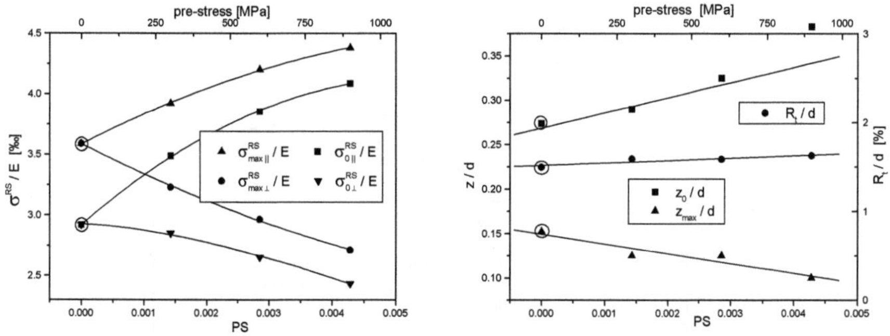

Fig. 8. Dimensionless output values vs. PS for constant $Si = Si_0$, $Ca = Ca_0$, and $\alpha = \alpha_0$

the course of the results determined by FEM very well. With this set of 24 formulae, that are not explicitly listed here, and the approach of equation 6 it is possible to get an instantaneous estimation of the desired output values for an arbitrary combination of the input parameters in a reasonable range, without the necessity to run a special FE-calculation with this very set of parameters. In order to test the product ansatz for evaluating the influence of changes in multiple input parameters the output values were compared with experimentally determined results and the results of an explicit FE-calculation for the same set of parameters. Therefore all four process parameters were set to a value significantly different from the standard. The chosen parameters are shown in Table 3. The dimensionless output values were determined according to Eq. 6.

Table 3. Parameter sets for method verification

standard parameters		parameters for verification	
$Si_0 = 1.0953 \cdot 10^9$	$(d = 0.56\ \text{mm})$	$Si = 2.135 \cdot 10^9$	$(d = 1.1\ \text{mm})$
$Ca_0 = 6.770 \cdot 10^{-3}$	$(v = 35\ \text{m/s})$	$Ca = 1.118 \cdot 10^{-2}$	$(v = 61\ \text{m/s})$
$\alpha_0 = 0°$		$\alpha = 30°$	
$PS_0 = 0$	$(\sigma_{pre} = 0\ \text{MPa})$	$PS = 1.429 \cdot 10^{-3}$	$(\sigma_{pre} = 300\ \text{MPa})$

In Fig. 9 the residual stress profiles of an ordinary FE-simulation and of an experiment are confronted with the characteristic values determined by the product ansatz. The results of the interesting output values are given in Table 4 and show remarkable accordance for the normalized residual stress components with a relative error of less than 12 %. The relative differences of the depth values and the roughness may still be improved for accurate predictions in the late dimensioning process, but don't effect severely because of the quite small residual stress gradients near the residual stress maxima.

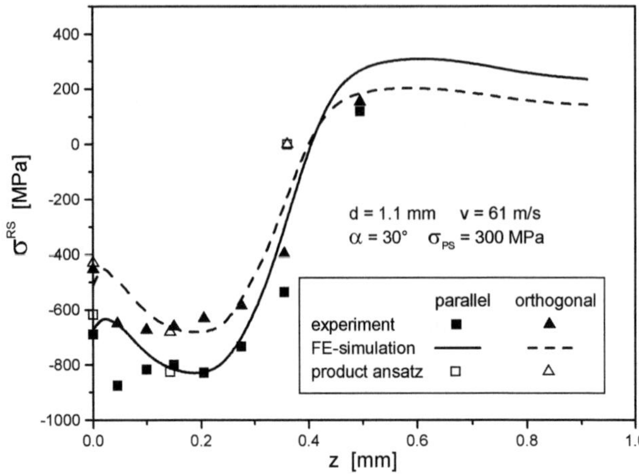

Fig. 9. Simulated and experimentally determined residual stress profiles

Anyhow, it is obvious that the product ansatz is a powerful method for a qick estimation of surface characteristics after shot peening processes.

Table 4. Comparison of the results determined by similarity rules and a separate FEA with experimental data

	product ansatz	**rel. error prod. ans. vs. exp.**	experiment	**rel. error exp. vs. FEM**	FE-simulation
$\sigma^{RS}_{max\|\|}/E$	$3.972 \cdot 10^{-3}$	**0.6 %**	$3.948 \cdot 10^{-3}$	**0.3 %**	$3.960 \cdot 10^{-3}$
$\sigma^{RS}_{max\perp}/E$	$3.236 \cdot 10^{-3}$	**1.0 %**	$3.205 \cdot 10^{-3}$	**1.0 %**	$3.199 \cdot 10^{-3}$
$\sigma^{RS}_{0\|\|}/E$	$2.939 \cdot 10^{-3}$	**11.5 %**	$3.276 \cdot 10^{-3}$	**2.4 %**	$3.238 \cdot 10^{-3}$
$\sigma^{RS}_{0\perp}/E$	$2.052 \cdot 10^{-3}$	**4.9 %**	$2.157 \cdot 10^{-3}$	**11.1 %**	$2.427 \cdot 10^{-3}$
z_{max}/d	0.1300	**43.4 %**	0.1864	**25.9 %**	0.1481
z_0/d	0.3280	**21.0 %**	0.4150	**12.7 %**	0.3683
R_t/d	0.0255	**21.1 %**	0.0309	**4.5 %**	0.0296

4 Computational Requirements and Computing Time

The material model presented in Sect. 2.2 is implemented into ABAQUS as an user subroutine written in Fortran code. The use of this subroutine requires the Intel Fortran and C++ compiler for its compilation and integration into ABAQUS.

Table 5. Computing time

Host name	wk1nc02	XC6000
Processor type	Opteron	Itanium
Clock frequency [MHz]	2200	1600
ABAQUS version	6.5-3	6.5-1
Parallelization mode	mpi	threads
computational time (1 CPU)	7 h 04 min	7 h 44 min
computational time (2 CPUs)	3 h 39 min	4 h 21 min
computational time (8 CPUs)		1 h 25 min

Beside the need for a sufficient high processor clock speed the simulation jobs require about 1500 MB of physical memory.

The shot peening FEM calculations were performed on an in house dual processor opteron workstation and on the scientific supercomputer XC6000. Table 5 shows the computing times of a typical shot peening simulation job for both computers and for parallel and non-parallel execution. The parallel execution was done using the domain-level method of ABAQUS splitting the model into a number of topological domains being computed individually by each CPU involved in the analysis. It can be seen that computational time of the shot peening simulation can be notable reduced taking advantage of the parallelization capability of the software. This leads in the case of a job execution on 8 CPUs on the XC6000 to a major time reduction of more than a factor 5 in comparison to a non-parallel execution.

The numerical data base necessary for the realization of the application of similarity mechanics presented above could only be produced within a reasonable short time taking into advantage the high computational capacity of the XC6000 supercomputer in combination with the good scalability of the simulation jobs.

References

[1] E. Buckingham. On physically similar systems; illustrations of the use of dimensional equations. *Physical Review*, 4:345–376, 1914.

[2] J.A. del Valle, R. Romero, and A.C. Picasso. Bauschinger effect in age-hardened inconel x-750 alloy. *Materials Science and Engineering*, A311:100–107, 2001.

[3] J. Kotschenreuther, L. Delonnoy, T. Hochrainer, J. Schmidt, J. Fleischer, V. Schulze, D. Löhe, and P. Gumbsch. Modelling, simulation and experimental tests for process scaling of cutting processes with geometrically defined edge. In Ferdinand Hollmann Frank Vollertsen, editor, *Process Scaling, Strahltechnik*, volume 24, pages 121–136, Bremen, 2003. BIAS.

[4] J. Lemaitre and J.L. Chaboche. *Mechanics of solid materials.* Cambridge University Press, 1990.

[5] R. Menig. *Randschichtzustand, Eigenspannungsstabilität und Schwingfestigkeit von unterschiedlich wärmebehandeltem 42 CrMo 4 nach modifizierten Kugelstrahlbehandlungen.* PhD thesis, Universität Karlsruhe, 2002.
[6] V. Schulze. *Modern Mechanical Surface Treatment: States, Stability, Effects.* John Wiley & Sons, 2006.
[7] V. Schulze and O. Vöhringer. Influence of alloying elements on the strain rate and temperature dependence of the flow stress of steel. *Metallurgical and Materials transactions*, 31A:825–830, 2000.
[8] J. Schwarzer, V. Schulze, and O. Vöhringer. Finite element simulation of shot peening - a method to evaluate the influence of peening parameters on surface characteristics. In Lothar Wagner, editor, *Shot Peening*, pages 507–515. WILEY-VCH Verlag GmbH & Co. KGaA, Garmisch-Partenkirchen 2002.

Computer-Aided Destruction of Complex Structures by Blasting

Steffen Mattern, Gunther Blankenhorn, and Karl Schweizerhof

Institut für Mechanik, Universität Karlsruhe (TH), D-76131 Karlsruhe
mechanik@ifm.uni-karlsruhe.de

1 Introduction

The controlled destruction of buildings at the end of their average life cycle has become more and more important during the last years. An economic way of dismounting is a demolition of a partially dismountable structure with explosives. Within this method, several structural elements of a building's load carrying system are removed by explosives to initiate a particular collapse. The strategy of blasting, i.e. the selection of the structural elements, which are destroyed by the explosives, is performed by engineers with special knowledge in this field of demolition. However the prediction of the collapse supported by fairly simple analysis models requires a great amount of experience and is very error sensitive if the building construction is complex.

In this contribution a 'complex construction' means a construction including several hundred structural elements like beams, plates, columns, etc. for which it is rather difficult to predict the collapse kinematics with simple mechanical tools. In order to develop a suitable strategy of blasting for complex structures, a reliable 'a-priori-simulation' of the collapse of the complete structure is mandatory. The goal of the investigations is to obtain a good prediction of the collapse mechanism by a numerical simulation of a complete real building. The analyses are performed using the Finite Element Method (FEM) combined with explicit time integration, which has already been approved as simulation tool for structural analyses involving impact and contact. Such rather costly simulations – costly with respect to the computational effort – are meant to support the development of a reliable and efficient simulation tool for demolition by blasting that will be developed throughout the research project FOR500 (funded by DFG). Complete FE-simulations of the failure process of a complex building require enormous computational resources and parallelization of such problems is inevitable.

Within the contribution, first some general informations about the state of the art in building demolition is given. Then the methods used for the numerical analyses are discussed and the simulated reference models are presented.

All computations and parametrical studies which were necessary to develop and improve the models were performed on the HP XC6000-Cluster of the University of Karlsruhe [2].

2 Building Demolition – State of the Art

Compared to methods such as destruction with dredgers or systematic deconstruction, blasting demolition definitely won recognition on braking down brickwork, reinforced concrete and steel constructions because of economic and technical reasons. Demolition by blasting, together with special possibilities of local weakening (e.g. cutting of reinforcement, deconstruction of supporting parts with special machines), allows the teardown of a building at a moderate time of preparation, proper planning of the blasting strategy and accurate positioning of the explosive charge presumed. Annoyance of residents is kept to a temporal minimum, business as usual and traffic around the object are disturbed only for a short while. The effort for necessary safety arrangements is considerably reduced compared to mechanical deconstruction, because the zone of danger can be evacuated during the collapse. For those reasons, popularity of blasting demolition raised in the last few years, especially in inner-city areas. General information about planning and accomplishment of a building demolition is given e.g. in [9].

Nevertheless, blasting demolition can be very dangerous, used for complex and non-trivial buildings, for which precise information about the construction details and the used materials is sometimes rare or even not available at all. The consequences can be recognized, when e.g. dynamic effects of the structure have not been considered correctly. This sometimes leads to uncontrolled collapses, which cause high damage. Dangerous situations can also arise, if a planned collapse is not completed and the remaining parts of the building have to be removed manually. In such situations expensive procedures are necessary to finish the demolition. Currently only several rudimental mechanical tools are used to predict the collapse, but in order to avoid accidents and the corresponding high costs, it is necessary to realize a reliable simulation of the complete collaps, which requires modern mechanical methods.

3 Numerical Analysis

3.1 Basic Idea

The safe execution of the destruction of a building using controlled explosives requires detailed and reliable knowledge about the kinematics of the collapse with a special blasting strategy. In order to realize collapse simulations e.g. with different locations of the explosive loads, the Finite Element Method with explicit time integration is used within this project. Applying this method makes it possible to analyse the entire building from the time of blasting till the end of the breakdown. The knowledge won by those simulations is used to support and validate an alternative method based on rigid body models which requires far less computational effort. Such an efficient way of simulation allows the concerning of uncertainties e.g. with fuzzy algorithms [10], which require many deterministic solutions of one structure. The final goal here is to extract such models of reduced size based on certain modeling or decision rules by rigidizing parts of the structure during the simulation [1]. In order to gain proper insight and in particular, for validation purposes many analyses of different structures, i.e. extensive parametrical studies are necessary.

3.2 Numerical Algorithms

First the numerical algorithms, used to solve this highly non-linear problem, have to be chosen. The problem is driven by finite deformations, finite rotations, non linear material behavior and multiple contact possibilities. On the other hand, the spatial discretization must be done by continuum elements to catch the contact surfaces in a better way, in comparison to structural elements. This leads to a large number of elements. Considering all requirements, a combination of FE-analysis with explicit time integration [3, 8, 5] is chosen. The small time step size, which is needed for numerical robustness, is compensated by highly efficient element formulations and the highly developed capabilities for dividing the problem to several computation units. Also the contact possibilities have to be taken into account which naturally reduce the possible time step size. As a conclusion the disadvantage of a small time step size in explicit methods – here the central difference scheme [11] – is abolished by its efficient implementation and parallelization [6].

4 Reference Models

4.1 General Information

In order to cary out the analyses, in this project the commercial FE-program LS-DYNA is used, which uses a central difference method for the time integration [7, 6]. The code is highly parallelized and perfectly suited to run on clusters such as the *HP XC6000*. For discretization of the structural parts, one-point under-integrated hexahedral elements are chosen, which perform excellently without showing any locking. However a stabilization against unphysical element kinematics, the hourglass modes, is required. The chosen hourglass stabilization – Belytschko-Bindemann assumed strain co-rotational stiffness form, which performs extremely well – is described in detail in [4].

For all concrete parts, a piecewise linear plasticity material model is used. The parameters, necessary for concrete like behavior of this on purpose simplified material law were obtained by calibrating with rather simple experimental examples. However, though the material model does not allow detailed modification concerning e.g. reinforcement, the reached approximation for the investigated mass dominated problems was fairly good. The possibility of element failure, which is necessary to simulate the appearance of local zones of accumulated damage (hinges) during the collapse event is also implemented in the material model. E.g. every time an element reaches a specific plastic strain, it is removed from the computation. Especially this mechanism helps to support the development of rigid body models as mentioned in Sect. 3.1. Concerning accuracy, it is acceptable to use the same material model for each simulation since the spare available documentations of the objects contain little information about the used concrete and steel. Investigations with more detailed material models on local submodels of the simulated structure are carried out in another subproject involved in the Research Unit 500 [1].

Every time contact appears during the collapse event, the kinematical configuration of the simulated structure changes abruptly. Hence correct determination of contact within parts of the building and between building and ground is also very important to obtain realistic collapse behavior of a model. For this reason fast automatic contact search algorithms are implemented in LS-DYNA. Though these algorithms show very good performance concerning accuracy and computation time, the search for contact requires a considerable amount of CPU-time of the whole simulation as each surface segment of the FE-mesh has to be considered. The chosen contact formulation is a penalty based node-to-segment algorithm for building to baseplate contact and a segment-to-segment algorithm between the building parts in all investigations. The base plate was modeled with four-node shell elements which were assumed to be rigid in each of the three presented models representing the contact segments for the base plate.

4.2 Storehouse in Weida/ Thüringen

As first example for a blast simulation, a storehouse in Thüringen was chosen, which has been demolished by blast in 1998. The framework of the seven storys was from reinforced concrete with masonry outer walls in the first and thin concrete walls in the upper floors. The building was 22 m high, 22 m long and 12 m wide with an approximate overall mass of 1900 tons. In the simulation, the collapse was reached by two steps of blasting as shown in Fig. 1. With the first explosion, two rows of columns were removed, after four seconds, the third row was destroyed. The weakening, caused by the first explosion was not sufficient to start the collapse of the building, so it stayed for four seconds on the two remaining rows of columns. After the second explosion, the building started bending forward which lead finally to the collapse. First the complete upper six storys began to rotate, after the cuboid got into contact to the ground, it started to break into pieces.

The discretized structure as depicted in Fig. 1, consists of 82867 hexahedral finite elements. Further informations about the simulation are given in Table 1. The results of a simulation can be seen in Fig. 4.2. Unfortunately,

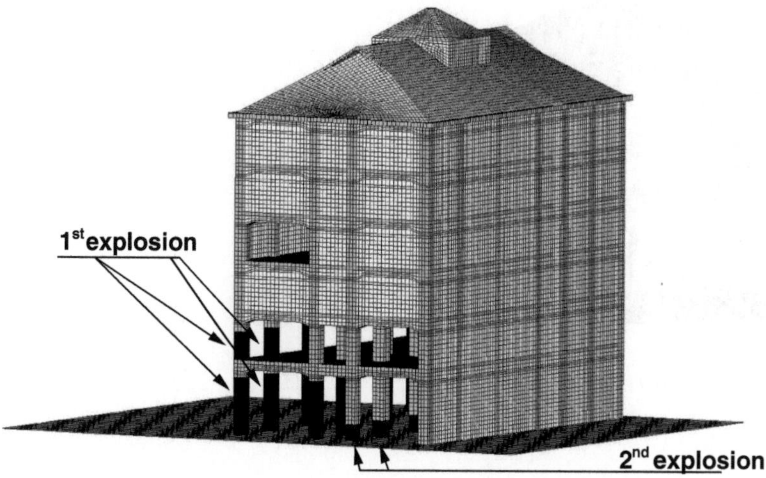

Fig. 1. Finite Element model of storehouse in Weida/ Thüringen – Removal of columns by blast in two steps

the documentation of the real collapse of 1998 is limited to one movie, which makes the validation of the simulation difficult. However, the beginning of the collapse is fairly realistically captured, compared to the available data.

Table 1. Comparison of the reference models – Weida (Sect. 4.2), Borna (Sect. 4.3) and Hagen (Sect. 4.4)

	Weida	**Borna**	**Hagen**
number of elements	82867	77079	392481
simulated time	9s	8s	—
number of processors	8	8	—
total CPU-time	65086 s	124756 s	—
CPU-time for contact search	30%	65%	—

4.3 Silo-building in Borna/ Sachsen

The second reference model, computed on the *HP-XC6000* Cluster is a silo building of about 25 m height with a base area of 36 m x 12 m and an approx-

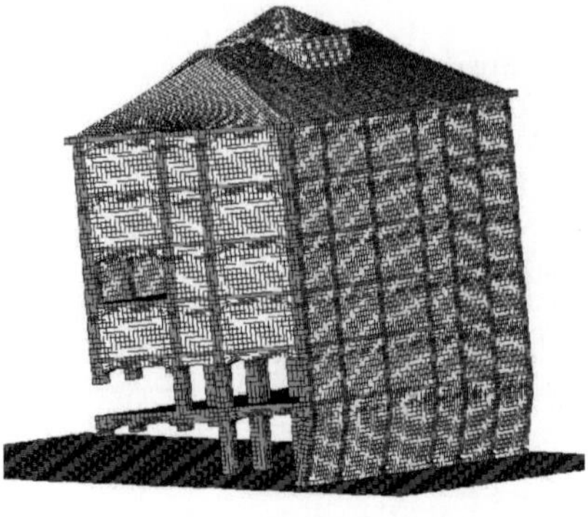

(a) $t_1 = 4.6\,\mathrm{s}$

(b) $t_1 = 5.0\,\mathrm{s}$

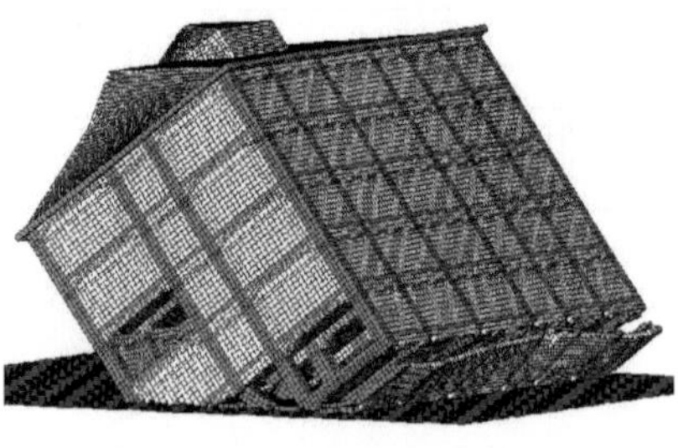

(c) $t_1 = 5.6\,\mathrm{s}$

Fig. 2. Three States of the collapse simulation

imate mass of 3600 tons. All columns and girders, as well as the six collecting bins were from reinforced concrete. Also a few concrete walls were included for stiffening reasons, however, most of the outer walls were from masonry. Before the start of the blasting, the structure was weakened by removing wedge-shaped parts of the inner concrete walls as shown in Fig. 3. The collapse in form of a bending movement to the front was realized by one explosion, removing the first row of columns completely. During the rotation and even after the first contact, the entire upper part stayed almost undeformed and was just turned over within the destruction.

The construction was discretized with 77079 hexahedral finite elements. As before, a rigid ground plate was modeled with shell elements, providing the segments for contact analysis. The discretized geometry is depicted in Fig. 3. In Table 1 it can be seen, that the CPU-time needed for the a complete simulation is much higher, as for the simulation of the model from Sect. 4.2, though the number of elements and the simulated collapse time is smaller. The reason for this effect is the huge amount of CPU-time required for the contact search algorithm. The problematic part of this structure is at the funnels, were fairly small elements are necessary for discretization. After getting into contact with the ground plate, here the contact search algorithm needs a lot of time at every time step. In the future, alternative models have to be investigated in detail to achieve a more efficient analysis without sacrificing accuracy.

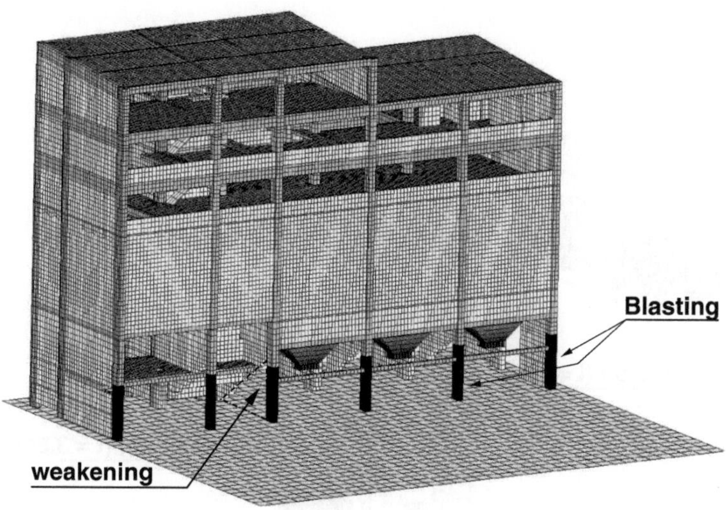

Fig. 3. Finite Element model of silo-building in Borna/ Sachsen

4.4 Sparkasse in Hagen/ Nordrhein-Westfalen – Highrise with 22 Storys

The third and largest example, simulated within the project is a 22 store building in Hagen/ Nordrhein-Westfalen. The building with a mass of about 26500 tons is 93 m high, 37 m long and 19 m wide. The main part of the structure is a frame construction of reinforced concrete, stiffened against shear with concrete walls. The building had been broken down in 2004. Before the blast process, all parts but the main load carrying construction were removed with machines or manually, also the facade was completely demounted. The planned collapse kinematic was reached by a combination of three explosions at 0 s, 2.5 s and 3.5 s, where as shown in Fig. 4 each time wedge-shaped parts were cut from the construction. The location of the weakened zones lead to a folding of the complete structure for reasons of minimizing of the space required for the debris.

In order to realize a rather regular mesh with fairly equal element sizes necessary for the proper simulation of the wave propagation process, 392481 solid elements were used for the discretization of the complete structure. Settings concerning contact or material were chosen with respect to the experience, gained by the simulation of the smaller buildings from Sects. 4.2 and 4.3. A complete simulation of the collapse of the building has not been carried out yet, so CPU-times are not known up to now. However the model is finished and the first complete run will be performed within the near future.

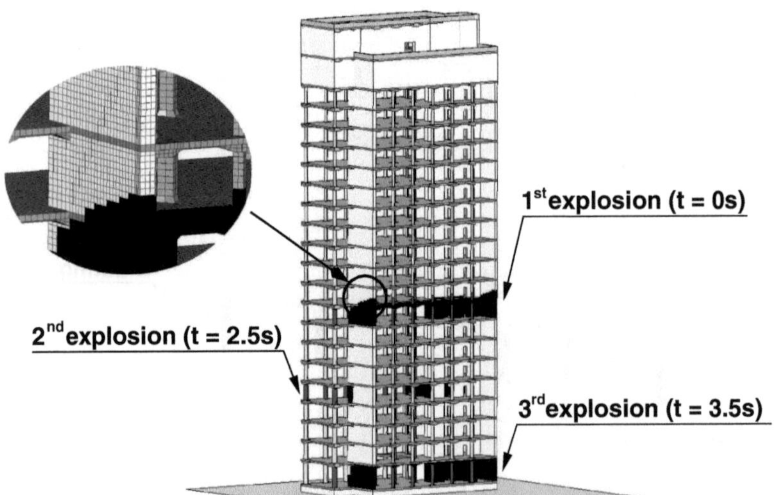

Fig. 4. Finite Element model of the highrisebuilding in Hagen

5 Closure

An insight into a part of the project work of the Reseach Unit 500 currently funded by DFG was given in this contribution. A number of fairly large Finite Element analyses have to be carried out in order to support the development of an alternative simulation process based on multi rigid body system analysis with comparable accuracy, but far less numerical effort. The development of the Finite Element models which is the focus of the current presentation itselves requires parametrical studies, to learn about e.g. contact or material parameters. These kinds of studies, especially with such large models are only possible, using high performance computers as the *HP XC6000*-Cluster. However the project is not finished yet – as mentioned in Sect. 4.4 one model has not been simulated yet – but a lot of experience about the simulation of blasting demolition have already been reached during the last year. This knowledge is indispensable for the further investigations of the Research Unit 500.

Acknowledgements

The financial support of the German Research Foundation (DFG) (Project FOR 500 - Computer aided destruction of complex structures using controlled explosives) is greatfully acknowledged.

Providing every available data such as plans and video material is very important for the success of this project. For this support the authors like to thank Dr.-Ing. Rainer Melzer.

References

1. DFG Research Unit 500. Computer aided destruction of complex structures using controlled explosives. http://www.sprengen.net/.
2. Höchstleistungsrechner-Kompetenzzentrum Baden-Württemberg. http://www.hkz-bw.de/.
3. K.-J. Bathe. *Finite-Elemente-Methoden.* Springer, 2002.
4. T. Belytschko and L.P. Bindemann. Assumed strain stabilization of the eight node hexahedral element. *Computer Methods in Applied Mechanical Engineering*, 105:225–260, 1993.
5. Ted Belytschko, Wing Kam Liu, and Brian Moran. *Nonlinear finite elements for continua and structures.* Wiley, 2000.
6. J.O. Hallquist. *LS-DYNA Theoretical Manual.* Livermore Software Technology Corporation, 1991-1998.
7. J.O. Hallquist. *LS-DYNA Keyword User's Manual.* Livermore Software Technology Corporation, 1992-2005.
8. T.J.R. Hughes. *The finite element method.* Dover Publ., 2000.
9. J. Lippok and D. Ebeling. *Bauwerkssprengungen.* Weißensee-Verlag, 2006.
10. B. Möller and M. Beer. *Fuzzy randomness.* Springer, 2004.
11. W.L. Wood. *Practical time-stepping schemes.* Clarendon Pr., 1990.

5 Closure

Wave Propagation in Automotive Structures Induced by Impact Events

Steffen Mattern[1] and Karl Schweizerhof[1,2]

[1] Institut für Mechanik, Universität Karlsruhe (TH), D-76131 Karlsruhe
mechanik@ifm.uni-karlsruhe.de
[2] project manager

1 Introduction

The prediction and numerical simulation of the propagation of high frequency vibrations caused by impact is very important for practical purposes and proper modeling is an open issue. In present automobiles more and more technical devices, which are getting their informations from sensors, are used to improve the safety concerning passengers. For example the activation of airbags and safety belts is controlled by sensors. For this reason it is important to assure that sensors perform reliably at a specific frequency range.

For the development of sensors, it is very important to know the kind of signal, which has to interpreted. When an impact happens at a specific location of the structure, a wave starts to propagate immediately. At the time, when the oscillations reach the point where the senor is located, they have already been influenced by many structural and material parameters, e.g. edges, spotwelds, foam material. It is necessary to calibrate the sensor in a way that if and only if a sufficiently hard impact has occurred, e.g. the airbag is activated.

In order to reduce costs for complex and expensive experiments, it is useful to perform numerical simulations of suitable structural parts to obtain general informations about wave propagation mechanisms in complex structures and to find out about the proper modeling with FE-programs. In this project, the powerful, highly parallelized commercial finite element code LS-DYNA [7, 6] is used to simulate different models and evaluate the influence of several modeling modifications and of other simulation parameters. In this contribution, a partial structure is described in detail and some of the performed modifications are explained. Different contact formulations are used as well as different material models for special parts. Damping is applied in parts of the structure and spotwelds are modelled with different methods, which is also a very important issue of simulation in automotive industry. Because of the large number of numerical investigations, performed during the project, only a few results are shown in the following sections. The results of the simulations are compared to experimental data with a special focus on amplitudes and frequencies at the locations of the sensors.

The models contain about 25000 – 50000 solid and shell elements and the simulated time of the impact event was in all cases 0.1 s, which lead to over $2 \cdot 10^6$ time steps per computation, using the explicit time integration available in LS-DYNA. With LS-DYNA, such kind of problems are particularly suited for computing on a high performance parallel computer, such as the *HP-XC6000 Cluster* at the University of Karlsruhe. Only with parallelization, it is possible to perform parametrical studies, which are necessary to obtain general knowledge about wave propagation and their proper modeling in complex structures in an appropriate time. Within the project, also investigations on academic examples with rather few elements were processed in order to gain information about wave propagation mechanisms as e.g. discussed in [5, 1] and their simulation with finite elements. However, several important effects can only be recognized at sufficiently complex models, which motivates the costly studies, presented in the following.

The results of the experiments were made available by other sources, thus they will not be explained further in the following. The main issue will be the simulation of the numerical models and the influence of different parameters.

2 T-Shaped, Spotwelded Structure Impacted by a Rigid Ball

2.1 Description

The structure, discussed in this contribution is a steel construction of top-hat profiles and sheets, connected with spotwelds, which is impacted by a metal ball at the top. This is a typical part of an automobile containing all relevant structural issues. In the experiments, the displacements were measured at one (S_1) and the accelerations at four sensors (locations $A_1 - A_4$), as shown in Fig. 1. The finite element model (Fig. 2) consists of about 33500 shell elements [3, 4] and 145 spotwelds, each modeled with one 8-node solid element. The steel plates are modeled with an elastic-plastic material. For the spotwelds, the specific material `*MAT_SPOTWELD` in LS-DYNA, which consists of an isotropic hardening plasticity model coupled to four failure models [7] is used. The metal ball, the impactor, is assumed to be a rigid body. Contact between impactor and plate is realized by an automatic penalty based segment-to-segment contact formulation. In order to connect two metal plates with spotwelds, the `*CONTACT_SPOTWELD` is applied, which ties the nodes of a solid element to the two neighboring deformable shell surfaces with constrains [7].

The results from the simulation of this basic finite element model, depicted in Fig. 1, compared to experimental results provided to the authors, are shown in Fig. 3. This is then taken as a basis for several parametrical studies, computed on the *HP-XC6000 Cluster*.

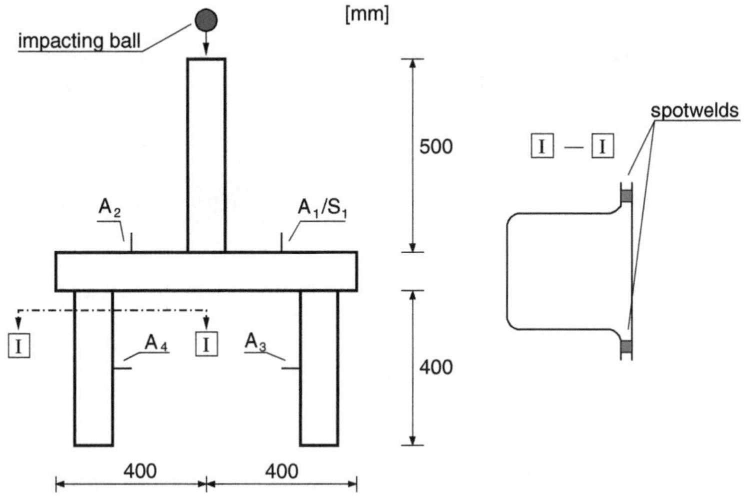

Fig. 1. Geometry of the T-shape structure, sensor location denoted by S_1 and A_1–A_4

Fig. 2. FE-Model of the T-shaped structure

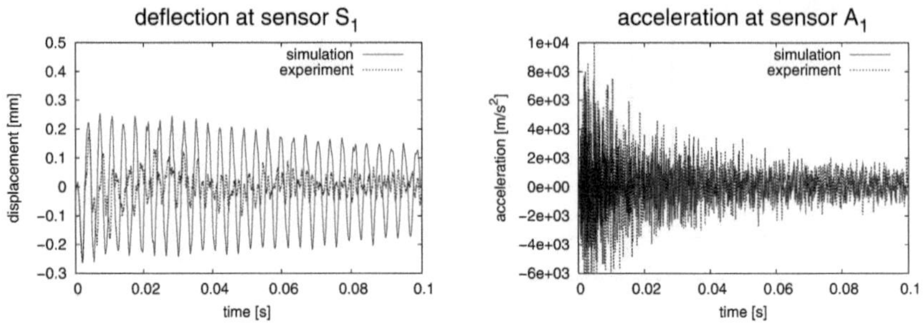

(a) full simulated time

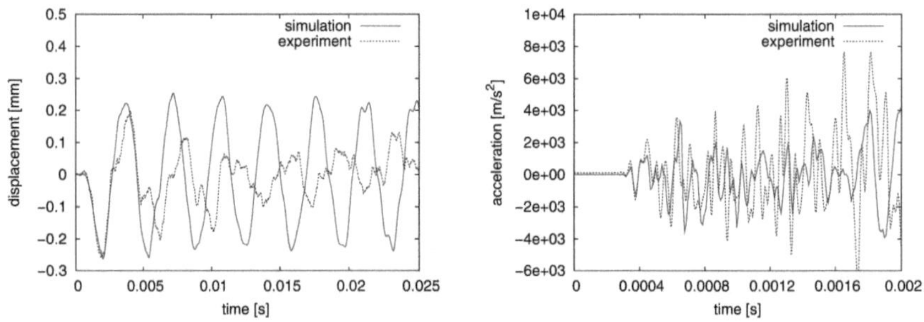

(b) cutout – beginning of the simulation

Fig. 3. Experimental and simulated displacements and accelerations at sensor 1

2.2 Parametrical Studies

2.2.1 Stabilization of Spotwelds

As already mentioned, the spotwelds are modeled with a standard eight-node hexahedral finite element using one-point under-integration combined with a stabilization against unphysical hourglass modes, as shown exemplarily in Fig. 4. The chosen hourglass stabilization – Belytschko–Bindemann assumed strain co-rotational stiffness form – is described in detail in [2, 4] and

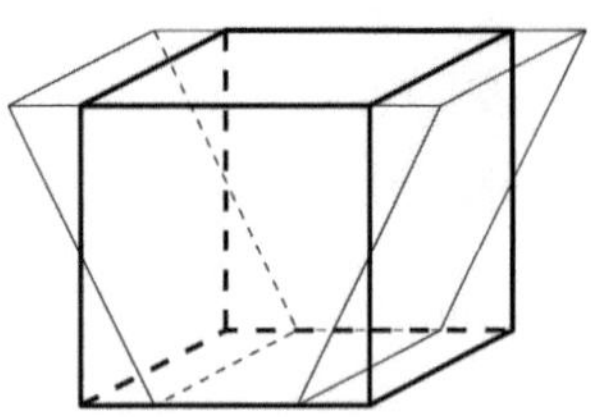

Fig. 4. Example for an hourglass mode

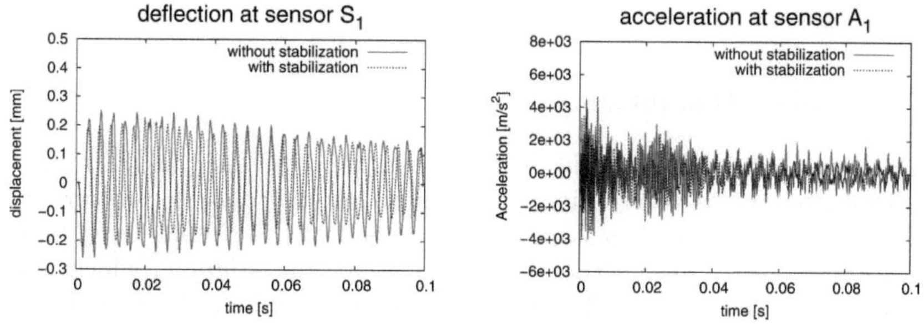

(a) full simulated time

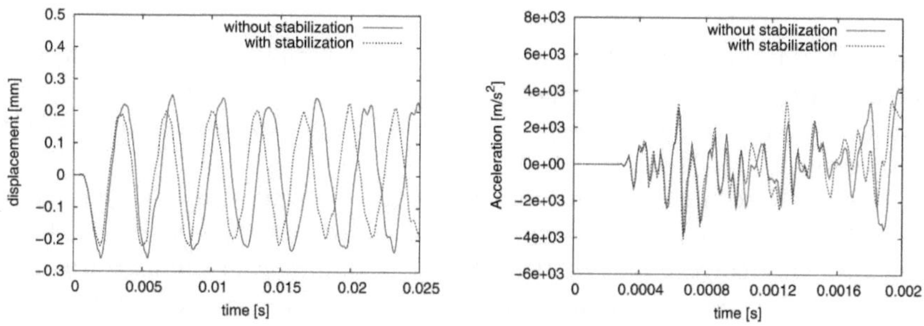

(b) cutout – beginning of the simulation

Fig. 5. Simulated displacements and accelerations with standard viscous- and with stiffness-controlled hourglass stabilization of the spotwelds

works properly for all applications performed in the project. By default, i.e. if no hourglass stabilization is chosen, a standard viscous hourglass control is used [7].

In Fig. 5, the results of the simulations with standard viscous and with Belytschko–Bindemann stiffness stabilization of the spotwelds are given. As expected the structure reacts slightly stiffer with stiffness controlled stabilization of the spotwelds, which leads to higher frequencies in the displacements. Also a reduction of the amplitudes can be noticed. The right side of Fig. 5 shows also only small changes in the accelerations. The properties concerning propagation of waves through the spotwelds have slightly changed as a consequence of stiffness controlled stabilization, which shows that there is a small influence of unphysical hourglass modes on the results of the original simulation. For problems with larger displacements and higher amplitudes, the effect would be more obvious, which means that for such kind of problems (e.g.

crash simulations), hourglass stabilization of under-integrated finite elements is mandatory.

2.2.2 Applied Damping

As can be seen in Fig. 3, the displacements decay very quickly in the experiment, but in the simulation, the amplitudes stay almost constant. This leads to far overrated displacements at the end of the simulation. The reason for the energy loss in the experiment is system damping, which is due to material damping as well as due to dissipation at boundaries and joints. Mechanically, this effect can be described with Rayleigh damping, where the damping matrix can be defined as

$$\mathbf{C} = \alpha \mathbf{M} + \beta \mathbf{K}. \tag{1}$$

Mass proportional and stiffness proportional damping is possible, which can be defined for the whole structure, or only for specific parts. Fairly problematic is the validation of the damping parameters α and β, thus parametrical studies are necessary.

First, only stiffness proportional damping is applied, which leads to damping of high frequencies. As coefficients, values from 0.10 to 0.25 were chosen, which roughly corresponds to 10% to 25% of damping in the high frequency domain [7]. For better visualization, in Fig. 6 only the results with maximum damping (25%) are plotted. It can be seen in Fig. 6, that damping of high frequencies has almost no influence on the displacements, and leads to a rather small reduction of amplitudes of the accelerations.

Then pure mass proportional damping was applied. Mass weighted damping is used to damp all motions including rigid body motions, this means damping in the lower frequency range. It is also possible to damp only special parts, or chose different damping coefficients for different parts. The results in Fig. 7 show – as expected – strong influence of the mass weighted damping on the amplitudes of displacements and acceleration.

Comparing the results of both described formulations, mass proportional damping leads to results, which are closer to the experimental data. Obviously, the main part of energy loss happens in the lower frequency domain. Another reason for dissipation of energy is the impact event. In the simulation, the ball is modeled as a rigid body, hence here no loss of energy is implied. To obtain results, closer to reality, also a contact damping could be applied, which means a dissipation of energy at the impact event. Further investigations concerning the combination of different ways of modeling damping are necessary.

2.2.3 Influence of Shell Element Formulation

In order to investigate the influence of the element formulation on the simulation result, besides the standard shell elements with hourglass-stabilization

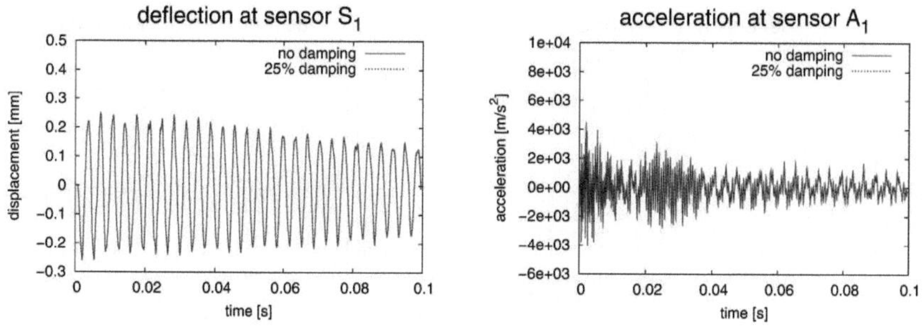

(a) full simulated time

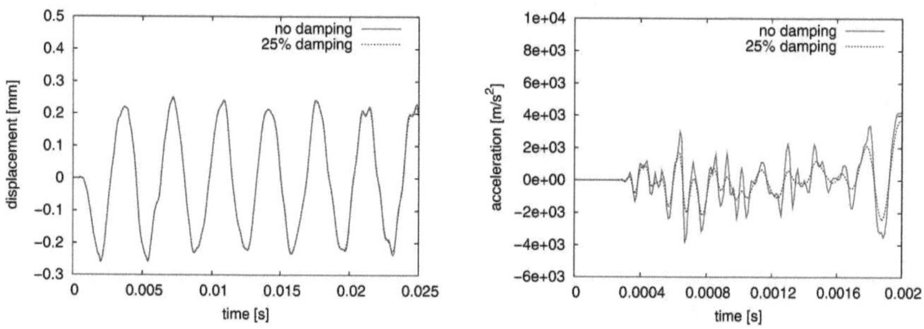

(b) cutout – beginning of the simulation

Fig. 6. Studying the influence of stiffness proportional damping

also fully integrated shell elements with an assumed strain formulation for the shear terms [7] are used to model all metal sheets. Spotwelds, impactor and contact formulation, as well as all boundary conditions are kept unmodified, so differences in the results can be associated exclusively to the element formulation. As expected, the fully integrated shell elements react much stiffer, which is depicted in Fig. 8. Although the displacements show only slightly higher frequencies, the differences in the accelerations can be clearly recognized. Frequencies as well as amplitudes are considerably higher than in the simulations with hourglass stabilized elements, which indicates a stiffer structure. In addition we have to note, that the choice of the element formulation strongly affects the efficiency. The computation with fully integrated shell elements takes about three times as long as the computation with under-integrated elements. Concerning the use of different shell element formulations, we have to conclude that, fully integrated elements are better suited for predicting the higher frequency domain. However the higher stiffness of this formulation leads to overrated amplitudes especially regarding the accelerations.

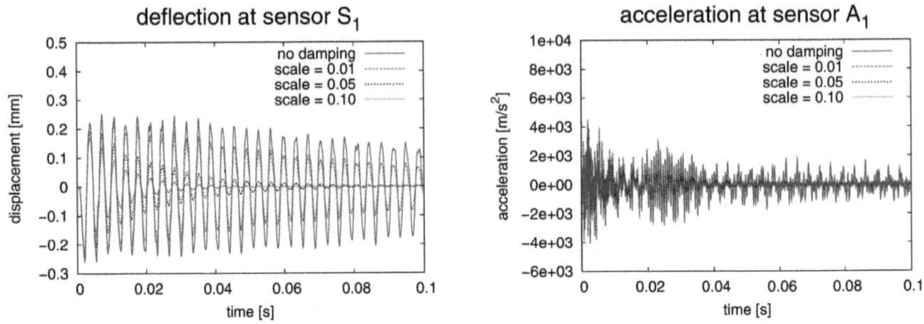

(a) full simulated time

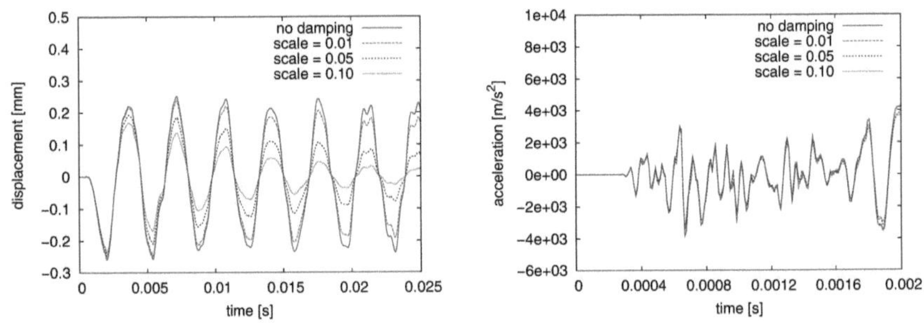

(b) cutout – beginning of the simulation

Fig. 7. Studying the influence of mass proportional damping

2.2.4 Refinement of the Mesh

As described in Sect. 2.1, the original discretization, depicted in Fig. 1 contains 33086 shell elements. The average element length in this model is 5 mm. In order to investigate the influence of mesh refinement on the results, the element length in both directions was reduced to about 2.5 mm, which lead to 132344 shell elements. All computations of these models were performend on 8 processors of a so called fat-node on the *HP-XC6000 Cluster*. The CPU-time per processor of this problem was about $3 \cdot 10^4$ s, which leads to a simulation time of approx. 8.5 h for the complete analysis. The results in Fig. 9 show higher frequencies and amplitudes in the accelerations and quickly decreasing amplitudes in the displacements. Compared to the experimental data, given in Fig. 3, frequencies and amplitudes correlate much better for the refined mesh. However, in the first few cycles, both simulations show very similar results, which is clearly visible for the displacements.

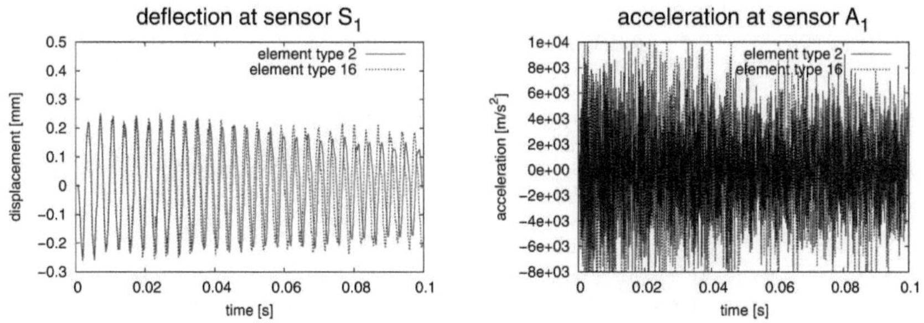

(a) full simulated time

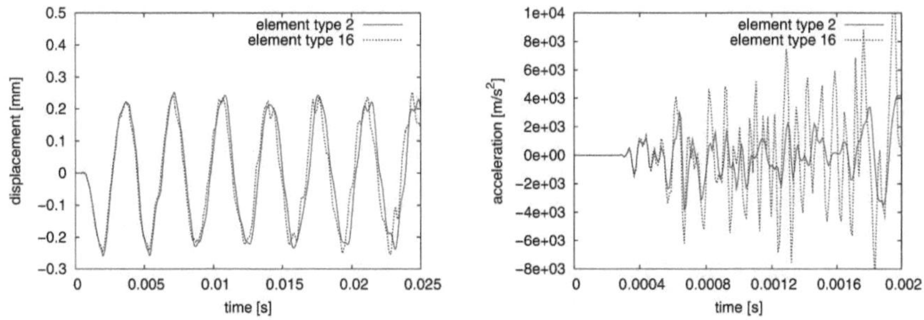

(b) cutout – beginning of the simulation

Fig. 8. Comparison of standard (type 2) shell element with viscous hourglass control and fully integrated (type 16) shell element

Obviously the lower frequency response is well captured already by the coarse mesh. However, the capability of the finer mesh to model more high frequency content – important in wave propagation – appears to be extremely important. A closer view on this partial issue by a Fourier decomposition is depicted in Fig. 10. It shows that lower frequencies can be simulated correctly even with the coarse mesh, but higher frequencies require finer meshes, even beyond the currently used refined mesh. The experimental results show also that there is a fairly high energy content between 500 and 1000 Hz which must be represented by the simulation model. This is a fairly surprising result, as studies with a ball impacting a plate had shown very reasonable correspondence with experimental observations.

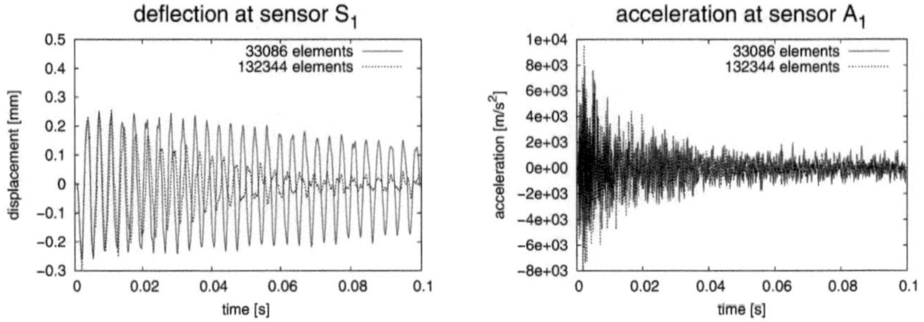

(a) full simulated time

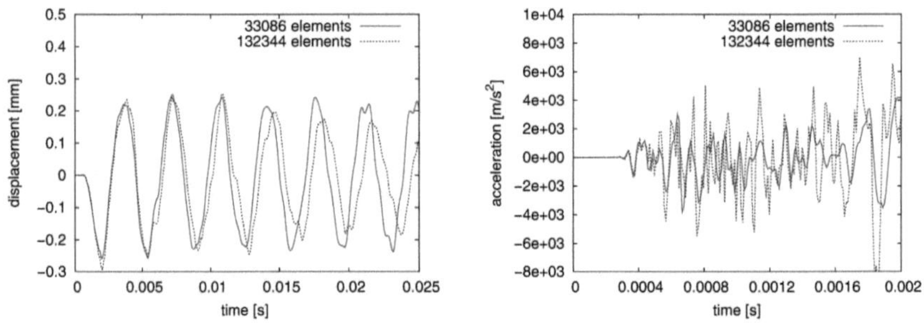

(b) cutout – beginning of the simulation

Fig. 9. Two levels of discretization of the steel structure

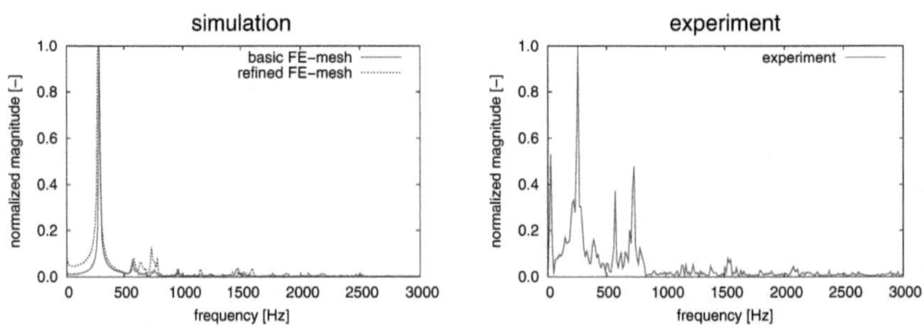

Fig. 10. Frequency amplitude spectra of displacements – simulation and experiment

3 Conclusions

The simulation results presented in Sect. 2 show only a part of the project work on the *HP-XC6000 Cluster*. The goal of the project was to generate knowledge about the proper FE-simulation of complex structures by modifying several relevant model parameters. In another part of the project, numerical examples with rather small numbers of elements are computed, to gain general information about wave propagation simulation with finite elements. Some of the performed modifications do not advance the results at all, compared to the experimental results, but general rules concerning the usage of different tools and procedures of the Finite Element Method can be obtained from these simulations. However the mesh refinement has shown that the computation of very large models on parallel computers is absolutely necessary. Though in many investigations it is not required, to aim at an absolute realistic simulation of the experimental results, because the numerical effort is much higher than the obtained advantage, it has proven that a constantly refined mesh is of vital importance. As described in Sect. 2.2.4, it may be often possible to simulate e.g. lower frequencies with rather coarse meshes. In order to realize the presented parametrical studies in an appropriate time, they were all carried out first with the rather coarse mesh. As the goal of the investigations was, to find out the general influence of the discussed parameters, for which the used discretization was sufficient.

The project is continued and computations on other complex structures with rather fine meshing will follow, which presumes the availability of high performance parallel computers, such as the *HP-XC6000 Cluster*. In further investigations, also the application of damping on the refined mesh from Sect. 2.2.4 with the experience from Sect. 2.2.2 will be analyzed. Other interesting issues are a) the influence of the boundary conditions on the experimental results and how accurate these can be captured in the simulation, and b) how the simulation reacts on changes in the joining achieved with a contact formulation used for the spotwelds. It is important to investigate all these parameters separately, which requires again many parametrical studies with high numerical effort. The final achievement of the study will be to develop modeling rules for wave propagation simulation in complex structures such as automotives under impact as found in crashworthiness events.

References

1. J.D. Achenbach. *Wave propagation in elastic solids.* North-Holland, 1973.
2. T. Belytschko and L.P. Bindemann. Assumed strain stabilization of the eight node hexahedral element. *Computer Methods in Applied Mechanical Engineering*, 105:225–260, 1993.
3. T. Belytschko and C.S. Tsay. A stabilization procedure for the quardilateral plate element with one-point quadrature. *International Journal for Numerical Methods in Engineering*, 19:405–419, 1983.

4. Ted Belytschko, Wing Kam Liu, and Brian Moran. *Nonlinear finite elements for continua and structures.* Wiley, 2000.
5. K.F. Graff. *Wave motion in elastic solids.* Ohio State Univ. Press, 1975.
6. J.O. Hallquist. *LS-DYNA Theoretical Manual.* Livermore Software Technology Corporation, 1991–1998.
7. J.O. Hallquist. *LS-DYNA Keyword User's Manual.* Livermore Software Technology Corporation, 1992–2005.

Miscellaneous Topics

Wolfgang Schröder

Aerodynamisches Institut, RWTH Aachen, Wüllnerstr. zw. 5 u. 7,
(D) – 52062 Aachen,
office@aia.rwth-aachen.de

This section documents the extremely broad field of numerical simulation. It pushes research in many other fields than just the customary topics like fluid dynamics, aerodynamics, structure mechanics and so forth. The following five articles show that today's computational approaches are by far not complete from a numerical or from a modeling point of view. However, when applied with the physically correct focus the novel numerical techniques can be used as reliable prediction methods to substantiate new theories.

The first paper is a continuation of the work reported in previous volumes. It is a cooperation between the Institute of Geosciences of the Friedrich-Schiller-University of Jena in Germany and the Department of Earth and Planetary Science of the University of California at Berkeley. The contribution focuses on the numerical simulation of the chemical differentiation of the Earth's mantle which induces the generation and growth of the continents and the formation and augmentation of the depleted mid-oceanic ridge basalt mantle. This problem is tackled in a three-dimensional compressible spherical-shell mantle via an integrated theory in conjunction with the problem of thermal convection. The entire evolution was computed starting with the formation of the solid-state primordial silicate mantle. The number, size, and form of the continents are not subject to any restrictions. The paper is based on a vast variation of parameters such as the viscosity-level parameter, the yield stress, and the temporal average of the Raleigh number.

The second contribution from the Institute of Geodesy of the University of Stuttgart focuses on the recovery of the geopotential. Since several satellite missions for gravity field recovery do already provide global and highly resolved data this challenging task can be tackled today. In this paper two new solution algorithms for successful geopotential recovery are developed. The first method is an iterative LSQR solver tailored for cluster systems. The alternative approach is a brute-force inversion, which is extremely efficient on SM architectures.

The subsequent article deals with molecular modeling of hydrogen bonding fluids. The authors are from the Institute of Technical Thermodynamics nad Thermal Process Engineering of the University of Stuttgart. Since the experimental data on thermophysical properties of pure fluids and mixtures are often narrow molecular modeling and simulation can be considered a useful alternative approach. Due to the requirement to resolve hydrogen bonds that are generated by strong intermolecular forces the simulations of fluids are quite costly which is why high-performance computing is a must to achieve findings in an acceptable time frame.

The fourth contribution from the Institute of Scientific Computing at Karlsruhe shows an error estimate to possess an invaluable advantage to solve partial differential equations. First, the quality of the Finite Difference Element Method to estimate the error is evidenced for some academic examples. Then, the numerical simulation is used to fuel-cell and fluid-structure interaction problems. Also in these applied problems the error estimates proved the quality for all components of the vector of solution. To obtain a similar quality control of the solution using customary grid refinement computations would be extremely costly.

Continental Growth and Thermal Convection in the Earth's Mantle

Uwe Walzer[1], Roland Hendel[1], and John Baumgardner[2]

[1] Institut für Geowissenschaften, Friedrich-Schiller-Universität, Burgweg 11, 07749 Jena, Germany
u.walzer@uni-jena.de
[2] Dept. Earth Planet. Science, University of California, Berkeley, CA 94720, USA

Abstract. The main subject of this paper is the numerical simulation of the chemical differentiation of the Earth's mantle. This differentiation induces the generation and growth of the continents and, as a complement, the formation and augmentation of the depleted MORB mantle. Here, we present for the first time a solution of this problem by an integrated theory in common with the problem of thermal convection *in a 3-D compressible spherical-shell mantle.* The whole coupled thermal and chemical evolution of mantle plus crust was calculated starting with the formation of the *solid-state* primordial silicate mantle. No restricting assumptions have been made regarding number, size and form of the continents. It was, however, implemented that moving oceanic plateaus touching a continent are to be accreted to this continent at the corresponding place. The model contains a mantle-viscosity profile with a usual asthenosphere beneath a lithosphere, a highly viscous transition zone and a second low-viscosity layer below the 660-km mineral phase boundary. The central part of the lower mantle is highly viscous. This explains the fact that there are, regarding the incompatible elements, chemically different mantle reservoirs in spite of perpetual stirring during more than 4.49×10^9 a. The highly viscous central part of the lower mantle also explains the relatively slow lateral movements of CMB-based plumes, slow in comparison with the lateral movements of the lithospheric plates. The temperature- and pressure-dependent viscosity of the model is complemented by a viscoplastic yield stress, σ_y. The paper includes a comprehensive variation of parameters, especially the variation of the viscosity-level parameter, r_n, the yield stress, σ_y, and the temporal average of the Rayleigh number. In the r_n–σ_y plot, a central area shows runs with realistic distributions and sizes of continents. This area is partly overlapping with the r_n–σ_y areas of piecewise plate-like movements of the lithosphere and of realistic values of the surface heat flow and Urey number. Numerical problems are discussed in Sect. 3.

1 Introduction

Investigations of the isotopic compositions of SNC meteorites and lunar rocks show a relatively quick chemical differentiation of Mars and Moon within the first 200 Ma (Norman et al., 2003; Nyquist et al., 2001). It is probable that not only the iron cores of these planetary bodies were formed rather quickly but also an early silicate crust developed from a magma ocean. A similar mechanism can be expected for the early Earth. In the case of the Earth, however, there are three additional modes of growth of the total continental mass (Hofmann, 2004) which continue to add juvenile mass in batches distributed over the whole geological history. The continental crust (CC) grows episodically (Condie, 2003). U, Th, K and other incompatible elements will be enriched in CC leaving behind parts of the mantle depleted in these elements: This reservoir of the mantle is called depleted MORB mantle (DMM) where MORB stands for mid-oceanic ridge basalt. These processes are in fact much more complex (Porcelli and Ballentine, 2002; Hofmann, 2004; Bennett, 2004; Rudnick and Gao, 2004; Fitton et al., 2004; Stracke et al., 2005; Walzer et al., 2006). Furthermore, there is still some doubt as to whether the rich part of the mantle below DMM is quasi-primordial or processed or mainly processed and to a lower degree quasi-primordial. Rich means rich in incompatible elements. It is only clear that this non-DMM part of the mantle must be richer in U,

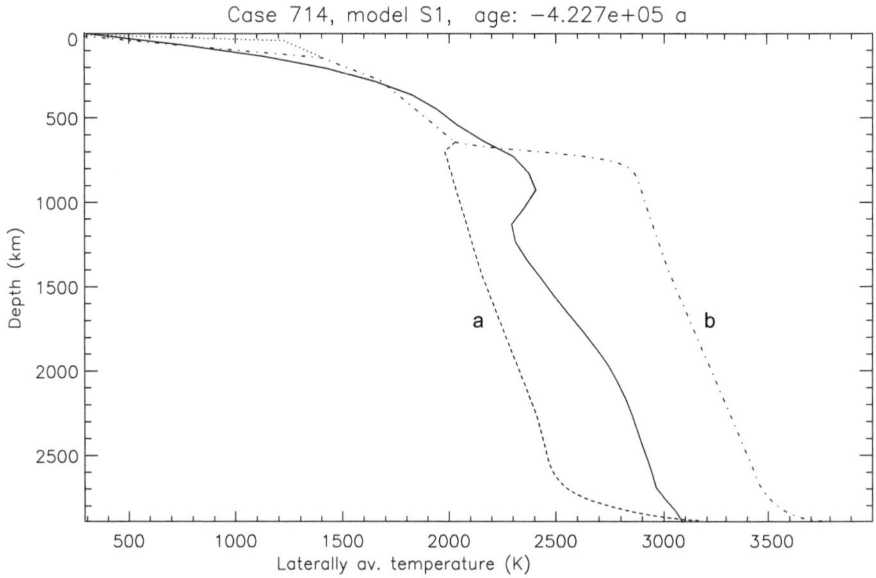

Fig. 1. The *solid curve* represents the laterally averaged temperature of the geological present time for a reference run with a viscoplastic yield stress of $\sigma_y = 135$ MPa and a viscosity-level parameter $r_n = -0.6$. Cf. Eqs. (1) and (2). The CMB temperature, T_c, is laterally constant but variable in time according to the parameterized heat balance of the Earth's core. The range of possible mantle geotherms according to parameterized models given by Schubert et al. (2001) is shown for comparison. *Label a* and *b* denote geotherms of whole-mantle and partially layered convection, respectively. The dotted line stands for a mid-oceanic ridge geotherm

Th and K, richer than DMM. Otherwise the requirements of energy supply would not be fulfilled.

The lateral movements of lithospheric plates, the collisions of continents, the collisions of continents and oceanic plates as well as orogenesis are caused by thermal solid-state convection of the mantle (Schubert et al., 2001). The energy sources of this convective motor are the kinetic energy of the Earth's accretion, the potential energy of the gravitational differentiation of a possibly homogeneous primordial Earth into an iron core and a silicate primordial mantle, and the nuclear bonding energy that is released by the decay of the radionuclides ^{238}U, ^{235}U, ^{232}Th and ^{40}K. Minor energy sources are the heat of tidal friction and the latent heat due to the growth of the inner core at the expense of the outer core. This heat is released by the cooling of the Earth.

2 Theory

We solve the balance equations of mass, momentum and energy using the anelastic liquid approximation in the formulation given by Walzer et al. (2003b). Furthermore, the balance equations of the mentioned radionuclides plus their corresponding daughter nuclides are taken into account. We supplement the viscous constitutive equation by a viscoplastic yield stress, σ_y, for

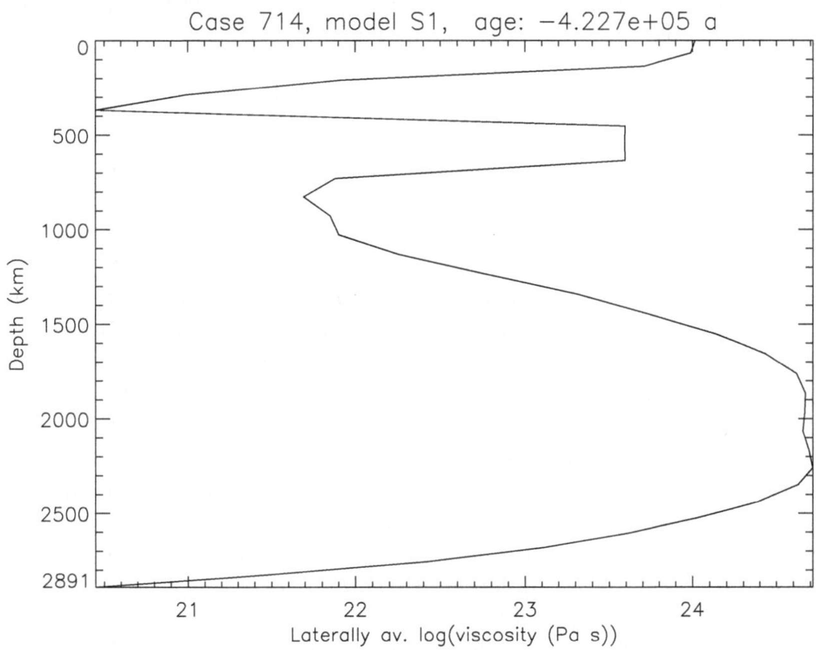

Fig. 2. The laterally averaged shear viscosity of the reference run for the geological present time

the uppermost 285 km and by a special viscosity profile for the initial temperature. This viscosity profile has been derived by Walzer et al. (2004a). In the present paper, the equations of chemical differentiation of oceanic plateaus plus the above mentioned equations are simultaneously solved. The new features of the theory are given in Walzer et al. (2006).

Here are only some remarks on the viscosity law. The shear viscosity, η, is calculated by

$$\eta(r,\theta,\phi,t) = 10^{r_n} \cdot \frac{\exp(c\,\overline{T_m/T_{av}})}{\exp(c\,\overline{T_m/T_{st}})} \cdot \eta_3(r) \cdot \exp\left[c_t \cdot T_m \left(\frac{1}{T} - \frac{1}{T_{av}}\right)\right] \quad (1)$$

where r is the radius, θ the colatitude, ϕ the longitude, t the time, r_n the viscosity-level parameter, T_m the melting temperature, T_{av} the laterally averaged temperature, T_{st} the initial temperature profile, T the temperature as a function of r, θ, ϕ, t. The quantity $\eta_3(r)$ is the viscosity profile at the initial temperature and for $r_n = 0$. So, $\eta_3(r)$ describes the dependence of the viscosity on pressure and on the mineral phase boundaries. The second factor of the right-hand side of Eq. (1) describes the increase of the viscosity profile with the cooling of the Earth. For $MgSiO_3$ perovskite we should insert c=14, for MgO wüstite c=10 according to Yamazaki and Karato (2001). So, the lower-mantle c should be somewhere between these two values. For numerical reasons, we are able to use only c=7. In the lateral-variability term, we inserted $c_t = 1$. For the uppermost 285 km of the mantle (plus crust), an effective viscosity, η_{eff}, was implemented where

$$\eta_{eff} = \min\left[\eta(P,T), \frac{\sigma_y}{2\dot{\varepsilon}}\right] \quad (2)$$

The pressure is denoted by P, the second invariant of the strain-rate tensor by $\dot{\varepsilon}$.

Oceanic plateaus are driven on the surface of the moving oceanic lithosphere near to the continents like on a conveyor belt. They will be subducted to a depth of about 100 km and suffer a further chemical differentiation. The emerging andesitic magmas will be added to the continent. So, the total mass of the continents increases generating a growing DMM or MORB source reservoir of the mantle, that is depleted in incompatible elements, as a countermove. The present numerical model depends only slightly on the question whether, except DMM, only FOZO, HIMU, EM1 and EM2 exist in the mantle (see Hofmann, 2004; Stracke et al., 2005) or whether there are additionally nearly primordial parts of the mantle (Bennett, 2004).

Although mantle convection is working by solid-state creeping, we calculate the process using the differential equations of a compressible anelastic liquid that is heated from within by the decay of ^{238}U, ^{235}U, ^{232}Th and ^{40}K. The mantle is additionally slightly heated from the core-mantle boundary (CMB). For this purpose, we use a parameterized cooling-core model.

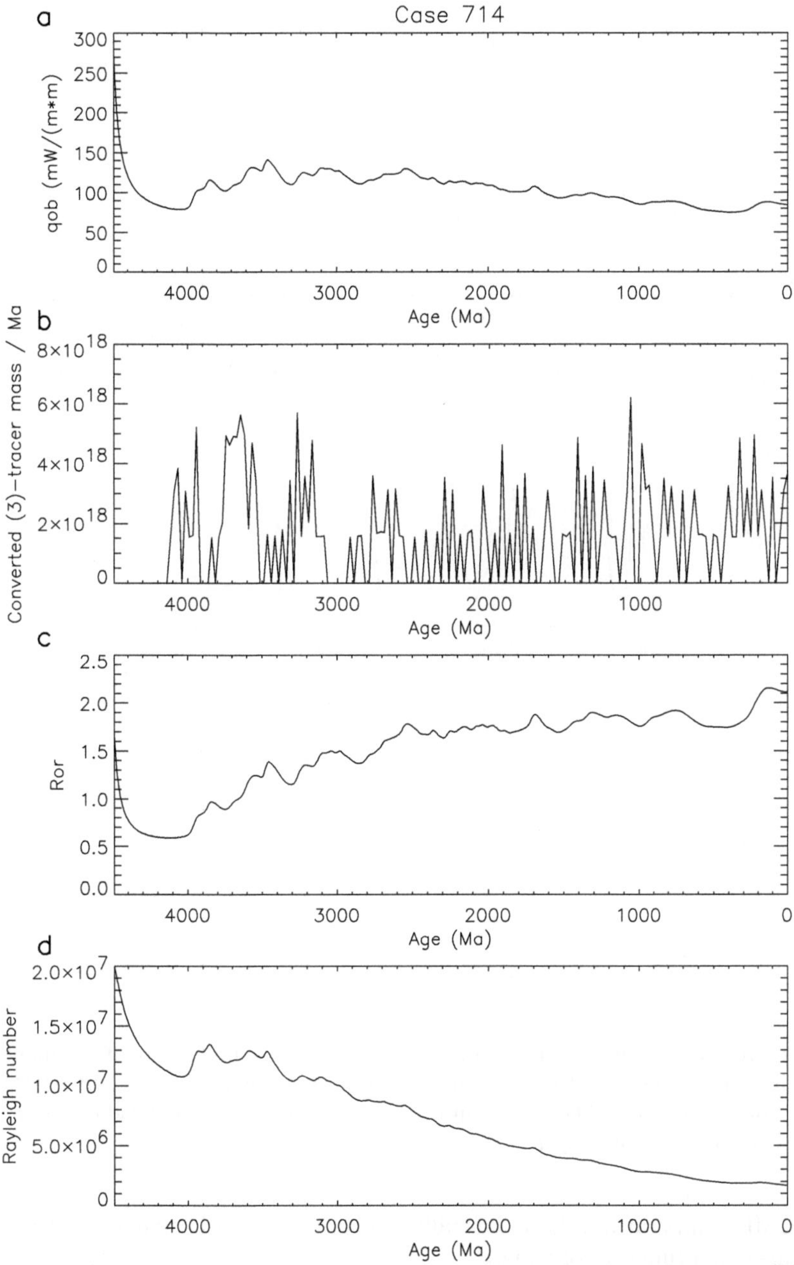

Fig. 3. **(a)** The evolution of the laterally averaged surface heat flow, *qob*. **(b)** Juvenile contributions to the total mass of the continents, expressed as converted (3)-tracer mass per Ma, as a function of time. **(c)** The evolution of the reciprocal value of the Urey number. *Ror* means *r*atio of the surface heat *o*utflow per unit time to the mantle's *r*adiogenic heat production per unit time. **(d)** The evolution of the Rayleigh number

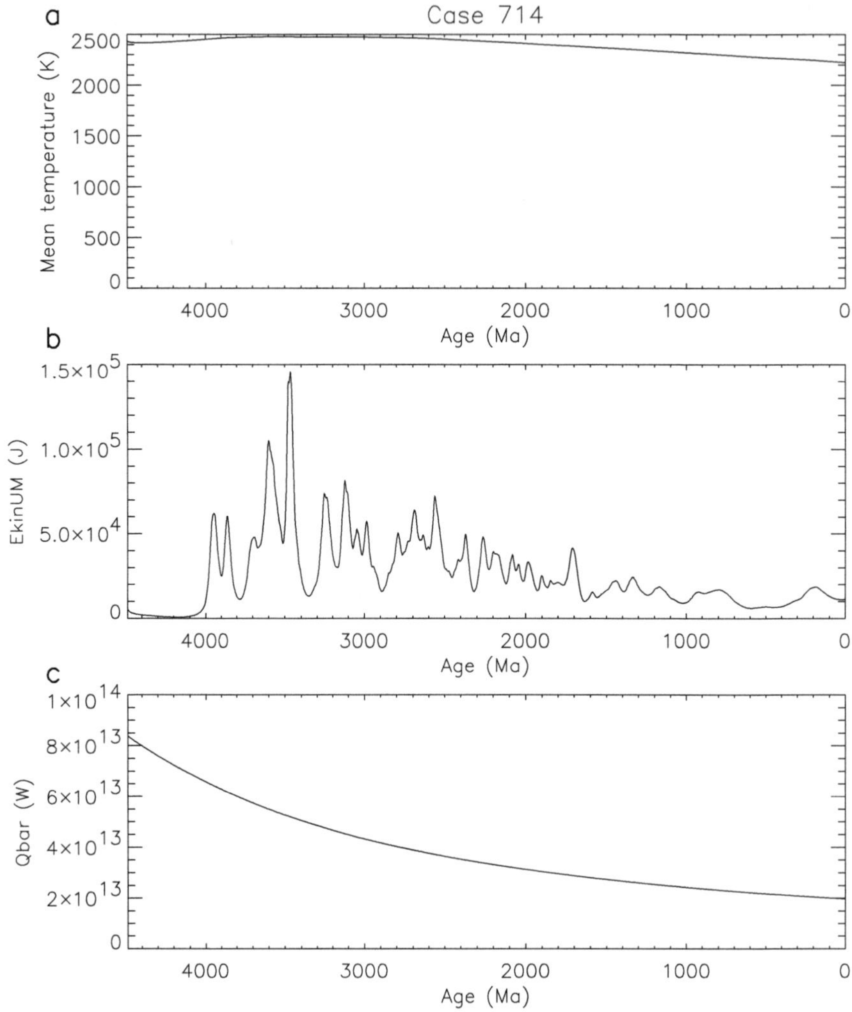

Fig. 4. (a) The evolution of the volume-averaged mean temperature of the mantle. **(b)** The kinetic energy of the thermal convection in the upper mantle, *EkinUM*, as a function of time. **(c)** The evolution of the power of the internal radiogenic heat generation of the mantle, *Qbar*

The CMB temperature, T_c, is laterally constant at a particular time. But T_c decreases as a function of time.

There is an open principal question: Why can we, at present, observe reservoirs with different abundances of the mentioned radionuclides in the interior of the mantle in spite of the continual stirring due to convection enduring for more than 4.49×10^9 a of the existence of the solid-state silicate mantle? A mathematical description of the segregation mechanism is given by Walzer et al. (2006).

3 Numerical Aspects

The total pressure dependence and the radial temperature dependence of viscosity is completely taken into account. For numerical reasons, however, only a part of the lateral temperature dependence of the viscosity could be used. At the mineral phase boundaries in the interior of the Earth's mantle, there are not only discontinuities of the seismic velocities and of the density but also jumps of activation volumes, activation energies and, therefore, of activation enthalpies. Since the viscosity depends exponentially on the activation enthalpy of the prevailing creeping process, the conclusion is inescapable that there are considerable viscosity jumps at the upper and lower surfaces of the transition zone. These jumps cause numerical problems in the solution of the balance equations. The problems have been solved. Nevertheless, our group

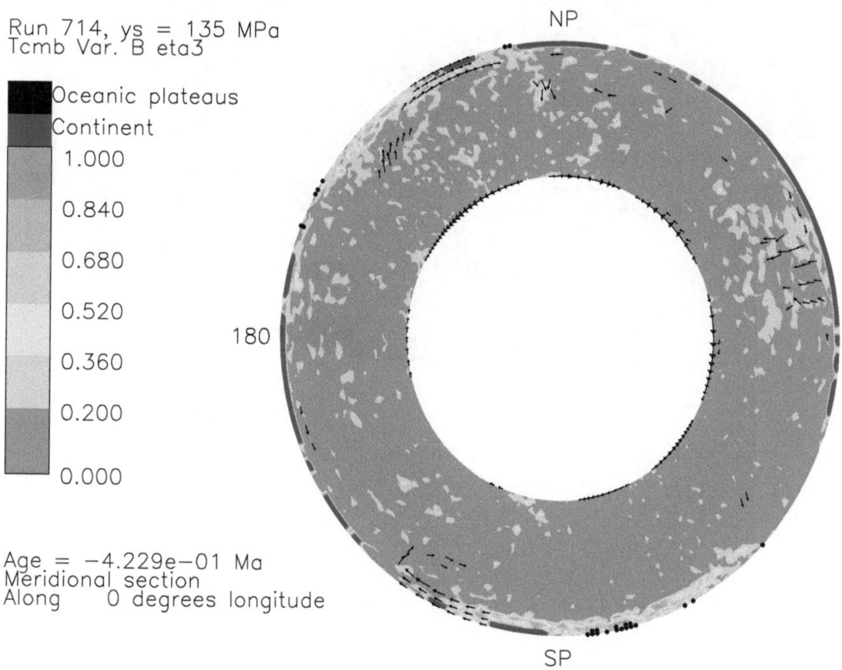

Fig. 5. The result of the chemical evolution of the silicate spherical shell of the Earth for the geological present time. Strongly depleted parts of the mantle are denoted by *yellow* or *greenish areas*. Less depleted and rich parts of the mantle have *orange colors*. Depleted means with a low content of incompatible elements like U,Th,K etc. Continents are depicted in *red*. *Black dots* represent oceanic plateaus

is searching for more effective solutions of the numerical jump problem. The minor discontinuity at a depth of 520 km has been neglected.

The mantle has been treated as a thick spherical shell. The discretization is made by projection of the edges of a concentric regular icosahedron (situated in the core) onto spherical shell surfaces with different radial distances from the center. These surfaces subdivide the mantle into thin shells. A first step of grid refinement is the bipartition of the edges of the spherical triangles into equal parts. Connecting the new points by great circles, we obtain four smaller triangles from each starting triangle. The refinement can be augmented by succesive steps of this kind. We can use different formulae for the distribution of the radial distances of the spherical grid surfaces. In this paper, we used exclusively a grid with isometric grid cells, as good as possible. The grid is non-adaptive. The Navier–Stokes equations as well as pressure and creeping velocity are discretized by finite elements. Piecewise

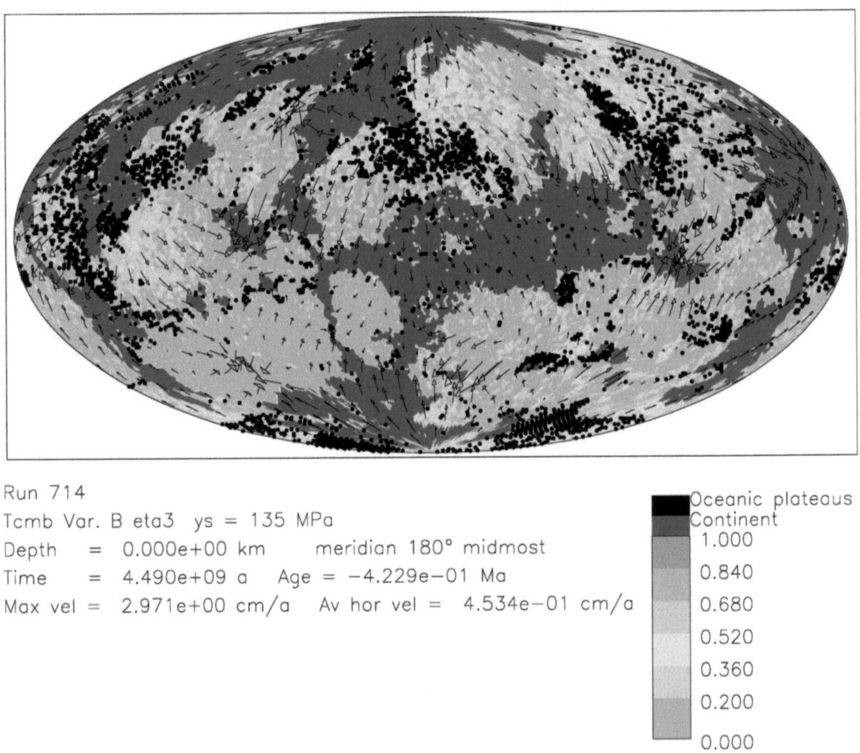

Fig. 6. The distribution of the red continents and the black oceanic plateaus at the Earth's surface for the geological present time of a reference run with a yield stress $\sigma_y = 135$ MPa and a viscosity-level parameter $r_n = -0.6$. Their are no prescriptions regarding number, size or form of continents in this model S3. *Arrows* represent the creeping velocities. The oceanic lithosphere is shown in *yellowish colors*

linear basis functions have been applied for the creeping velocity, piecewise constant or also piecewise linear basis functions are used for the pressure. The equations for pressure and creeping velocity have been simultaneously solved by the Ramage–Wathen procedure. This is an Uzawa algorithm. The energy equation has been solved using an iterative multidimensional positive-definite advection-transport algorithm with explicit time steps (Bunge and Baumgardner, 1995). In the Ramage–Wathen procedure, the corresponding equation systems are solved by a multigrid procedure and a Jacobi smoother. In the multigrid procedure, prolongation and restriction are matrix-dependently executed. Only in this way, it was possible to handle the strong variations and jumps of the coefficients that mimic the strong viscosity gradients (Yang, 1997). Radial and horizontal lines are used in the Jacobi smoother. For the formulation of chemical differentiation, we used a tracer modul developed by Dave Stegman. This modul contains a second-order Runge–Kutta procedure to move the tracers in the creeping-velocity field. Each tracer conveys, e.g., the abundances of the radionuclides. In this case, it is an active tracer since

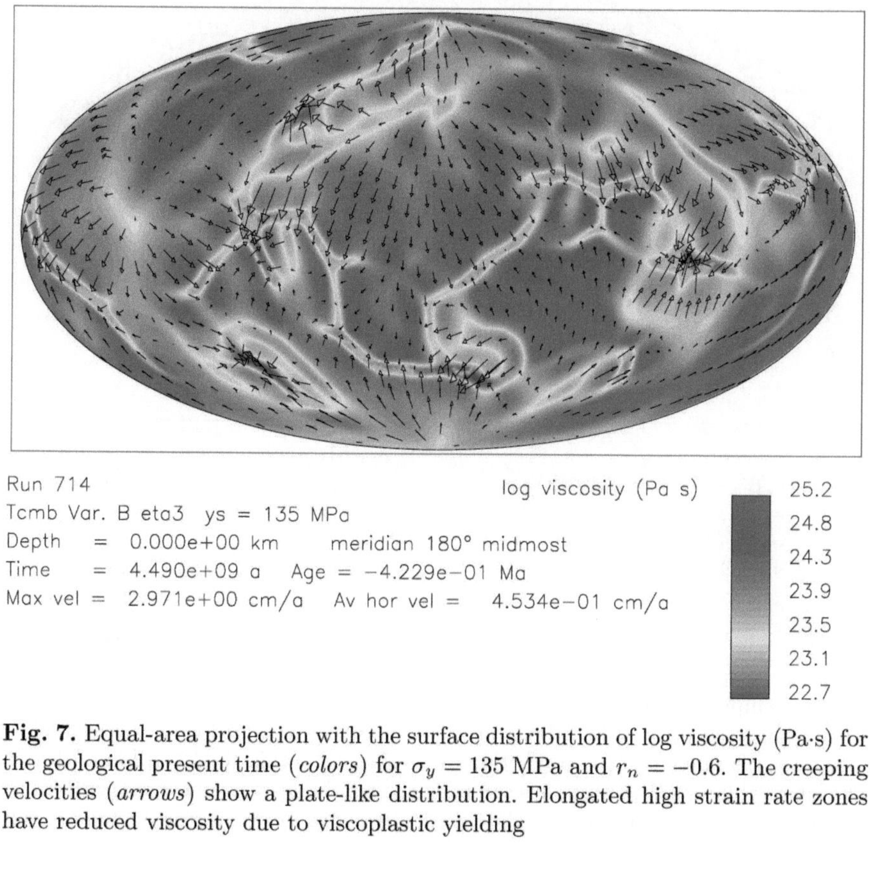

Fig. 7. Equal-area projection with the surface distribution of log viscosity (Pa·s) for the geological present time (*colors*) for $\sigma_y = 135$ MPa and $r_n = -0.6$. The creeping velocities (*arrows*) show a plate-like distribution. Elongated high strain rate zones have reduced viscosity due to viscoplastic yielding

the multitude of tracers determines the heat production rate per unit volume dependent on time and location vector.

The FORTRAN code is parallelized by domain decomposition and explicit message passing (MPI)(Bunge, 1996).

4 The Developing Stages of the Present Thermal and Chemical Evolution Model of the Earth's Mantle

Our latest papers, [25], [26], second part of [27], [28], [29], use a viscosity profile of the mantle that is derived from the seismic model PREM (Dziewonski and Anderson, 1981) and solid-state physics. The latter term denotes not only theoretical-physics papers but also the experimental results of the Karato group. Meanwhile, the steep viscosity gradients at the upper and lower surface of the transition zone could be handled. Formerly [24], it was necessary

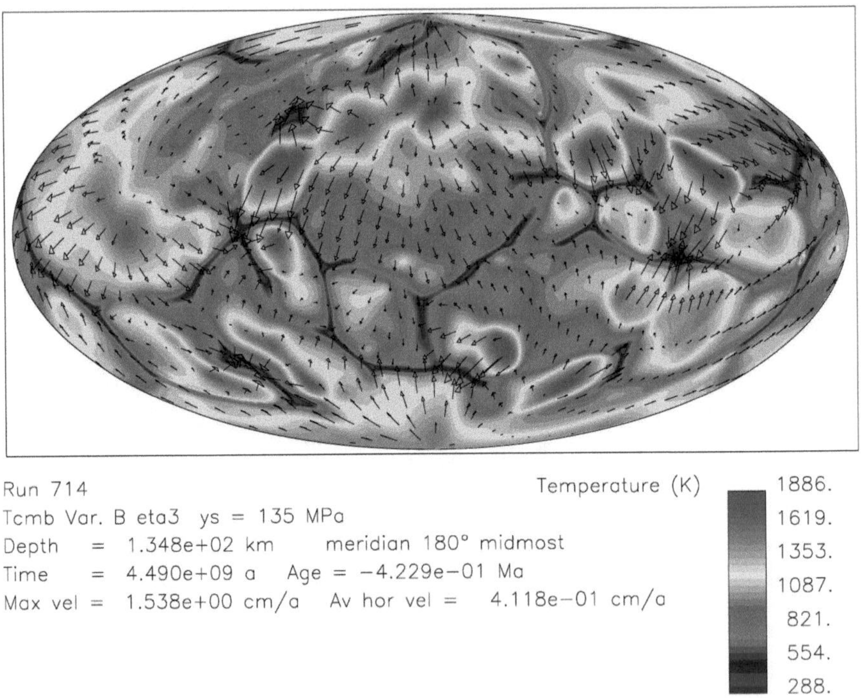

Fig. 8. Temperature distribution (*colors*) and creeping velocities (*arrows*) for a reference run with $\sigma_y = 135$ MPa and $r_n = -0.6$, for the geological present. The shown area is an equal-area projection of a spherical surface in 134.8 km of depth. The narrow blue subducting zones can be found also in deeper equal-area projections of temperature distribution. The slab-like features are rather narrow in comparison with the broad upwelling zones

to slope the viscosity jumps stronger. In [24], a constant CMB heat flow of 28.9 mW/m^2 was assumed using Anderson (1998). Regarding the constancy with respect to time, this assumption was founded on Stevenson et al. (1983), Stacey (1992) and Schubert et al. (2001). In [25], [26], second part of [27], [28], [29], however, it is assumed that the CMB is *laterally* isothermal. A core-cooling model was implemented: the CMB temperature was determined after each time step, in dependence from the computed CMB heat flow (Steinbach et al., 1993; Honda and Iwase, 1996). Now, both, CMB temperature and CMB heat flow, are dependent on time. As a third innovation, the Newtonian rheology has been supplemented by a *viscoplastic yield stress,* σ_y. We made a variation of the parameters and found *stable plate-like solutions on a sphere* for a central σ_y–$Ra(2)$ region where $Ra(2)$ is the temporal average of the Rayleigh number, averaged over the last 2000 Ma [25], [26]. This model, S2, had, however, only oceanic lithospheric plates but no continents. We used exclusively models of whole-mantle convection with two interior phase boundaries and si-

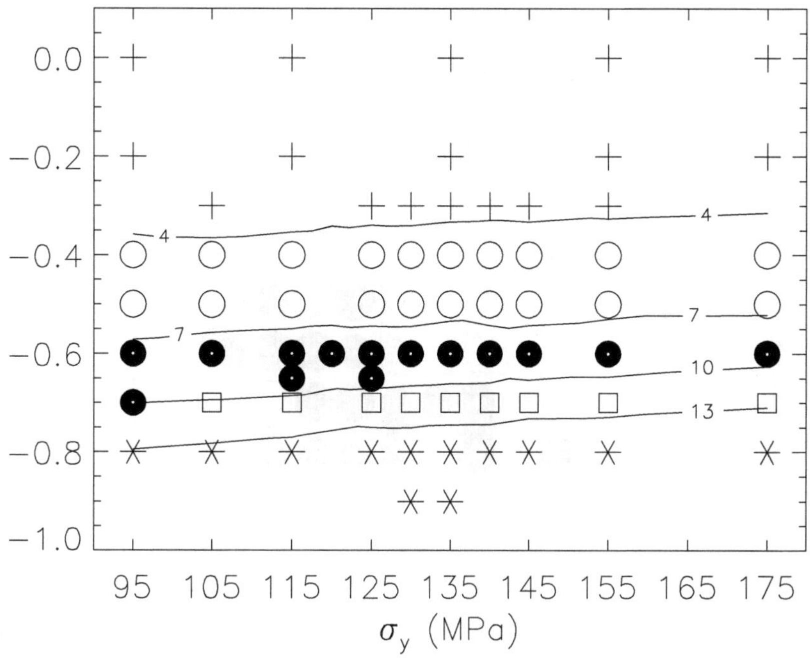

Fig. 9. The time average of the Rayleigh number, *Ra*, as a function of yield stress, σ_y, and viscosity-level parameter, r_n. Each sign represents one run. *Asterisks* stand for runs with $13 \times 10^6 \leq Ra$. *White squares* represent runs with $10 \times 10^6 \leq Ra < 13 \times 10^6$. *Little black disks with a white center* typify runs with $7 \times 10^6 \leq Ra < 10 \times 10^6$. *White circles* denote runs with $4 \times 10^6 \leq Ra < 7 \times 10^6$. Finally, *plus signs* stand for runs with $Ra < 4 \times 10^6$

multaneously the long-term thermal evolution from the very beginning of the existence of the solid silicate mantle. The modelled mantle was compressible with variable viscosity and time-dependent heating from within plus a slight heating from below. The dynamical results using our own mantle viscosity profile have been compared with our dynamical results using viscosity profiles of Kaufmann and Lambeck (2002) and King and Masters (1992) [26].

In [24], another viscosity profile with two low-viscosity layers and a *purely viscous* constitutive equation is applied. This assumptions lead to reticularly connected very thin plate-like downwellings that dip perpendicularly into the mantle. It is no surprise that there are neither plates at the surface nor transform faults. If $Nu_{(2)}$ stands for the temporal average of the Nusselt number, averaged over the last 2000 Ma, and $Ra_{(2)}$ is the corresponding mean for the Rayleigh number then we find $Nu_{(2)} = 0.120 Ra_{(2)}^{0.295}$ for a wide parameter range.

In [28] that is based on [25] and [26], the growing values of the viscosity profile due to cooling of the Earth are taken into account. Therefore, the laterally averaged surface heat flow as a function of time, the Urey and Rayleigh

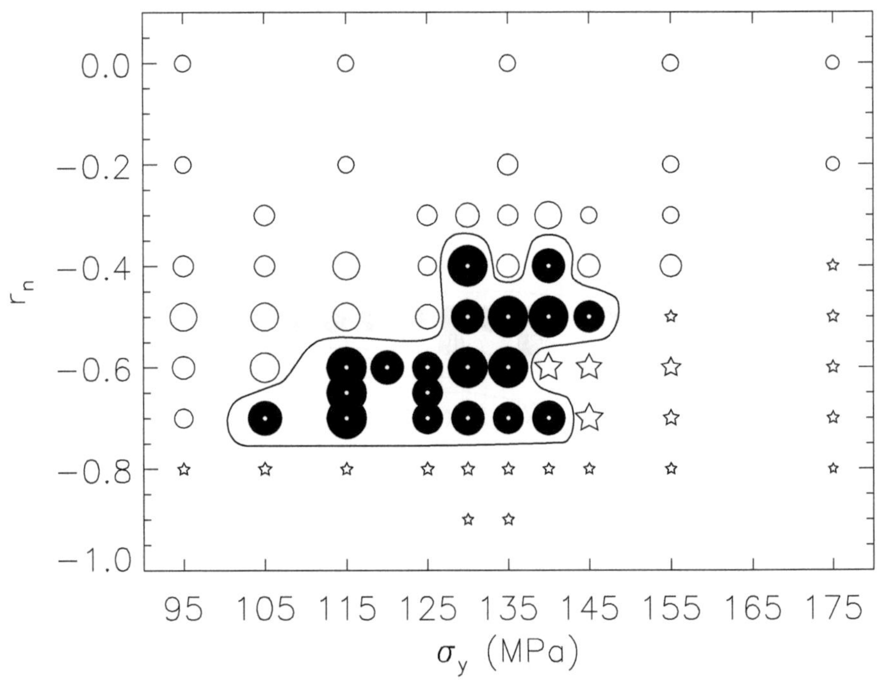

Fig. 10. Type of continental distribution in the σ_y–r_n plot. *Little black disks with a white center* stand for Earth-like distributions where the size of the disk is a measure of the quality. *Five-pointed stars* represent an unrealistic multitude of tiny continents. *White circles* depict runs with reticularly connected, narrow stripe-like continents

numbers and the volume-averaged temperature as functions of time correspond in the general features with parameterized cooling models. Contrary to the parameterized curves, the curves of [28] show temporal variations. This is more realistic for geological reasons. A well-developed plateness is stable and shows realistic configurations at the surface. The stability of the Earth-like distributed, very thin downwellings is influenced by the pressure dependence of the viscosity connected with the existence of two low-viscosity layers: One layer is the conventional asthenosphere just below the lithosphere. The other one is situated immediately beneath the 660-km phase boundary but the central part of the lower mantle is highly viscous. The latter feature guarantees the relative lateral stability of those plumes which came from the CMB.

Walzer et al. (2006) present an integrated theory, S3, of mantle convection plus chemical differentiation of the growing continental crust and the complementary depleted MORB mantle. Plateness and realistic cooling features of the former step [28] of the model are conserved. The mathematical tools of the

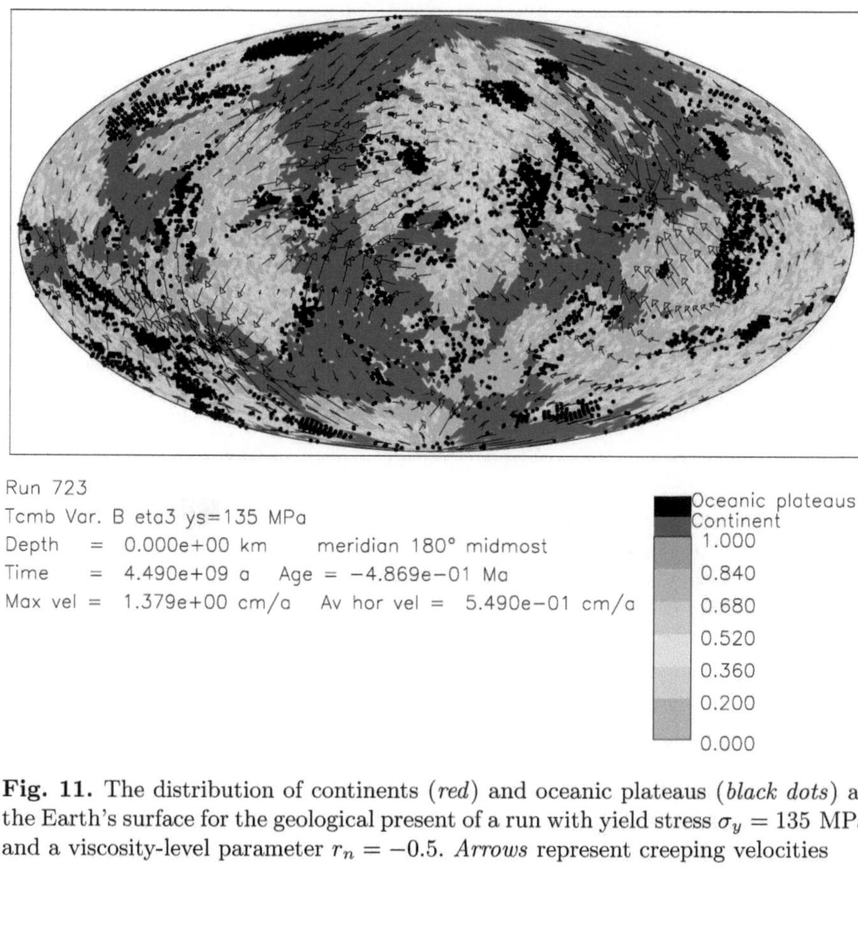

Fig. 11. The distribution of continents (*red*) and oceanic plateaus (*black dots*) at the Earth's surface for the geological present of a run with yield stress $\sigma_y = 135$ MPa and a viscosity-level parameter $r_n = -0.5$. *Arrows* represent creeping velocities

theory are given in [29]. The complicated geochemical features are discussed there. Some particular items of the chemistry are indicated in Chap. 1 of the present paper.

5 Results and Discussions

5.1 Thermal and Chemical Evolution using a Reference Run

In this paper, only such numerical results of the new model, S3, are presented that are *not* given in Walzer et al. (2006). To come to the most important item first, the results of this subsection apply also to other runs that are situated in a certain central area of the r_n–σ_y plot. The selected reference run is determined by a yield stress $\sigma_y = 135$ MPa and a viscosity-level parameter $r_n = -0.6$. Now we consecutively present the figures. The corresponding

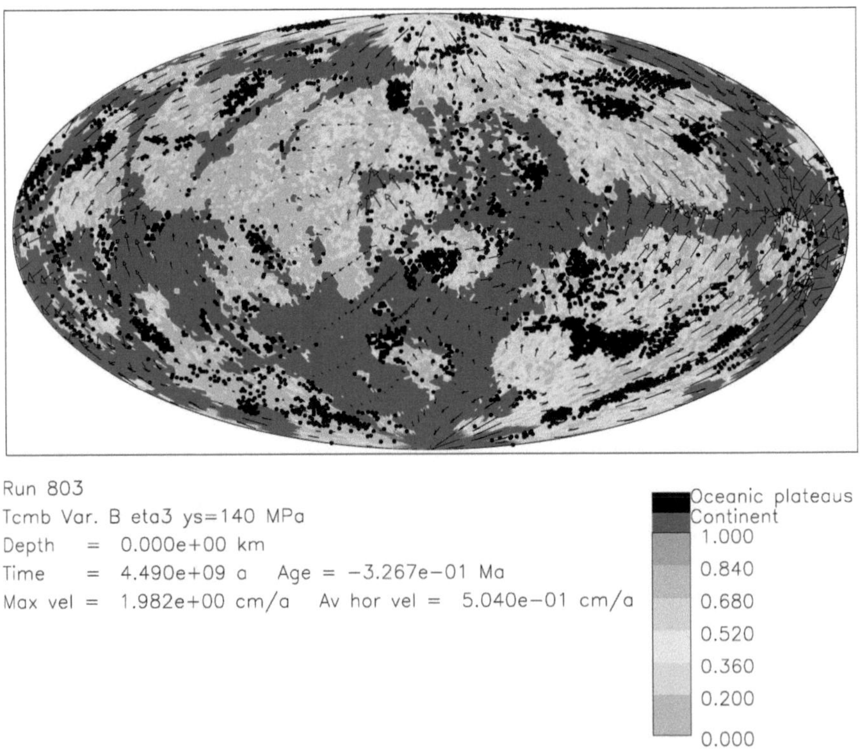

Fig. 12. The distribution of continents (*red*) and oceanic plateaus (*black dots*) at the Earth's surface for the geological present of a run with yield stress $\sigma_y = 140$ MPa and a viscosity-level parameter $r_n = -0.5$. *Arrows* denote creeping velocities

discussion is immediately after each presentation. In Fig. 1, the laterally averaged temperature is plotted as a function of depth for the geological present time. Its curve is nearer to the parameterized geotherm of whole-mantle convection than to the corresponding temperature curve of layered convection. This is quite understandable since we observe whole-mantle convection in the results of S3. However, the flow is somewhat hindered by the highly viscous transition zone and by the 660-km phase boundary. Therefore the temperature is somewhat elevated but not to such a degree that would be expected for a purely thermal coupling at the 660-km boundary.

Figure 2 shows the laterally averaged viscosity for the present. The derivation of this type of viscosity profile is given in [26]. We will discuss the profile from below to above: The low viscosity in the D″ layer is determined by the strong temperature gradient at CMB. Therefore the temperature influence on viscosity is larger than the pressure influence in this special zone. The middle part of the lower mantle is highly viscous due to the gradual pres-

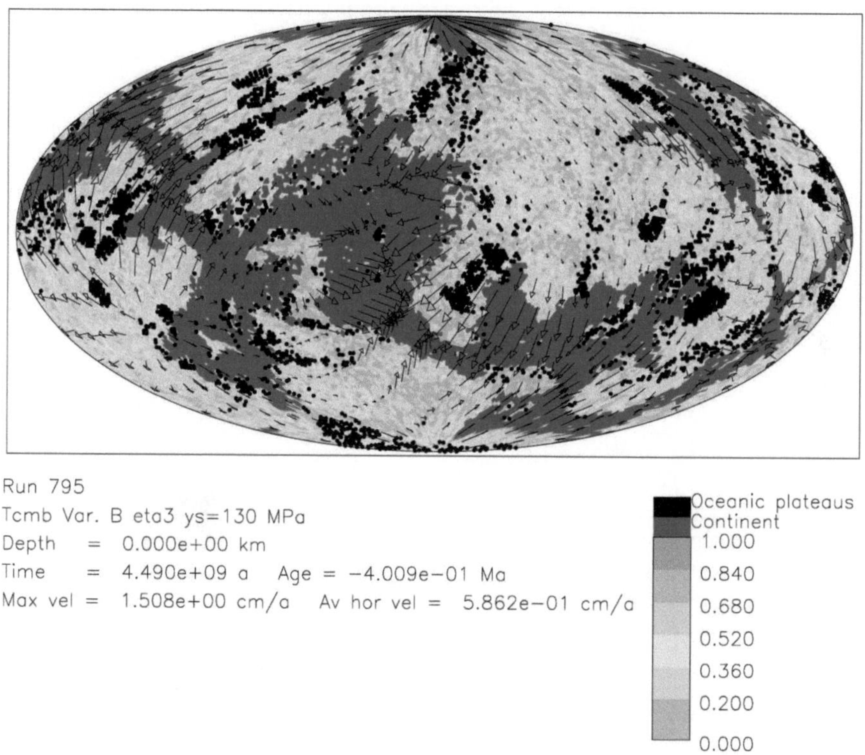

Fig. 13. The distribution of continents (*red*) and oceanic plateaus (*black dots*) at the Earth's surface for the geological present of a run with yield stress $\sigma_y = 130$ MPa and a viscosity-level parameter $r_n = -0.4$. *Arrows* represent creeping velocities

sure increase. This causes two effects: *Firstly*, CMB-based plumes use always the same pipe if they have eaten through to the surface. It is energetically more advantageous to use always the same duct instead to burn a new hole in this high-viscosity layer. This is a plausible explanation for the slow lateral movement of the plumes. They move orders of magnitude slower than the lithospheric plates. That's why the plumes can be used as a reference frame for plate motions. Of course, the sluggish lower-mantle bulk convection moves also the tubes of the plume. *Secondly*, chemical inhomogeneity is relatively well conserved in this central part of the lower mantle. It is unimportant for S3 whether this inhomogeneity is primordial or whether it is created by subduction. But if we look for a source region of ^{3}He, ^{20}Ne and ^{22}Ne, than it could be that not only the remains of old and new subduction slabs are conserved in this middle layer of the lower mantle but also parts of a quasi-primordial mantle. The above-said isotopes are not produced in the Earth. On the other hand, it is not very convincing that the mentioned *light* noble gases came to

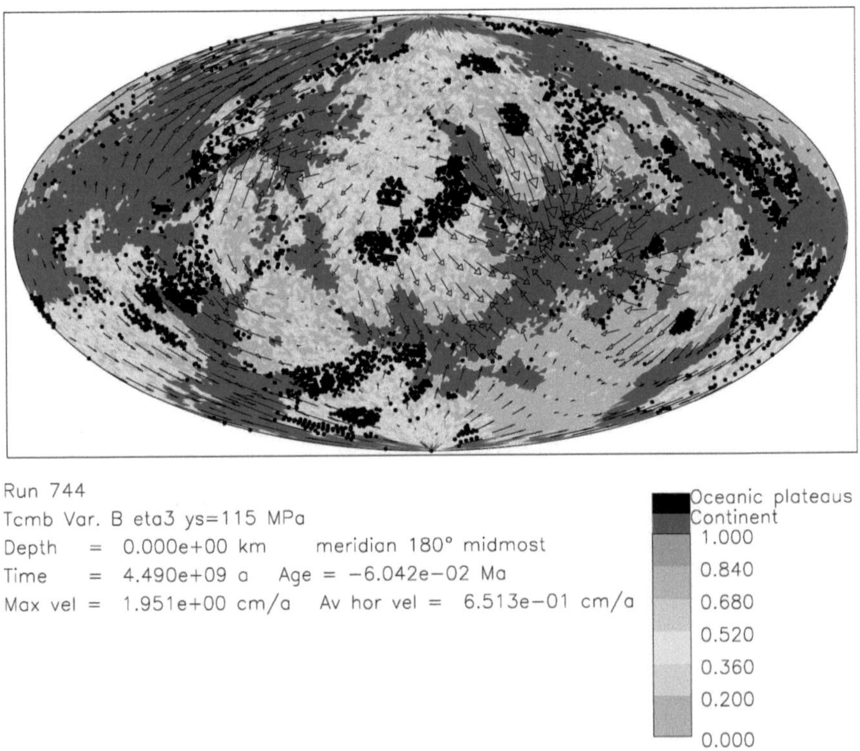

Fig. 14. The distribution of continents (*red*) and oceanic plateaus (*black dots*) at the Earth's surface for the geological present of a run with yield stress $\sigma_y = 115$ MPa and a viscosity-level parameter $r_n = -0.7$. *Arrows* denote creeping velocities

the interior by subduction. There are proposals that these nuclides could be primordially stored in the Earth's core. This doesn't seem to be very plausible since the metallic fluid of the outer core moves with 10–30 km/a. So, each part of this iron alloy was very often in contact with the CMB during the Earth's evolution. Therefore we conclude that if the mantle is not able to store the noble gases than the core by no means whatever. However, this discussion is not relevant for the calculations of S3. Figure 2 indicates a high-viscosity transition zone. It could be that this is true only for a layer between 520 and 660 km depth. The augmentation of the observed seismicity in the transition zone could point to high viscosity. The two low-viscosity layers above and beneath the transition layer cause a strong stirring of the material there. This is a good explanation for the tendency to homogeneity of DMM. Near the surface, Fig. 2 displays a highly viscous lithosphere. Runs with viscosity profiles of other structure show that the two low-viscosity layers below the lithosphere

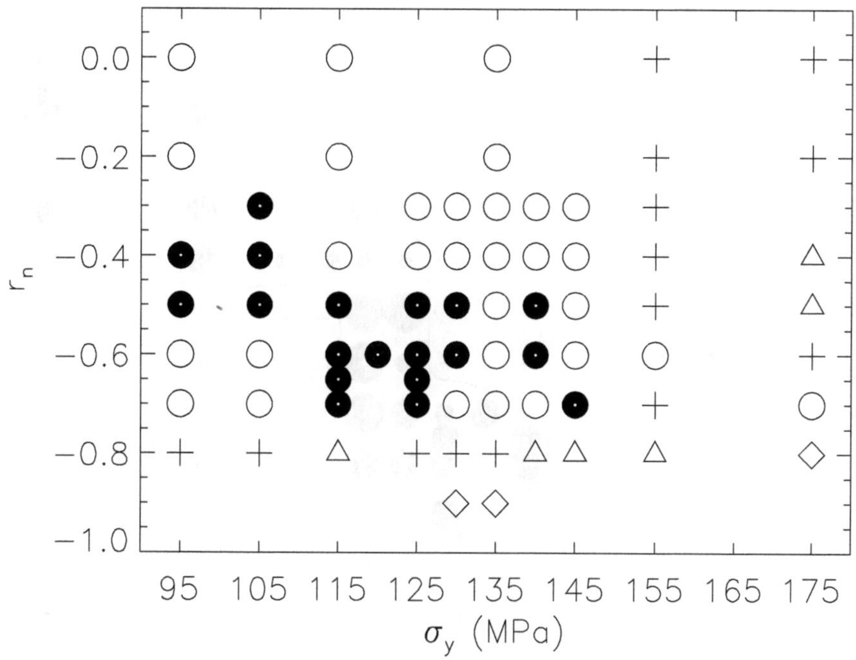

Fig. 15. A classification of the runs using the deviation of the calculated size of present-day continents from the observed size. The quantity d_c is the observed surface percentage of the continents (= 40.35%) minus the computed surface percentage of the continents. *Little black disks with a white center* stand for slight deviations, namely $-4.5 \leq d_c < 4.5$ percent, *white circles* for $4.5 \leq d_c < 13.5$, plus signs represent $13.5 \leq d_c < 22.5$. *White triangles* denote runs with $22.5 \leq d_c < 31.5$, *white diamonds* stand for $31.5 \leq d_c < 40.5$ The quantity d_c of the runs is shown as a function of yield stress, σ_y, and of viscosity-level parameter, r_n

and below the 660-km discontinuity are essential for the generation of *very* thin cold sheet-like downwellings.

Figure 3 represents the temporal dependence of some relevant quantities of the reference run: The first panel exhibits the evolution of laterally averaged heat flow at the Earth's surface. For the present time, the curve arrives at a realistic value: The observed mean global heat flow is 87 mW/m^2 (Pollak et al., 1993). The second panel shows the growth of the total continental mass as a function of time. The behaviour of the model corresponds to the observation that the mantle grew in batches in the geological history (Condie, 2003). The third panel of Fig. 3 displays the temporal dependence of *Ror*. The quantity *Ror* is the reciprocal value of the Urey number. The curve is roughly similar to that of corresponding parameterized models except the smaller variations, of course. The fourth panel of Fig. 3 represents the decrease of the Rayleigh number as a function of time.

Figure 4 shows similar evolution diagrams: The first panel exhibits the subsiding of the mean temperature. From an age of $\tau = 3000$ Ma, it decreases

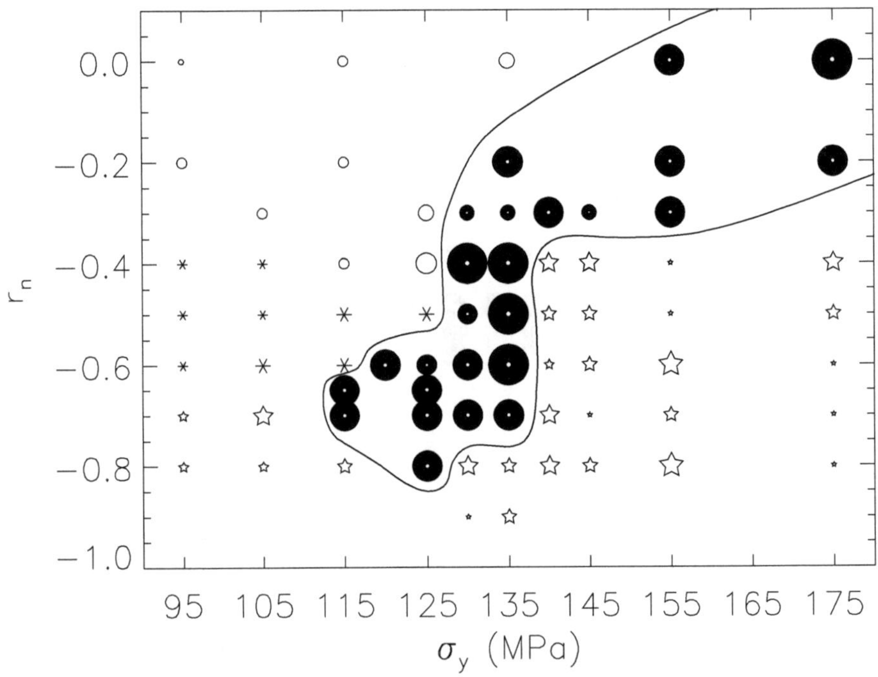

Fig. 16. Plateness in the σ_y–r_n plot. Plate-like solutions with narrow subducting zones are represented by *little black disks with a white center*. Its surface area is a measure of plateness. *White circles* stand for runs with broad downwellings and minor plateness. *White five-pointed stars* denote unrealistic runs with local subduction only. *Asterisks* represent runs with a rather complex behavior with lots of small, but not narrow downwellings

until the present time with a gradient of 250 K per 3000 Ma. This is somewhat less than our expectation: From komatiite research we expect 300 K decrease per 3000 Ma. The second panel of Fig. 4 shows the time dependence of the kinetic energy of the upper-mantle flow, *EkinUM*. It is a measure of the power transmission to the laterally moving oceanic lithospheric plates. The third panel of Fig. 4 demonstrates the decrease of the total radiogenic heat production of the mantle.

Figure 5 displays the distribution of chemical heterogeneity at the end of the chemical evolution of the mantle in a meridional section. There are no pure unblended reservoirs. However, DMM prevails immediately beneath the continents (red) and below the oceanic lithosphere. This is a realistic feature of the model since, where the real lithosphere is ripped, MORB magma flows out if there is no plume in the surrounding. The MORB source (DMM) is depleted and relatively homogenized. This is demonstrated by the low standard deviation of the isotope ratios and of chemical quantities (Allègre and Levin, 1995)

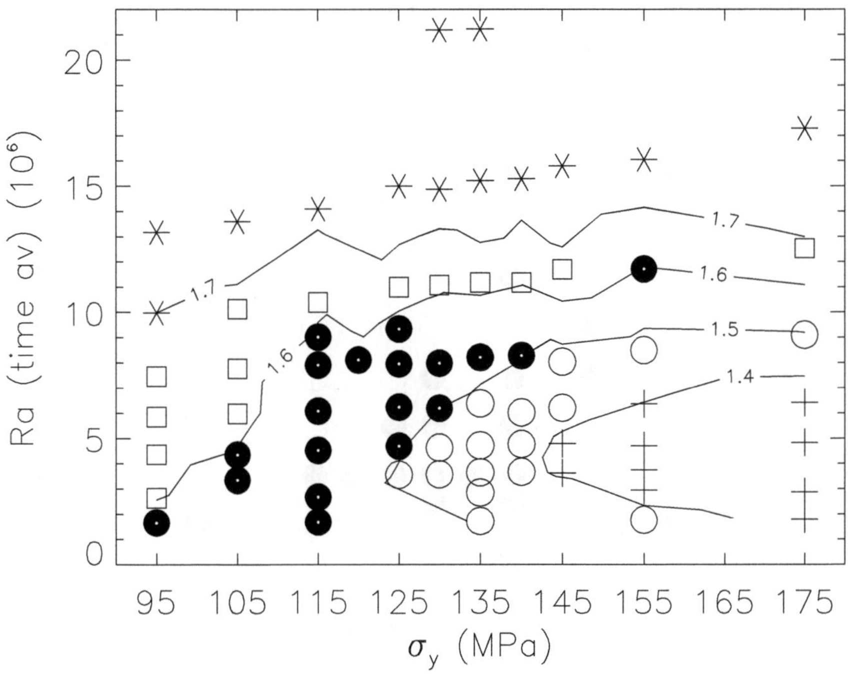

Fig. 17. The time average of the *r*atio of the surface heat *o*utflow per unit time to the mantle's *r*adiogenic heat production per unit time, *Ror*, as a function of the time average of the Rayleigh number, *Ra*, and the yield stress, σ_y. *Asterisks* denote $1.7 \leq Ror$. *White squares* stand for $1.6 \leq Ror < 1.7$. *Little black disks with a white center* represent the realistic range of $1.5 \leq Ror < 1.6$. *White circles* depict runs with $1.4 \leq Ror < 1.5$. Finally, plus signs stand for $Ror < 1.4$

although Hofmann (2004) modifies this statement. Figure 5 shows a marble-cake mantle but reversed than up to now proposed: The *depleted* mantle parts are disconnected and distributed like raisins. It is an important result that we did not obtain simply connected areas for one geochemical reservoir. However, the depleted areas tend to be in the upper parts of the mantle. This is not amazing since they develop below the lithosphere, and the conventional asthenosphere promotes the lateral propagation of DMM because of its low viscosity.

Figure 6 shows the distribution of continents (red) and of oceanic plateaus (black dots) at the Earth's surface for the geological present time. The moving oceanic lithosphere carries along the oceanic plateaus. If the plateaus touch a continent they will be connected with the continent. This is the only additional implementation. Else, there are no charges or special rules. Neither number nor form nor size of the continents are prescribed. The configuration simply results from the numerical solution of the system of equations and the corresponding initial and boundary conditions. At first, the comparison with the observed continents was done simply visually, then by a development into spherical harmonics using a function of the coefficients, A_n^m and B_n^m, that does not depend on the location of the pole of the coordinate system. We are not

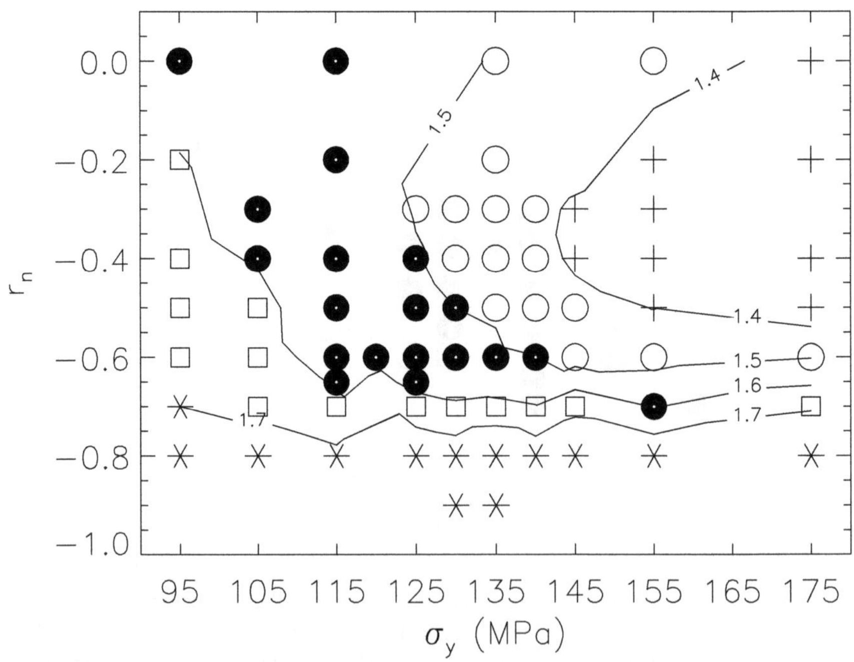

Fig. 18. The magnitude *Ror* as a function of the viscosity-level parameter, r_n, and the yield stress, σ_y. For the description of the symbols, see Fig. 17.

aware of former papers on spherical-shell mantle convection with continents that evolve due to physical laws and which are *not* simply put on the surface.

Figure7 exhibits a plate-like motion of the lithospheric pieces at the surface. The colors denote the logarithm of the shear viscosity in Pa·s. Figure 8 displays the present-time temperature on an equal area projection of an interior spherical surface in 134.8 km of depth. The blue lines depict downwelling zones beneath collision lines at surface.

5.2 Variation of Parameters: The Generation of Continents

A multitude of runs were systematically investigated in order to point out that the selected reference run is by no means an exceptional case but it is typical for a certain area of parameters. Furthermore, it is to be demonstrated that the systematic part of the results is not too far from the real Earth. Figure 9 shows a relatively simple result: The parameter r_n serves for a (single) shift of the starting profile of the viscosity at the beginning of a run. As expected,

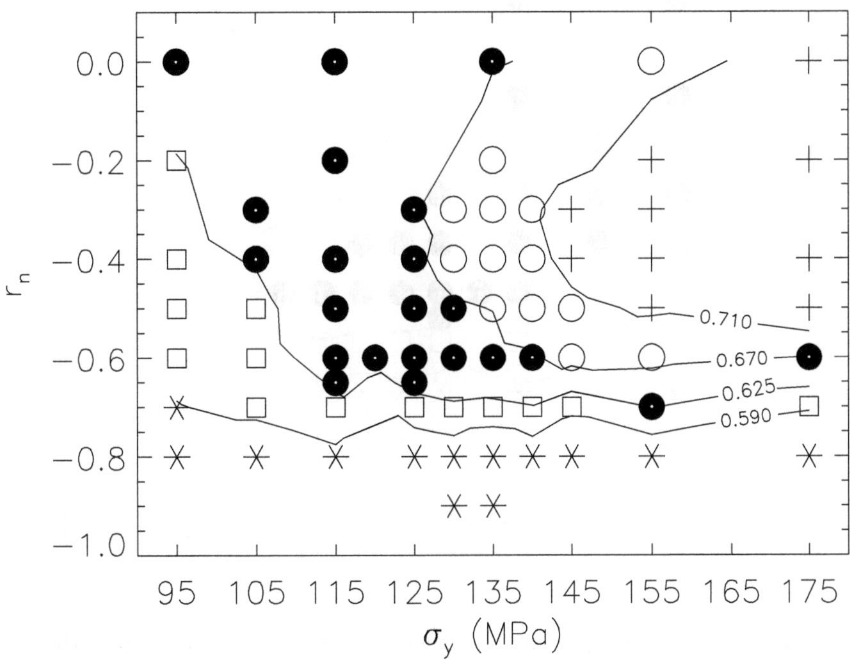

Fig. 19. The time average of the Urey number, Ur, as a function of the viscosity-level parameter, r_n, and the yield stress, σ_y. *Asterisks* stand for runs with $Ur \leq 0.59$. *White squares* denote $0.59 < Ur \leq 0.625$. *Little black disks with a white center* represent runs with $0.625 < Ur \leq 0.67$. *White circles* signify runs with $0.67 < Ur \leq 0.71$. *Plus signs* depict runs with $0.71 < Ur$

Fig. 9 demonstrates that the quantity Ra grows with decreasing r_n because of Eq. (1). There is , however, also a *small* influence of the yield stress.

Size and form of the computed continents have been systematically compared with the observed continents. Figure 10 exhibits Earth-like continent solutions in the center of an r_n-σ_y plot. These realistic runs are signified by black disks with a white center. A Ra-σ_y plot of the continent quality resembles Fig. 10 since Ra monotonously decreases with increasing r_n. So, the new figure stands on its head. Figures 11 through 14 show final solutions of the distribution of continents for other input parameters r_n and σ_y, different from those of Fig. 6. Further investigations have been done to restrict the realistic r_n-σ_y area. Figure 15 displays the distribution of differences, namely observed final continental area minus calculated final continental area. Evidently, a run is only realistic if it has a black disk in Fig. 10 as well as in Fig. 15. But also in this case, a certain middle r_n-σ_y area remains for the favourable solutions.

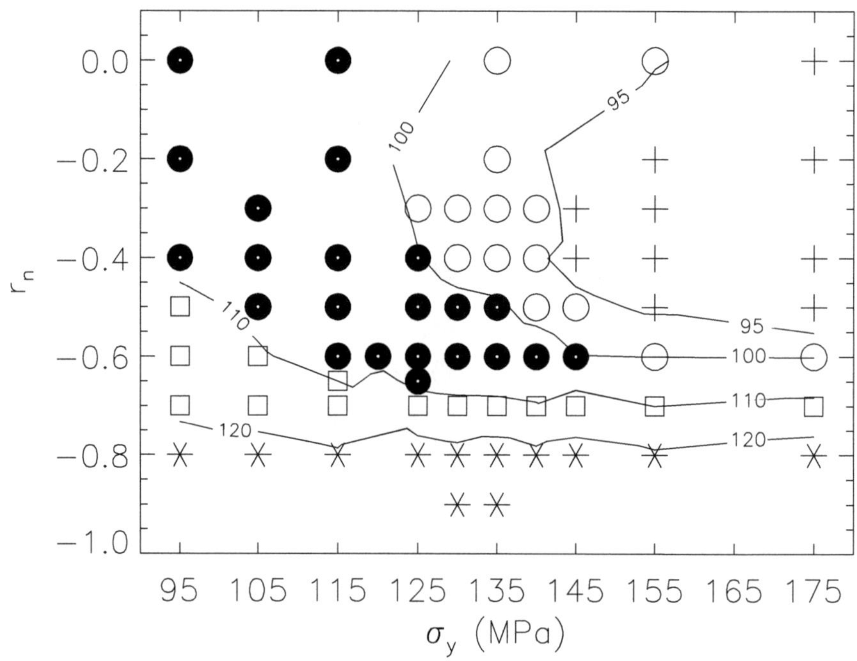

Fig. 20. The *symbols* represent the classes of the temporal average of the surface average of the heat flow, *qob*, in a r_n–σ_y plot. The following numbers are given in mW/m^2. *Asterisks* depict runs with $120 \leq qob$. *White squares* stand for $110 \leq qob < 120$. *Little black disks with a white center* denote runs with $100 \leq qob < 110$, *white circles* are for $95 \leq qob < 100$. *Plus signs* depics runs with $qob < 95$

5.3 Variation of the Parameters: Generation of Plate-Like Solutions and the Distribution of Thermal Parameters

Figure 16 shows a classification of runs regarding the planforms of flows near the surface. Plate-like solutions are represented by black disks. It is indicative of Model S3 that, again, an overlap of the black disks is observed with those of Figs. 10 and 15. Figure 17 represents the distribution of the time average of the ratio of the heat outflow at the surface to the radiogenic heat generation in the interior, *Ror*, as a function of the time average of the Rayleigh number, *Ra*, and of the yield stress, σ_y. Runs with realistic values of the ratio *Ror* are depicted by black disks. Figure 18 shows an r_n-σ_y plot of the ratio *Ror*. The Urey number is represented by Fig. 19 in an r_n-σ_y diagram. Finally, the time average of mean surface heat flow for each run are presented in an r_n-σ_y plot. The realistic values are again signified by black disks. We observe a partial covering with the favourable field of continent distribution of Fig. 10. – The conclusions can be found in the Summary.

Acknowledgements

We gratefully acknowledge the supply of computing time by the Höchstleistungsrechenzentrum Stuttgart (HLRS) and by the NIC Jülich.

References

1. Allègre, C.J., and Levin, E. (1995), Isotopic systems and stirring times of the Earth's mantle, *Earth Planet. Sci. Lett.*, *136*, 629–646.
2. Anderson, O.L. (1998), The Grüneisen parameter for iron at outer core conditions and the resulting conductive heat and power in the core, *Phys. Earth Planet. Int.*, *109*, 179–197.
3. Bennett, V.C. (2004), Compositional evolution of the mantle, in *Treatise on Geochemistry*, Vol.2, *The Mantle and the Core*, edited by R.W. Carlson, 493–519, Elsevier, Amsterdam.
4. Bunge, H.-P. (1996), Global mantle convection models, PhD Thesis, University of California, Berkeley.
5. Bunge, H.-P., and Baumgardner, J.R. (1995), Mantle convection modelling on parallel virtual machines, *Computers in physics*, *9*, no. 2, 207–215.
6. Condie, K.C. (2003), Incompatible element ratio in oceanic basalts and komatiites: tracking deep mantle sources and continental growth rates with time, *Geochem. Geophys. Geosyst.*, *4 (1)*, 1005, doi:10.1029/2002GC000333.
7. Dziewonski, A.M., and Anderson, D.L. (1981), Preliminary reference Earth model. *Phys. Earth Planet. Int.*, *25*, 297–356.
8. Fitton, G., Mahoney, J., Wallace, P., and Saunders, A. (eds.) (2004), *Origin and Evolution of the Ontong Java Plateau*, *Geol. Soc. London Spec. Publ.*, *229*.
9. Hofmann, A.W. (2004), Sampling mantle heterogeneity through oceanic basalts: Isotopes and trace elements, in *Treatise an Geochemistry*, Vol. 2, *The Mantle and the Core*, edited by R.W. Carlson, 61–101, Elsevier, Amsterdam

10. Honda, S. and Iwase, Y. (1996), Comparison of the dynamic and parameterized models of mantle convection including core cooling. *Earth Planet. Sci. Lett.*, *139*, 133–145.
11. Kaufmann, G., and Lambeck, K. (2002), Glacial isostatic adjustment and the radial viscosity profile from inverse modeling. *J. Geophys. Res.*, *107*, (B11), 2280,doi:10.1029/2001JB000941.
12. King, S.D., and Masters, G. (1992), An inversion for radial viscosity structure using seismic tomography, Geophys. Res. Lett., *19*, 1551–1554.
13. Norman, M., Borg, L., Nyquist, L. and Bogard, D. (2003), Chonology, geochemistry, and petrology of a ferroan noritic anorthosite clast from Descartes breccia 67215: clues to the age, origin, structure, and impact history of the lunar crust, *Meteorit. Planet. Sci.*, *38*, 645–661.
14. Nyquist, L.E., Bogard, D.D., Shih, C.-Y., Greshake, A., Stoffler, D., and Eugster, O. (2001), Ages and geological histories of martian meteorites, *Space Sci. Rev.*, *96*, 105–164.
15. Pollak, H.N., Hurter, S.J., and Johnson, J.R. (1993), Heat flow from the Earth's interior: analysis of the global data set, *Rev. Geophys.*, *31*, 267–280.
16. Porcelli, D., and Ballentine, C.J. (2002), Models for the distribution of terrestrial noble gases and evolution of the atmosphere, in *Noble Gases in Geochemistry and Cosmochemistry*, edited by D. Porcelli, C.J. Ballentine, and R. Wieler, *Rev. Min. Geochem.*, *47*, 411–480.
17. Rudnick, R.L., and Gao, S. (2004), (eds.) *Treatise on Geochemistry*, Vol. 3, Elsevier, Amsterdam.
18. Schubert, G., Turcotte, D.L., and Olson, P. (2001), *Mantle Convection in the Earth and Planets*, 940 pp., Cambridge Univ. Press, Cambridge, UK.
19. Stacey, F.D., (1992), *Physics of the Earth*, 3rd edn., Brookfield Press, Brisbane, 513 pp.
20. Steinbach, V., Yuen, D.A. and Zhao, W.L. (1993) Instabilities from phase transitions and the timescales of mantle thermal evolution, *Geophys. Res. Lett.*, *20*, 1119–1122.
21. Stevenson, D.J., Spohn, T., and Schubert, G. (1983), Magnetism and thermal evolution of the terrestrial planets. *Icarus*, *54*, 466–489.
22. Stracke, A., Hofmann, A.W., and Hart, S.R. (2005), FOZO, HIMU, and the rest of the mantle zoo, *Gochem. Geophys. Geosyst.*, *6 (5)*, Q05007, doi:10.1029/2004GC000824.
23. Walzer, U., Hendel, R., Baumgardner, J. (2003a), Variation of non-dimensional numbers and a thermal evolution model of the Earth's mantle. In: Krause, E., Jäger, W.(Eds.), High Performance Computing in Science and Engineering '02. Springer-Verlag, Berlin Heidelberg New York. pp. 89–103. ISBN 3-540-43860-2.
24. Walzer, U., Hendel, R., Baumgardner, J. (2003b), Viscosity stratification and a 3D compressible spherical shell model of mantle evolution. In: Krause, E., Jäger, W., Resch, M., (Eds.), High Performance Computing in Science and Engineering '03. Springer-Verlag, Berlin Heidelberg New York. pp. 27–67. ISBN 3-540-40850-9. Model S1
25. Walzer, U., Hendel, R., Baumgardner, J. (2003c), Generation of plate-tectonic behavior and a new viscosity profile of the Earth's mantle. In Wolf, D.,Münster, G.,Kremer, M. (Eds.), NIC Symposium 2004. NIC Series 20, pp. 419–428., ISBN 3-00-012372-5.

26. Walzer, U., Hendel, R., Baumgardner, J. (2004a), The effects of a variation of the radial viscosity profile on mantle evolution. *Tectonophysics*, Volume 384, Issues 1–4, 55–90. Model S2
27. Walzer, U., Hendel, R., Baumgardner, J. (2004b), Toward a thermochemical model of the evolution of the Earth's mantle. In: Krause, E., Jäger, W., Resch, M., (Eds.), High Performance Computing in Science and Engineering '04. Springer-Verlag, Berlin Heidelberg New York. pp 395–454. ISBN 3-540-22943-4.
28. Walzer, U., Hendel, R., Baumgardner, J. (2005), Plateness of the oceanic lithosphere and the thermal evolution of the Earth's mantle. In: Krause, E., Jäger, W., Resch, M., (Eds.), High Performance Computing in Science and Engineering '05. Springer-Verlag, Berlin Heidelberg New York. pp. 289–304. ISBN-10-540-28377-3.
29. Walzer, U., Hendel, R., Baumgardner, J. (2006), An integrated theory of whole-mantle convection, continent generation, and preservation of geochemical heterogenity, *J. Geophys. Res.*, submitted.
30. Yamazaki, D., and Karato, S.-I. (2001), Some mineral physics constraints on the rheology and geothermal structure of the Earth's lower mantle, *American Mineralogist*, *86*, 385–391.
31. Yang, W.-S. (1997), Variable viscosity thermal convection at infinite Prandtl number in a thick spherical shell. *Thesis*, Univ. of Illinois, Urbana-Champaign, 185 pp.

Efficient Satellite Based Geopotential Recovery

O. Baur, G. Austen, and W. Keller

Universität Stuttgart, Institute of Geodesy, Geschwister-Scholl-Str.24D,
D-70174 Stuttgart,
baur@gis.uni-stuttgart.de

Abstract. This contribution aims at directing the attention towards the main inverse problem of geodesy, i.e. the recovery of the geopotential. At present, geodesy is in the favorable situation that dedicated satellite missions for gravity field recovery are already operational, providing globally distributed and high-resolution datasets to perform this task. Due to the immense amount of data and the ever-growing interest in more detailed models of the Earth's static and time-variable gravity field to meet the current requirements of geoscientific research, new fast and efficient solution algorithms for successful geopotential recovery are required.

1 Introduction

The task of geodesy or more precisely of physical geodesy can be summarized as follows: determination of the physical shape of Earth, the so-called geoid, and the external terrestrial gravity field. In the following we refer hereunto as geopotential recovery. Before the availability of modern observational methods such as airborne or space-borne techniques this task has been completed through approaches making use of ground gravity observations. Traditionally, these data had to be observed pointwise in field campaigns. In fact, conducting measurements manually resulted into an amount of data well manageable. On the other hand, since the data was far from being continuous or globally distributed, it was only possible to achieve a rather unsatisfying geopotential solution. Remedy was found October 1957 with the launch of Sputnik I, the world's first artificial satellite. Shortly after, already a number of other spacecraft were orbiting around the Earth, influenced primarily by the Earth's mass attraction. Observing the satellite orbits from ground tracking stations provided semi-continuous and more globally distributed data leading to an improved geopotential modeling. This progress involved an increase in the amount of observations and the computational demand. With the advent of the Global Positioning System (GPS) and the advances in receiver technology in the 1990s, continuous satellite tracking became available. By means of dedicated geoscientific, GPS-tracked satellite missions such as GRACE and GOCE a temporary climax in geopotential mapping is reached. These satellite missions are characterized by low and near polar orbits delivering global and continuous high quality science data. Thus they provide the best possible basis for modern gravity field determination, yet giving rise to high demands on hardware and numerical algorithms.

The discussion on the geopotential recovery problem in this contribution is organized into the following sections. General remarks on geopotential parameterization and on the satellite missions GRACE and GOCE follow next. A paragraph on the importance of using High Performance Computing (HPC) for gravity field determination concludes the introductory section. Sections. 2 and 3 outline concepts and methodologies for geopotential recovery with GRACE and GOCE. Of major importance is the aspect of adopting HPC for solving the parameters of high-resolution geopotential models, which is addressed in Sect. 4 together with a description of the used HPC platforms, typical job sizes and runtime results. A last section giving some conclusions and an outlook completes the paper.

1.1 Geopotential Parameterization

To get started, a first grasp of the physical and mathematical nature of potential theory is outlined. Typically, the geopotential of the Earth $V(\lambda, \varphi, r)$ is given in terms of spherical coordinates (λ, φ, r) by means of a spherical harmonics series expansion model

$$V(\lambda, \varphi, r) = \frac{GM}{R} \sum_{l=0}^{\infty} \sum_{m=-l}^{l} \left(\frac{R}{r}\right)^{l+1} e_{lm}(\lambda, \varphi)\, \bar{v}_{lm}\,, \tag{1}$$

$$e_{lm}(\lambda, \varphi) = \begin{cases} \bar{P}_{lm}(\sin\varphi)\cos m\lambda & 0 \leq m \leq l \\ \bar{P}_{l|m|}(\sin\varphi)\sin |m|\,\lambda & -l \leq m < 0 \end{cases}. \tag{2}$$

$V(\lambda, \varphi, r)$ reflects the integrated effect of all infinitesimal mass elements forming the Earth's body on a test particle located in the Earth's exterior and is referred to as Newtonian gravitational potential. $V(\lambda, \varphi, r)$ mirrors irregularities in the mass distribution inside the Earth. Together, the gravitational potential and the centrifugal potential, which is related to the rotation of the Earth, add up to the gravity potential. The double sum expression in (1) can be compared with a two-dimensional Fourier expansion. In theory an infinite series is necessary to fully characterize the gravitational field, but practically the double sum is truncated at a maximum degree $l_{\max} = L$ determined by the satellite's observation principle and the desired resolution of the field. As a matter of course a more detailed resolution results in a higher value for L. The geopotential series representation consists of the attenuation factor $(R/r)^{l+1}$ and orthonormal base functions, the so-called surface spherical harmonics $e_{lm}(\lambda, \varphi)$, cf. (2), with $\bar{P}_{lm}(\sin\varphi)$ being the normalized Legendre functions of the first kind. From the attenuation factor follows directly one of the prevailing physical properties of Earth's gravity field, i.e. its decline with increasing distance from the geocenter. In fact $V \to 0$ holds for $r \gg R$ and $r \to \infty$. This property explains the fact that a geoscientific satellite mission has to be placed in an orbit as low as possible to achieve high sensitivity with respect to the Earth's attraction. Both the geocentric constant GM and the mean Earth radius R are fixed. The coefficients $\bar{v}_{lm}$ are unknown parameters, describing the deviation of the terrestrial gravitational potential from its first order spherical approximation. Their estimation can be performed best by globally distributed observation data such as provided by GRACE and GOCE. The field equation to be solved in the mass-free external space of the Earth is Laplace's differential equation

$$\Delta V = 0\,, \tag{3}$$

$$\Delta = \frac{\partial^2}{\partial r^2} + \frac{2}{r}\frac{\partial}{\partial r} + \frac{1}{r^2}\frac{\partial^2}{\partial \varphi^2} + \frac{\tan\varphi}{r^2}\frac{\partial}{\partial \varphi} + \frac{1}{r^2\cos^2\varphi}\frac{\partial^2}{\partial \lambda^2}\,. \tag{4}$$

Equation (3) is a homogeneous, second-order partial differential equation of elliptic type which results from Newtonian mechanics, neglecting Earth rotation, tidal and other geodynamic effects and (4) resembles the definition of the Laplace operator in spherical coordinates. It can be shown by means of a separation approach that the terrestrial potential $V(\lambda, \varphi, r)$ given in terms of the spherical harmonics series expansion of (1) satisfies (3).

1.2 Satellite Missions GRACE and GOCE

To resolve the static and time-variable gravity field of the Earth, the GRACE (Gravity Recovery And Climate Experiment) mission [4] was initiated by cooperation of the German Aerospace Center (DLR) and the National Aeronautics and Space Administration (NASA). GRACE was launched in March 2002 and will be in orbit for five to seven years. This twin-satellite mission is composed of two identical spacecraft, one satellite following its companion in the same orbit. The public therefore commonly refers to the satellites as Tom and Jerry. Both spacecraft are tracked by GPS and their relative motion is measured precisely by a highly sensitive microwave link. Relative variations in range and range-rate reflect the inhomogeneity of the Earth's mass distribution. The analysis of the relative motion allows the resolution of the long- to medium-wavelength (i.e. 40 000 km – 400 km) parts of the terrestrial gravity field with a much higher accuracy than possible so far. Furthermore, the time-variability of the gravity field can be detected by analyzing series of monthly static solutions.

Within the Living Planet program of the European Space Agency (ESA) the Earth explorer core mission GOCE (Gravity field and steady-state Ocean Circulation Explorer) is planned to be launched in 2007 and will be the first satellite mission in space applying three-dimensional gradiometry [3]. Satellite gradiometry is highly sensitive for the medium- to short-wavelength parts of the gravitational spectrum, thus will allow to determine the static Earth gravity field down to features of 200 km – 150 km in terms of spatial resolution. A GOCE mission duration of 20 months is planned, subdivided in two operational phases of six months each and an hibernation phase of eight months in between.

1.3 Advanced Geocomputations

As becomes obvious from Sects. 1.1 and 1.2 the complexity of an Earth gravity field model depends on different factors. First of all the observation principle (e.g. inter-satellite link, satellite gravity gradiometry, ...) determines the maximal possible spatial resolution of the geopotential, which corresponds to a certain degree L of the spherical harmonics series model, und thus the number of unknowns M. Furthermore, the normal matrix $\mathbf{N}$ involved within the context of a least squares solution by means of a Gauß-Markov model (minimizing the L_2-norm) and which has to be kept in the memory, is of size $M \times M$. These considerations are reflected in Table 1. For example, up to 63 000 unknown

Table 1. Memory requirement for normal matrix storage

resolution L	number of unknowns M	memory requirement for $\mathbf{N}$ (MByte)
150	22 798	2079
200	40 398	6528
250	62 998	15 875
300	90 598	32 833

parameters have to be estimated by means of least squares techniques from observations provided by the low Earth orbiting GOCE satellite. The corresponding upper half of the symmetric normal matrix $\mathbf{N}$ amounts to ~16 GB, which exceeds by far the amount of memory available on a standard personal computer. On the other hand, to fully assess the problem dimension mission lifetimes and data sampling rates have to be considered. The GRACE mission with its design lifetime of 5–7 years and a data sampling rate of 5 s will collect up to 40–50 millions observations. The GOCE mission, with a sampling rate of up to 1 s, will deliver in a one-year observation phase approximately twice the number of observations.

Due to the substantial dimension of the numerical problem to be solved coming up with the GOCE and its likely follow-up missions, only the use of HPC infrastructure and adoption of parallel programming standards together with optimized numerical libraries will allow to overcome the challenge arising from the computational point of view. Thus, HPC is the key technology to a successful accomplishing of the demands of modern and efficient satellite based geopotential recovery.

2 GRACE Gravity Field Recovery

The first question to be settled is how geopotential recovery is achieved with GRACE. As already seen in Sect. 1.2 the GRACE mission features a so-called leader-follower-configuration. This concept will be considered more closely in the next paragraph. A paragraph describing the GRACE observational model and the resulting system of equations follows. Exemplary GRACE results conclude this section.

2.1 Mission Concept

The GRACE mission layout is illustrated in Fig. 1. The inter-satellite measurement device, the so-called K-Band microwave link to determine relative satellite motion with an accuracy of $\sim 1\,\mu\text{m/s}$, and the on-board GPS receivers to absolutely position the satellite twins with an accuracy of $\sim 1\,\text{cm}$ are the key scientific instruments to perform geopotential recovery. The concept of inter-satellite measurements between two low flying spacecraft is denoted as low–low satellite-to-satellite tracking (LL-SST), whereas high–low (HL) SST refers to the inter-satellite distance measurements of the high flying GPS satellites and the low flying GRACE space segment (cf. Fig. 1). The separation of the two satellites is $220 \pm 50\,\text{km}$. The initial orbit height was $480\,\text{km}$ decaying down to $\sim 300\,\text{km}$ due to atmospheric influences. Moreover, the orbit is polar (with an inclination of $i \sim 89.5°$) and near circular (with an eccentricity of $e < 0.005$), which guarantees compliance with scientific data prerequisites. Typical GRACE geopotential solutions aim at gravity field resolution degrees of $L = 100 - 150$.

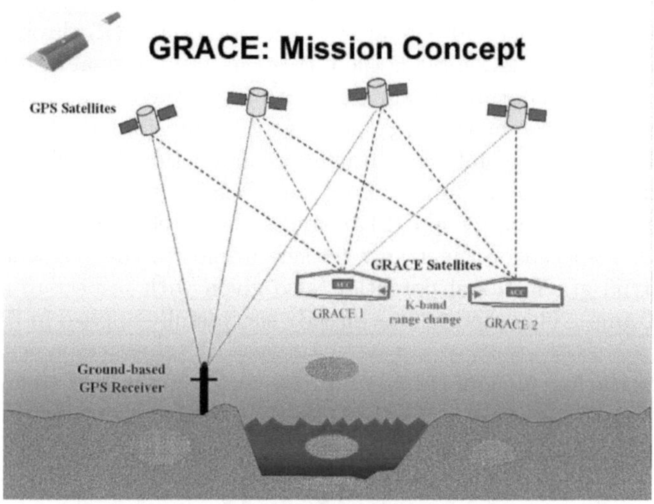

Fig. 1. GRACE mission concept

2.2 Mathematical Methodology

The GPS derived HL-SST position vector of the i-th GRACE satellite ($i = 1, 2$) with respect to a geocentric inertial frame of reference reads as follows

$$\mathbf{X}_i = [X_i \quad Y_i \quad Z_i]^T \, . \tag{5}$$

Differentiating the inter-satellite range ρ

$$\rho = \|\mathbf{X}_2 - \mathbf{X}_1\| = \|\Delta\mathbf{X}\| = \sqrt{\Delta\mathbf{X} \cdot \Delta\mathbf{X}} \tag{6}$$

twice with respect to time results in

$$\dot{\rho} = \frac{1}{\rho}\Delta\mathbf{X} \cdot \Delta\dot{\mathbf{X}} \, , \tag{7}$$

$$\ddot{\rho} = \frac{1}{\rho}\left(\Delta\mathbf{X} \cdot \Delta\ddot{\mathbf{X}} + \Delta\dot{\mathbf{X}} \cdot \Delta\dot{\mathbf{X}} - \frac{\dot{\rho}}{\rho}\Delta\mathbf{X} \cdot \Delta\dot{\mathbf{X}}\right) . \tag{8}$$

Equation (8) relates the GRACE range-accelerations $\ddot{\rho}$ to HL-SST derived vectorial position $\Delta\mathbf{X}$, velocity $\Delta\dot{\mathbf{X}}$ and acceleration $\Delta\ddot{\mathbf{X}}$ differences as well as to LL-SST inter-satellite range ρ and range-rate $\dot{\rho}$ data. The idea is now to associate the accelerations of the satellites with the Earth's gravitational field in order to introduce the mathematical relationship of observables and unknown field parameters. This is done by taking advantage of Newton's Law of Motion

$$\ddot{\mathbf{X}} = \mathbf{\Gamma} \, , \tag{9}$$

with $\mathbf{\Gamma}$ being subject to

$$\mathbf{\Gamma} = \mathbf{R}_{\mathbf{Xx}}\gamma = \mathbf{R}_{\mathbf{Xx}}\nabla V(\mathbf{x}) \, . \tag{10}$$

In (10), γ represents the gravitational acceleration vector caused by the mass attraction of the Earth and is related to the potential $V(\mathbf{x})$ through the gradient operator $\gamma = \nabla V(\mathbf{x})$. Since the terrestrial gravitational field is rotating with the Earth, γ is naturally expressed with respect to a body-fixed reference frame. Given that the matrix $\mathbf{R}_{\mathbf{Xx}}$ performs the transformation of the Earth-fixed to the space-fixed reference frame, it follows that $\mathbf{\Gamma} = \mathbf{R}_{\mathbf{Xx}}\gamma$ describes the specific gravitational force vector with respect to an inertial reference frame of reference. Expression (9) balances the satellite acceleration vector (given with respect to an inertial reference frame) to the gradient vector of the terrestrial gravitational potential (also given with respect to an inertial reference frame). This concept was already successfully applied in [12] and [2].

By defining the line-of-sight (LOS) unit base vector and its first time derivative

$$\mathbf{b}^{\mathrm{LOS}} = \frac{\Delta\mathbf{X}}{\rho} \, , \tag{11}$$

$$\dot{\mathbf{b}}^{\mathrm{LOS}} = \frac{\Delta\dot{\mathbf{X}}}{\rho} - \frac{\dot{\rho}\Delta\mathbf{X}}{\rho^2} \, , \tag{12}$$

equation (8) reads as follows

$$\ddot{\rho} = \mathbf{b}^{\mathrm{LOS}} \cdot \Delta\mathbf{\Gamma} + \dot{\mathbf{b}}^{\mathrm{LOS}} \cdot \Delta\dot{\mathbf{X}}\,, \tag{13}$$

with $\Delta\mathbf{\Gamma}$ being subject to

$$\Delta\mathbf{\Gamma} = \mathbf{R}_{\mathbf{Xx}}\Delta\gamma = \mathbf{R}_{\mathbf{Xx}}\left(\nabla V(\mathbf{x}_1) - \nabla V(\mathbf{x}_2)\right). \tag{14}$$

The second term on the right-hand side of (13) is referred to as velocity correction. It accounts for the fact that the GRACE assembly represents a moving observational system with respect to inertial space. The resulting observational model

$$\ddot{\rho} + \frac{1}{\rho}\left(\dot{\rho}^2 - \|\Delta\dot{\mathbf{X}}\|^2\right) = \mathbf{b}^{\mathrm{LOS}} \cdot \Delta\mathbf{\Gamma} \tag{15}$$

is obtained by re-ordering (13) and re-writing the velocity correction term. It equates the difference of the gravitational potential gradients of the two satellites projected along their LOS direction, right-hand side of (15), with the GRACE pseudo-observables, left-hand side of (15), which depend on range, range-rate, range-acceleration and satellite velocity differences. The pseudo-observables form the observation vector $\mathbf{y}$. The unknown coefficients $\bar{v}_{lm}$ of the spherical harmonic series expansion are collected in the vector of unknowns $\mathbf{x}$ and appear linearly. Denoting the matrix which defines the functional relationship of unknowns $\mathbf{x}$ and observations $\mathbf{y}$ by $\mathbf{A}$ leads to the corresponding linear and over-determined system of equations

$$\mathbf{A}\mathbf{x} = \mathbf{y} + \mathbf{e}\,. \tag{16}$$

2.3 Exemplary Results

Probably the most outstanding scientific outcome of the GRACE mission is its capability of detecting time-dependent variations in the geopotential. This is displayed in Fig. 2, where 24 monthly solutions are compared to a mean solution for the same period of time. (For more details on static and time-variable gravity field solutions see e.g. http://www.gfz-potsdam.de/grace/). In particular around the Amazon basin and the Himalayas region geoid variations of up to 1 cm can be determined. As mentioned before, such variations reflect mass transport on or below Earth surface. Due to the distinct occurrence of mass surpluses and mass deficiencies with respect to a certain time of the year, the geoid variations in the case under consideration can be associated in majority to surface water variations. Taking known precipitation values into account this assumption can be approved. Thus, GRACE plays an important role in hydrology. It should be emphasized that the repeated and near realtime evaluation of monthly GRACE solutions is a challenging task from the computational point of view.

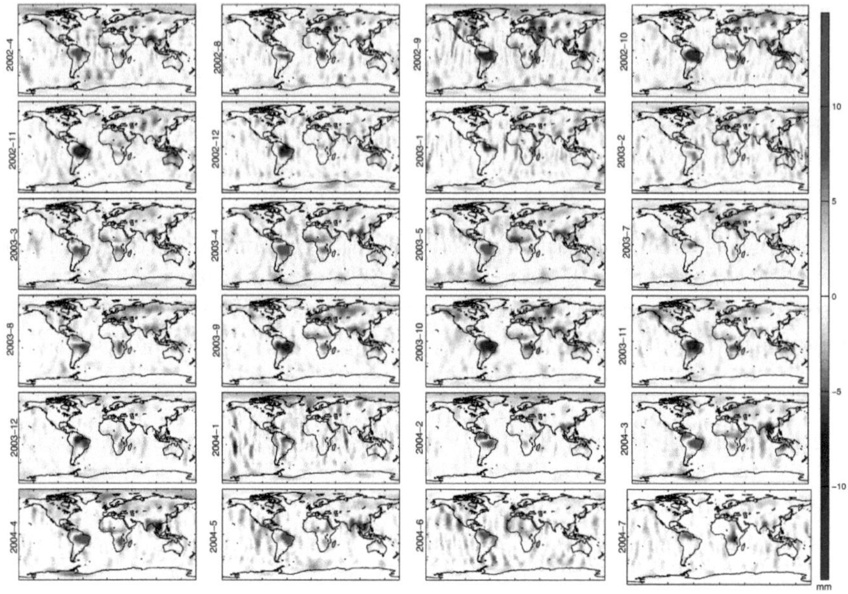

Fig. 2. Monthly geoid variations detected by GRACE

3 GOCE Gravity Field Recovery

Complementary to the GRACE mission objectives as outlined in Sect. 2, GOCE observations will allow for high-resolution static geopotential modeling. To meet this challenge satellite gravity gradiometry (SGG) will be realized for the first time in space geodetic applications [3].

3.1 Mission Concept

To recover short-scale geopotential features it is inescapable to provide high-sensitive observations requiring the satellite orbit to be as low as possible. However, low altitude requests high energy availability to keep the spacecraft in a free fall environment. To attain a balance between the conflictive specifications, an altitude of about 240–250 km has been chosen for GOCE. Actually, energy supply is realized by a sun-synchronous satellite orbit together with solar panels mounted on the whole satellite surface. Non-gravitational effects such as air drag and solar radiation pressure are compensated in situ. The Earth-pointing spacecraft is equipped with a star tracker for absolute orientation. Moreover, attitude control is achieved by a thruster assembly. Similarly to GRACE, a GPS receiver allows precise orbit determination. Figure 3 presents a simplified illustration of the satellite with its most important on-board components. The main instrument is a tri-axial capacitive gradiometer [8]. The principle of gradiometry is providing gravitational gradients (GGs)

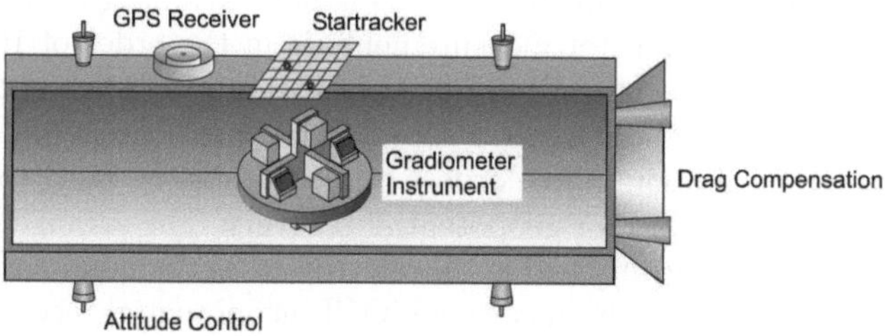

Fig. 3. Fundamental GOCE payload (GOCE Project Office, Munich)

as observables for gravity field determination [13]. From the technical point of view these quantities can be obtained by scaled differential gravitational acceleration measurements. In particular, for GOCE the gradiometer instrument consists of six three-dimensional accelerometers. They are mounted pairwise on the gradiometer axes as outlined in Fig. 4. Let $\frac{d}{2}$ denote the distance between the origin of the gradiometer and an arbitrary accelerometer, than the observed gradients read

$$\varGamma_{ij} = \frac{s_j(p_{2,i}) - s_j(p_{1,i})}{d}; \quad i,j = 1,2,3. \tag{17}$$

Therein, $s_j(p_{*,i})$ denotes the measurement of the $*$-th accelerometer on the axis i in direction of the axis j. Due to gradiometer frame rotations with respect to the inertial space, the measurements $\varGamma_{ij}$ consist of both the GGs V_{ij} and rotational effects, i.e. the centrifugal and Euler term:

$$\varGamma_{ij} = V_{ij} + \varOmega_{ik}\varOmega_{kj} + \dot{\varOmega}_{ij}. \tag{18}$$

Both rotational effects have to be subtracted from $\varGamma_{ij}$ to finally obtain the single GGs. They are arranged in the so-called gravitational tensor. Its coef-

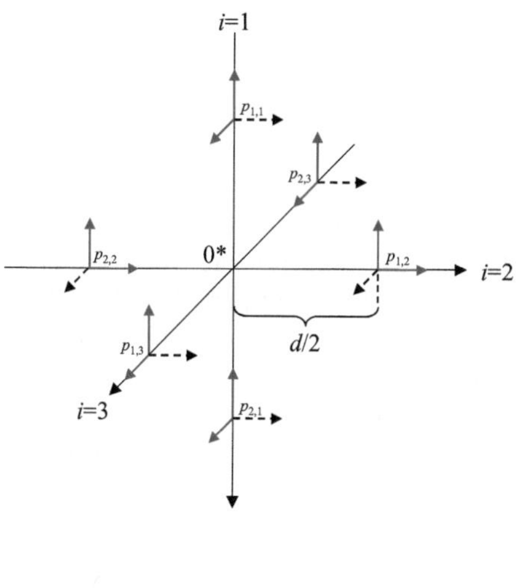

Fig. 4. GOCE gradiometer diamond configuration (low-sensitive axes labeled by *dotted lines*)

ficient matrix $\mathbf{V} = [V_{ij}]$ of dimension 3×3 is both symmetric and trace-free. The accuracy of accelerometer measurements is in the order of $10^{-12}\,\mathrm{m\,s^{-2}}$. Unfortunately, due to technical reasons only two axes of each accelerometer can be realized with highest accuracy level. The third one is limited in sensitivity by a factor of about three orders of magnitude. In SGG the elements V_{ii} are predominately used for geopotential recovery. The arrangement of the accelerometer axes according to Fig. 4, referred to as diamond configuration, ensures the optimal assembly in case of GOCE geopotential recovery [3]. The main diagonal elements of $\mathbf{V}$ can be provided with an accuracy in the range of 6–8 $\mathrm{mE\,Hz^{-1/2}}$ ($1\,\mathrm{E} = 1\,$Eötvös $= 10^{-9}\,\mathrm{s^{-2}}$).

3.2 Mathematical Methodology

Regarding gravity field modeling, GGs correspond to second order derivatives of the geopotential (1). They are by far more sensitive to the underlying force field then gravitational accelerations (first order derivatives). Actually, SGG is the state-of-the-art technology to perform high-resolution gravity field research. Starting from the gravitational acceleration vector

$$\nabla V(\mathbf{x}(\lambda, \varphi, r)) = \mathbf{e}_i v_i \tag{19}$$

with ∇ denoting the gradient operator as outlined in Sect. 2.2 the application of the gradient operator once more leads to

$$\nabla \ \ (\nabla V(\mathbf{x}(\lambda, \varphi, r))) = \mathbf{e}_i \otimes \mathbf{e}_j V_{ij}. \tag{20}$$

Equation (20) yields an analytical expression of the gravitational tensor in terms of the gravity field model parameters $\bar{v}_{lm}$. Each single component V_{ij} constitutes one type of observation. Exemplary for the component V_{33}

$$V_{33} = \frac{GM}{R^3} \sum_{l=0}^{L} \sum_{m=-l}^{l} \left(\frac{R}{r}\right)^{l+3} (l+2)(l+1) e_{lm}(\lambda, \varphi) \bar{v}_{lm} \tag{21}$$

holds. The linear functional relation between the model parameters and the GGs from measurements according to (18) is straightforward. Hence, again a linear system of equations in the same line as described in Sect. 2.2, cf. (16), results.

Commonly, SGG analysis is performed on the level of single GGs, in particular the main diagonal elements of the gravitational tensor. Details can be found in e.g. [14, 15, 5, 16, 11]. A completely different approach is based on the rotational invariants of the gravitational tensor [13]. The motivation for this method comes along with the absence of any knowledge about the gradiometer frame orientation with regard to the orbit frame, the inertial space respectively. The invariants approach has been adopted to the GOCE mission scenario in [2].

3.3 Exemplary Results

Figure 5 presents the signal content of each single GG. Obviously, V_{11} (also denoted as V_{xx}) predominantly images east–west features, the component V_{22} (or V_{yy}) north–south properties accordingly. Radial gravitational features are mainly mapped on V_{33} (or V_{zz}). The combined signal content of all the three main diagonal GGs reflects the total structure of the Earth gravitational field. However, the radial component contains the major information opposite to the other GGs. Closed-loop numerical case study results are illustrated in Fig. 6. They are based on a simulated GOCE data set covering one month of observation data with a sampling rate of $\Delta t = 5\,\mathrm{s}$. The Earth gravity model EGM96 [7] up to degree and order 300 is used for data simulation. V_{33}-only analysis is performed including a realistic measurement noise model [3]. The blue line shows the coefficients estimate for a maximum resolutions of $L = 300$ in terms of degree-error root mean square (DE-RMS) values relative to EGM96. Obviously, with one month of observation data geopotential recovery is limited to $l \sim 230$. This improves for expanded data coverage. The sun-synchronous GOCE orbit causes the inclination of the satellite orbit to be in the range of 96.6°. Thus, a double polar cap each of the area of $1.7 \cdot 10^6\,\mathrm{km}^2$ is not covered by observations. Polar regions reflect in the low-order spectral domain in terms of harmonic coefficients. This becomes apparent in Fig. 7 where the empirical relative errors of the parameter estimate are shown. Additional information such as e.g. delivered by terrestrial, respectively airborne, gravimetry and the GRACE satellite mission is necessary to finally provide a gravity field model covering the whole spectral domain.

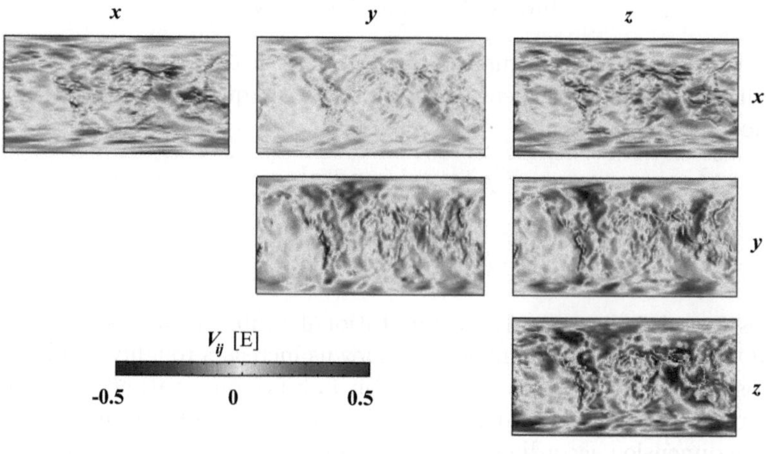

Fig. 5. Signal content of GGs (GOCE Project Office, Munich)

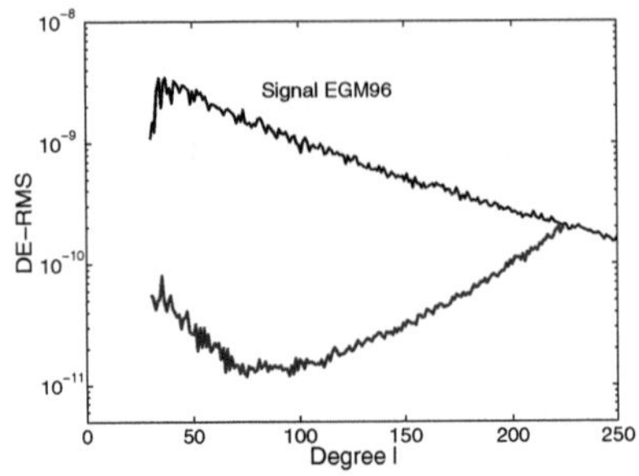

Fig. 6. DE-RMS values from noisy GOCE data analysis

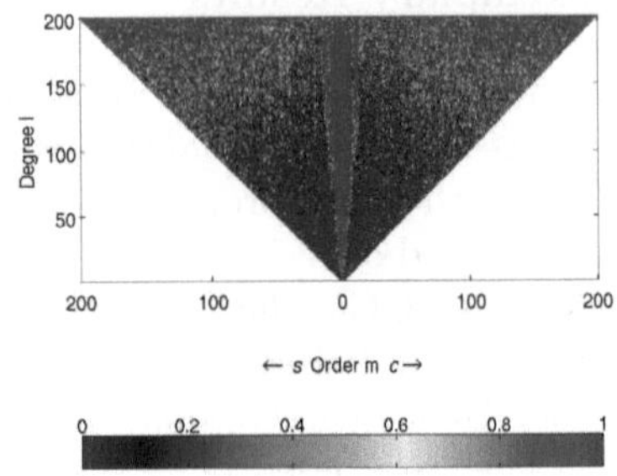

Fig. 7. Empirical relative errors based on noisy data

4 High Performance Computing

Both GRACE and GOCE data analysis yields an overdetermined (ill-posed) linear system of equations. It reads in terms of a standard Gauß-Markov model [6]

$$E\{\mathbf{y}\} = \mathbf{Ax}, \quad \mathbf{Ax} = \mathbf{y} + \mathbf{e}, \quad E\{\mathbf{ee}^T\} = D\{\mathbf{y}\} = \mathbf{\Sigma} = \sigma^2\mathbf{P}^{-1}. \tag{22}$$

In (22) the design matrix $\mathbf{A}$ relates the vector of unknowns $\mathbf{x}$ to the vector of observations $\mathbf{y}$. The observation variance-covariance information $D\{\mathbf{y}\} = \mathbf{\Sigma}$ is the product between the variance of unit weight σ^2 and the inverse positive definite weight matrix $\mathbf{P}$. L_2-norm minimization of the residual vector $\mathbf{e}$ according to

$$\min \|\mathbf{e}\|^2 = \min_{\mathbf{x}} \|\mathbf{Ax} - \mathbf{y}\|^2_{\mathbf{\Sigma}^{-1}} \tag{23}$$

finally provides the least squares (LS) estimate of the unknowns $\hat{\mathbf{x}}$. Writing out full minimization problem (23) yields the (uniquely determined) normal equation system

$$\mathbf{A}^T\mathbf{PA}\hat{\mathbf{x}} = \mathbf{A}^T\mathbf{Py}, \tag{24}$$

$$\mathbf{N}\hat{\mathbf{x}} = \mathbf{b}, \tag{25}$$

$$\hat{\mathbf{x}} = \mathbf{N}^{-1}\mathbf{b}. \tag{26}$$

Figures 8 and 9 illustrate the computational problem dimension regarding the GOCE mission. The number of unknowns increases roughly quadratically with rising maximum degree of resolution L. Since observations are collected uniformly, their number depends linearly on the mission lifetime. The peak problem dimension according to the considered scenario ($\Delta t = 5\,\mathrm{s}$) is achieved with approximately $N = 20$ millions of observations solving for about $M = 63\,000$ unknown geopotential model parameters.

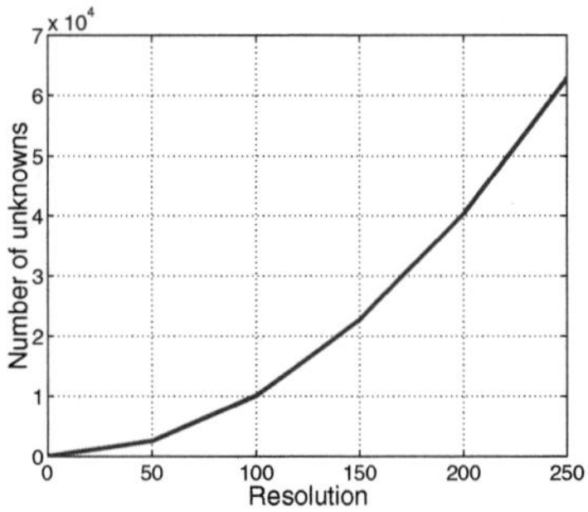

Fig. 8. Number of unknowns dependent on maximum resolution L

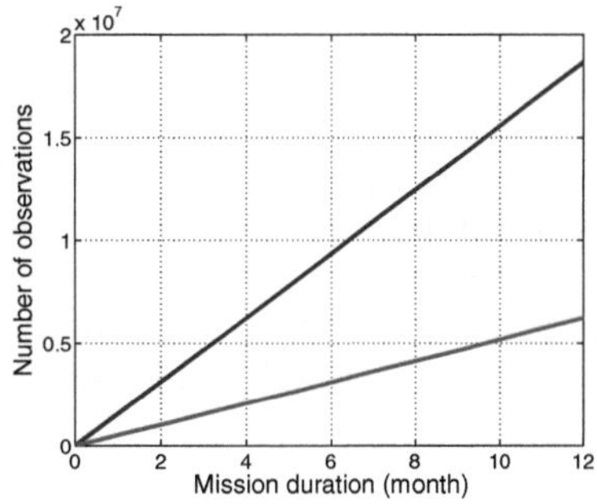

Fig. 9. Number of GOCE observations dependent on mission duration (Δt = 5 s). *Blue line*: V_{11}, V_{22} and V_{33}. *Red line*: V_{33}-only

4.1 Implementation of Tailored Algorithms

From the computational point of view the strategy to estimate geopotential model parameters is twofold. In a first approach (23) is treated by means of an iterative solver, namely the LSQR method [9, 10]. Concerning preconditioning and regularization we adapt the basic algorithm for its application in satellite geodesy. Notice that within the iterative process the major computational costs occur for the setup of the design matrix. Due to the character of the LSQR solver matrix-matrix and matrix-vector multiplications are avoided by means of repeated vector-vector operations. Thus, memory requirement is small since the design matrix can be handled for each row separately. Additionally, observation processing can be performed independently from each other. An effective parallel programming implementation of the LSQR method is not limited to a certain computing architecture. Since communication between successive processing stages is small, shared memory (SM), distributed memory (DM) as well as hybrid platforms are suited well for the iterative approach. However, array processors perform poorly. Preferably we run LSQR on cluster systems.

Opposite to the iterative method, alternatively brute-force inversion of the normal equation system (25) is performed. This approach is predominately motivated for exact variance-covariance information evaluation of the parameter estimate which is of utmost importance for a multitude of further investigations. It can be obtained with

$$D(\hat{\mathbf{x}}) = \hat{\sigma}^2 \mathbf{N}^{-1}, \quad \hat{\sigma}^2 = \frac{(\mathbf{A}\hat{\mathbf{x}} - \mathbf{y})^T \mathbf{P} (\mathbf{A}\hat{\mathbf{x}} - \mathbf{y})}{N - M}. \tag{27}$$

For the brute-force approach the whole normal matrix, its inverse respectively, has to be stored (at least one triangle since $\mathbf{N}$ is symmetric). According to Table 1 memory requirement rises very fast with increasing number of unknowns.

Thus, efficient normal equation system inversion is limited to shared memory systems. Otherwise communication effort increases considerably. Moreover, our findings indicate that optimal runtime scaling on SM systems is only achievable using a moderate number of CPUs, i.e. eight to twelve processors [1]. With respect to vectorization it can be stated that depending on the problem dimension the brute-force approach is well suited for array processors. Notice that avoiding normal matrix assembly (as done for iterative solvers such as LSQR) only approximate variance-covariance estimation is possible.

4.2 Requirements and Results

Concluding from the statements in Sect. 4.1 we perform the iterative LSQR method on the CRAY Opteron cluster (Strider), the brute-force approach on the NEC SX-8 platform respectively. The last one is considered as a pure SM system with OpenMP programming, i.e. so far we use only one single node. Table 2 presents typical user times for some selected analysis scenarios. Actually, in the current level of processing we exclusively perform code

Table 2. User time values for satellite based geopotential recovery ($\Delta t = 5\,\mathrm{s}$)

	L	observation period	#observations (10^6)	user time (d)
GRACE	160	1 month	0.5	1
		1 year	6	12[1]
		5 years	30	60[1]
GOCE	200	1 month	1.5	6
		6 months	9	36[1]
		(2 × 6 months	18	72[1])

[1] extrapolated value

development and test production runs based on simulated but close to reality scenarios. Data sets covering one month are sufficient for most studies. However, real data analysis will be a hot topic in the near future. Anyway, in Table 2 runtimes exceeding one month of observation data are extrapolated. Linear extrapolation has been proven to be justified. To come along with the challenges according to Table 2, computing resources for satellite geopotential recovery arise as outlined in Table 3. Concerning simulated case studies run times should not exceed a few hours up to one day since often repeated calculations are necessary to adjust different parameters. The same holds for monthly GRACE real data processing. On the other hand, extensive analysis of both GRACE and GOCE data occur to a lesser extent. Thus, they can take up to one week or even longer. Since GRACE data processing requires moderate hardware availability compared to acceptable runtime both

Table 3. Resources requirement for satellite based geopotential recovery

	kind of processing	NEC SX-8 (nodes)	CRAY Strider (nodes)
GRACE	simulation	1	≤ 8
	real data	$1-2$	$16-32$
GOCE	simulation	$1-2$	$16-32$
	real data	$2-4$	$32-64$

SM and DM systems can be used optionally. This does not hold for GOCE. In particular, pure SM processing is limited to small dimensioned problems. Otherwise, cluster systems are preferred. These can be both massive parallel architectures or clusters of SM nodes.

5 Conclusions and Outlook

The computational challenges regarding high-resolution, respectively repeated, terrestrial gravity field estimation based on large-scale data sets increased dramatically in recent years. We successfully adopted HPC facilities for satellite based geopotential recovery against the background of the complementary missions GRACE and GOCE. Dependent on hardware availability and scientific analysis output we realized two independent tailored implementations. The first one is the iterative LSQR solver suited best for cluster systems. Alternatively brute-force inversion of the resulting normal equation system is very efficient on SM architectures. In particular, in that case array processors perform very well. In its current development stage our implementations are ready for essential scientific case studies. Moreover, GRACE real data analysis will be prepared in the near future.

Acknowledgements

The authors thank the High Performance Computing Center Stuttgart (HLRS) for the opportunity to use their computing facilities, most notably for the helpful technical support. In addition, the authors thank the Center for Computing and Networking Services (SARA) in Amsterdam for excellent cooperation within the HPC-Europa Transnational Access Program.

References

1. Austen, G., Baur, O., Keller, W.: Use of High Performance Computing in Gravity Field Research. In: Nagel, W.E., Jäger, W., Resch, M. (eds.): High Performance Computing in Science and Engineering '05. Springer, pp. 305–318 (2006)

2. Baur, O., Grafarend, E.W.: High-Performance, GOCE Gravity Field Recovery from Gravity Gradients Tensor Invariants and Kinematic Orbit Information. In: Flury, J., Rummel, R., Reigber, C., Rothacher, M., Boedecker, G., Schreiber, U. (eds.): Observation of the Earth System from Space. Springer Berlin Heidelberg New York, pp. 239–253 (2006)
3. European Space Agency (ESA): Gravity Field and steady-state ocean circulation mission. ESA Publications Division, Reports for Mission Selection of the four candidate earth explorer missions, ESA SP-1233(1), ESTEC, Noordwjik (1999)
4. Jet Propulsion Laboratory (JPL): GRACE science and mission requirements document. 327-200, Rev. B, Jet Propulsion Laboratory, Pasadena, CA (1999)
5. Klees, R., Koop, R., Visser, P., van den IJssel, J.: Efficient gravity field recovery from GOCE gravity gradient observations. J. Geod., 74, 561–571 (2000)
6. Koch, K.-R: Parameter estimation and hypothesis testing in linear models. Springer, 2nd Edition (1999)
7. Lemoine, F.G., Kenyon, S.C., Factor, J.K., Trimmer, R.G., Pavlis, N.K., Chinn, D.S., Cox, C.M., Klosko, S.M., Luthcke, S.B., Torrence, M.H., Wang, Y.M., Williamson, R.G., Pavlis, E.C., Rapp, R.H., Olson, T.R.: The Development of the Joint NASA GSFC and NIMA Geopotential Model EGM96. NASA Goddard Space Flight Center, Greenbelt, Maryland, USA (1998)
8. Müller, J.: Die Satellitengradiometermission GOCE. DGK, Series C, No. 541, Munich (2001)
9. Paige, C.C., Saunders, M.A.: LSQR: An algorithm for sparse linear equations and sparse least squares. ACM T. Math. Software, 8, pp. 43–71 (1982a)
10. Paige, C.C., Saunders, M.A.: LSQR: Sparse linear equations and least squares problems. ACM T. Math. Software, 8, pp. 195–209 (1982b)
11. Pail, R., Plank, G.: Assessment of three numerical solution strategies for gravity field recovery from GOCE satellite gravity gradiometry implemented on a parallel platform. J. Geod., 76, pp. 462–474 (2002)
12. Reubelt, T., Austen, G., Grafarend, E.W.: Harmonic analysis of the Earth's gravitational field by means of semi-continuous ephemeris of a Low Earth Orbiting GPS-tracked satellite. Case study: CHAMP. J. Geod., 77, pp. 257–278 (2003)
13. Rummel, R.: Satellite Gradiometry. In: Sünkel, H. (ed.): Mathematical and Numerical Techniques in Physical Geodesy. Lect. Notes Earth Sci., 7, Springer Berlin (1986)
14. Rummel, R., Sansò, F., van Gelderen, M., Brovelli, M., Koop, R., Miggliaccio, F., Schrama, E., Scerdote, F.: Spherical harmonic analysis of satellite gradiometry. Netherlands Geodetic Commission, New Series, 39 (1993)
15. Schuh, W.D.: Tailored Numerical Solution Strategies for the Global Determination of the Earth's Gravity Field. Mitteilungen der Geodätischen Institute der TU Graz, 81, Graz (1996)
16. Sneeuw, N.: A semi-analytical approach to gravity field analysis from satellite observations. DGK, Series C, No. 527, Munich (2000)

Molecular Modeling of Hydrogen Bonding Fluids: Monomethylamine, Dimethylamine, and Water Revised

Thorsten Schnabel, Jadran Vrabec, and Hans Hasse

Institut für Technische Thermodynamik und Thermische Verfahrenstechnik, Universität Stuttgart, D-70550 Stuttgart, Germany
vrabec@itt.uni-stuttgart.de

1 Introduction

In chemical engineering, the knowledge on thermopysical properties of pure fluids and mixtures is important for the design and optimization of processes. As the experimental data base is often narrow, methods are required that predict thermophysical properties quantitatively. Usually, equations of state or G^{E} models are used for that purpose. They are known as excellent correlation tools, but they lack in predictive power and hold only little promise for further improvement.

Molecular modeling and simulation is an alternative approach for pure fluids and mixtures with excellent predictive power, a high potential for further development and various applications for technical problems. Especially for hydrogen bonding and associating fluids, phenomenological models have a poor description of thermophysical properties compared to molecular modeling. However, simulations of fluids forming hydrogen bonds are computationally quite expensive. The reason for that lies in the resolution of hydrogen bonds which involves the occurrence of very strong intermolecular forces and the formation of clusters.

2 Pure Component Molecular Models

New rigid molecular models for monomethylamine, dimethylamine and water based on Lennard–Jones potentials were developed in this work. These models account for hydrogen bonding and polarity of the molecules by using point charges. In the following, the procedure for the molecular model development of the amines is described.

The nuclei positions of all atoms were determined with the quantum chemistry software package GAMESS (US) [12] for a single molecule in the vacuum.

The basis set 6-31G and the Hartree–Fock method were applied for geometry optimization. Starting from these positions, for the methyl group the AUA4 parameters of Ungerer et al. [13] were applied. The AUA4 parameters were optimized by Ungerer et al. for vapor-liquid equilibria of alkanes. Following their approach for anisotropic united-atom LJ sites, a small offset of the nitrogen and hydrogen sites in the direction of the hydrogen nuclei positions was allowed for optimization. Five parameters were fitted to yield optimal saturated liquid densities and vapor pressures. The parameters optimized are: the offset, the point charges of the amine group and the hydrogens, as well as the Lennard–Jones size and energy parameters of this group. The point charges at the methyl Lennard–Jones sites were set in such a manner to yield overall neutrality of the molecular model.

The potential energy u_{ij} between two molecules i and j is given by

$$u_{ij}\left(r_{ijab}\right) = \sum_{a=1}^{m}\sum_{b=1}^{m} 4\epsilon_{ab}\left[\left(\frac{\sigma_{ab}}{r_{ijab}}\right)^{12} - \left(\frac{\sigma_{ab}}{r_{ijab}}\right)^{6}\right] + \frac{q_{ia}q_{jb}}{4\pi\varepsilon_0 r_{ijab}}, \qquad (1)$$

where m is the number of sites of the molecular model, a is the site index of molecule i and b the site index of molecule j, respectively. The site–site distance between molecules i and j is denoted by r_{ijab}. σ_{ab}, ϵ_{ab} are the Lennard–Jones size and energy parameters, q_{ia} and q_{jb} are the point charges located at the sites a and b on the molecules i and j, respectively. Finally, $\varepsilon_0 = 8.854188 \cdot 10^{-12}$ F/m is the permittivity of vacuum. The interaction between unlike Lennard–Jones sites of two molecules is defined by the Lorentz–Berthelot combining rules [1, 7]

$$\sigma_{ab} = \frac{\sigma_{aa} + \sigma_{bb}}{2}, \qquad (2)$$

$$\epsilon_{ab} = \sqrt{\epsilon_{aa}\epsilon_{bb}}\,. \qquad (3)$$

2.1 Monomethylamine

A new rigid monomethylamine model was developed with the aim to describe the vapor-liquid equilibria accurately with low computational cost for this primary amine. It uses nuclei off-center Lennard–Jones united atoms for the methyl (CH_3) and the amine (NH_2) group, accounting for repulsion and dispersion. Point charges are located on the methyl and the amine Lennard–Jones sites as well as on the two hydrogen site positions of the amine group. Hence, the new molecular model consists out of two Lennard–Jones and four point charge sites, i.e. $m = 4$ in Eq. (1). The geometry and the potential parameters for the new monomethylamine model are given by Fig. 1 and Table 1.

Figure 2 depicts the saturated densities and Fig. 3 shows the vapor presure of the present model together with the experimental data of monomethylamine and simulation data of a recently published transferable potential by

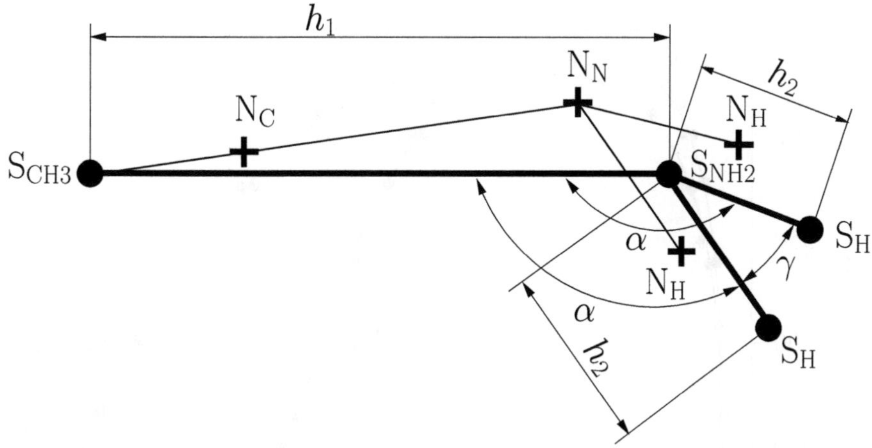

Fig. 1. Geometry of the present monomethylamine model: S_i indicates the model site i and N_j the nucleus position of atom j obtained from quantum chemistry calculations. Note, that both the real molecule and the model are not planar. Hence, α and h_2 look different here only due to the two-dimensional projection

Table 1. Lennard–Jones, point charge and geometry parameters of the present monomethylamine model, cf. Eq. (1) and Fig. 1; electronic charge $e = 1.602177 \cdot 10^{-19}$ C

Site	σ_{aa} Å	ϵ_{aa}/k_B K	q_{ia} e
S_{CH3}	3.6072	120.150	+0.19525
S_{NH2}	3.3151	142.147	−0.88653
S_H	0	0	+0.34564

h_1 Å	h_2 Å	α deg	γ deg
1.78615	1.01778	111.994	105.920

Wick et al. [15]. This much more complex, but tranferable molecular model uses an all atom approach for every nucleus and it considers internal degrees of freedom. Hence, that model is approximately five times computationally more expensive than the present model. The agreement between the present monomethylamine model as well as the transferable potential and the experimental data is excellent. However, the present molecular model yields overall slightly better agreement with experimental data in both saturated densities and vapor pressure compared to the transferable model.

The simulaltion results of the present monomethylamine model yield mean unsigned errors compared to experimental data [4] in vapor pressure, saturated liquid density and heat of vaporization of 6.5, 0.6 and 0.7 %, respectively,

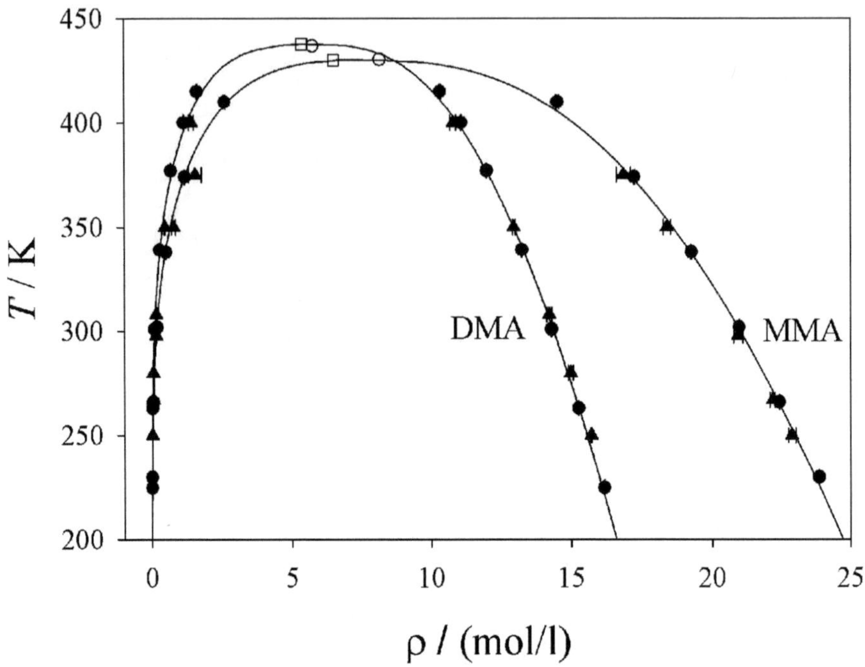

Fig. 2. Saturated densities of mono- and dimethylamine (MMA and DMA): *Bullet*, present simulation; *empty bullet*, critical point derived from present simulated data; *triangle up*, simulation data from Wick et al. [15]; *line*, experimental data [4]; *empty square*, experimental critical point

in the temperature range from 210 to 410 K, which is about 50 to 95 % of the critical temperature. Following the procedure suggested by Lotfi et al. [8], the critical temperature, density and pressure were determined. The results compare favorably to experimental data (numbers in parentheses): $T_c = 430.29$ (430.05) K, $\rho_c = 8.16$ (6.49) mol/l and $p_c = 7.71$ (7.46) MPa.

2.2 Dimethylamine

The present molecular model for dimethylamine was developed in a similar way as the monomethylamine model described in Sect. 2.1. Dimethylamine belongs to the category of secondary amines. The repulsion and dispersion of the dimethylamine molecule is considered with off-center Lennard–Jones united atom sites at the two methyl groups (CH_3) and the amine group (NH). Point charges were located at the Lennard–Jones positions and at the shifted hydrogen site. Hence, the new molecular model consists out of three Lennard–Jones and four point charge sites, i.e. $m = 4$ in Eq. (1).

In the optimization procedure, the point charge at the amine and the hydrogen sites, the Lennard–Jones size and energy parameters of the amine

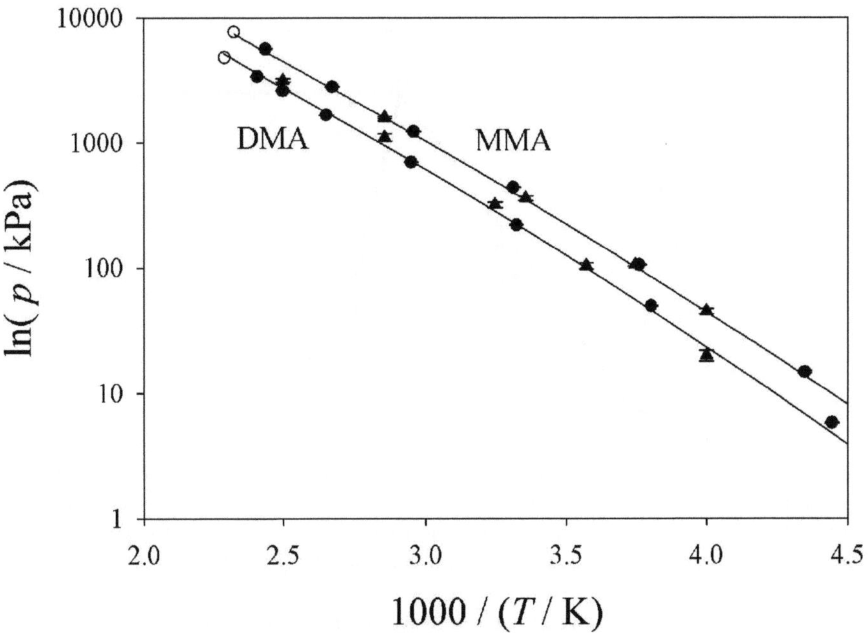

Fig. 3. Vapor pressure of mono- and dimethylamine (MMA and DMA): *Bullet*, present simulation; *empty bullet*, critical point derived from present simulated data; *triangle up*, simulation data from Wick et al. [15]; *line*, experimental data [4]

group as well as the offset of the amine group in direction to the hydrogen bonding atom were adjusted to the experimental saturated liquid density and the vapor pressure [4]. The point charges on the CH_3 Lennard–Jones centers were set in such a manner to yield overall neutrality of the molecular model.

The geometry and the potential parameters for the present dimethylamine molecular model are given by Fig. 4 and Table 2.

Figures 2 and 3 also include the saturated densities and the vapor pressure of dimethylamine of the present molecular model and the transferable potential of Wick et al. [15] as well as experimental data [4]. The agreement between the present as well as the transferable potential model and the experimental data is excellent. Again, the present molecular model yields overall better agreement with experimental data in both saturated densities and vapor pressure when compared to the transferable model. The simulaltion results of the present dimethylamine model yield mean unsigned errors compared to experimental data [4] in vapor pressure, saturated liquid density and heat of vaporization of 6.2, 0.4 and 2.4 %, respectively. This holds for the whole temperature range from 220 to 415 K, which is again about 50 to 95 % of the critical temperature. Following the procedure suggested by Lotfi et al. [8], the critical temperature, density and pressure were determined. The results

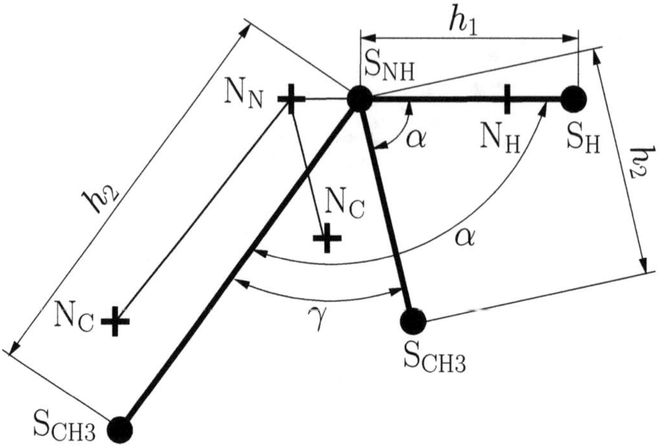

Fig. 4. Geometry of the present dimethylamine model: S_i indicates the model site i and N_j the nucleus position of atom j obtained from quantum chemistry calculations. Note, that both the real molecule and the model are not planar. Hence, α and h_2 look different here only due to the two-dimensional projection

Table 2. Lennard–Jones, point charge and geometry parameters of the present dimethylamine model, cf. Eq. (1) and Fig. 4.

Site	σ_{aa} Å	ϵ_{aa}/k_B K	q_{ia} e
S_{CH3}	3.6072	120.150	+0.03774
S_{NH}	3.4800	72.856	−0.45959
S_H	0	0	+0.38411

h_1 Å	h_2 Å	α deg	γ deg
1.01844	1.70196	111.368	108.924

also compare favorably to experimental data (numbers in parentheses): T_c = 436.73 (437.65) K, ρ_c =5.74 (5.35) mol/l and p_c = 4.81 (5.21) MPa.

2.3 Water

Many molecular models for water based on the Lennard–Jones and the exponential-6 potential, with rigid and flexible point charges as well as point polarizablities have been published. But none of them describes the thermophysical properties over a wide range of state points favorably. The exponential-6 model of Errington and Panagiotopoulos [5] is the best molecular water model in the literature to describe vapor-liquid equilibrium data which uses state independent potential parameters. However, the exponential-6 model is difficult to be used in mixtures since it can not be reasonably combined with other Lennard–Jones based molecular models.

In previous work, the molecular water model TIP4P of Jorgensen et al. [6], which uses one Lennard–Jones site and three off-center point charges, was reparameterized [3, 10]. Since the optimized TIP4P model still has a poor description of the vapor pressure, two new molecular model types were parameterized with the aim to describe saturated liquid densities and vapor pressures in the full vapor-liquid equilibrium temperature range.

The first new molecular model type consists out of one Lennard–Jones site and four point charges. For the parameterization, the TIP5P water model of Mahoney and Jorgensen [9] was taken as a starting point. The second molecular model parameterized in this work is an all atom molecular model since it uses two additional Lennard–Jones sites at the hydrogen positions.

Figures 5 and 6 show saturated densities and vapor pressure of the previously optimized TIP4P [3, 10], the two new models developed here, the exponential-6 model of Errington and Panagiotopoulos [5] as well as experimental data [4].

The description of the saturated densities by the new optimized TIP5P model is favorable. However, the description of the vapor pressure is signifi-

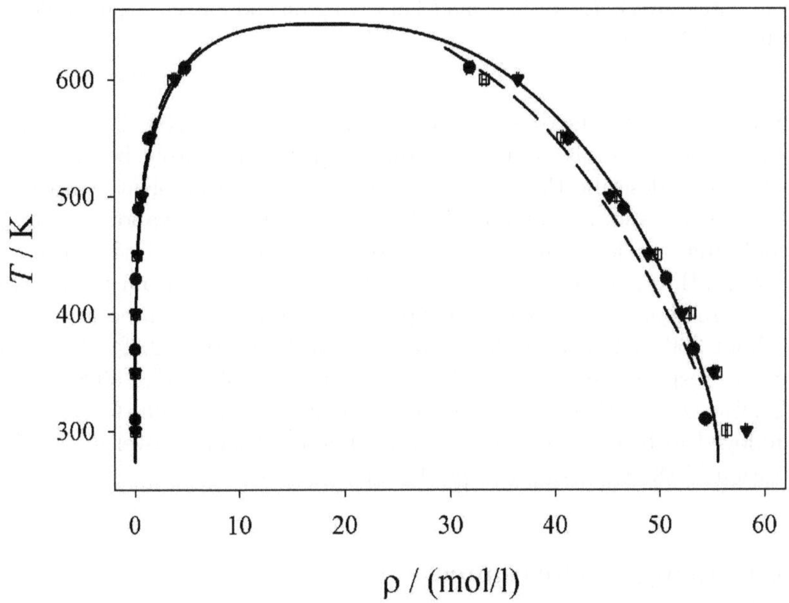

Fig. 5. Saturated densities of water. *Bullet*: previously optimized TIP4P model [3, 10]; *empty square*: recently optimized TIP5P model; *triangle down*: recently optimized all atom water model; *dashed line*: model of Errington and Panagiotopoulos [5]; *line*: experimental data [4]

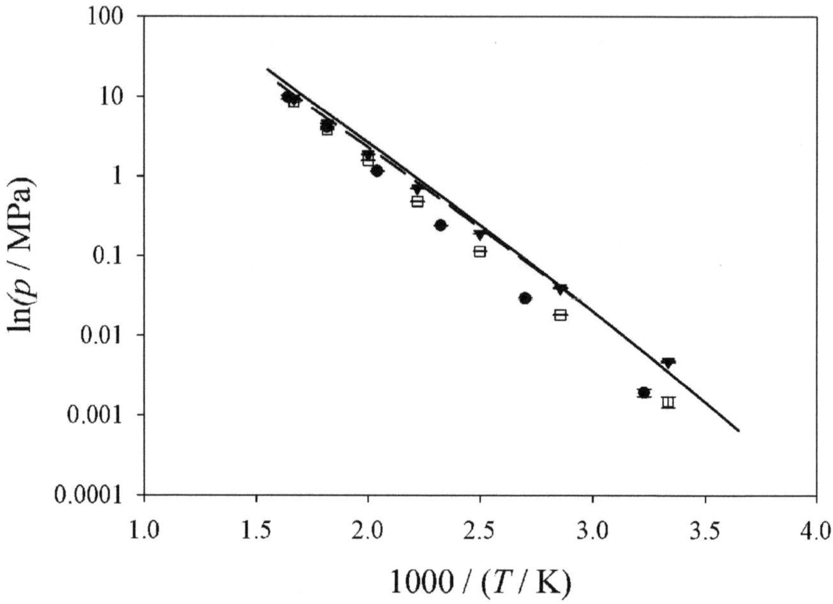

Fig. 6. Vapor pressure of water. *Bullet*: previously optimized TIP4P model [3, 10]; *empty square*: recently optimized TIP5P model; *triangle down*: recently optimized all atom water model; *dashed line*: model of Errington and Panagiotopoulos [5]; *line*: experimental data [4]

cantly worse than by the exponential-6 model. The optimized all atom molecular model describes the saturated liquid densities favorably but it can not reproduce and describe the saturated densities around ambient temperature and the density anomaly of water. The description of the vapor pressure by this molecular model is overall favorable. The exponential-6, the previously optimized TIP4P, the two new optimized TIP5P and all atom models yield mean unsigned errors compared to experimental data in a temperature range from about 300 to 620 K for the saturated liquid density of 2.3%, 2.0%, 3.0% and 1.5%, respectively, and for the vapor pressure of 23.1%, 50.8%, 45.5% and 22.3%. Due to these results, other molecular model types for water will be employed in future work and parameterized to yield an overall satisfying description of the vapor-liquid equilibrium.

3 Computing Performance

The computations of the vapor-liquid equilibria for the new molecular models described in Sects. 2.1 to 2.3 and for their development were carried out on the NEC SX-8, the NEC Xeon EM64T (Cacau) and the Cray Opteron cluster (Strider) with the MPI molecular simulation program *ms*2 developed in our group. The molecular dynamics part of the program yields on the NEC SX-8 in a single processor job a vectorization ratio of 99.7 % and a sustained performance of 5.1 GFlops. Due to MPI communication overheads and shorter vector lengths in a parallel job, the sustained performance reduces to 4.1 GFlops per processor on a 8 CPU (one node) job and the total vectorization ratio to 99.6 %. A flow trace analysis of a parallel one node job on the NEC SX-8 is given in Table 3. Since the diameter of the cut-off sphere is close to the simulation box length and a list of interacting particles is obtained before the calculation of the interaction, the vectorization ratio and sustained performance is favorable. Note, that the parallelization of the molecular dynamics part of *ms*2 is obtained using Plimptons particle-based decomposition algorithm [11].

A performance investigation on the NEC SX-8, Cacau and Strider regarding the speed-up and scale-up of a molecular dynamics run is shown in Fig. 7.

On the NEC SX-8, the jobs were carried out using one node with 8 CPUs where such a job lasts approximately 4 hours depending on the molecular model used. On Cacau and Strider, typically 8 and 16 CPUs were used for a job where it lasts approximately 24 and 16 hours, respectively.

Table 3. Flow trace on one node of the NEC SX-8 for a parallel molecular dynamics simulation run

EXCL. TIME [%]	AVER. TIME [msec]	MFLOPS	V.OP RATIO	AVER. V.LEN	I-CACHE MISS	O-CACHE MISS
ms2_potential.tpotljlj_force						
15.2	2.696	4710.2	99.83	233.0	0.0089	0.0413
ms2_potential.tpotcc_force						
13.1	3.616	4621.1	99.69	233.1	0.0048	0.0262
ms2_component.tcomponent_atom2mol						
9.1	40.476	19.0	35.57	48.0	0.0341	0.0185
ms2_potential.tpotcq_force						
6.9	3.811	5375.3	99.88	232.2	0.0036	0.0158
ms2_potential.tpotcd_force						
6.3	3.506	5170.1	99.88	232.3	0.0034	0.0159
ms2_potential.tpotqc_force						
6.1	3.362	6212.5	99.81	233.3	0.0032	0.0363
ms2_potential.tpotdc_force						
5.7	3.176	5831.1	99.83	233.5	0.0033	0.0359
ms2_component.tcomponent_mol2atom						
3.7	16.237	45.9	47.82	71.2	0.0194	0.0154
⋮	⋮	⋮	⋮	⋮	⋮	⋮
total						
100.0	2.602	4098.9	99.58	231.9	0.4158	1.0459

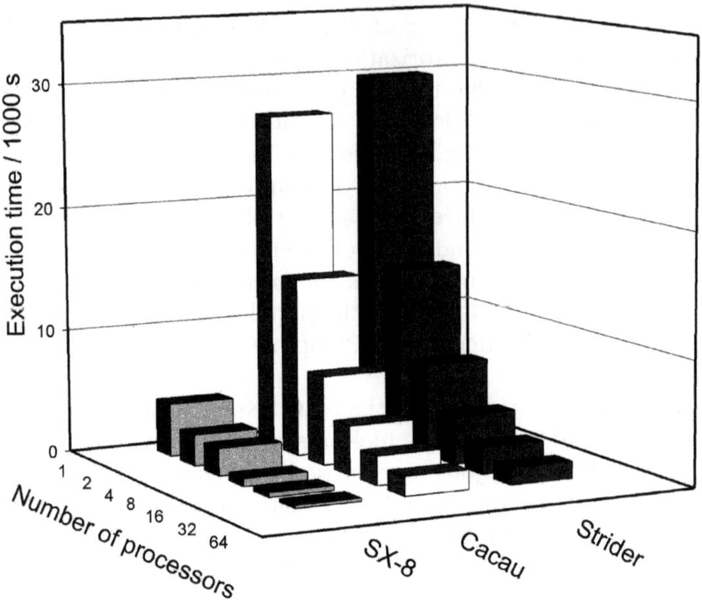

Fig. 7. Execution times of a molecular dynamics run on NEC SX-8, Cacau and Strider over number of CPUs

References

1. Berthelot, D.: Sur le Mélange des Gaz. Comptes Rendus de l'Académie des Sciences Paris, **126**, 1703 (1889).
2. Chen, B., Potoff, J.J., Siepmann, J.I.: Monte Carlo calculations for alcohols and their mixtures with alkanes. Transferable potential for phase equilibria. 5. United atom description of primary, secondary, and tertiary alcohols. J. Phys. Chem. B, **105**, 3093 (2001).
3. Derbali, Y.: Molekulare Wassermodelle zur Vorhersage thermophysikalischer Stoffdaten. Diploma Thesis, University of Stuttgart, Stuttgart (2005)
4. Daubert, T.E., Danner, R.P.: Data Compilation Tables of Properties of Pure Compounds. AIChE (1984).
5. Errington, J.R., Panagiotopoulos, A.Z.: Phase equilibria of the modified exponential-6 potential from Hamiltonian scaling grand canonical Monte Carlo. J. Chem. Phys., **109**, 1093 (1998).
6. Jorgensen, W.L., Chandrasekhar, J.D., Madura, R.W., et al.: Comparison of Simple Potential Functions for Simulating Liquid Water. J. Chem. Phys., **79**, 926 (1983).
7. Lorentz, H.A.: Über die Anwendung des Satzes vom Virial in der kinetischen Theorie der Gase. Annalen der Physik, **12**, 127 (1881).
8. Lotfi, A., Vrabec, J., Fischer, J.: Vapour liquid equilibria of the Lennard–Jones fluid from the *NpT* plus test particle method. Mol. Phys., **76**, 1319 (1992).

9. Mahoney, W.M., Jorgensen, W.L.: A five-site model for liquid water and the reproduction of the density anomaly by rigid, nonpolarizable potenial functions. J. Chem. Phys., **112**, 8910 (2000).
10. Nagel, W.E., Jäger, W., Resch, M.: High Performance Computing in Science and Engineering '05. Springer, Berlin (2005).
11. Plimpton, S.: Fast Parallel Algorithms for Short-Range Molecular Dynamics. J. Comp. Phys., **117**, 1 (1994).
12. Schmidt, M.W., Baldridge, M.W., Boatz, J.A., et al.: General atomic and molecular electronic structure system. J. Comput. Chem., **14**, 1347 (1993).
13. Ungerer, P., Beauais, C., Delhommelle, J., et al.: Optimization of the anisotropic united atoms intermolecular potential for n-alkanes. J. Chem. Phys., **112**, 5499 (2000).
14. Widom, B.: Some topics in the theory of fluids. J. Chem. Phys., **39**, 2808 (1963).
15. Wick, C.D., Stubbs, J.M., Rai, N., Siepmann, J.I.: Tranferable Potentials for Phase Equilibria. 7. Primary, Secondary, and Tertiary Amines, Nitroalkanes, Nitriles, Amides, Pyridine, and Pyrimidine. J. Chem. Phys. B, **109**, 18974 (2005).

The Application of a Black-Box Solver with Error Estimate to Different Systems of PDEs

Torsten Adolph and Willi Schönauer

Institut für Wissenschaftliches Rechnen, Forschungszentrum Karlsruhe
torsten.adolph@iwr.fzk.de
willi.schoenauer@iwr.fzk.de

Happy are those people that do not see the errors.

Abstract. At first a brief overview of the Finite Difference Element Method (FDEM) is given, above all how an explicit estimate of the error is obtained. Then for some academic examples the estimated and exact error are compared showing the quality of the estimate. The PDEs for fuel cells of PEMFC and SOFC type with extremely nonlinear coefficients are solved and the error estimate shows the quality of the solution. Finally for a complicated fluid/structure interaction problem of a high pressure Diesel injection pump, where the domain of solution has 3 subdomains with different PDEs and where a nested iteration procedure is needed, the PDEs are solved and the global error estimate shows the quality of the solution. For all these examples it would be very difficult to obtain a quality control of the solution by conventional grid refinement tests.

1 Introduction

The development of the Finite Difference Element Method (FDEM) at the computer center of the University of Karlsruhe has been supported by the German Ministry of Research (BMBF). The application of FDEM to the numerical simulation of fuel cells (FCs) has been supported by the Research Alliance Fuel Cells of the state Baden-Württemberg. In this paper we present a compilation of results of these projects.

Never before such problems have been solved with error estimates. So the emphasis of this paper will be on the error estimate: together with the solution we present values or plots for the error estimates. Because of the limited accorded space of the paper we cannot present all the details of FDEM and of the examples. However, we will give the precise information where these details are in the corresponding reports. As these reports are in the Internet, the reader can immediately have a look at them at his computer.

2 The Finite Difference Element Method (FDEM)

FDEM is an unprecedented generalization of the FDM on an unstructured FEM mesh. It is a black-box solver for arbitrary nonlinear systems of 2-D and 3-D elliptic or parabolic PDEs. If the unknown solution is $u(t,x,y,z)$ the operator for PDEs and BCs (boundary conditions) is (2.4.1) and (2.4.2) in [1]:

$$Pu \equiv P(t,x,y,z,u,u_t,u_x,u_y,u_z,u_{xx},u_{yy},u_{zz},u_{xy},u_{xz},u_{yz}) = 0 \,. \tag{1}$$

For a system of m PDEs u and Pu have m components:

$$u = \begin{pmatrix} u_1 \\ \vdots \\ u_m \end{pmatrix} , \quad Pu = \begin{pmatrix} P_1 u \\ \vdots \\ P_m u \end{pmatrix} . \tag{2}$$

Because we have a black-box solver, the PDEs and BCs and their Jacobian matrices of type (2.4.6) in [1] must be entered as Fortran code in prescribed frames.

The geometry of the domain of solution is entered as a FEM mesh with triangles in 2-D and tetrahedra in 3-D. The domain may be composed of subdomains with different PDEs and non-matching grid. From the element list and its inverted list we determine for each node more than the necessary number of nodes for difference formulas of a given consistency order q. By a sophisticated algorithm from this set the necessary number of nodes is selected, see Sect. 2.2 in [1]. From the difference of formulas of different consistency order we get an estimate of the discretization error. If we want e.g. the discretization error for u_x and $u_{x,d,q}$ denotes the difference formula of consistency order q, the error estimate d_x is defined by

$$d_x := u_{x,d,q+2} - u_{x,d,q} \,, \tag{3}$$

i.e. by the difference to the order $q+2$. This has a built-in self-control: if this is not a "better" formula the error estimate shows large error.

With such an error estimate we can explicitly compute the error of the solution by the error Eq. (2.4.8) in [1]. The knowledge of the error estimate allows a mesh refinement and order control in space and time (for parabolic PDEs), see Sect. 2.5 in [1].

A special problem for a black-box solver is the efficient parallelization because the user enters his domain by the FEM mesh. We use a 1-D domain decomposition with overlap to distribute the data to the processors, see Sect. 2.8 in [1]. We use MPI. A detailed report on the parallelization is [2]. The resulting large and sparse linear system is solved by the LINSOL program package [3] that is also efficiently parallelized for iterative methods of CG type and (I)LU preconditioning.

3 Academic Examples

The purpose of these academic examples is to check the quality of the error estimate (which is one level higher than the usual check for the quality of the solution), see Sect. 2.10 in [1]. We define the global relative error for a component l of the solution and the global relative error by

$$\frac{\|\Delta u_{d,l}\|}{\|u_{d,l}\|}, \quad \frac{\|\Delta u_d\|}{\|u_d\|} = \max_l \frac{\|\Delta u_{d,l}\|}{\|u_{d,l}\|}, \tag{4}$$

where $\Delta u_{d,l}$ is computed from the error equation (component l of (2.4.8) in [1]). The norm $\|\cdot\|$ is the maximum norm. We generate from the original PDE $Pu = 0$ a "test PDE" for a given solution $\bar{u}$ by $Pu - P\bar{u} = 0$ that has $\bar{u}$ as solution. This prescription holds also for the BCs. The exact global relative error then is

$$\frac{\|\bar{u} - u_d\|}{\|u_d\|}. \tag{5}$$

We compute on the HP XC6000 with Itanium2 processors of 1.5 GHz (University of Karlsruhe). As exact solution $\bar{u}$ we select either a polynomial of a given order or a sugar loaf type function (2.10.16) in [1]. We solve the Navier–Stokes equations in velocity/vorticity form (2.10.13) in [1] with the unknown functions velocity components u, v and vorticity ω, and Reynolds number $Re = 1$. We solve on a circle with radius $= 1$ on a grid with 751 nodes, 1410 elements that has been generated by the commercial mesh generator I-DEAS. We compute with 8 processors. The given CPU time is that of the master processor 1.

Table 1 shows the results. Here are two remarks: for $\bar{u}$ polynomial of order 6 and consistency order $q = 6$ we should reproduce $\bar{u}$ exactly which is expressed by the small errors. For the sugar loaf function and consistency order $q = 6$ we get a large error estimate. This shows the built-in self-control: near the top of the sugar loaf the grid is too coarse for the consistency order $q + 2 = 8$ that is used for the error estimate, the order 8 is "overdrawn" (higher order may not be better).

Table 1. Results for the solution of the Navier–Stokes type equations on a circle with 751 nodes for different consistency orders q and test function $\bar{u}$

	order $q = 2$		order $q = 4$		order $q = 6$	
type $\bar{u}$	error exact error estim.	CPU sec.	error exact error estim.	CPU sec.	error exact error estim.	CPU sec.
pol. order 6	0.154 0.155	0.158	0.914E-02 0.367E-01	0.175	0.108E-10 0.109E-08	2.131
sugar loaf	0.694E-01 0.642E-01	0.168	0.238E-01 0.220E-01	0.184	0.457E-02 0.736*	1.853

* here the order 8 for the error estimate is overdrawn (too coarse grid)

Table 2. Results for the self-adaptation of mesh and order for sugar loaf test function for prescribed global relative error 0.25×10^{-2} (0.25%)

cycle	no. of nodes	no. of elem.	no. of nodes ref.	no. of nodes with order 2	4	6	global relat. error exact	estimated	sec. for cycle
1	751	1410	132	427	320	4	0.305E-01	0.280E-01	1.021
2	1332	2493	345	180	1144	8	0.109E-01	0.950E-02	3.604
3	2941	5469	—	360	2556	25	0.179E-02	0.174E-02	10.086

For the demonstration of the self-adaptation we solve the same problem with the sugar loaf function again with 8 processors, but now we switch on the mesh refinement and order control for a global relative error of 0.25%. The results are shown in Table 2. The requested accuracy needs 3 refinement cycles. "no. of nodes ref." is the number of refinement nodes that determine the refinement elements from which then follows the new number of nodes. Observe the excellent error estimate that results from the optimal local order. Figure 1 shows the grid after the 3rd cycle, the refinement is clearly visible.

As mentioned above we wanted to demonstrate the quality of the error estimate. The estimate is the better the smaller the error is, this is a natural consequence of (3). What we have seen in Table 1 is also part of our test technique for each new problem: we at first create from the new problem a test PDE $Pu - P\bar{u} = 0$ and check with polynomial test solutions $\bar{u}$ the error estimate like in Table 1.

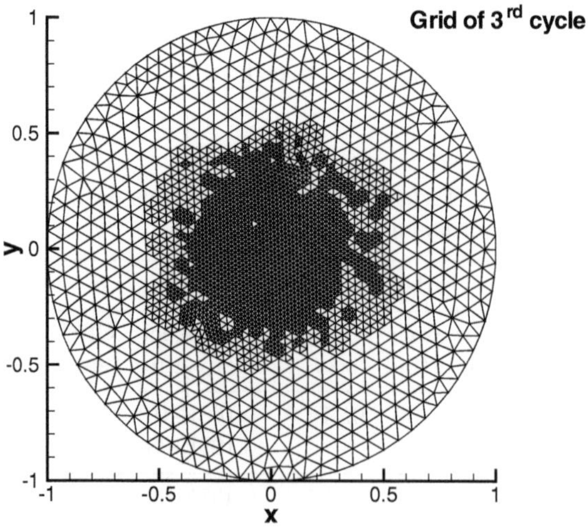

Fig. 1. Refined grid after 3rd cycle of Table 2

4 The Numerical Simulation of Fuel Cells (FCs)

This project was supported by the Research Alliance Fuel Cells (FABZ) of the state of Baden-Württemberg. The corresponding report [4] was written by ZSW Ulm for the equations of the PEMFCs (Part I), by the IWE of the University of Karlsruhe for the equations of the SOFCs (Part II) and by the RZ of the University of Karlsruhe for the numerical solution of the corresponding PDEs (Part III). These problems are characterized by the extreme nonlinearity of their coefficients that depend in a complicated way on the variables themselves. Quite naturally we will report here only on that part of [4] that deals with the numerical solution of the PDEs. Up to now nobody had solved these PDEs with an error estimate.

4.1 Numerical Solution of the PDEs for PEMFCs

The PEMFCs (Polymer–Electrolyte–Membrane FC or Proton Exchange Membrane FC) are the "cold" FCs, operating at about 300 K. The domain of solution for the used model is the GDL (gas diffusion layer) that is at its lower left half open to the oxygen channel and at its lower right half closed by a rib, see Fig. 2. How this is cut out of a whole cell can be seen at Fig. 1 on page III.4 in [4].

The variables for this problem are the molar flux densities of oxygen in x-direction $\dot{n}_o^x$ and y-direction $\dot{n}_o^y$, similarly for water vapor $\dot{n}_w^x$ and $\dot{n}_w^y$, and for nitrogen $\dot{n}_n^x$ and $\dot{n}_n^y$, the partial pressure for oxygen p_o, for water p_w and for nitrogen p_n, the total pressure p and as a special variable that has physical meaning only at the reaction layer the current density i. So we have 11 variables and need a system of 11 PDEs. These PDEs are given in Part I of [4],

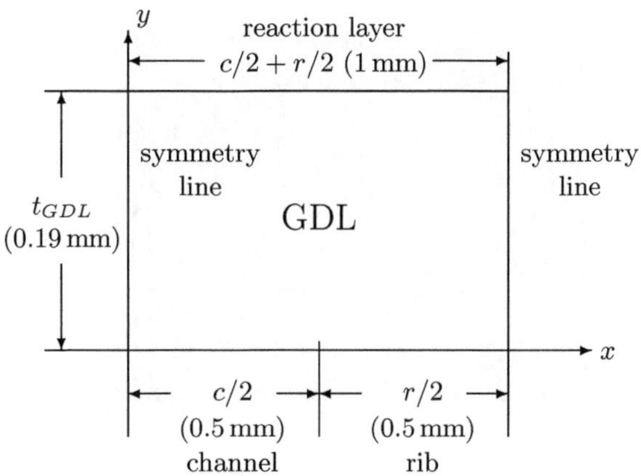

Fig. 2. Domain of solution for the PEMFC

and in Table 1 on page III.6 is shown which equation is used for which variable position. We show below as an example a typical species transport equation, but without explanation of the notations:

$$\begin{aligned}&\frac{\dot n_o^x}{DKn_o} + \frac{p_w \dot n_o^x - p_o \dot n_w^x}{D_{ow}p} + \frac{p_n \dot n_o^x - p_o \dot n_n^x}{D_{on}p} + \frac{1}{RT}\frac{\partial p_o}{\partial x} - \frac{p_o}{RTp}\frac{\partial p}{\partial x}\\ &+ \left[\frac{B_o}{DKn_o} + \frac{B_o p_w}{D_{ow}p}\left(1-\frac{B_w}{B_o}\right) + \frac{B_o p_n}{D_{on}p}\left(1-\frac{B_n}{B_o}\right)\right]\frac{p_o}{RTp}\frac{\partial p}{\partial x} = 0 .\end{aligned} \tag{6}$$

The effective permeabilities B_j depend in a complicated nonlinear way from the pressure p and the partial pressures p_i. We do not discuss here the BCs, they are given on pages III.9–III.11 in [4].

We computed with 32 processors of the HP XC6000 with 1.5 GHz Itanium2 processors on a grid of 200 × 201 nodes in x,y-direction, we used consistency order $q = 4$. Because we have 11 variables per node we have 442200 unknowns. The computation time was 4123 sec on the master processor 1.

Figure 3 shows a typical result for $\dot n_w^y$ and its error. There is a quasi-singularity at the lower boundary where the BCs change from channel to rib.

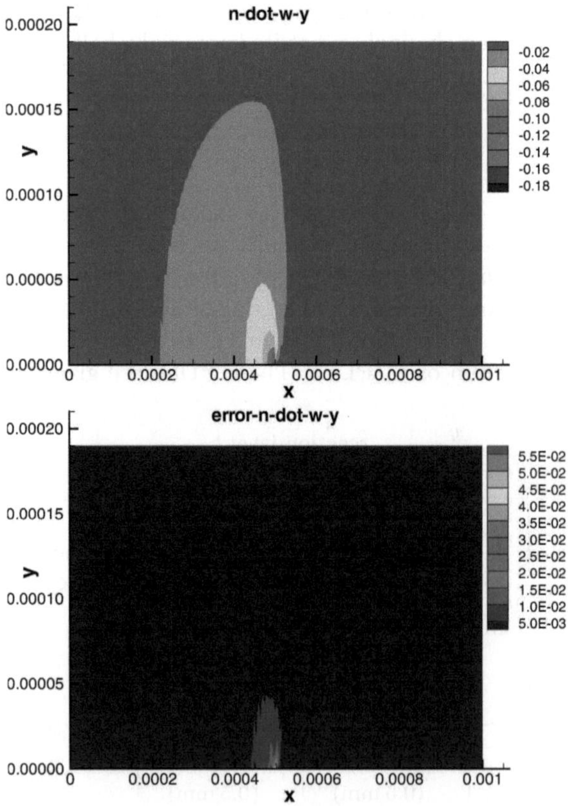

Fig. 3. Contour plot of molar flux density of water vapor in y-direction $\dot n_w^y$ and its error

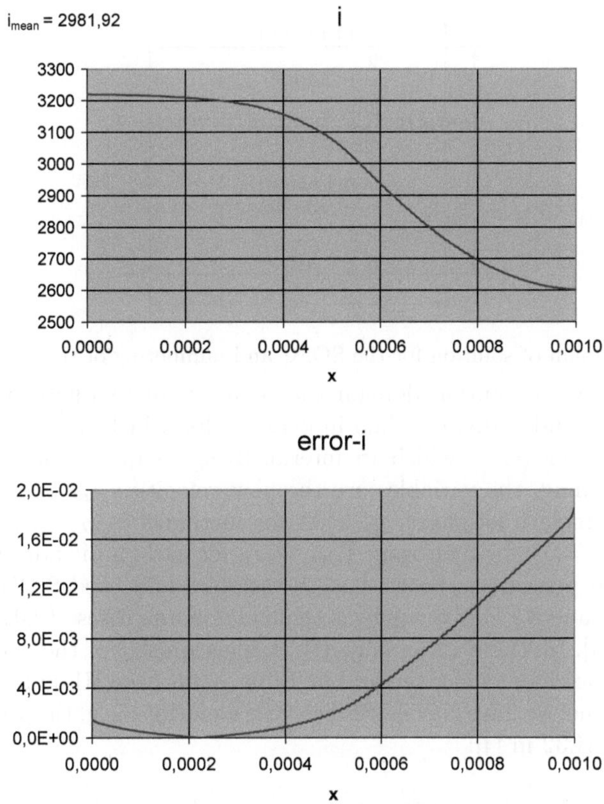

Fig. 4. Current density i along the reaction layer and its error

Figure 4 shows the current density i and its error along the reaction layer. This is the most interesting quantity for the engineers.

In [4] on pages III.15–III.24 are figures of type Fig. 3 for all variables. So the engineer can see if he can trust the solution. If there would be no error estimate he had to do grid refinement tests and observe for all variables and nodes (here 442200 values) how they change with finer grid. So the error estimate that consumes only a fraction of the total computation time is an invaluable feature in the solution process.

4.2 Numerical Solution of the PDEs for SOFCs

The SOFCs (Solid Oxide FCs) are the "hot" FCs, operating at about 1200 K. The domain of solution for the used model is the anode with flow in porous media and the gas channel with Navier–Stokes equations, see Fig. 5.

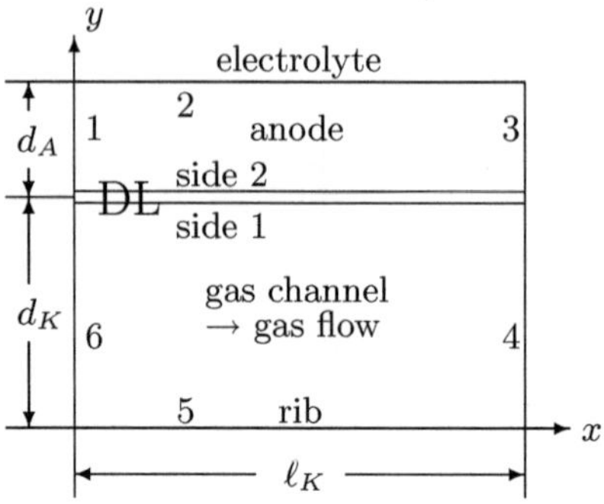

Fig. 5. Domain of solution for the SOFC and numbering of the boundaries

Here we have a solution domain that consists of two subdomains with different PDEs and a dividing line in-between for which we must prescribe coupling conditions (CCs) which are interior BCs. Because the model includes methane reforming, the variables are the flow velocities u_x, u_y in x- and y-direction, the mole fractions $Y_{CH_4} = Y_3$ for methane, $Y_{CO} = Y_4$ for carbon monoxide, $Y_{H_2} = Y_5$ for hydrogen, $Y_{CO_2} = Y_6$ for carbon dioxide, $Y_{H_2O} = Y_7$ for steam, and pressure p. So we have 8 variables and need 8 PDEs for the anode domain and 8 PDEs for the gas channel domain. These PDEs are given in Part II of [4]. In Table 12 on page III.31 the sequence of the variables and equations in the channel is given and in Table 15 on page III.36 in the anode.

In the channel we have Navier–Stokes-type equations, e.g. the y-momentum, (18) on page III.32 in [4]:

$$\begin{aligned} &\varrho u_x \frac{\partial u_y}{\partial x} + \varrho u_y \frac{\partial u_y}{\partial y} + \frac{\partial p}{\partial y} - \frac{\partial \mu}{\partial y}\left(\frac{4}{3}\frac{\partial u_y}{\partial y} - \frac{2}{3}\frac{\partial u_x}{\partial x}\right) \\ &- \mu\left(\frac{4}{3}\frac{\partial^2 u_y}{\partial y^2} + \frac{1}{3}\frac{\partial^2 u_x}{\partial x \partial y} + \frac{\partial^2 u_y}{\partial x^2}\right) - \frac{\partial \mu}{\partial x}\left(\frac{\partial u_x}{\partial y} + \frac{\partial u_y}{\partial x}\right) = 0\,, \end{aligned} \tag{7}$$

and we have transport equations like that for methane, (19) on page III.32 in [4]:

$$\begin{aligned} &-\frac{\partial p}{\partial x} u_x Y_3 - \frac{\partial u_x}{\partial x} p Y_3 - \frac{\partial Y_3}{\partial x} p u_x - \frac{\partial p}{\partial y} u_y Y_3 - \frac{\partial u_y}{\partial y} p Y_3 - \frac{\partial Y_3}{\partial y} p u_y \\ &+ \frac{\partial D_{3,gas}}{\partial y}\left(\frac{\partial p}{\partial y} Y_3 + \frac{\partial Y_3}{\partial y} p\right) + D_{3,gas}\left(\frac{\partial^2 p}{\partial y^2} Y_3 + 2\frac{\partial p}{\partial y}\frac{\partial Y_3}{\partial y} + \frac{\partial^2 Y_3}{\partial y^2} p\right) = 0\,. \end{aligned} \tag{8}$$

In the anode the Navier–Stokes equations are replaced by Darcy's law. The species transport equations have additional terms by the y-dependence of p and by the chemical reactions. The transport equation for methane, (34) on page III.37 is now

$$\begin{aligned} &-\frac{\partial p}{\partial x} u_x Y_3 - \frac{\partial u_x}{\partial x} p Y_3 - \frac{\partial Y_3}{\partial x} p u_x - \frac{\partial p}{\partial y} u_y Y_3 - \frac{\partial u_y}{\partial y} p Y_3 \\ &\quad - \frac{\partial Y_3}{\partial y} p u_y + D_{3,gas}\left(\frac{\partial^2 p}{\partial y^2} Y_3 + 2\frac{\partial p}{\partial y}\frac{\partial Y_3}{\partial y} + p\frac{\partial^2 Y_3}{\partial y^2}\right) \\ &+ \frac{\partial D_{3,gas}}{\partial y}\left(\frac{\partial p}{\partial y} Y_3 + \frac{\partial Y_3}{\partial y} p\right) + D_{3,gas}\left(\frac{\partial^2 p}{\partial x^2} Y_3 + 2\frac{\partial p}{\partial x}\frac{\partial Y_3}{\partial x} + p\frac{\partial^2 Y_3}{\partial x^2}\right) \\ &\quad + \frac{\partial D_{3,gas}}{\partial x}\left(\frac{\partial p}{\partial x} Y_3 + \frac{\partial Y_3}{\partial x} p\right) - \frac{RT}{d_A} r_3 = 0\,. \end{aligned} \tag{9}$$

These equations are extremely nonlinear because μ in (7) and the $D_{i,gas}$ depend nonlinearly on the Y_j, similarly the reaction rates r_k are depending.

Here we must define the BCs for the 6 boundaries of Fig. 5 and the CCs for side 1 and side 2 of the dividing line DL. These conditions are given in Tables 17–19 on pages III.40 and III.41 in [4].

The numerical solution of these equations was much more critical than that of the PEMFCs because of the extreme nonlinearity of the coefficients. We computed with a grid with 80 nodes in x-direction and 41 nodes in the channel and 41 nodes in the anode in y-direction, resulting in 52480 unknowns. We used consistency order $q = 4$. The computation time on 8 processors of the HP XC6000 with processors Itanium2, 1.5 GHz, was 510 sec on the master processor 1. The colour plots of the results and error estimates for all 8 variables in the channel and anode are presented in [4], pages III.47–III.62. We present here only the mole fraction Y_{CO} and its error in Fig. 6 for the anode.

The ampleness of the generated information can be estimated only if one looks at all the nice colour plots of the report [4]. Here the engineer can

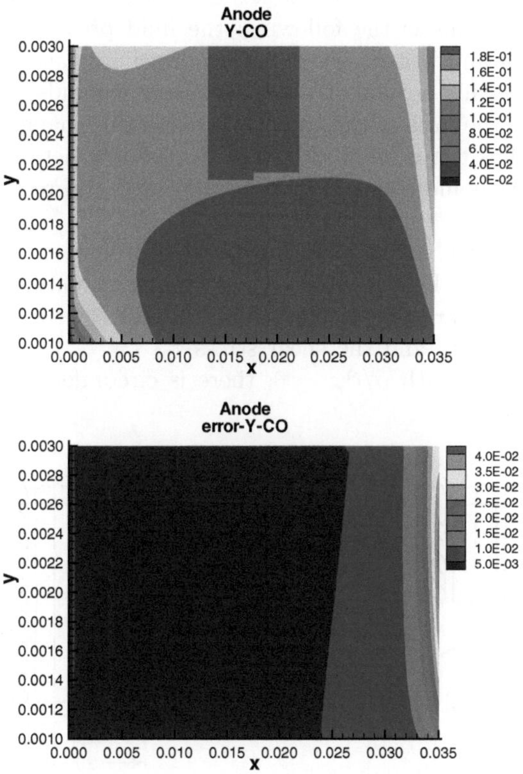

Fig. 6. Contour plot of mole fraction Y_{CO} and its error for the anode

immediately see the quality of the solution from the error plots, this is an unprecedented gain in information.

5 Fluid/Structure Interaction for a High Pressure Diesel Injection Pump

A detailed presentation of this problem (cooperation with Bosch) is given in Sect. 3.3 in [1]. In a high pressure Diesel injection pump the housing extends under the injection pressure of 2000 bar and the piston is compressed so that the lubrication gap between housing and piston changes its form and consequently the leakage flow changes. This is a fluid structure interaction problem. The problem is simplified by replacing the complicated shape of the housing by a tube or bush, see Fig. 7. The piston does not move, so we have a static configuration.

Our domain of solution now has 3 subdomains with different PDEs: in the housing and piston we must solve the elasticity equations of steel, in-between we have the gap with the Navier–Stokes equations for Diesel. The coupling between these domains is the following: the fluid pressure p is the normal stress for housing and piston, so we have a direct interaction of the flow on the structure. By this normal stress the housing expands and the piston is compressed, thus the form of the gap changes and this changes the flow which changes p and thus the normal stress etc. So the interaction of housing and piston on the flow is indirect and more complicated and requires an iterative procedure.

We solve the problem in axisymmetrical coordinates, then x in Fig. 7 becomes the radius r. For the elasticity equations in housing and piston the dependent variables are the displacements w and u in z- and r-direction, the stresses σ_z, σ_r, σ_φ and the shear stress τ_{rz} $(= \tau_{zr})$. Although we have rotational symmetry with $\partial/\partial\varphi = 0$, there is circumferential stress σ_φ. So

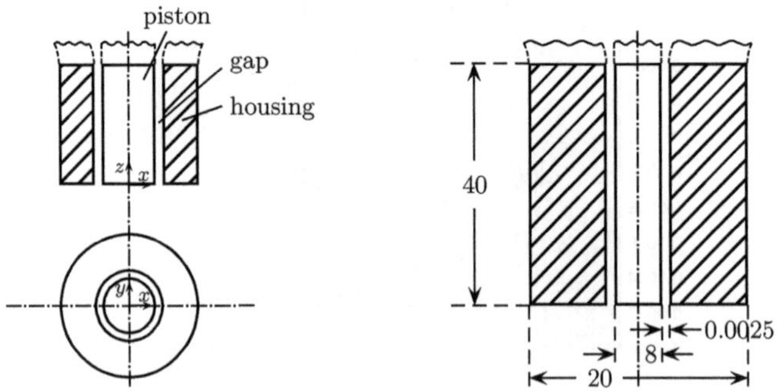

Fig. 7. Symbolic configuration and dimensions in mm. In reality the gap is extremely thin

we have 6 variables and need 6 PDEs. They are given in (3.3.4.1)–(3.3.4.6) on page 129 in [1]. Here we show the first and last of these equations in incremental form (index "old" is for the last solution):

$$\frac{1}{E}\left[\sigma_z - \sigma_{z,old} - \nu(\sigma_\varphi - \sigma_{\varphi,old}) - \nu(\sigma_r - \sigma_{r,old})\right] - \frac{\partial w}{\partial z} = 0\,, \tag{10}$$

$$\frac{\partial \tau_{rz}}{\partial r} + \frac{\partial \sigma_z}{\partial z} + \frac{\tau_{rz}}{r} = 0\,. \tag{11}$$

Here E is the elasticity module and ν is Poisson's ratio. Table (3.3.4.7) in [1] gives the information which equation is used in which position (for which variable) in the system of 6 equations.

In the lubrication gap we must solve the Navier–Stokes equations. The variables are the velocity components w and u in z- and r-direction, and the pressure p. So we need a system of 3 equations. Here we show the momentum equation in r-direction and the continuity equation, (3.3.4.17) and (3.3.4.19) on page 123 in [1]:

$$u\frac{\partial u}{\partial r} + w\frac{\partial u}{\partial z} + \frac{1}{\varrho}\frac{\partial p}{\partial r} - \frac{\eta}{\varrho}\left(\frac{\partial^2 u}{\partial r^2} + \frac{1}{r}\frac{\partial u}{\partial r} - \frac{u}{r^2} + \frac{\partial^2 u}{\partial z^2}\right) = 0\,, \tag{12}$$

$$\frac{\partial u}{\partial r} + \frac{u}{r} + \frac{\partial w}{\partial z} = 0\,. \tag{13}$$

Here η and ϱ are the dynamical viscosity and the density. But now we have a problem: the whole domain has 3 subdomains where two of them have 6 variables and PDEs, and one has 3 variables and PDEs. However, the code is designed for the same number of variables in the whole domain. So we add in the fluid domain 3 dummy variables with variable $= 0$ as PDEs and BCs.

Table (3.3.4.20) on page 133 in [1] shows which equation is used in which position of the system, i.e. for which preferred variable: for w we take the continuity equation, for u the r-momentum equation and for p the z-momentum equation. This is a quite natural ordering. If we want to discuss the BCs for the flow we immediately meet a serious problem: at the entry we have a prescribed pressure of 2000 bar $=$ 200 N/mm^2 and at the exit of 0 bar. Because in the Navier–Stokes equations there is only $\partial p/\partial z$ we can prescribe for incompressible flow the pressure only at one position, e.g. 2000 bar at the entry. Then the pressure at the exit is the result of the Navier–Stokes equations. At the high pressure entry we prescribe a parabolic velocity profile for w: $w(r) =$ parabola with w_{max} in the middle of the entry. The choice of w_{max} determines the pressure at the exit, i.e. the pressure drop in the gap. This introduces what we call $\boldsymbol{w_{max}}$**-iteration**: we start with an appropriate value of w_{max}, determine p_{exit} and if it is too large we increase w_{max} (gives larger pressure drop), if it is too small, we reduce w_{max}. This is made by a sophisticated iteration procedure.

The detailed discussion of the BCs for housing, piston and fluid is given in Sect. 3.3.4 in [1] and would be too lengthy for this paper. However, the

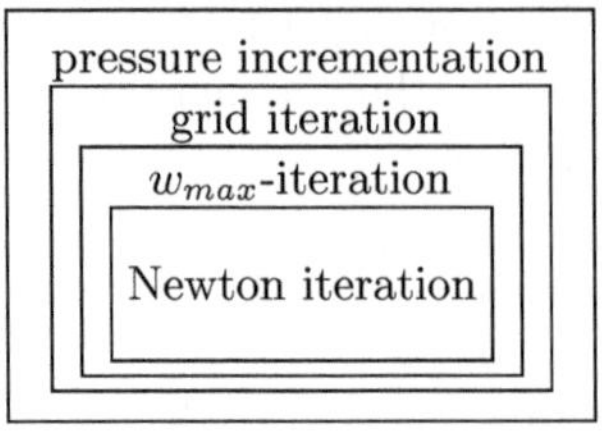

Fig. 8. The nested iterations for the solution process

essential part is the coupling between fluid and structure: if we have computed displacements for the pressure p given by the fluid flow, we apply in an incremental form these displacements by a shift algorithm that is presented in Sect. 3.3.3 in [1]. The index "old" in (10) denotes the values before the shifting (on the old grid). Then we observe if the grid still shifts. This induces what we call **grid iteration**. If the grid does no longer move we have the solution of our problem. Thus we have the nested iterations shown in Fig. 8. The innermost iteration is the Newton iteration for the solution of the PDEs, then we must determine w_{max} for the exit pressure zero, then we must apply the displacements until the grid does no longer move. The outermost iteration gives the possibility to increase gradually the entry pressure.

When we made the first numerical experiments we had serious difficulties caused by the extreme differences in length scales: housing and piston in cm, lubrication gap in micrometer, see Fig. 7. After some trials we decided to use cm as length scale. The discretization errors of housing and piston for the used grid caused the surfaces to be "rough" in the micrometer scale. Therefore we applied a smoothing of these surfaces.

We used a grid of 401 × 80 in z,r-direction for the housing, 401 × 40 for the fluid and 401 × 81 for the piston. We computed with 16 processors of the HP XC6000 with 1.5 GHz Itanium2 processors. The CPU time on the master processor 1 was 3354 sec, of which 3296 sec were needed for the linear solver LINSOL with full LU preconditioning. Table 3 gives some results for 2000 bar entry pressure for housing and fluid. In Table 3.3.5.1 on page 138 of [1] further values for entry pressures of 1500 bar to 3000 bar are given. Figure 9 shows the form of the lubrication gap for 2000 bar entry pressure. The bold lines show the original channel. It is amazing how the high injection pressure changes the lubrication gap from the manufacturing dimension of 2.5 micrometer to up to 11.5 micrometer. The error estimates in Table 3 show the quality of the solution. For the fluid there is a maximal error of w of 87%, but the (arithmetic) mean error is 1.4%, so large errors occur only locally.

Figure 10 shows the contour plot of the velocity w in z-direction which is responsible for the leakage flow that is in this case 2.40 cm^3/s. Figure 10 also shows its error plot. Here we can see that the large errors occur only locally. There a much finer grid should be used. As the mean error of w is 1.4% we can conclude that the volume flow is also accurate to this error level. Note that this is a global error estimate that includes all the errors of all the equations in the coupled domains. Here again the error estimate gives us the certainty

Table 3. Maximum value of solution component, max. relat. error and mean relat. error of solution component and volume flow through the gap for entry pressure of 2000 bar

Housing

		max. solution	max. error	mean error
w	cm	0.4143E−02	0.10E−03	0.11E−04
u	cm	0.7584E−03	0.32E−03	0.82E−04
σ_z	N/cm^2	0.2244E+05	0.30E−02	0.12E−04
σ_r	N/cm^2	0.2000E+05	0.28E−02	0.22E−04
σ_φ	N/cm^2	0.2772E+05	0.65E−03	0.67E−04
τ_{rz}	N/cm^2	0.9837E+03	0.24E−01	0.87E−04

Fluid

		max. solution	max. error	mean error	Volume [cm^3/s]
w	cm/s	0.3339E+04	0.87E+00	0.14E−01	2.40
u	cm/s	0.3810E+00	0.96E+02	0.10E+01	
p	N/cm^2	0.2000E+05	0.94E−01	0.60E−02	

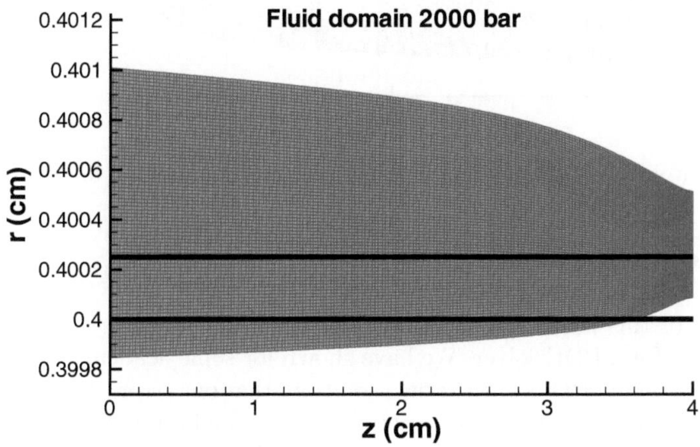

Fig. 9. Fluid domain with computational grid for 2000 bar entry pressure and original channel (*bold lines*)

that we can trust our solution. The reader may think how he could get this certainty by other methods for this complicated fluid/structure interaction problem.

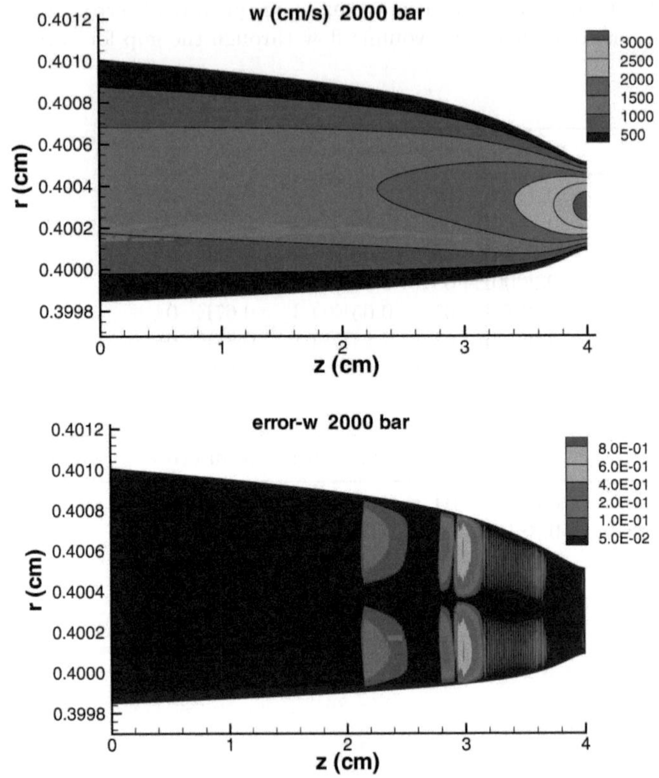

Fig. 10. Contour plot for the velocity w in z-direction for 2000 bar and its error

6 Concluding Remark

The purpose of this paper was to show that an error estimate is an invaluable advantage for a PDE solver. We have shown for some academic examples where we constructed PDEs with known solution, that our error estimate gives an excellent approximation for the exact error. It is the better the smaller the error is. Then we solved the PDEs for the numerical simulation of fuel cells of PEMFC and SOFC type and the error estimates showed the quality of the solution for all components of the systems. Finally, we solved a fluid/structure interaction problem for a high pressure Diesel injection pump. The high pressure of 2000 bar bends up the housing that the lubrication gap widens from 2.5 micrometer up to 11.5 micrometer. The solution algorithm is a complicated nested iteration. Nevertheless we compute a global error estimate for the coupled domains of housing, piston and fluid that tells us that we can trust our solution. The conventional way would be to do a sequence of grid refinements. The reader may imagine what amount of work would be necessary to obtain a comparable information like our error estimate.

References

1. W. Schönauer, T. Adolph, FDEM: The Evolution and Application of the Finite Difference Element Method (FDEM) Program Package for the Solution of Partial Differential Equations, 2005, available at www.rz.uni-karlsruhe.de/rz/docs/FDEM/Literatur/fdem.pdf
2. T. Adolph, The Parallelization of the Mesh Refinement Algorithm in the Finite Difference Element Method, Doctoral Thesis, 2005, available at www.rz.uni-karlsruhe.de/rz/docs/FDEM/Literatur/par_mra_fdem.pdf
3. LINSOL, see www.rz.uni-karlsruhe.de/rd/linsol.php
4. The Numerical Simulation of Fuel Cells of the PEMFC and SOFC type with the Finite Difference Element Method (FDEM), collective paper by three groups, 2005, available at www.rz.uni-karlsruhe.de/rz/docs/FDEM/Literatur/fuelcells.pdf

References

Scalable Parallel Suffix Array Construction

Fabian Kulla[1] and Peter Sanders[2]

[1] Forschungszentrum Karlsruhe, 76344 Eggenstein-Leopoldshafen, Germany,
Fabian.Kulla@iwr.fzk.de

[2] Universität Karlsruhe, 76128 Karlsruhe, Germany,
sanders@ira.uka.de

Abstract. Suffix arrays are a simple and powerful data structure for text processing that can be used for full text indexes, data compression, and many other applications in particular in bioinformatics. We describe the first implementation and experimental evaluation of a scalable parallel algorithm for suffix array construction. The implementation works on distributed memory computers using MPI, Experiments with up to 128 processors show good constant factors and make it look likely that the algorithm would also scale to considerably larger systems. This makes it possible to build suffix arrays for huge inputs very quickly. Our algorithm is a parallelization of the linear time DC3 algorithm.

1 Introduction

The suffix array, a lexicographically sorted array of the suffixes of a string, has numerous applications, e.g., in string matching, genome analysis, and text compression. For example, one can use it as full text index: To find all occurrences of a pattern P in a text T, do binary search in the suffix array of T, i.e., look for the interval of suffixes that have P as a prefix. A lot of effort has been devoted to efficient construction of suffix arrays, culminating recently in three direct linear time algorithms. One of the linear time algorithms, *DC3* [3] is very simple and can also be adapted to different models of computation. An external memory version of the algorithm [1] already makes it possible to construct suffix array for huge inputs. However, this takes many hours and hence a scalable *parallel algorithm* might be more interesting. This is the subject of the present paper. We describe the algorithm, *pDC3*, in Sect. 2 and experimental results in Sect. 3. Section 4 concludes with an outline of possible additional questions.

Related Work: A longer version of this paper is published at EuroPVM/ MPI [4]. There are numerous theoretical results on parallel suffix *tree* construction. Suffix trees can be easily converted to suffix arrays. However, these algorithms are fairly complicated. We are not aware of any implementations.

Recently, a trend is to use simpler suffix array construction algorithms even as a means of constructing suffix trees. The basic ideas for parallel suffix array construction based on the DC3 algorithm are already given in [3] for several theoretical models of parallel computation. Here, we outline a practical algorithm with particular emphasis on implementation and experimental evaluation. We are only aware of a single implemented parallel algorithm for suffix array construction [2]. This algorithm is practical but based on string sorting and thus needs quadratic work in the worst case. Furthermore it seems that all processing elements (PEs) need access to the complete input.

2 The pDC3 Algorithm

We use the shorthands $[i, j] = \{i, \ldots, j\}$ and $[i, j) = [i, j-1]$ for ranges of integers and extend to substrings as seen below. The **input** of a suffix array construction algorithm is a *string* $T = T[0, n) = t_0 t_1 \cdots t_{n-1}$ over the alphabet $[1, n]$, that is a sequence of n integers from the range $[1, n]$. For convenience, we assume that $t_j = 0$ for $j \geq n$. For $i \in [0, n]$, let S_i denote the *suffix* $T[i, n) = t_i t_{i+1} \cdots t_{n-1}$. The goal is to sort the sequence $\langle S_0, \ldots, S_{|}n-1\rangle$ of suffixes of T. The **output** is the *suffix array* $SA[0, n)$ of T, a permutation of $[0, n)$ satisfying $S_{SA[0]} < S_{SA[1]} < \cdots < S_{SA[n-1]}$. Let p denote the number of processors (PEs). PEs are numbered from 0 to $p - 1$.

At the most abstract level, the DC3 Algorithm is very simple and completely independent of the model of computation: It first constructs the suffix array of the suffixes starting at *sample* positions $i \bmod 3 \neq 0$. This is done by reduction to the suffix array construction of a string of two thirds the length, which is then solved recursively. The reduction works in two steps. First, the suffixes are sorted by their first three characters. Then these triples are replaced by *lexicographic names* respecting the relative order of the triples. The ranks of the sample suffixes are used to annotate the original input. With this annotation, two *arbitrary* suffixes S_i and S_j can be compared by looking at $T[i, i+2]$ and the annotations at positions $[i, i+2]$. For a more detailed explanation refer to [3, 4].

In the parallel algorithm, input, output, and intermediate tuple sequences are uniformly distributed over all PEs. This makes most operations easy to parallelize. The algorithm has considerable demands on communication bandwidth however, since the necessary sorting and permutation operations require several all-to-all communications involving total data volume $\Theta(n)$. The only somewhat nontrivial observation is that a sequential scanning operation needed for lexicographic naming in the sequential algorithm can be replaced by a parallel prefix computation. An analysis of the algorithm reveals that the parallel sorting operations of $\Theta(n)$ constant size tuples are the limiting factor to scalability. We use quicksort for local sorting and a simple variant of comparison based sample sort [6] for parallel sorting.

3 Experiments

We have implemented pDC3 using C++ and MPI. Measurements were done at the Rechenzentrum of Universitt Karlsruhe on a HP Integrity rx2620 with 64 dual 1.5 GHz Itanium 2 nodes (6 MB Cache). The machine has 64 × 12 GB of memory. Nodes are connected by a Quadrics QSNet II network with 800 MByte/s bandwidth. Our implementation uncovered a bug in the implementation of `MPI_alltoallv`. This bug is fixed now.

We have used the big real world inputs from [1]: The human genome, 3.125 GByte of books from the Gutenberg project, and 522 MByte of source code. In addition, we use the artificial inputs a^n and $(abc)^{n/3}$. Timing is started when all PEs have initialized MPI and hold n/p characters of the input each.

Figure 1 shows the work performed for the Gutenberg input using 16–128 PEs using one or two CPUs on each node. We see that sorting and merging takes most of the time. Communication time (mainly all-to-all) takes only a small fraction of the total execution time. However, it should be noted that low cost machines with switched Gigabit Ethernet have an order of magnitude smaller communication bandwidth than our machine. On such machines, communication would take about half of the time. (Which might still be ac-

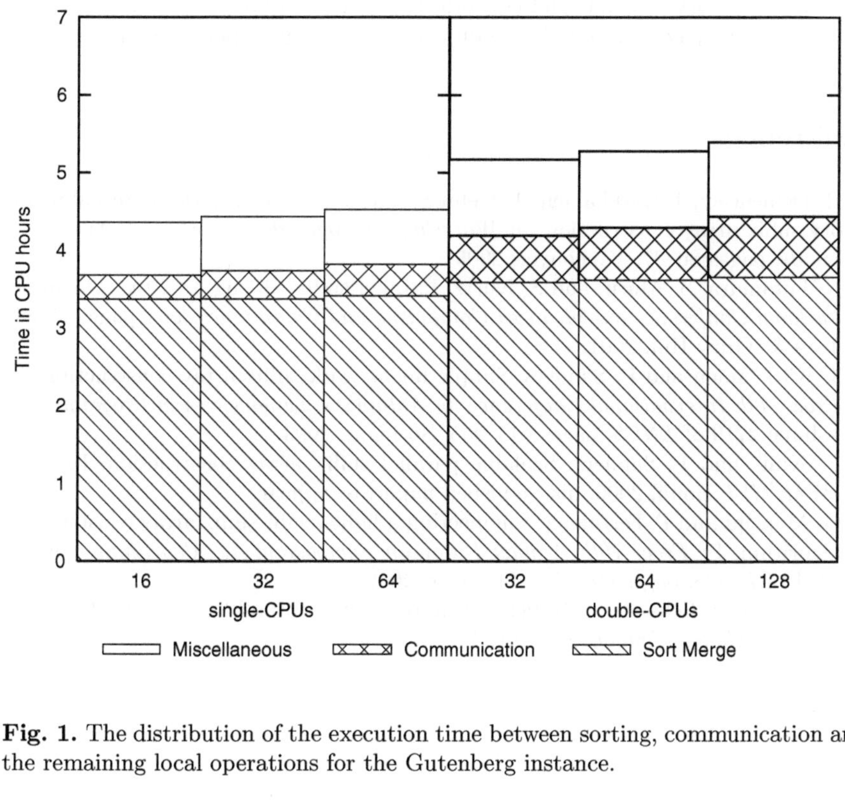

Fig. 1. The distribution of the execution time between sorting, communication and the remaining local operations for the Gutenberg instance.

ceptable considering that such machines are much cheaper). The overall work increases only slightly when increasing the number of processors. This indicates good scalability. As to be expected, using both CPUs increases internal work and total communication time since the CPUs have to share the main memory and the network interface.

For the biggest instance that we can process sequentially (source code), we have made comparisons with the simple sequential linear time implementation from [3] and with the fastest practical suffix array construction algorithm [5]. With the minimal number of two processors our parallel algorithm already outperforms the simple sequential algorithm significantly. The break even point to [5] is at four processors. The work per processor is about half as much as for the external algorithm from [1] on a 2 GHz Intel Xeon processor.

4 Conclusions

We have demonstrated that pDC3 is a practicable and scalable way to build huge suffix arrays. Several improvements could be considered. For example, pDC3 might scale to machines with thousands of processors if we would sort samples in parallel. For inputs that are so large that they do not even fit in the main memory of a parallel computer, a parallel external algorithm could be developed by combining the results of the present paper with [1].

References

1. R. Dementiev, J. Kärkkäinen, J. Mehnert, and P. Sanders. Better external memory suffix array construction. In *Workshop on Algorithm Engineering & Experiments*, pages 86–97, Vancouver, 2005.
2. N. Futamura, S. Aluru, and S. Kurtz. Parallel suffix sorting. In *Proc. 9th International Conference on Advanced Computing and Communications*, pages 76–81. Tata McGraw-Hill, 2001.
3. J. Kärkkäinen and P. Sanders. Simple linear work suffix array construction. In *Proc. 30th International Conference on Automata, Languages and Programming*, volume 2719 of *LNCS*, pages 943–955. Springer, 2003.
4. F. Kulla and P. Sanders. Scalable parallel suffix array construction. In *Euro PVM/MPI*, Bonn, 2006. distinguished paper, to appear.
5. G. Manzini and P. Ferragina. Engineering a lightweight suffix array construction algorithm. In *Proc. 10th Annual European Symposium on Algorithms*, volume 2461 of *LNCS*, pages 698–710. Springer, 2002.
6. H. Shi and J. Schaeffer. Parallel sorting by regular sampling. *Journal of Parallel and Distributed Computing*, 14(4):361–372, 1992.

MIX
Papier aus verantwortungsvollen Quellen
Paper from responsible sources
FSC® C105338

If you have any concerns about our products, you can contact us on
ProductSafety@springernature.com

In case Publisher is established outside the EU, the EU authorized representative is:
Springer Nature Customer Service Center GmbH
Europaplatz 3, 69115 Heidelberg, Germany

Printed by Libri Plureos GmbH
in Hamburg, Germany